T0177710

SEMICONDUCTOR NANOPHOTONICS

Semiconductor Nanophotonics

Prasanta Kumar Basu
Bratati Mukhopadhyay
Rikmantra Basu

OXFORD
UNIVERSITY PRESS

OXFORD
UNIVERSITY PRESS

Great Clarendon Street, Oxford, OX2 6DP,
United Kingdom

Oxford University Press is a department of the University of Oxford.
It furthers the University's objective of excellence in research, scholarship,
and education by publishing worldwide. Oxford is a registered trade mark of
Oxford University Press in the UK and in certain other countries.

Impression: 2

Published in the United States of America by Oxford University Press
198 Madison Avenue, New York, NY 10016, United States of America.

British Library Cataloguing in Publication Data
Data available

Library of Congress Control Number: 2022933845

ISBN 978–0–19–878469–2

DOI: 10.1093/oso/9780198784692.001.0001

Printed and bound by
CPI Group (UK) Ltd, Croydon, CR0 4YY

Preface

Light is an essential component for the creation of living species, their growth, sustenance, and evolution on the planet Earth. Human society has made continuous efforts to extract light energy and other forms of energy derived from the main source, the Sun, for their livelihood, improvement of their living conditions, and survival against enemies. Curious minds at all times, in all civilizations, and in all countries tried to understand the mysteries of light and find new ways of utilizing light.

Scientific and systematic investigation about light was initiated by Newton and pursued by many giants like Snell, Huygens, Fraunhoffer, Fresnel, Fermat, Clausius, Young, to name a few. Maxwell through his famous equations laid a strong foundation for electromagnetic waves.

Till the turn of the twentieth century, the wave nature of light and electromagnetic radiation, apart from a few earlier efforts to consider the particulate nature of light, known as the corpuscular theory, was firmly believed. The wave theory and classical statistics faced a stumbling block while explaining the blackbody radiation spectra. Finally, Planck came with his revolutionary idea of quanta and thus a new era began. The wave particle duality concept soon followed, and the name photon was coined to signify the particulate nature of light. Formulation of Bose-Einstein and Fermi-Dirac statistics successfully explained the behaviour of bosons and fermions, respectively.

A breakthrough in the field of science and technology occurred with the announcements of solid state and gas lasers and thereafter the miniaturization of lasers was initiated by the inventions of semiconductor lasers. Further downscaling of the size of semiconductor lasers was possible with the introduction of double heterostructures, followed by quantum nanostructures in the form of Quantum Wells, Quantum Wires, and Quantum Dots. The remarkable progress in optical fiber communication owes much to these tiny semiconductor nanostructured devices.

Although the term photon has been in use over several decades, the word **Photonics** is somewhat new. The photonic devices and associated physics had been the subject of interest over a long period by members of the Institute of Electrical and Electronics Engineers (IEEE) Lasers and Electro Optic Society. The name of this society has been changed to the IEEE Photonics Society in 2008–2009. A few other societies and institutions still prefer to publish papers in the area titled optics or optoelectronics.

The term Nanophotonics or Nanooptics is still younger. The subject deals with novel physical phenomena and related application areas using devices and structures having dimensions in the nanometric range. The subject is already vast and full of novel physical phenomena and myriads of applications in communication, sensing, medicine, pollution monitoring, light manipulation with optical tweezers, and in future quantum computing, and quantum information processing, to list a few.

The aim of the book is to cover some of the novel physical phenomena and introduce a few application areas using semiconductors and their nanostructures. The targeted readers are graduate students and researchers entering into this fascinating new field. Focus on semiconductors only somewhat restricts the coverage, but the emphasis stems out of the authors experience as well as the need to limit the size of the book.

The book is more or less divided into three parts. Part I contains four chapters: an introduction, basic semiconductor theory, optical processes based on electromagnetic theory, and photons and their quantization. Part II embodies five chapters: Electron photon interactions in bulk, in Quantum Wells, and Excitonic processes in bulk and Quantum Wells, in Quantum Wires or nanowires and then in Quantum Dots or nanoparticles. After developing the theoretical background, the book delves into the actual subject. Part III introduces the structure and properties of microcavities, which are essential elements for observing various nanophotonic phenomena. The strong interaction between light and matter is first introduced in Chapter 11, in which the cavity quantum electrodynamics phenomena is discussed. Chapter 12 gives an introduction to Bose–Einstein condensation in semiconductor nanostructures and basic theory of polariton laser. Chapter 13 deals with surface plasmon polariton: phenomena and applications, followed by Chapter 14 titled 'Spasers and plasmonic nanolasers'. Chapter 15 covers optical metamaterials: phenomena and applications. Chapter 16 covers the evolution of small lasers or nanolasers, some interesting features and current and emerging applications of nanolasers.

It may be mentioned that the last four chapters point out the importance of metals in the Nanophotonics arena. The relevant theories are first developed considering metal as the conductor. Thereafter, due coverage is given to the replacement of metals by heavily doped semiconductors, properly modifying the underlying theories, the wavelength range covered by use of semiconductors and the role of semiconductors in surface plasmonics and its amplification in surface plasmon amplification by stimulated emission of radiation (SPASER)s and plasmonic nanolasers.

There are many interesting phenomena that need their proper place in a textbook or an advanced textbook like the present one. To name a few, slow and fast light, single photon emission and detection, laser cooling, nanostructures in quantum computing, and cryptography are important topics calling the attention of the beginners and creating their awareness in the emerging fields. However, these topics could not be included in the present treatise for want of space.

Another important limitation of the present book is its lack of rigour in the basic theoretical treatment of strong light-matter interaction. The needed theory must be based on the language of second quantization used in Quantum Optics. Instead, the whole book makes use of semi-classical theory, with the only exception of introducing Jaynes-Cummins formula. In most of the cases, phenomena related to strong light-matter interaction are described by using classical analogues like coupled harmonic oscillators. The approach is partly due to the authors' inexperience and mostly to create a first-hand awareness amongst the targeted audience, comprising graduate students and researchers in physics, electrical engineering, materials science, and practicing engineers. There are a few textbooks and research monograms containing excellent articles by lead experts giving detailed exposures. These sources however, though cited in the present book, are of interest to advanced level researchers.

The authors record their indebtedness to many people, a list of which is given in the accompanying *Acknowledgement* page. The authors are thankful to the editorial staff of Oxford University Press for their constant encouragements and cooperation and patience to publish this book.

Prasanta Kumar Basu
Institute of Radio Physics and Electronics, University of Calcutta, India
Bratati Mukhopadhyay
Institute of Radio Physics and Electronics, University of Calcutta, India
Rikmantra Basu
ECE Department, National Institute of Technology Delhi, India

Acknowledgements

The authors acknowledge with thanks helpful comments and useful references received from Professor Pallab Bhattacharyya of University of Michigan in preparing the manuscript. Professor B M Arora, formerly with Tata Institute of Fundamental Research (TIFR) and the Indian Institute of Technology (IIT) Bombay, kindly reviewed a few chapters and provided useful suggestions for which the authors feel thankful. Dr Tapajyoti Dasgupta, an alumnus of the Institute of Radio Physics and Electronics, now at the Instrumentation and Applied Physics department of the Indian Institute of Science (IISc) Bangalore, not only provided a large number of softcopies of publications cited in this work, but also went through some chapters, providing useful comments. The authors feel indebted to him. Professor Guo En Chang, National Chung Cheng University (CCU), Chita Yi, Taiwan, kindly arranged several visits of P K Basu to his laboratory, which provided a very conducive and hospitable environment for the preparation of the book. He thankfully acknowledges his kind help as well as support from AIM-Hi of CCU, and from the Ministries of Science & Technology (MOST) and of Education (MoEd) of the Republic of Taiwan. The authors are also thankful to Dr. Harshvardhan Kumar, the then Senior Research Fellow at the National Institute of Technolgoty (NIT) Delhi and now a faculty of the Laxmi Niwas Mittal Institute of Information Technology (LNM IIT), Jaipur, for his help in the project in various ways. Ms. Namrata Shaw, Senior Research Fellow at the Institute of Radio Physics and Electronics, is to be thanked for carefully reading some portions of the manuscript. The Executive Committee of IEEE Photonics Society, Kolkata, arranged a few lectures/tutorials on some topics of the book, that provided useful feedback from the audience. The authors are thankful to the committee members.

Prasanta is indebted to his wife Mrs. Chitrani Basu and his sister late Dr. Kaberi Basu for all encouragements and support received and for their sustained efforts to make their beloveds' academic pursuit remarkably free from family duties.

Bratati would like to convey her heartfelt gratitude to her mentor Professor Prasanta Kumar Basu who enkindled her inquisitiveness and made her familiar with the infinite mysteries of Semiconductor Physics. She is also much privileged to enjoy unstinted moral support from her husband Dr. Saibal Bhattacharyya and their daughter Sucheta Bhattacharyya during every struggle and success of her life. They have been constant sources of inspiration throughout the present venture also.

Rikmantra acknowledges with thanks the fruitful collaboration with Professor Guo En Chang of CCU and for arranging several visits to his laboratory under support from AIM-Hi, which expanded his knowledge base. He is also thankful to the authorities of National Institute of Technology Delhi for academic and administrative support and encouragements.

Finally, Rikmantra is indebted to his parents for all support received in his life. No words will be sufficient to express his gratitude to his mother Chitrani and his late aunt Kaberi for sharing their ecstasy with him in his bright periods and for providing moral support and boosting his spirits in the dark periods. They gave all support for his pursuit of academic career.

Contents

List of Figures

List of Tables

1

Introduction

1.1 Introduction to Nanophotonics

The term Nanophotonics means Photonics at the nanometre scale. The term Photonics has been coined to replicate the term Electronics. Therefore, it will not be out of place to briefly discuss first the definition of Electronics, its development, scope, and application areas, which are of course well-known. The definition, aim, and scope of Photonics and Nanophotonics are introduced thereafter.

1.1.1 Electronics: Some milestones

The subject we now call Electronics found its identity when Lee de Forest developed the triode valve in 1906. Earlier the vacuum diode was developed and the behaviour of electrons in vacuum tubes became the subject of investigation by many workers. The word 'Electronics' means the study of the motion of electrons in a medium, control of the motion to produce amplification of electrical signals to generate and shape electrical signals and to detect and process the signals. This naturally led to the development of electronic circuits.

Though early Electronics knew only vacuum tubes, electronic circuits and systems using tubes created wonders. The primary application area was communication, particularly, telephony and later radio broadcasts. In addition to the development of transmitters and receivers, many new methods related to transmission, propagation, and radiative properties of electromagnetic waves cropped up. The electronic properties of materials became the subject of intensive study. Revolutionary concepts about communication techniques, modulation schemes, coding, etc., were put forward. Based on these concepts, the frequency spectrum of electromagnetic (EM) waves, the carrier for communication, went beyond the radio frequency range, to encompass microwaves. That brought in newer devices and passive components in the microwave range. Electronic devices and circuits also showed the ability to control large motors and other electrical equipment and developed production engineering.

The subject Electronics crossed several milestones during World War II. Some noteworthy examples are radio detection and ranging (RADAR), control of aircrafts,

Semiconductor Nanophotonics. Prasanta Kumar Basu, Bratati Mukhopadhyay, and Rikmantra Basu, Oxford University Press.
© P.K. Basu, B. Mukhopadhyay, R. Basu (2022). DOI: 10.1093/oso/9780198784692.003.0001

instruments' landing systems, development of digital computers like electronic numerical integrator and calculator (ENIAC), undersea communication, and many other such.

After World War II, it was the start of the Electronics revolution ushered in by miniaturization of devices and systems. The first breakthrough came with the announcement of transistors, and then after a decade or so, came the integrated circuit (IC). With miniaturization of devices and circuits offering the benefit of low power consumption, there was no looking back. Today, almost all electronic gadgets contain ICs, made of semiconductors, mostly silicon (Si). Remarkable progress in computers, communication, and ICs ushered in the information age.

1.1.2 Photonics

The Greek word *photos* means light. Light had been considered as EM waves until the turn of the twentieth century. However, in order to explain blackbody radiation, Planck had to introduce the corpuscular nature of light. His idea got support from Einstein's explanation of photoelectric effect. The wave-particle duality concept put forward by de Broglie lent support to the concept that light can be thought of as waves as well as particles. The particulate nature of light is described by the term 'photon'. Photons obey Bose statistics, which differ from the statistics of electrons as follows.

As stated already, the word Photonics has been coined to mimic the term Electronics. It is concerned with study of motion of photons in a medium, their generation and detection, as well as control of emission, transmission, modulation, signal processing, switching, amplification, signal shaping, and sensing. In other words, photons are expected to perform all the functions done by electrons. Like electronic circuits, efforts are also being made to realize photonic or all-optical circuits. At present most photonic applications are done by using visible and infrared light, though efforts are being made to extend the spectrum to ultraviolet region as well as in the THz range.

In comparison to Electronics, Photonics is rather a young field. The real progress in the area started in 1960s with the development of laser, first as a solid state (Ruby) laser, then in the form of semiconductor lasers. The development of continuous wave (CW) room-temperature semiconductor lasers and of low loss silica optical fibres led to the rapid progress in the field. Both are responsible for the deployment of optical fibre communication links and networks. An erbium (Er) doped fibre amplifier is also responsible for the development that ultimately paved the way for telecommunications revolution of the late twentieth century and provided the infrastructure for the Internet.

The word 'Photonics' came into the common vocabulary much later. Though the term was coined in the early 1960s and scientists, mostly from Bell Labs, used the term, the word 'optoelectronics' was more in use initially, as in most of the devices and systems the electronic properties at optical wavelengths were exploited. Terms like 'electro-optics' were also in use. In the 1980s telecom data operators adopted fibre-optic data transmission. Around that time IEEE Lasers and Electro Optic Society (LEOS) introduced a journal entitled *IEEE Photonics Technology Letters*. The LEOS has been rechristened as IEEE Photonics Society since 2006.

1.1.3   Coverage of photonics

Photonics encompasses both fundamental studies and application areas. In this section, we shall be mostly concerned with devices and systems.

1.1.3.1   *Photonics in communication*

The most important application area is communication using optical fibres. Fig. 1.1 shows the block diagram of a dense wavelength division multiplexed (DWDM) system that is currently in operation in almost all long haul communication links.

As shown in Fig. 1.1, a number (N) of lasers are used in the link. Each laser emitting at a particular wavelength (for example λ_1, λ_2....) is modulated with voice, picture, and computer data, usually in digital format. The lasers may be directly modulated, or external modulators are used to impress the signal on the light wave emitted by a laser. All the signals coming from N number of sources are combined by a multiplexer, which is a passive waveguide device. The multiplexed signal is then fed to a fibre, a few hundred kms long. At the output end of the fibre, the signals become attenuated and material dispersion of the core of the fibre produces significant spreading of the digital pulses. A regenerator is used to reshape and regenerate the digital pulses in their original form (not shown in Fig. 1.1). A repeater or regenerator, which includes a detector, a laser, and different electronic circuits including a decision circuit, is then employed to reshape the pulses. In the regenerator, a photo detector first converts the weak and distorted optical signal into a stream of cleaned electrical pulses. A laser then converts the cleaned electrical pulses into optical pulses, which then propagate through another long section of the fibre.

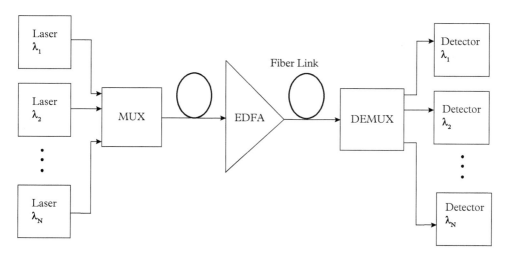

Figure 1.1 *Schematic diagram of a wavelength division multiplexing (WDM) point-to-point communication link.*

At the present time, a number of erbium-doped fibre amplifiers (EDFA) are inserted in the link at regular intervals to amplify the weak optical signals. An add-drop multiplexer (ADM) is used in the link to drop some of the signals at and add new signals from a nearby location. This process repeats over a few more sections of fibre. At the extreme right end of Fig. 1.1 the receiver section is shown. Different wavelengths are first demultiplexed and each signal is converted into electrical signal by a photodetector, which is then amplified electronically and finally processed to extract the originally sent information.

The block diagram and brief description of the operation of the link suggests that a number of active and passive devices are involved in building the whole system. These include laser diodes and associated electronic circuits to bias the lasers, bias control circuits, modulators, amplifiers, detectors and their bias circuits, signal processors, switches, optical fibre, waveguides, multiplexer/demultiplexer, add drop multiplexer(MUX/DEMUX, ADM)s, and many more.

1.1.3.2   *Other application areas of photonics*

Apart from communication, there are many different application areas. Fig. 1.2 summarizes the areas covered by Photonics.

1.1.4   Nanophotonics: A brief introduction

According to US National Academy of Sciences, Nanophotonics means 'the science and engineering of light matter interactions that take place on wavelength and subwavelength scales where the physical, chemical or structural nature of natural or artificial nanostructured matter controls the interactions.'

It may be mentioned at this stage that human beings had used Nanophotonics concepts in their daily life over many centuries, without knowing the principles, of course. An example is Lycurgus cup, kept now at the British Museum, London, probably dating back to the fourth century AD. The different colours originating from the glass are due to metal nanoparticles embedded in it. The glass appears red at places where light is transmitted through it, but at places where light is scattered near the surface, the scattered light appears greenish.

1.2   Nanophotonics: Scope

We now attempt to elaborate the definition and scope of Nanophotonics as given previously. Obviously, Nanophotonics deals with the physical phenomena induced by light rays or by a bunch of photons and exploitation of these phenomena to realize devices and systems of improved performance. The known physical phenomena observed in larger structures may also be present in nanostructures with altered characteristics. For example, absorption, spontaneous and stimulated emissions also take place in nanostructures, but their properties are modified. In addition, new phenomena are observed in nanostructures. In regard to devices, the fundamental working principle may remain the same, but the devices may show better performance or may work at wider range of wavelengths or may be modulated with higher frequency.

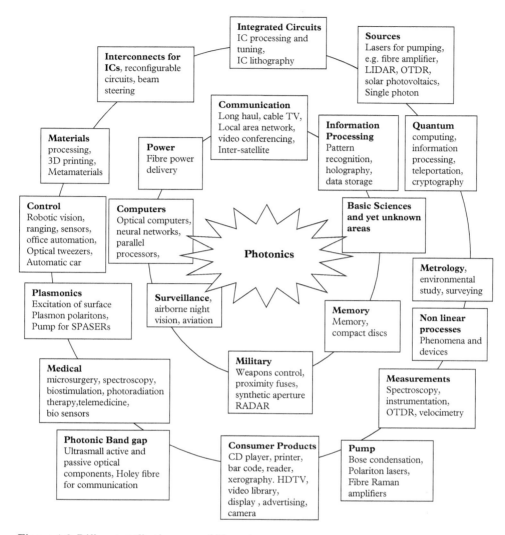

Figure 1.2 *Different application areas of Photonics.*

In the following paragraphs we give a brief outline of the scope of Nanophotonics. In general, Nanophotonics cover more or less four different areas. Our discussion is solely limited to solid state inorganic materials and devices.

1.2.1 Semiconductor nanostructures

The most widely studied area of Nanophotonics focusses on this structure. The development of semiconductor nanostructures is solely due to the advent of sophisticated epitaxial techniques like molecular beam epitaxy (MBE) and metal organic chemical vapour deposition (MOCVD). With these techniques it is possible to grow semiconductor multilayers, each layer having thickness as low as 1 nm. The structures are generally called Quantum Nanostructures, a brief idea of which will be given in Section 1.3. The first member of the family, a Quantum Well (QW), is made by sandwiching a gallium arsenide (GaAs) layer between two layers of aluminum gallium arsenide (AlGaAs) alloy. A rectangular potential well formed in the conduction (valence) band of GaAs quantizes the motion of electrons (holes) along the direction of growth of the GaAs QW.

Semiconductor Nanophotonics owes its origin to the absorption measurement in AlGaAs QWs made by Dingle et al (1974). As pointed out already, semiconductor Nanophotonics is concerned with study of electron-photon interactions in quantum nanostructures, discovery of new phenomena, and design and fabrication of new devices with improved performance. Some of the novel phenomena and devices will be mentioned in Section 1.4.

1.2.2   Photonic band gap structures

Photons, like electrons, possess momentum and have also wave-like properties. In bulk materials the wave propagates along all the three directions. However, when a waveguide is formed by sandwiching a higher refractive index (RI) dielectric (or semiconductor) between two layers of lower RI dielectrics (semiconductors), the EM wave is confined along the thickness of the waveguide layer, but the wave propagates freely along the other two dimensions. This one-dimensional confinement of light is analogous to electron confinement in a QW. Two-dimensional (2D) confinement of EM wave may be obtained in an optical fibre, in which the wave propagates along the axis. Complete confinement is accomplished in a photonic crystal.

In a photonic crystal (PhC) periodic arrangement of high and low RI materials gives rise to a gap in the frequency (ω)-wave vector (k) diagram, just like a band gap in a semiconductor. EM waves whose frequency falls in the gap region cannot propagate through the crystal. Defects can be introduced or states in the gap can be created by modulating the periodicity.

PhCs form an important branch of Nanophotonics. The structures and properties will be discussed briefly in Section 1.3.2.

1.2.3   Light-Matter interaction

One of the important branches of Nanophotonics is the study of light-matter interaction, principally electron-photon interaction, in nanoscale dimensions. In general, the strong confinement of photons in nanometre sized waveguides, leads to an altered density-of-states (DOS) for photons as well as considerable enhancement of the electric field in the region. These in turn give rise to novel phenomena not exhibited by bulk materials.

The altered DOS in waveguides can give rise to Purcell effect (Purcell 1946) predicted as early as in 1946. It alters the spontaneous emission rate in a semiconductor laser and enhances the spontaneous emission coupling factor in a lasing mode. It is possible then to achieve a threshold less laser or a laser with threshold current as low as 1 μA. The increased field strength may even convert the spontaneous emission, otherwise thought to be irreversible, into a reversible process (Yablonovitch 1987). These are some of the examples of novel optical phenomena representing a growing field known as Cavity Quantum Electrodynamics (CQED) in quantum optics.

1.2.4 Plasmonics

While all the three branches of Nanophotonics described previously rely on semiconductors and dielectrics, this emerging branch of Nanophotonics involves metals and dielectrics (Maier, 2006, 2007).

One of the problems of EM wave is the sizable length scale, which is the wavelength of light. For a wave of 1 μm wavelength the device size cannot be made less than this, which is dictated by the diffraction limit. The only parameter to shape light is the RI η of the material which is in the range $1.3 \leq \eta \leq 4.0$. PhCs provide the minimum length scale achievable. To go beyond this limit, that is to go in the subwavelength regime, novel strategies are required. A useful method is to create a plasma wave at a noble metal-dielectric interface. Noble metals like Au, Ag, etc., possess large negative RI at visible wavelengths and contain a large density of free electrons. An oscillating wave of charge carriers, known as plasma wave, is created by the oscillating EM waves. In fact, the propagating wave is a mixture of plasma waves and light waves, called in quantum mechanical language the surface plasmon-polariton. When two metal-insulator layers are brought in close proximity, forming a metal-insulator-metal structure, light is strongly confined in the dielectric gap, breaking the barrier of diffraction limit. Moreover, with reduced thickness of the insulator gap, not only the confinement increases, but also there is a significant reduction of the in-plane wavelength. The value has been as low as 50 nm for a free space wavelength in the range 450 to 650 nm.

A disadvantage of surface plasmon wave is its high absorption loss. This problem has called due attention from workers and schemes for proper amplification have been demonstrated (Bergman and Stockman 2003: Oulton et al 2009; Zheludev et al 2009; Oulton 2012).

The realization of EM waves working in the subwavelength regime opens up the prospect of obtaining tiny photonic devices making way for photonic integrated circuits (PICs) just like electronic ICs grown on Si platform.

1.3 Introduction to nanostructures

In Section 1.2, semiconductor nanostructures, PhCs, and plasmonic nanostructures have been introduced very briefly. In this subsection, preliminary ideas of these nanostructures will be given in some more details.

1.3.1   Semiconductor nanostructures

Electrons and holes behave as classical particles in bulk semiconductors. By assigning different effective masses to electrons and holes, the particle motion is governed by Newton's laws. However, when one or more dimensions in a bulk semiconductor material are reduced to a few nanometres, quantum mechanical principles are needed to describe the behaviour of the particles. The structures are then called Quantum Nanostructures. The dimension of such nanostructure should be below some critical length, which may be defined in several ways. One such length scale, the de Broglie wavelength of the particle, is more in use. It is expressed as $\lambda = h/p$, where h is Planck constant and p is the momentum of the electron or hole. If one particular dimension of the structure is less than the de Broglie wavelength, then the motion of the electron or hole gets quantized in that direction. The behaviour is governed by the wellknown example of the 'particle in a box' treated in all textbooks on quantum mechanics. In Fig. 1.3, an electron is, shown confined in a rectangular potential well of width L having infinite barrier heights at both the boundaries. In this situation, the electron wavefunction gets reflected at the boundaries and forms a standing wave pattern. Since the electron's free movement is now restricted along this direction, its motion gets quantized, and the quantized energy levels are given by

$$E_n = \frac{\hbar^2}{2m}\left(\frac{n\pi}{L}\right)^2, n = 1, 2, 3, \ldots \ldots \tag{1.1}$$

where m is the mass of the electron.

The concept can be applied to a semiconductor material the thickness L, along the z-direction should be less than the de Broglie wavelength. In a practical semiconductor structure, it is not possible to have infinite potential barrier at the two ends. However, in many cases the ideal situation is assumed, and the electron energy levels become quantized as given by Eq. (1.1) with mass m replaced by the electron effective mass m_e. A

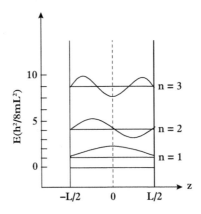

Figure 1.3 *Quantized energy levels and wavefunctions of a particle in a box problem.*

few quantized energy levels and the corresponding wavefunctions of electrons are shown in Fig. 1.3. The electrons are however free to move along two other directions, x and y, and the resulting electron gas is called the two-dimensional electron gas (2DEG). The structure is called a QW. It may be realized by sandwiching a thin layer of GaAs between two layers of $Ga_{1-x}Al_xAs$. There are, apart from the GaAs/GaAlAs combination, many different pairs of semiconductors, either lattice matched or mismatched to each other, that are used to fabricate QW structures.

One may move one step forward by considering a semiconductor having dimensions less than the de Broglie wavelength in two directions: the y and z directions (Arakawa and Sakaki 1982). The electrons cannot move freely in both these directions and have free motion along the x direction only. The structure is called a quantum wire (QWR) and it supports a one-dimensional electron gas (1 DEG). This structure is shown in Fig. 1.4 in a rectangular shape. However, cylindrical configuration is also possible, when the structure is called a nanowire (NWR).

Consider now that the dimension of the semiconductor is less than the de Broglie wavelength in all three directions (Arakawa and Sakaki 1982). The motion of the electron is then inhibited in all three dimensions. The electron energy levels are quantized and become very sharp like energy levels of atoms. The structure, as shown in Fig. 1.4 is box-like and is called a quantum box (QB) structure. The electron gas a QB supports is zero-dimensional (0 DEG) with no degrees of freedom. The QB structure is seldom realized, however. The practical structures, mainly grown via self-assembly, are either of pyramid shape or dome shaped, and are generally named as quantum dot (QD) structures.

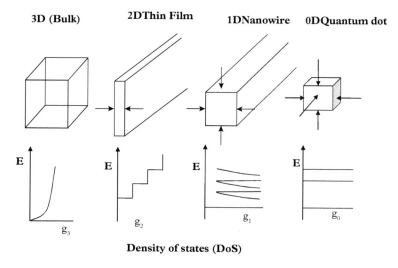

Figure 1.4 *Schematic diagrams of bulk, thin film QW, nanowire, and QD structures (upper part) and the corresponding DOS function (lower part).*

An interesting feature of quantum nanostructures is the change of DOS function for the electrons. Fig. 1.4 shows how this DOS function progressively alters from bulk to QW, to QWR, and finally to QB structures.

1.3.2   Photonic bandgap structures

Multiple layers of two dielectrics having different RIs have been in use in optoelectronics over a long period. These are Bragg gratings (BGRs) having a very high reflectivity (~ 99% or more) and are used in vertical cavity surface emitting lasers (VCSELs). In other words, the BGRs have sharply defined pass and stop bands. They indeed represent a one-dimensional photonic crystal structure. However, since grating structures like these have been used for many years before the advent of modern photonic crystals, it is not common to call them photonic crystals (John 1987; Joannopoulos et al 2007).

A periodic variation of RI in two- and three-dimensions may be created on a single chip by multiple coupled photon confining structures. The periodicity may be of the order of the wavelength of the photons. The resulting structure is a PhC. An example of two-dimensional (2D) PhC lattice is given in Fig. 1.5, in which (a) shows a triangular lattice and (b) shows a square lattice. Both the lattice constant a and the radius of perturbation Γ should be a few hundredths of nanometres for 1 μm free space wavelength of light.

The method of realizing 2D PhCs is illustrated in Fig. 1.6. In Fig. 1.6(a) a lattice of higher RI rods is produced in a layer of lower index material, say, air. Alternatively, as illustrated in Fig. 1.6(b), a lattice of air holes (lower RI) may be etched in a layer of higher index material, say Si. These lattice structures must possess a high aspect ratio, which is height divided by width.

Photons in a PhC behave in the same way as the electrons behave in a semiconductor crystal. Hence the term PhC is introduced. In a semiconductor, electrons move in a periodic potential. Their behaviour is governed by a Schrödinger equation, which is

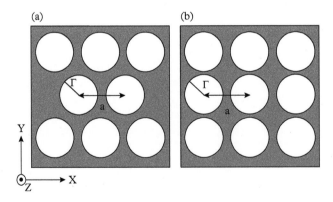

Figure 1.5 *(a,b) Diagram of common two-dimensional photonic crystal structures: (a) a triangular lattice and (b) a square lattice.*

(a)

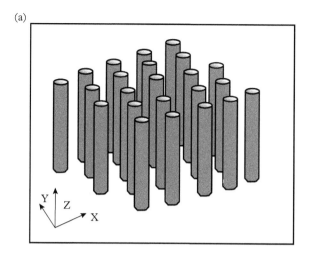

(b)

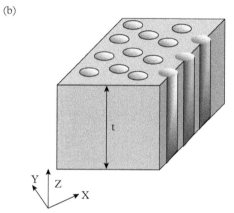

Figure 1.6 *(a,b) Two-dimensional photonic crystals:*
(a) dielectric rods in air and (b) holes in higher-index material.

solved by taking into account the periodicity. The solutions give rise to conduction and valence bands separated by a band gap. The results may be summarized in the E-k diagram, which is quite wellknown.

For photons, one needs to solve the Helmholtz equation in presence of a periodic variation of RI. The process yields a very similar dispersion relation: a plot of energy (frequency ω) vs. wavevector (k). A schematic of the ω-k relation is given in Fig. 1.7. A common point in the dispersion relations is the existence of a gap called the 'photonic band gap'. In this forbidden zone, no wave can propagate.

It may be pointed out that the solution of a wave equation in a periodic variation of RI is more complicated that its counterpart in a semiconductor, Schrödinger equation. While a closed form analytical equation may be derived for the latter, no such form

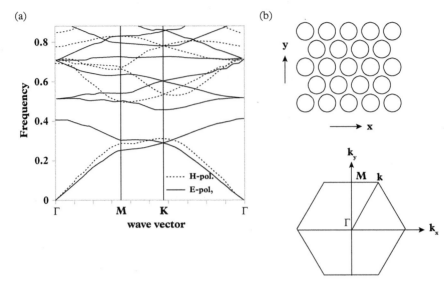

Figure 1.7 *(a) Photonic band structure of a 2D silicon photonic crystal with a hexagonal arrangement of cylindrical air holes; (b) the schematical diagram of the hexagonal lattice (top) and the corresponding first BZ of reciprocal lattice with high symmetry points (bottom) are shown in the right hand side.*

Source: Birner A, Wehrspohn R B, Gösele U M, and Busch K (2001) *Advanced Materials* 13(6): March 16: 377–388. Redrawn with permission; copyright: John Wiley (2001).

exists for PhCs. The solutions are obtained by computer-aided programmes based on finite-difference time-domain (FTTD) analysis.

The calculated ω-k relation for photons propagating in a hexagonal Si lattice with air holes in between (Birner et al 2001) is shown in Fig. 1.7. The dispersion relation is shown for the first Brillouin zone. The corner points of the zone Γ (= 0), X (= π/ax) and M (= π/x + y) are marked in the diagram. The bandgaps for TE photons are easily identified. If the frequency of EM waves falls within this forbidden zone, the wave cannot propagate. The plots in Fig. 1.7 resemble the E-k diagram for a semiconductor.

1.3.3   Metallic nanostructures

Surface plasmon waves are supported by various types of waveguides. A M-I-M waveguide has briefly been introduced in Section 1.2.3.

Surface plasmon polariton waveguides may be of different types (Berini and de Leon 2011). The simplest one is a one-dimensional (1D) waveguide consisting of a single metal-dielectric interface. A metal film sandwiched between two symmetric dielectric claddings and a dielectric film sandwiched between two symmetric metal claddings can also act as 1D waveguide. Examples of 2D waveguides are dielectric-loaded, metal stripe, gap, low-index hybrid, wedge, and channel waveguides. The structures are similar to 2D dielectric waveguides in common use.

1.3.3.1 *Nanoantenna*

Conventional antennas are widely used to transmit radio or TV signals. Nanoantennas working at optical frequencies have nanoscale dimensions and are plasmonic nanostructures used to generate surface plasmons. As the oscillation frequency of the surface plasmon matches that of the incident EM waves, a localized 'surface plasmon resonance' (LSPR) occurs. This creates an unprecedented level of light intensity in a very small space—around 100 nm^3. It may thus be possible to illuminate a single molecule by a single photon. The simplest nanoantenna is a single metallic nanoparticle, the free electrons in which can support LSPR at visible wavelengths. Any object brought into this so-called locally confined field—or 'nanofocus'—will affect the LSPR in such a way that it can then be detected using a technique called dark-field microscopy—a technique where only scattered light makes up an image. Moreover, an optically excited nanoparticle can radiate light in a controlled way.

A small, isolated nanoparticle can work as an electrical dipole and radiate with a broad angular distribution. However, using the concepts developed for radio frequency (RF) antenna, the directivity can be controlled, and scattering from an array of nanoantenna can also be monitored.

The large enhancement of light intensity in a nanoantenna leads to investigations in quantum optical phenomena as well as new applications. Nanoantennas coupled to single quantum emitter can give rise to highly directional spontaneous emission. Due to large light intensity, there is a large enhancement of Purcell factor, a quantitative measure of light-matter interaction, as high as 1000. The brightness of the fluorophore has been increased 1,000-fold or more, thus improving fluorescence microscopy in biological systems. The LSPR can be used for sensing flammable molecules and observation of single molecules and study of infrared (IR) vibrational spectroscopy by use of sharp metal tips to enhance field amplitude.

1.3.3.2 *Metamaterials and metasurfaces*

Metamaterials are artificial materials. Unlike natural materials, they must be prepared by special techniques. The remarkable property of metamaterials lies in their peculiar EM behaviour as their permittivity or permeability can be controlled. The negative values of μ and ε merit special mention in this respect, see (Solymar and Walsh (2009); Koenderink et al 2015).

Metamaterials at radio frequencies can be synthesized by using a resonant circuit comprising an inductance and a capacitance. This can be realized by using a split-ring in which the C-shaped ring acts as an inductance and the gap acts as a capacitor. Subsequently, many other structures, in particular, periodically stacking, split-ring, and thin wires are used as metamaterials at microwave frequencies.

Metallic nanoparticles can be used not only as optical resonators concentrating light within, but also to realize optical metamaterials. Such optical metamaterials can be formed by ordered or disordered arrangements of resonant nanoscale photonic scattering elements.

Optical metamaterials, like their microwave brethren, exhibit unusual EM properties. A metamaterial slab composed of 3D chiral helix antenna, offers large selectivity to circular polarization. The materials can refract light in unexpected directions. A very interesting phenomenon, the negative refraction, can be exploited to produce *invisibility cloaks*. An object covered by a metamaterial shell remains undetectable at a certain frequency range. However, the materials show strong frequency dispersion as well as large absorption, limiting the practical applicability of the passive metamaterials.

There is a need for reduction of loss, and at the same time the devices need be integrated on a large surface area. As a result, attention is now shifted from 3D metamaterials to 2D metasurfaces. Metasurfaces are planarized, ultrathin, and patterned artificial surfaces. It is possible to have functionalities of conventional optics and metamaterials in these metasurfaces. Since light does not propagate along one direction, absorption losses are reduced. Optical resonances induced by plasmonic nanomaterials are strongly localized. One can observe anomalous refraction, reflection, and control of light in the subwavelength region. Ultrathin lenses, beam steering devices, and generation of angular momentum of light are some examples of how the metasurfaces can be utilized. Use of metasurfaces as a substrate for all-optical circuitry needed for filtering, signal processing, and even computing has been investigated.

1.4   Novel phenomena in Nanophotonics: A brief outline

Many novel physical phenomena are exhibited by nanophotonic structures. Table 1.1 gives a list of such phenomena, which is by no means complete. As semiconductor nanostructures have been available since the mid-1970s, most of the entries in the table represent the physical processes observed in such structures.

1.5   Applications of Nanophotonics

All the application areas indicated in Fig. 1.1 pertain to Nanophotonics as well. In fact many areas as shown there have opened up with the availability of nanophotonic structures. A few such areas have already been mentioned in Sections 1.3 and 1.4.

In semiconductor nanophotonic devices, both light and electron are confined within the nanometre scale. The resulting strong light matter interaction brings about several advantages. For example, enhancement of spontaneous emission due to Purcell effect leads to very low threshold or almost thresholdless laser operation. The reduced power consumption helps improve the multi terabit global Internet traffic. Nanophotonic technologies are applied to materials processing and 3D printing, medical diagnostics,

Table 1.1 *Some physical phenomena observed in Nanostructures*

Phenomenon/ Process	Brief description	Structure used	Application
Absorption spectra in QWs, QWRs, QDs	Spectra follows the DOS	QW, QWR, QD	Detectors
Recombination spectra in QWs, QWRs, QDs	Enhanced stimulated recombination	As above	Lasers, amplifiers, switches
Excitonic absorption	Sharper and enhanced absorption with reduced dimension	As above	Characterization, emitters
Intersubband absorption	Absorption at longer wavelengths than band to band absorption in bulk	As above	QWIP, QDIP working at mid-IR and longer wavelengths
Intersubband emission	Emission at longer wavelengths, even at THz	As above	QCL at mid IR and longer wavelength, THz laser
Quantum Confined Stark Effect (QCSE)	Shift of excitonic absorption peak with electric field	As above	Electroabsorption modulator
Photovoltaic effect	Generation of voltage due to absorption of light	As above	Solar cells
Photorefractive effect	Light induced change in RI	As above	Dynamic holography
Nonlinear optics	Two photon absorption, degenerate four wave mixing, harmonic generation	As above; stronger effect with reduced dimension	Switch, wavelength converter
Cavity QED: Purcell effect	Inhibited or enhanced spontaneous and stimulated emission	Semiconductor nanostructures enclosed in high Q cavity (PhC)	Threshold less or ultralow threshold laser
Cavity QED: Rabi oscillation	Reversibility of spontaneous emission	As above	Fundamental studies

Continued

Table 1.1 *Continued*

Phenomenon/ Process	Brief description	Structure used	Application
Slow, fast, and stopped light	Change in group velocity	As above	Optical delay line, optical buffer, optical memory
Laser cooling	Cooling of a substance by a laser spot	Rare earth doped fibre, Semiconductor nanostructures	Optical refrigeration, cooling of ICs, cryogenic coolers for space-bourn equipment
Optical processes in PBG structures	Propagation of EM waves	PhC	Filters, mirrors, waveguides, bends, all-optical switches, all-optical memory and logic, modulator, detectors, lasers
PBG structures: Holey fibre		2D PhC	Communication
Surface Plasmon Polariton	Propagation of hybrid light-plasma waves at metal insulator interface	Plasmonic structures	Light propagation at subwavelength regime; interconnect for ICs, Photonic ICs, SPASERs, and other components, convergence of Electronics and Photonics
Plasmoelectric effect	Generation of electric potential by optical illumination	Plasmonic structure	Photovoltaic cells
Non-reciprocal flow of light	Flow of light along a forward direction	Plasmonic structure	All optical isolator
Strongly confined light intensity	Highly directional spontaneous emission, enhanced Purcell factor	Plasmonic nanoantenna	Detection of flammable molecules, biological sensors
Unusual optical phenomena	Negative refraction, generation of angular momentum of light	Optical metamaterials and metasurfaces	Ultrathin planar super lenses, beam steering, photonic circuitry

and sensing. The devices form key components quantum computing, cryptography, entangled photon emitters operating at high qbit rates, in cyber security and in many other emerging areas.

The tiny size of nanophotonic devices: both active and passive,opens up the possibility of realizing integrated optical circuits with less cost and space and increased functionalities. Infinite flexibility of design is possible with changing dimensions, shapes, spacings, and materials by changing dimensions of physical structure, composition of the materials, and combinations of the materials, as in a hetrostructure.

There are many areas needing optical circuits with reduced cost and size and more functionality other than the emerging areas as mentioned previously. These include consumer Electronics systems for optical data storage, digital imaging and display, industrial optics, e.g., sensors and control systems, and optical communication and networking.

Realization of a cost-effective integrated optoelectronic or photonic circuits is beset with many difficulties. First, different materials are used in making different components, unlike in Si-based ICs. Second is the size mismatch between passive optical components and active semiconductor-based photonic devices. Both these technological issues create major concern to the community and efforts are afoot globally to arrive at a viable solution to these problems.

Introduction of engineered structure like the photonic band gap or photonic crystal structures has expanded the horizon of Nanophotonics. It is now possible to realize passive components like waveguides with sharp bends conducive to miniaturized circuits. Another artificial structure is metamaterials or metasurfaces, the development of which has led to newer application areas, that could not even be dreamt earlier. Many other and novel application areas are being explored by using such materials.

To end this section, we should mention electronic-photonic integration by combining electronic systems and photonic system on the same chip, ideally on Si platform. Section 1.6 deals with this and points out the ultimate need, plasmonics replacing both Electronics and Photonics for tinier, speedier, and greener system.

1.6   Problems of integration

Si-based very large-scale integration (VLSI) technology has reached a remarkable height. The feature size of transistors has been reduced to 5 nm and even less, allowing a higher number of transistors within a processor. However, with shrinking of the transistor size, the interconnects should also be scaled down. This not only increases the delay of transfer of signal as the metal interconnect behaves as a resistance-inductance capacitance (RLC) circuit, but also increases the power dissipation of the circuit.

One solution to the 'interconnect bottleneck' is to use an optical link replacing metal interconnection. Use of optics increases the speed of the system and at the same time reduces the power dissipation. The method has been successfully applied for board-to-board and chip-to-chip connections. The ultimate aim is to use optics for intrachip interconnect. However, the links must be compatible with complementary metal-oxide semiconductor (CMOS) fabrication technology and at the same time tiniest active devices like lasers, modulators, and detectors need be used. The minimum size of photonic devices is however limited by fundamental laws of diffraction.

1.6.1   Ultimate Electronics and Photonics

The present information age owes its existence to the unprecedented development in computers and communication. These developments are possible due to advances in VLSI technology based on Si and also in the fibre-based communication. However, both Electronics and Photonics have inherent limitations that are now showing up. Electronic circuits based on semiconductors including Si and other materials enable realization of nanoscale elements for computation, signal processing, and information storage. Optical fibres facilitate information transfer at a high speed over a long distance. Unfortunately, however, the speed of electronic circuits tends to saturation due to interconnect delay. Photonic devices do not reach and are not expected to reach nanoscale dimensions due to the diffraction limit and therefore make realization of truly photonic ICs, making use of Electronics-Photonics integration on the same chip a remote possibility.

The real solution for both high-speed and nanoscale size lies in plasmonics. It offers both the size of present-day Electronics and the speed of photonic link.

The advantages of plasmonics or metal-based Nanophotonics over Electronics and Photonics are clearly indicated in Fig. 1.8, in which the operating speed versus device dimension of electronic and photonic devices as well as those of plasmonic devices are shown (Brongersma and Shalaev 2010; Hu et al 2011). The lower region in left signify an electronic domain dominated by semiconductors. The size for an individual device has reached 3 nm, but the speed is limited to about 10 GHz due to heat generation and interconnection delay. The top rectangle at the right represents the Photonics' domain, the materials of choice being insulators and semiconductors. While the speed of devices and systems approaches THz limit, the size is limited by fundamental law of diffraction. The size acts as a deterrent to realizing photonic ICs of a footprint comparable to electronic ICs. The top region in the left belongs to plasmonics, in which devices can break the diffraction limit and at the same time show high speed.

As already stated, electronic devices show smaller dimensions with lower speeds, and photonic devices show higher speeds but larger sizes. Both small size and high speed, coupled with less power dissipation, are characteristics of plasmonic devices. It can serve as a bridge between Photonics and Nanoelectronics. As a result of intense research in the area for over two decades, the loss process in metals in plasmonic structures has recently been taken care of (Azzam et al 2020).

While the realization of truly plasmonic systems is beyond today's capabilities and too futuristic, there are efforts to work with hybrid electronic-photonic-plasmonic systems (Obzay 2006; Notomi et al 2011). A useful review on this is presented by Hu et al (2011); Papaioannou et al (2012); and Pleros (2018). Plasmonics for

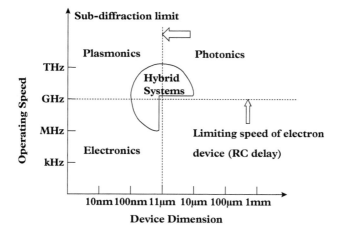

Figure 1.8 *Speed versus dimension of electronic, photonic, and plasmonic devices. The dashed lines indicate approximate physical limitations of different devices. The hybrid systems employ the combination of Electronics, Photonics, and plasmonics.*

telecommunication applications has been reviewed by Carvalho and Mejía-Salazar (2020). A monolithic bi-CMOS electronic-plasmonic transmitter has been fabricated and tested by Kohl et al (2020).

Reading List

Basu P K (1997) *Theory of Optical Processes in Semiconductors, Bulk and Microstructures.* Oxford, UK: Oxford University Press.

Gaponenko S V (2010) *Introduction to Nanophotonics.* Cambridge, UK: Cambridge University Press.

Hunsperger R G (2009) Nanophotonics. In: *Integrated Optics: Theory and Technology*, 6th edn. pp. 469–497. New York: Springer Sc. Business Media.

Joannopoulos J, Johnson S G, Meade R, and Winn J (2007) *Photonic Crystals, Moulding the Flow of Light*, 2nd edn. Princeton, NJ: Princeton University Press.

Koenderink A F, Alù A, and Polman A (2015) Nanophotonics: Shrinking light-based technology. *Science* 348(6234): 516–521.

Maier S A (2007) *Plasmonics: Fundamentals and Applications.* New York: Springer.

Prasad P N (2004) *Nanophotonics.* New York: Wiley-Interscience.

Solymar L and Walsh D (2010) Electrical properties of materials. In: *Artificial Materials and Metamaterials*, 8th edn. Chapter 15. Oxford: Oxford University Press.

Zalevsky Z and Abdulhalim I (2014) *Integrated Nanophotonic Devices*, 2nd edn. Amsterdam: Elsevier.

References

Arakawa Y and Sakaki H (1982) Multidimensional quantum well laser and temperature dependence of its threshold current. *Applied Physics Letters* 40(11): 77–78.

Azzam S I, Kildishe A V, Alexander V, Ma Ren-Min, Cun-Zheng Ning, Rupert Oulton, Vladimir M Shalaev, Mark I Stockman, Jia-Lu Xu, and Xiang Zhang (2020) Ten years of spasers and plasmonic nanolasers. *Light: Science & Applications* 9(90): 1–21.

Bergman D J and Stockman M I (2003) Surface plasmon amplification by stimulated emission of radiation: Quantum generation of coherent surface plasmons in nanosystems. *Physical Review Letters* 90: 027402.

Berini P and De Leon I (2011) Surface plasmon–polariton amplifiers and lasers. *Nature Photonics* 6: 16–24.

Birner A, Wehrspohn R B, Gösele U M, and Busch K (2001) Silicon based photonic crystals. *Advanced Materials* 13(6): 377–388.

Mark L Brongersma and Vladimir M Shalaev (2010) The case for plasmonics. *Science* 328(April): 440–441.

Carvalho W O F and Mejía-Salazar J R (2020) Plasmonics for telecommunications applications. *Sensors* 20(2488): 1–21.

Dingle R, Wiegmann W, and Henry C H (1974) Quantum states of confined carriers in very thin AlGa As-GaAs-Al Ga As heterostructures, *Physical Review Letters* 33: 827–830.

Hu E L, Brongersma M, and Baca A (2011) Applications: Nanophotonics and plasmonics. In: *Nanotechnology Research Directions for Societal Needs in 2020*, M C Roco, C A Mirkin, and M C Hersan (eds.) Chapter 9, pp. 417–444. Dodrecht: Springer.

John S (1987) Strong localization of photons in certain disordered dielectric superlattices, *Physical Review Letters* 58(23): 2486–2489.

Kohl U et al (2020) A monolithic bipolar CMOS electronic–plasmonic high-speed transmitter. *Nature Electronics* 3: 338–345.

Maier S A (2006) Plasmonics: Metal nanostructures for subwavelength photonic devices. *IEEE Journal of Selected Topics in Quantum Electronics* 12: 1214.

Notomi M, Shinuya A, Nozaki K, Tanabe T, Matsuo S, Kuramochi E, Sato T, Taniyama T, and Sumikura H (2011) Low-power nanophotonic devices based on photonic crystals towards dense photonic network on chip. *IET Circuits, Devices and Systems* 5(2): 84–93.

Oulton R F, Sorger V J, Thomas Zentgraf T, Ma R-M, Christopher Gladden C, Dai L, Bartal G, and Zhang X (2009) Plasmon lasers at deep subwavelength scale. *Nature* 461: 629–632.

Oulton R F (2012) Surface plasmon lasers: Source of nanoscopic light. *Materials Today* 15(12): 26–34.

Ozbay E (2006) Plasmonics: Merging photonics and electronics at nanoscale dimensions. *Science* 311: 189–193.

Papaioannou S, Vyrsokinos K, Kalavrouziotis D, Giannoulis G, Apostolopoulos D, Avramopoulos H, Zacharatos F, Hassan K, Weeber J -C, Markey L, Dereux A, Kumar A, Bozhevolnyi S I, Suna A, de Villasante O G, Tekin T, Waldow M, Tsilipakos O, Pitilakis A, Kriezis E, and Pleros N (2012) Merging Plasmonics and Silicon Photonics Towards Greener and Fast er Network-on-Chip. Solutions for Data Centers and High-Performance Computing Systems. In: *Plasmonics – Principles and Applications, 2012*, Papaioannou et al (eds.) Chapter 21. London, UK: InTech Open Ltd.

Pleros N (2018) Silicon photonics and plasmonics towards network-on-chip functionalities for disaggregated computing, 2018 Optical Fibre Communications Conference and Exposition (OFC). San Diego, CA, USA: IEEE.

Purcell E M (1946) Spontaneous emission probabilities at radio frequencies. *Physical Review* 69: 681.

Yablonovitch E (1987) Inhibited spontaneous emission in solid-state physics and electronics. *Physical Review Letters* 58(20): 2059–2062.

Zheludev N I, Prosvirnin S L, Papasimakis N, and Fedotov V A (2008) Lasing spaser. *Nature Photon* 2: 351–354.

2

Basic properties of semiconductors

2.1 Introduction

Electronic and optical properties of semiconductor nanostructures have been the subject of intensive studies since the early 1970s. Semiconductor based nanophotonic devices have been fabricated and used in practical applications almost at the same time.

In the present chapter, the basic properties of semiconductors and their nanostructures will briefly be discussed. The presentation in the chapter is rather sketchy, as a number of excellent text books and research monographs are available which discuss at length the basic physics and application areas. A list of such books and monographs is given at the end of this chapter to help readers understand the phenomena occurring in semiconductor bulk and nanostructures, and principles of operations of the devices. The present chapter gives a summary of the properties, most relevant equations without detailed derivation, and elementary ideas about how the structures are made. Important optical properties of nanostructures will be discussed in later chapters.

The present chapter introduces the basic properties of bulk semiconductor. It discusses the band diagram, density-of-states (DOS), carrier concentration, transport of carriers. The concept of excess carriers and recombination is then given, followed by a brief discussion of excitons. After introducing semiconductor heterojunctions, the concept of quantization and the formation of two-dimensional electron gas (2DEG) in a Quantum Well (QW) are discussed. The idea is then extended to the case of Quantum Wires (QWRs) and Quantum Dots (QDs). The formation of multiple QWs and superlattices is then discussed. Finally, the concept of two-dimensional (2D) excitons is introduced.

Most of the Nanophotonics related phenomena to be discussed in this book are observed in QDs and devices based on the principles employ semiconductor QDs. The electronic and optical processes in QDs are discussed in a later chapter.

2.2 Band structure

Semiconductors are generally single crystals consisting of regular repetition of a certain basic structural unit called unit cell. As most of the semiconductors have zincblende or

Semiconductor Nanophotonics. Prasanta Kumar Basu, Bratati Mukhopadhyay, and Rikmantra Basu, Oxford University Press.
© P.K. Basu, B. Mukhopadhyay, R. Basu (2022). DOI: 10.1093/oso/9780198784692.003.0002

diamond structures where each atom is surrounded by four of its neighbouring atoms, the concept of an energy band comes into the picture, instead of the discrete energy level.

The formation of energy bands in a crystal is described by considering a one-dimensional (1D) chain of atoms to represent the periodic structure. The simplest models to understand the formation of bands in a 1D chain of atoms are the Kronig–Penny model and the tight binding model. According to these models, there are several allowed energy bands where the electrons are free to move with a momentum $\hbar k$ and the allowed bands are separated by a forbidden energy bandgap. The allowed values for $\mathbf{k}$ are restricted within the reduced Brillouin zone.

However, a perfect crystal consists of many valence electrons and ions. The many electron problem is reduced to single electron problem by several levels of approximation and finally the one-particle Schrodinger equation is solved.

For an electron in a periodic potential

$$V(r) = V(r + R) \tag{2.1}$$

where $R = n_1 a_1 + n_2 a_2 + n_3 a_3$ and a_1, a_2, a_3 are the lattice vectors, and n_1, n_2, and n_3 are integers, the electron wavefunction satisfies the Schrodinger equation

$$H\psi(r) = \left[-\frac{\hbar^2}{2m_0} \nabla^2 + V(r)\psi(r) = E(r)\psi(r) \right] \tag{2.2}$$

The Hamiltonian remains constant under translation by the lattice vectors, $r \rightarrow r + R$. Thus $\psi(r + R)$ will be different from $\psi(r)$ by a constant of unity magnitude to prevent the wavefunction from growing infinitely by the repetitive translation of $\mathbf{R}$. The general solution of the previous equation is given by

$$\psi_{nk}(r) = exp(ik.r)u_{nk}(r) \tag{2.3}$$

where $u_{nk}(r + R) = u_{nk}(r)$ is a periodic function. The wavefunction $\psi_{nk}(r)$ is called Bloch function. The Bloch functions are eigen functions of the one-electron Schrodinger equation, and therefore they are orthogonal to one another. Thus,

$$\int \psi_{n'k'} \psi_{nk} d^3 r = \delta_{n'n} \delta_{k'k} \tag{2.4}$$

The energy is given by

$$E = E_n(k) \tag{2.5}$$

where n refers to the band and $\mathbf{k}$ the wave vector of electron. The plots of energy (E) as a function of wave vector ($\mathbf{k}$)are known as the dispersion relation.

In most cases, it is sufficient to consider the energy levels within a few times of the thermal energy from the band extrema. We therefore consider the properties of electrons and holes near the edges of the conduction and valence bands. The E-k relationship, near the band edges, can be approximated by a quadratic equation

$$E(\mathbf{k}) = \frac{\hbar^2 k^2}{2m^*} \tag{2.6}$$

where m^* is called effective mass. Its value is different from a free electron mass as the electrons in the crystal lattice are always under the effect of periodic potential of the lattice. As a result, the free electron mass must be modified according to the periodicity of crystal potential to consider the electrons in the crystal are essentially free. It is to be mentioned here that, the previous equation corresponds to a band of a single scalar mass which is absolutely independent on the direction of the **k**-vector. This corresponds to the spherical constant energy surfaces which are applicable for the conduction band of many direct band gap semiconductors like gallium arsenide (GaAs).

On the other hand, the dispersion relation for some semiconductors can be written as

$$E(\mathbf{k}) = \frac{\hbar^2}{2} \left[\frac{(k_x - k_{0x})^2}{m_x} + \frac{(k_y - k_{0y})^2}{m_y} + \frac{(k_z - k_{0z})^2}{m_z} \right] \tag{2.7}$$

This represents a band with ellipsoidal surfaces of equal energy and in this case, the effective mass depends on the **k**-vector and effective mass tensor with components is given by

$$m_{ij} = \frac{\hbar^2}{(\partial^2 E / \partial k_i \partial k_j)} \tag{2.8}$$

This can be illustrated considering six conduction band valleys in silicon (Si) having ellipsoidal constant energy surfaces. In this case there are six minima. The energy measured from the conduction band edge varies parabolically with the value of **k**. The effective mass for an electron is not isotropic but is a tensor for each valley. If the axes coincide with the highly symmetric axes, the E-k relation can be written as

$$E(\mathbf{k}) = E_{k0} + \frac{\hbar^2}{2} \left[\frac{(k_x - k_{0x})^2}{m_l} + \frac{(k_y - k_{0y})^2}{m_t} + \frac{(k_z - k_{0z})^2}{m_t} \right] \tag{2.9}$$

In this case, ellipsoidal constant energy surfaces are proportional to $m_l^{1/2}$ and $m_t^{1/2}$ respectively, where m_l and m_t being the longitudinal and transverse effective mass respectively. These ellipsoidal constant energy surfaces are shown in Fig. 2.1.

There are three valence bands; they are heavy-hole (HH), light-hole (LH), and split-off (SO) bands. For HH and LH bands, the E-k relationship near the band edges can be calculated by using k.p perturbation theory of Basu et al (2015)(Appendix A).

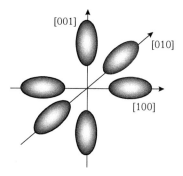

Figure 2.1 *Six degenerate conduction band valleys in Si.*

The dispersion relation for the valence band is now given (Luttinger and Kohn 1955).

$$E(\mathbf{k}) = Ak^2 \pm \left[B^2 k^4 + C^2 \left(k_x^2 k_y^2 + k_y^2 k_z^2 + k_z^2 k_x^2\right)\right]^{1/2} \tag{2.10}$$

A, B, and *C* are different constants related to the momentum matrix element (Appendix A). The + and − signs are for LH and HH bands, respectively. The constant energy surfaces described by Eq. (2.11) are not spherical for $C \neq 0$ and are known as warped surfaces.

The E-**k** relationship for the split-off (SO) band is given by

$$E(\mathbf{k}) = -\Delta + Ak^2 \tag{2.11}$$

Here, Δ is the amount of energy by which SO is shifted from HH or LH band as shown in Fig. 2.2 and constant A is same as that in Eq. (2.10).

The detailed band structures of some important semiconductors are given by Cohen and Chelikowsky (1989).

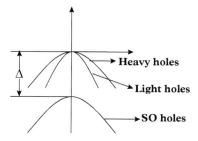

Figure 2.2 *Valence band structure.*

2.3   Density of states

Let us consider a cubic crystal of dimension L in each of the x-, y-, and z-directions. The electron wavefunction given by Eq. (2.3) will vanish at the crystal boundaries if $k_x L = 2\pi l$, $k_y L = 2\pi m$, and $k_z L = 2\pi n$, where l, m, and n are integers. Each allowed value of k occupies a volume $(2\pi/L)^3$ in $\mathbf{k}$ space as the difference between two adjacent k_x values is $2\pi/L$. Assume two concentric spheres of radii k and $k + dk$ in $\mathbf{k}$-space. Hence, the total number of quantum states in that volume, taking into account two possible values of spin, will be given by

$$dN = 2 \times \frac{4\pi k^2 \, dk}{(2\pi/L)^3} \tag{2.12}$$

Considering first an isotropic semiconductor having a constant spherical energy surface for an electron, the dispersion relation can be written as

$$E(\mathbf{k}) = E_C + \frac{\hbar^2 k^2}{2m_c^*} \tag{2.13}$$

where m_c^* is the effective mass of the electron. Therefore, one can obtain

$$k^2 = \frac{2m_c^* (E - E_C)}{\hbar^2} \tag{2.14}$$

and

$$k\, dk = \frac{m_c^*}{\hbar^2} \, dE \tag{2.15}$$

Using Eq. (2.15) and Eq. (2.16), one can easily obtain the DOS per unit volume per unit energy interval from Eq. (2.13) as

$$dN = \frac{1}{2\pi^2} \left(\frac{2m_c^*}{\hbar^2} \right)^{3/2} (E - E_C)^{1/2} \, dE \tag{2.16}$$

This relation for the DOS in the conduction band is valid only for a semiconductor having an isotropic electron effective mass m_c^*. This expression must be modified for those in which crystallographic anisotropy exists and effective masses are different in different directions of $\mathbf{k}$-vector.

In that case, the dispersion relation can be written as

$$E(\mathbf{k}) = E_C + \frac{\hbar^2}{2} \left[\frac{(k_x - k_{0x})^2}{m_x} + \frac{(k_y - k_{0y})^2}{m_y} + \frac{(k_z - k_{0z})^2}{m_z} \right] \tag{2.17}$$

Using the transformation

$$k_x^* = \left(\frac{m_0}{m_x}\right)^{1/2}(k_x - k_{0x}), \; k_y^* = \left(\frac{m_0}{m_y}\right)^{1/2}(k_y - k_{0y}), \; k_z^* = \left(\frac{m_0}{m_z}\right)^{1/2}(k_z - k_{0z})$$

The previous equation can be reduced to the isotropic mass m_0. As a result, the dispersion relation becomes

$$E(k^*) = \frac{\hbar^2 k^{*2}}{2m^*} \tag{2.18}$$

In this case, the volume element is to be noted as

$$dk_x^* dk_y^* dk_z^* = \left(\frac{m_0^3}{m_x m_y m_z}\right) dk_x dk_y dk_z \tag{2.19}$$

Transforming back into **k**-space and considering the valley degeneracy in conduction band as g_{vc}, one can obtain the DOS function for conduction band as

$$N(E)dE = \frac{g_{vc}}{2\pi^2}\left(\frac{2m_{de}^*}{\hbar^2}\right)^{3/2}(E - E_C)^{1/2}dE \tag{2.20}$$

where $m_{de}^* = (m_x m_y m_z)^{1/3}$ is the DOS effective mass of the electron in the conduction band.

A similar equation holds for the DOS for holes in the valence band.

The DOS functions for the conduction band and valence bands are shown in Fig. 2.3.

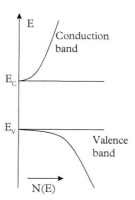

Figure 2.3 *Density-of-state (DOS) functions in the conduction and valence band.*

2.4 Doping

Pure or intrinsic semiconductors contain electrons and holes. Electrons from the valence band (VB) acquire enough energy to go to the conduction band (CB), thereby leaving a hole in the VB. The densities of electrons (n_i) and holes (p_i) in such a pure semiconductor are equal, $n_i = p_i$. Introduction of impurities in a pure semiconductor increases the electron or hole densities by several orders of magnitude. A common example is to introduce pentavalent P in Si. The four outermost electrons of P that replace a silicon (Si) atom take part in the covalent bonding process. The fifth electron is loosely bound to the P nucleus. A small amount of energy, called the donor binding energy E_d can make this electron free and available in the conduction band. The expression for donor binding energy as calculated by using Bohr's model for an H atom is given as

$$E_d = \left(\frac{\varepsilon_0}{\varepsilon_s}\right)^2 \left(\frac{m_0}{m_e}\right) \frac{e^4 m_0}{32\pi^2 \varepsilon_0^2 \hbar^2} = 13.6 \left(\frac{\varepsilon_0}{\varepsilon_s}\right)^2 \left(\frac{m_0}{m_e}\right) (eV) \tag{2.21}$$

Example 2.1 *Let us assume $\varepsilon_s = 16$ and $m_e = 0.12\, m_0$, which correspond approximately to germanium (Ge). The calculated value of the donor binding energy is 6.4 meV.*

The donor binding energy is small compared to the thermal energy $k_B T$ at or near room temperature. Therefore, almost all impurities like P donate their fifth electron to the CB. It is well-known that group V impurities act as donors and group III impurities act as acceptors in Si or Ge. It follows easily that in a III–V semiconductor group, II impurities like carbon (C), beryllium (Be), magnesium (Mg), zinc (Zn), etc., act as acceptors when they replace the group III atoms, and group VI impurities act as donors by replacing group V atoms. However, certain impurity atoms, like Si, act as donors if they replace gallium (Ga) atoms in GaAs but act as acceptors if they occupy the sites of arsenic (As) atoms. Such impurities are called amphoteric and show complicated behaviour.

2.5 Carrier concentration

The number of electrons in an occupied conduction band level can be evaluated by the total number of states, $N(E)$, multiplied by the occupancy, $f(E)$, integrated over the conduction band.

$$n = \int_{E_C}^{\infty} N(E) f(E) dE \tag{2.22}$$

where E_C represents the bottom of the conduction band and

$$f(E) = \frac{1}{1 + exp\left[(E - E_F)/k_B T\right]} \tag{2.23}$$

where E_F is known as the Fermi level.

In most of the cases, a Boltzmann approximation is valid and Eq. (2.23) takes the form as

$$f(E) = exp\left[(E - E_F)/k_B T\right] \tag{2.24}$$

Using Eqs. (2.20) and (2.24) and introducing a variable $x = (E - E_C)/k_B T$, Eq. (2.22) can be expressed as

$$n = \frac{g v_c}{2\pi^2}\left(\frac{2m_{de}^* k_B T}{2\pi \hbar^2}\right)^{3/2} exp\left[-(E_C - E_F)/k_B T\right] \int_0^\infty x^{1/2} e^{-x} dx \tag{2.25}$$

The integral is in the form of a Gamma function and equal to $\frac{1}{2}\pi^{1/2}$. The number of electrons per unit volume is then expressed as

$$n = 2g v_c \left(\frac{m_{de}^* k_B T}{2\pi \hbar^2}\right)^{3/2} exp\left[-(E_C - E_F)/k_B T\right] = N_C exp\left[-(E_C - E_F)/k_B T\right] \tag{2.26}$$

The DOS effective mass for holes in all cubic semiconductors is expressed as

$$m_{dh}^* = (m_{hh}^{3/2} + m_{lh}^{3/2})^{2/3} \tag{2.27}$$

where m_{hh} and m_{lh} are, respectively, the heavy hole and light hole effective masses. The hole density in the valence band can be obtained by replacing m_{de}^* by m_{dh}^*, $g v_c$ by $g v_h$, and N_C by N_V in Eq. (2.26).

Example 2.2   *The perfector N_C in Eq. (2.26) is called the effective DOS. Let $m_{de}^* = 0.067m_0$ and $m_{dh}^* = 0.5m_0$ for GaAs. As there is no valley degeneracy, $g v_c = 1$, the effective DOS for conduction and valence bands are respectively $4.3\times10^{17} cm^{-3}$ and $8.87\times10^{18} cm^{-3}$ at room temperature.*

Eq. (2.26) is modified for Fermi–Dirac distribution of carriers (see Problem 5). The expressions contain an integral $F_{1/2}(\eta)$ which may be evaluated numerically. Approximate formulas are also available to calculate carrier densities for the degenerate case (see, for example Joyce and Dixon (1977)).

2.6   Scattering mechanisms and transport

The electron wavefunction given by Eq. (2.3) has been obtained by solving a Schrödinger equation in a perfect periodic crystal that assumes that the motion of lattice atoms is frozen. The lattice atoms at a finite temperature however oscillate around their positions of equilibrium. The lattice vibration gives rise to a vibrating potential and perturbs the motion of the electrons by changing its wavevector from **k** to **k'**. In quantum mechanical language the lattice vibrations are quantized, and the quantum of the energy is called a phonon. The vibrational spectra have broadly two branches: acoustic phonon and optical phonons, each having transverse and longitudinal modes.

The change of motion of electrons due to the lattice vibration is due to electron-phonon interaction or scattering. In addition, the presence of impurities, both ionized donors and acceptors, as well as unionized (neutral) impurities introduce a perturbing potential in an otherwise perfect crystal and lead to electron-impurity scattering. Scattering may also occur in the presence of traps, defects, dislocations, and in semiconductor alloys, due to structural imperfections and many other agencies.

The electronic and optoelectronic properties in semiconductors are altered due to all the scattering mechanisms mentioned previously. In this section, we shall be concerned about how the scattering mechanisms control the electron transport, particularly the conductivity.

2.6.1   Scattering probability and relaxation time

As stated already, the effect of scattering is to alter the initial state of an electron represented by $|k\rangle$ to a newer state $|k'\rangle$. The probability of such a transition is given for the case of electron-phonon scattering by the Fermi Golden Rule as

$$S(k, k') = \frac{2\pi}{\hbar} |\langle k'| V |k\rangle|^2 \delta(E_k \pm \hbar\omega_q - E_{k'}) \qquad (2.28)$$

where $|\langle k'| V |k\rangle|^2$ is called the squared matrix element for scattering, V is the scattering potential, $E_k (E_{k'})$ denotes the electron energy in the initial (final) state, $\hbar\omega_q$ is the phonon energy, the Dirac delta function represents energy conservation, and $+(-)$ signs denote phonon absorption (emission).

The relaxation time may be calculated from the scattering probability and is expressed as

$$\frac{1}{\tau(k)} = \sum_{k'} S(k, k')(1 - \cos\theta) \qquad (2.29)$$

where the summation runs over all the final states having wavevector k' satisfying momentum and energy conservation and θ is the angle between k and k'.

For the calculation of the matrix elements for different scattering mechanisms the reader is referred to several texts (Nag 1980; Singh 1993) in Reading list.

2.6.2   Low field mobility

Electrons and holes in a semiconductor possess a random thermal velocity $v_{th}^2 = 3k_B T/m$, at a finite temperature T, where m denotes the effective mass of the carrier. As mentioned earlier, the carriers make frequent collisions with vibrating lattice ions, impurities: both ionized and neutral, other charge carriers, defects, and so on. A single electron travels some distance due to its thermal motion, and then a scattering event changes its direction of motion and thus the processes of free travel and scattering repeat. After a large number of collisions, there is no net displacement of the electrons. The current density $\mathcal{J}$ of an electron is expressed as

$$\mathcal{J} = -nev_d \tag{2.30}$$

where n is the electron density and v_d is its drift velocity. In absence of an electric field, the net displacement and drift velocity are zero. This leads to zero current density, as expected.

When an electric field F is applied, the force deviates slightly in the electron paths between collisions and produces a net displacement of the electrons in a direction opposite to the direction of the electric field. The effective displacement of carriers per unit time is the drift velocity v_d. The applied electric field exerts a force $P = -eF$ on the electrons which gives rise to an acceleration expressed as $P = m_e(dv/dt)$. For a small electric field, the acceleration may be written as v_d/τ, where τ is called the relaxation time. Thus

$$-eF = m_e v_d/\tau \tag{2.31}$$

This leads to the following expression for the drift velocity:

$$v_d = -\frac{e\tau}{m_e}F = -\mu_n F \tag{2.32}$$

The proportionality constant $\mu_n = e\tau/m_e$ between v_d and F: the drift velocity per unit electric field is called the mobility of electrons. A similar relation exists for holes.

The relaxation time τ, given by Eqs. (2.29) and (2.31) is alternatively termed as the mean scattering time and it represents the mean time between two successive collisions. Various scattering processes have already been introduced in this section. Assuming that the total probability of scattering is the sum of the individual probabilities, one may write for the total probability as

$$\frac{1}{\tau} = \frac{1}{\tau_L} + \frac{1}{\tau_{imp}} + \frac{1}{\tau_{ee}} + \dots \tag{2.33}$$

where the suffixes L, imp, and ee stand, respectively, for the scattering time for lattice, impurity and electron-electron scattering. Using the relation between the mobility and relaxation time, one may arrive at the following simple rule, known as the Matthiessen's rule:

$$\frac{1}{\mu} = \sum_{i=1}^{N} \frac{1}{\mu_i} \tag{2.34}$$

where μ_i is the mobility limited by the ith scattering mechanism and μ is the overall mobility.

The temperature dependencies of various relaxation times and of the related mobility may be known from Eq. (2.29). However, the temperature behaviour can be qualitatively understood by the following arguments. Since the lattice scattering is due to the interaction of electrons with the vibrating atoms, with increased vibration amplitude with increasing temperature, the frequency of such collisions increases. This makes the relaxation time shorter and therefore the lattice-scattering limited mobility decreases with the increasing temperature. At low temperatures, lattice scattering is less significant, but the thermal velocity of the carriers is also lower. Since a slower electron is more strongly influenced by the Coulomb field of an ionized impurity, the impurity scattering becomes more frequent at lower temperatures making the mobility lower. The temperature variation of lattice scattering mobility is $\mu_L \propto T^{-x}$, $0 < x < 3$, and that of impurity-scattering mobility is $\mu_{imp} \propto T^{3/2}$. Considering these two scattering mechanisms, it may be shown that the overall mobility increases with temperature, attains a maximum value, and then decreases due to dominance of lattice scattering (see Problem 2.11). At a constant temperature, the mobility at low doping level is limited by the lattice scattering. However, with an increased doping level the mobility decreases, due to the additional contribution of the ionized impurity scattering.

Example 2.3 *Assuming that, in a GaAs sample, there are two scattering mechanisms of relaxation time of 0.1 ps and 0.2 ps. The values of the mobility due to these two scattering mechanisms are 2624.2 cm²/V-sec, and 5248.4 cm²/V-sec respectively. Using the Mathiessen rule the net mobility is 1749.5 cm²/V-sec.*

The simplified expression for the mobility given after Eq. (2.32) should be modified by considering the electrons distribution function in the following manner:

$$\mu = \frac{\langle e\tau \rangle}{m_e} \tag{2.35}$$

The average relaxation time may be expressed as

$$\langle \tau \rangle = \frac{\int \tau(E) E^{3/2} \, (\partial f_0/\partial E) \, dE}{\int E^{3/2} \, (\partial f_0/\partial E) \, dE} \tag{2.36}$$

where f_0 is the equilibrium distribution function (Fermi–Dirac or Boltzmann).

2.6.3   High field effect

From Eqs. (2.30) and (2.32), $\mathcal{J} = \sigma F$, and $\sigma = ne\mu$ is the conductivity. This is Ohm's law, in which the mobility is assumed to be independent of the applied field F. However, in most semiconductors when the field is large, ~ 1 kV/cm, the mobility becomes a function of the field. In this situation, the drift velocity of the carriers becomes comparable to the thermal velocity. The total energy of the electrons, the sum of thermal energy, and the energy gained from the field, no longer equals $(3/2)k_B T_L$, but can be expressed as $(3/2)k_B T_e$, where T_L is the temperature of the lattice, and T_e is termed as 'electron temperature'. It is easy to note that $T_e > T_L$. The electrons are hot. Similar is the case for holes. The effect is termed as the 'hot carrier effect'.

The variations of the drift velocity of electrons and holes in Si and Ge are a linear increase with the field, satisfying Ohm's law. The velocities then increase sublinearly and finally attain a saturation value after a critical field. The drift velocity in n-type GaAs shows a linear increase at low fields, but after a critical field, v_d decreases with field, showing a negative differential resistance region. It then increases almost linearly with a lower slope and finally assumes a saturation value.

The following empirical expressions are used to give the values of the electron and hole drift velocities at different temperatures:

$$v_d = v_l \frac{F}{F_c} \left[\frac{1}{1 + \left(\frac{F}{F_c} \right)^\beta} \right]^{1/\beta} \tag{2.37}$$

The values for Si at 300 K are:

	Electron	Hole
v_l cm/s	1.07×10^7	8.34×10^6
F_c V/cm	6.91×10^3	1.45×10^4
B	1.11	2.637

2.7   Excess carriers and recombination

So far, we have determined the thermal equilibrium values of electron and hole concentration in the conduction band and valence band respectively. It is possible to create excess carriers, that is, to make the carrier concentration higher than its thermal equilibrium values. Excess carriers can be created by optical excitation, or electron bombardment, or they can be injected across a forward biased p-n junction.

2.7.1   Absorption and recombination

When a bunch of photons with energy $\hbar\omega \geq E_g$ falls on a semiconductor, there is an insignificant amount of absorption, that is, electrons from the valence band are excited to the many unoccupied states available in the conduction band. The excited electrons in the conduction band created holes in the valence band and may primarily have much higher energy than the band edge energy. But subsequently, by the process of scattering, electrons and holes lose their energy until their equilibrium energies become equal to the band edge energies. These electron-hole pairs (EHPs) are called excess carriers and, as they are out of balance, they will recombine eventually by different process of recombination observed in a semiconductor.

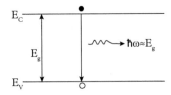

Figure 2.4 *Direct band-to-band recombination processes annihilating an excess electron-hole pair (EHP)*

For a direct band gap semiconductor, an electron from the conduction band falls to an empty state in the valence band spontaneously, thus annihilating the EHP and in this case the probability that an electron and a hole will recombine is constant in time. Assuming low-level injection, the excess carrier concentration at any instant of time can be expressed as

$$\delta n(t) = \Delta n\, exp\left(-\frac{t}{\tau_n}\right) \tag{2.38}$$

where Δn represents the initial excess carrier concentration and τ_n is called the re-combination lifetime or often called the minority carrier lifetime. Direct band to band recombination is shown in Fig. 2.4.

The situation is different for indirect band gap semiconductors as both momentum and energy change are involved during recombination. In this case, recombination takes place with the assistance of defect states created by impurity atoms, either deliberately introduced or present spontaneously due to the semiconductor fabrication technology. These defect states act as the recombination centers. However, recombination occurs in two steps: the defect state first captures a carrier from one of the bands and then captures the opposite type of carrier from the other band, thus annihilating the EHP. The extended theory of this type of recombination was proposed by Shockley and Read (1952); and by Hall (1952). So, this type of recombination in an indirect band gap semiconductor is known as Shockley–Read–Hall (SRH) recombination process.

Another interesting recombination process is observed in heavily doped semiconductor material, where an electron and a hole recombine directly without involving any defect level and the released energy is transferred to another carrier. In other words, this is a three-carrier process. For a heavily doped n-type semiconductor, as the number of electrons in the conduction band is higher than the number of holes in the valence band, after recombination of an EHP, the released energy is captured by another electron in the conduction band, thereby moving to a higher energy state in that band. So the process is known as e-e-h process. The reverse is the case for p-type semiconductor where the re-leased energy is captured by hole, thereby moving to a higher energy state in the valence band and subsequently known as h-h-e process. However, it is clear from the previous discussion, that no electromagnetic radiation is emitted by this recombination process, instead the released energy is absorbed by a third carrier. This process of recombination is called auger recombination and schematically described in Fig. 2.5.

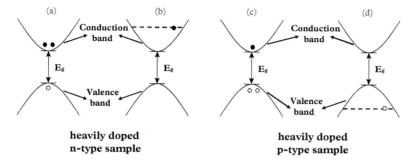

Figure 2.5 *e-e-h process (a) before recombination (b) after recombination; and h-h-e process (c) before recombination (d) after recombination.*

The recombination may take place at the semiconductor surface due to the presence of surface states. The surface states are introduced due to the discontinuity of the lattice structure at the semiconductor surface. As a result, a large number of energy states in the forbidden energy gap, called surface states, enhance the recombination rate at the surface. Moreover, there may be adsorbed ions, molecules, or damage in the surface layer to increase the recombination rate.

So far as we have discussed, it is clear that the recombination of EHPs can occur in different ways: both radiative and non-radiative. All these phenomena can be captured by following the expression for the recombination rate:

$$R(n) = An + Bn^2 + Cn^3 + R_{st}N_{ph} \tag{2.39}$$

In Eq. (2.39), the linear term An is due to the nonradiative process, Bn^2 is due to spontaneous recombination process where an electron in the conduction band recombines with a hole in the valence band, the cubic term Cn^3 is due to Auger recombination. The last term is due to stimulated recombination and is proportional to the number of photons N_{ph}.

Example 2.4 *Assume that the values of the coefficients A, B, and C are 3×10^7, 0.12×10^{-10} $cm^3\ s^{-1}$ and $9.6 \times 10^{-29}\ cm^6\ s^{-1}$ respectively. The value of τ becomes 31 ns with $n = 1 \times 10^{17}\ cm^{-3}$.*

2.7.2 Diffusion process

A spatial variation of an electron or hole concentration can exist in a semiconductor. This may be due to non-uniform doping over space. Also, when excess carriers are generated nonuniformly in a semiconductor, the electron and hole concentrations vary with the position in the sample. In the presence of such spatial variation in carrier concentration, carriers move from the high concentration region to the low concentration region. This type of motion is called *diffusion* and, due to carrier charge, it constitutes a current called diffusion current. For electrons the expression reads

$$\mathcal{J}_n = eD_n \frac{dn}{dx} \tag{2.40}$$

D_n is the proportionality constant and is known as the *diffusion coefficient* or *diffusivity of electrons.*

Similarly for holes, the diffusion current is

$$\mathcal{J}_p = -eD_p \frac{dp}{dx} \tag{2.41}$$

The diffusivity and mobility are related, and for electrons

$$\frac{D_n}{\mu_n} = \frac{k_B T}{e} \tag{2.42}$$

and for holes, $\frac{D_p}{\mu_p} = \frac{k_B T}{e}$ is valid.

This important equation is known as Einstein relation. However, this relation is only valid for non-degenerate semiconductors.

Example 2.5 *The electron mobility of GaAs is 8000 cm²/V-sec. So, the diffusion coefficient of the electron at 300 K will be 206 cm²/sec.*

2.7.3   Concept of a quasi-fermi level

In the presence of excess carriers, the law of mass action $np = n_i^2$ does not hold good, and in this non-equilibrium condition the concept of equilibrium Fermi level is no longer valid. However, the electron and hole concentrations are expressed in the same exponential form, by defining separate quasi-Fermi level F_e and F_h for electrons and holes respectively for non-equilibrium cases. Eq. (2.26) and a similar equation for holes are accordingly modified to yield the following expressions for the non-equilibrium case in steady state:

$$n = n_i exp\left(\frac{F_e - E_i}{k_B T}\right) = N_c exp\left(-\frac{E_c - F_e}{k_B T}\right) \tag{2.43}$$

$$p = n_i exp\left(\frac{E_i - F_h}{k_B T}\right) = N_v exp\left(-\frac{F_h - E_v}{k_B T}\right) \tag{2.44}$$

Taking product of these two equations yields

$$np = n_i^2 exp\left(\frac{F_e - F_h}{k_B T}\right) \tag{2.45}$$

Three different situations now arise: when $F_e > F_h$, $np > n_i^2$ and excess carrier injection takes place, when $F_e < F_h$ $np < n_i^2$, carrier extraction occurs, and when $E_{Fn} = E_{Fp}$, equilibrium situation is obtained with $np = n_i^2$.

Example 2.6 *Consider a p type semiconductor at T = 300K with carrier concentration of p_0 = 10^{15} cm^{-3}, n_i = 1.5 × 10^{10} cm^{-3}, and n_0 = 2.25 × 10^5 cm^{-3}. In nonequilibrium, assume that the excess carrier concentrations are $\delta n = \delta p = 5 \times 10^{12}$ cm^{-3}. So the position of the quasi-Fermi level for electron will be 0.151 eV above the intrinsic level and that for hole will be 0.289 eV below the intrinsic level.*

Example 2.7 *Assume a GaAs sample with n = p = 1×10^{18} cm^{-3}. The quasi-Fermi level for electrons will be 0.041eV above the conduction band and 0.055eV below the valence band using Joyce–Dixon relationship at T = 300 K.*

2.7.4 The continuity equation

In this section, we consider the overall effect of drift, diffusion, generation as well as recombination of carriers occurring in a semiconductor. The combined effect is expressed by the *continuity equation*. The derivation is given in standard texts (see for example Streetman and Banerjee 2015; Neamen 2012, in the Reading list).

The 1D continuity equations for minority carriers (n_p for electrons in p-type semiconductor and p_n for holes in n-type semiconductors) are written as

$$\frac{\partial n_p}{\partial t} = n_p \mu_n \frac{\partial \Xi}{\partial x} + \mu_n \Xi \frac{\partial n_p}{\partial x} + D_n \frac{\partial^2 n_p}{\partial x^2} + G_n - \frac{n_p - n_{p0}}{\tau_n} \tag{2.46}$$

$$\frac{\partial p_n}{\partial t} = -p_n \mu_p \frac{\partial \Xi}{\partial x} - \mu_p \Xi \frac{\partial p_n}{\partial x} + D_p \frac{\partial^2 p_n}{\partial x^2} + G_p - \frac{p_n - p_{n0}}{\tau_p} \tag{2.47}$$

In the previous two equations, Ξ is the electric field, G is the generation rate, subscript 0 refers to equilibrium concentration of carriers, and the other symbols are already defined. The last term in each equation represents recombination rate. There are three unknown quantities to be evaluated (n, p and Ξ) in the previous two continuity equations. In principle, these two equations along with the Poisson's equation with appropriate boundary conditions may be solved simultaneously to obtain a unique solution. However, as the equations are nonlinear, analytical solutions are only possible with some simplifying assumptions. Assuming that there is no electric field ($\Xi = 0$) and no other generation of carriers excepting the thermal generation ($G_n = G_p = 0$), the continuity equations, Eqs. (2.46) and (2.47), may be simplified as

$$\frac{\partial n_p}{\partial t} = D_n \frac{\partial^2 n_p}{\partial x^2} - \frac{n_p - n_{p0}}{\tau_n} \tag{2.48}$$

$$\frac{\partial p_n}{\partial t} = D_p \frac{\partial^2 p_n}{\partial x^2} - \frac{p_n - p_{n0}}{\tau_p} \tag{2.49}$$

Writing the excess electron concentrations as $\delta n = n_p - n_{po}$ and a similar expression for excess holes, we obtain

$$\frac{\partial \delta n}{\partial t} = D_n \frac{\partial^2 \delta n}{\partial x^2} - \frac{\delta n}{\tau_n} \tag{2.50}$$

$$\frac{\partial \delta p}{\partial t} = D_p \frac{\partial^2 \delta p}{\partial x^2} - \frac{\delta p}{\tau_p} \tag{2.51}$$

Under steady state distribution of excess carriers, the diffusion equations become

$$\frac{\partial^2 \delta n}{\partial x^2} = \frac{\delta n}{D_n \tau_n} = \frac{\delta n}{L_n^2} \tag{2.52}$$

$$\frac{\partial^2 \delta p}{\partial x^2} = \frac{\delta p}{D_p \tau_p} = \frac{\delta p}{L_p^2} \tag{2.53}$$

where $L_n = \sqrt{D_n \tau_n}$ and $L_p = \sqrt{D_p \tau_p}$ are, respectively, the diffusion lengths of electrons and holes.

Assume that excess electrons are somehow injected at the surface ($x = 0$) of a semi-infinite bar of a semiconductor. Solving the steady state continuity equation Eq. (2.52), for electrons, the distribution of excess electron concentration in the bar may be expressed as

$$\delta n(x) = \Delta n \exp(-x/L_n) \tag{2.54}$$

where $\Delta n = \delta n|_{x=0}$. The injected excess electron concentration therefore dies out exponentially along x due to recombination and at a distance equal to the diffusion length L_n the excess electron distribution is reduced to $1/e$ of its value Δn at the point of injection. The steady state distribution of excess electrons causes diffusion, and therefore an electron current, in the direction of increasing concentration. The electron diffusion current can be expressed as

$$J_n(x) = -\frac{eD_n}{L_n} \delta n(x) \tag{2.55}$$

In a similar manner, the hole diffusion current may be written as

$$J_p(x) = -\frac{eD_p}{L_p} \delta p(x) \tag{2.56}$$

2.8 Excitons

As already discussed, when a semiconductor material is excited, an electron from the fully occupied valence band is transferred to the one of the unoccupied available states in the conduction band. As a result, EHPs are created and they are normally considered as uncorrelated. But in actual situation, an attractive Coulomb interaction is assumed to occur between them as they are oppositely charged. Consequently, a hydrogen-like bound pair must be thought of instead of a free EHP. These complex elementary excitations in solids are called *excitons*.

Since two particles are involved, the problem is solved by using two-particle Schrodinger equation

$$\left[-\frac{\hbar^2}{2m_e}\nabla_e^2 - \frac{\hbar^2}{2m_h}\nabla_h^2 - \frac{q^2}{4\pi\varepsilon|\mathbf{r}_e - \mathbf{r}_h|}\right]\psi_{ex} = E\psi_{ex} \tag{2.57}$$

where m_e (m_h) is the effective mass of electron (hole) and $|\mathbf{r}_e - \mathbf{r}_h|$ is the separation between the positions of the two particles.

To transform the two-body problem into one-body problem, the following transformations have been introduced:

$$r = r_e - r_h; \quad K = k_e - k_h; \quad k = \frac{m_e k_e + m_h k_h}{m_e + m_h}; \quad R = \frac{m_e r_e + m_h r_h}{m_e + m_h} \tag{2.58}$$

The transformed Hamiltonian will be

$$\mathcal{H} = \frac{\hbar^2 K^2}{2(m_e + m_h)} + \left[\frac{\hbar^2 k^2}{2m_r} - \frac{q^2}{4\pi\varepsilon|r|}\right] \tag{2.59}$$

where m_r is the reduced mass.

The first part of the previous Hamiltonian is due to the motion of center of mass of the bound pair and second part accounts for the relative motion of electron and hole. From the first part, we get the plane wave solution as

$$\psi_{cm} = e^{(iK.R)} \tag{2.60}$$

The envelope function $\phi(\mathbf{r})$ related to the second part satisfies the relation

$$\left[\frac{\hbar^2 k^2}{2m_r} - \frac{q^2}{4\pi\varepsilon|r|}\right]\phi(r) = E\phi(r) \tag{2.61}$$

Eq. (2.61) is the equation for hydrogen atom problem and the eigenvalues are given by

$$E_n = \frac{1}{2}\frac{m_r q^4}{(4\pi\varepsilon)^2}\frac{1}{n^2\hbar^2} = \frac{1}{n^2}\left(\frac{m_r}{m_0}\right)\left(\frac{\varepsilon_0}{\varepsilon}\right)^2 \times 13.6eV \tag{2.62}$$

Therefore the complete wavefunction may be written as

$$\psi_{nK} = e^{(iK.R)}\phi(r)\varphi_e(r_e)\varphi_h(r_h) \tag{2.63}$$

Example 2.8 *The electron effective mass in GaAs is 0.067 m_0 and the hole effective mass is 0.5 m_0. The reduced mass $m_r = 0.059 m_0$. Taking $\varepsilon = 13.1 \varepsilon_0$, the binding energy for lowest exciton state (1s) is 4.67 meV.*

2.9   Alloys and heterojunctions

Semiconductor alloys constitute an important class of materials and have been developed due to the need to develop materials with a chosen band gap and a chosen lattice constant. Growth of good quality alloys paved the way for realizing heterojunctions and afterwards quantum nanostructures. Novel electronic and photonic devices have been fabricated and used in practical systems.

2.9.1   Alloys

Alloys may be classified as binary, ternary, and quaternary depending upon the number of different atoms involved in making the materials. Two atoms A and B may have the composition $A_x B_{1-x}$ making it a binary alloy and a practical example is $Si_{1-x}Ge_x$. A well-known example of a ternary alloy in the form $A_{1-x}B_x C$ using three atoms A, B, and C is $Ga_{1-x}Al_x As$: an alloy of GaAs and aluminum arsenides (AlA)s. An alloy represented generally by $A_{1-x}B_x C_y D_{1-y}$, as for example $In_{1-x}Ga_x As_y P_{1-y}$, is called a quaternary alloy.

The important parameters of an alloy like lattice constant, band gap, etc., are dependent on the parameters of the constituent materials. In its simplest form, the lattice constant of the alloy $A_{1-x}B_x C$ obeys the following linear relation, known as Vegard's law:

$$a_{ABC} = (1 - x)a_{AC} + x a_{BC} \tag{2.64}$$

Where the subscripts correspond to the constituent binaries. In most cases, the relationship is nonlinear and the parameter P_{ABC} of a ternary alloy $A_{1-x}B_x$ C may be expressed in terms of the corresponding parameters of binary compounds AC and BC as

$$P_{ABC} = (1 - x)P_{AC} + x P_{BC} + x(1 - x)P_{AB} = a + bx^2 + cx^3 \tag{2.65}$$

where $a = P_{AC}, b = P_{BC} - P_{AC} + P_{AB}, c = -P_{AB}$ The parameter, c, called the bowing parameter, arises due to lattice disorder created by intermixing atoms A and B on the lattice site normally occupied by one kind of atom in a binary compound.

2.9.2   Heterojunctions

A heterojunction is formed by joining two semiconductors having different band gaps, permittivities, electron affinities, etc. Modern growth techniques ensure an abrupt change in the band gap, and other characteristics of the constituent semiconductors at the heterointerface. The two semiconductors may or may not have similar lattice constant. For lattice mismatch the layers experience strain. We simplify our discussion by assuming an abrupt heterojunction and lattice matched pairs.

An important issue for heterojunctions is to know the band lineup at the heterointerface (see Fig. 2.6). Though many theoretical models are available in the literature, not a single model can explain the results obtained for all the materials systems. We present an empirical rule to illustrate the band discontinuity in a single heterojunction.

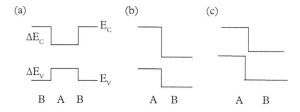

Figure 2.6 *Different types of band alignment: (a) Type I,*
(b) type II, and (c) broken-gap type II band alignments.

2.9.2.1 Empirical rule for band alignment

In the absence of an accepted theory for band line-up in heterojunctions, experimental results are used to determine band offsets ΔE_c and ΔE_v. It is now accepted that the ratio $\Delta E_c/\Delta E_v$ = 60/40 or 65/35 for a gallium aluminum arsenide (GaAlA)s/GaAs system. On the other hand, this ratio is approximately 40/60 for indium gallium arsenide phosphide (InGaAsP)/indium phosphate (InP) pairs.

Example 2.9 *For $Ga_{0.7}Al_{0.3}As/GaAs$ system, the band gap difference is $1.798 - 1.424 = 0.374\,eV$. Taking the 65/35 ratio, $\Delta E_c = 0.243\,eV$ and $\Delta E_v = 0.131\,eV$. The band line-up is illustrated in Fig. 2.6(a).*

A simple theory, called the model solid theory (van de Walle 1989; van de Walle and Martin), is useful in predicting the band offsets between various semiconductor pairs

2.9.2.2 Different types of band line-up

The band line-up for a double heterojunction with lower gap semiconductor A sand-wiched between two layers of higher gap semiconductor B is shown in Fig. 2.6(a). A typical example is an aluminum gallium arsenide (AlGaAs)-GaAs-AlGaAs double heterojunction. There are discontinuities ΔE_c and ΔE_v respectively, in the CB and VB at the heterointerfaces. It is seen from Fig. 2.6(a) that a square well is formed in both the conduction and valence bands of the lower gap semiconductor B in the BAB structure. This type of band line-up is commonly termed as type-I. In this type of band alignment both electrons and holes occupy the same layer, the lower gap A layer.

Two other types of band line-up are also shown in Fig. 2.6. In Fig. 2.6(b) the staggered line-up, called the type II line-up, is shown. This occurs in GaInAs/GaAsSb or in Si/SiGe systems. As may be seen from Fig. 2.6(b), both the band edges of one semiconductor are shifted in the same direction relative to those of the other. Contrary to type I case, in type II alignment, electrons and holes accumulate in different materials, electrons in B and holes in A. The extreme case as shown in Fig. 2.6(c) is known as type II misaligned. This is found in InAs/(GaSb) pair. In this case, the conduction band edge of one semiconductor, for example InAs, is below the valence band edge of the other semiconductor GaSb.

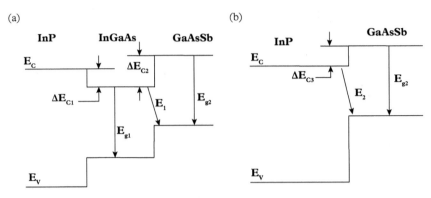

Figure 2.7 *Band alignment in (a) InP-In$_{0.52}$Ga$_{0.48}$As-GaAs$_{0.5}$Sb$_{0.5}$ and (b) InP-GaAs$_{0.5}$Sb$_{0.5}$ heterostructures*

Example 2.10 *Two examples of Type II alignment are given in Fig. 2.7 (Hu et al 1998). Here $E_1 = 0.453$ eV and $E_2 = 0.630$ eV are the values for the spatially indirect gap in (a) and (b) respectively. The band gap for GaAsSb is $E_{g2} = 0.813$ eV. The conduction band offsets are $\Delta E_{c2} = E_{g2} - E_1 = 0.36$ eV, and $\Delta E_{c3} = E_{g2} -; E_2 = 0.18$ eV.*

2.10   Quantum structures

A brief introduction to semiconductor nanostructures has been given in Chapter 1. Most of the present electronic and photonic devices are based on such nanostructures in which at least one dimension of the layer of interest is only a few nanometres thick. The motion of particles is quantum mechanically confined along the concerned direction. The confinement gives rise to electron or hole gas of lower dimensions. We consider in this section the conditions necessary for quantum confinement and then a few basic electronic properties.

2.10.1   Condition for quantum confinement

The electrons in a bulk semiconductor have free motion in all the three directions: x, y, and z, as indicated by the E-**k** relationship expressed by

$$E(k) = \frac{\hbar^2 k^2}{2m_e}, \quad k = k_x, k_y, k_z. \tag{2.66}$$

where m_e is the effective mass of electrons. The condition for quantum confinement is that the electron must be confined in a 1D potential well of width comparable to its de Broglie wavelength. In that situation, the free motion of the electron is inhibited in that direction. The electron waves are reflected at the two boundaries of the potential well and a standing wave is formed leading to quantized energy levels. The electron is however free to move along the other two dimensions, and accordingly, a 2DEG forms. The electron wavelength $\lambda = 2\pi/k$ is related to its energy E by

$$\lambda = \frac{2\pi\hbar}{\sqrt{2m_eE}} \tag{2.67}$$

Example 2.11 *Let us assume that the energy of the electron is 25 meV and the effective mass of electron is $m_e = m_l = 0.91\ m_0$: the longitudinal mass in Si conduction band. Then λ $\approx 8\ nm$. The value for GaAs will be $\approx 30\ nm$ using $m_e = 0.07\ m_0$.*

2.10.2　Quantum wells

Present electronic and optoelectronic devices are based on nanostructures where at least one dimension of the layer of interest is only few nanometres thick. The behaviour of the particles in such nanostructure is not like those in the bulk material and the motion is quantum mechanically confined along concerned direction. If the electron motion is confined in a 1D potential well having width comparable to its de Broglie wavelength, the electron waves are then standing wave in nature and energy becomes quantized along that particular direction. The electrons, however, are free to move along the other two dimensions and accordingly, a 2DEG results.

The 1D potential well may be realized in a double heterojunction. Consider the simplest case of a double heterojunction made by GaAs and its alloy Al_xGa_{1-x} As, as shown in Fig. 2.8. In the figure, rectangular potential well with abrupt walls at the two heterointerfaces exists in the valence, and the conduction bands of the lower gap GaAs Are sandwiched between two AlGaAs layers. Here, the simplest approximation has been made by considering that the difference in the band gap of two semiconductors is consumed by the discontinuities in the conduction and valence bands in the two heterointerfaces. If the thickness d of GaAs along the growth direction (z-direction) is comparable to the de Broglie wavelength, the quantum confinement is expected along

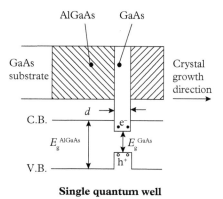

Single quantum well

Figure 2.8 *The QW structure formed by sandwiching GaAs layer in between two AlGaAs layers*

that direction. Such structures are known as QWs. In the given figure GaAs is the QW and AlGaAs is called the barrier.

2.10.2.1 *Simplified energy levels and density of states*

The energy levels in the conduction and valence bands can be calculated by using effective mass theory. For simplifications, the lattice constant and the effective mass of electrons in the two materials are assumed to be equal. In the given diagram, the motion of electron is confined along the z-direction, and they are free to move along (x-y) plane. The effective mass Schrodinger equation in the presence of a potential along the z-direction, $V(z)$, may be written as

$$\left[-\frac{\hbar^2}{2m_{||}}\frac{\partial^2}{\partial r^2} - \frac{\hbar^2}{2m_z}\frac{\partial^2}{\partial z^2} + V(z)\right]\Psi(r,k) = E\Psi(r,k) \tag{2.68}$$

Here, $m_{||}$ and m_{zz} are, respectively, the electron effective mass along the layer and along the z-direction, $V(z)$ denotes the energy at the bottom of the conduction band, and E is the total energy of the electron. The total wavefunction is written in product form as

$$\Psi(r,k) = \chi(r)\phi_n(z) \tag{2.69}$$

Using this in Eq. (2.68), one may obtain the following relation for different quantized energy levels:

$$\left[-\frac{\hbar^2}{2m_z}\frac{d^2}{dz^2} + V(z)\right]\phi_n(z) = E_n\phi_n(z) \tag{2.70}$$

Under an infinite well approximation

$$V(z) = 0 \quad |z| < d/2 \tag{2.71a}$$

$$= \infty \quad |z| > d/2 \tag{2.71b}$$

Since the electron cannot penetrate into a semi-infinite region of infinite potential energy, the envelope function must satisfy the condition

$$\phi_n\left(\frac{d}{2}\right) = \phi_n\left(-\frac{d}{2}\right) = 0 \tag{2.72}$$

The following two equations are two independent envelope functions satisfying the effective mass equations:

$$\phi_n(z) = \sqrt{\frac{2}{d}}sin(k_n z) \tag{2.73a}$$

$$\phi_n(z) = \sqrt{\frac{2}{d}} \cos(k_n z) \qquad (2.73b)$$

where

$$k_n = \sqrt{\left(\frac{2 m_z E_n}{\hbar^2}\right)} \qquad (2.74)$$

According to Eq. (2.73a and b),

$$k_n = \frac{n\pi}{d} \qquad (2.75)$$

where n is a positive integer, with even values applying to Eq. (2.73a) and odd values to Eq. (2.73b). The energy eigenvalues are given by

$$E_n = \frac{\hbar^2}{2m_z} \left(\frac{n\pi}{d}\right)^2 \qquad (2.76)$$

A diagram showing the first few energy levels, or subbands, as they are called, and envelope functions is given in Fig. 2.9. The total energy including the translational energy parallel to the interface is now expressed as

$$E(k, n) = E_n + \frac{\hbar^2 k^2}{2m_{\parallel}} \qquad (2.77)$$

where k is the in-plane (2D) wave vector. The energy dispersion relation is shown in Fig. 2.10. The dispersion relation for holes is expressed identically as Eq. (2.77) by using the appropriate hole effective mass.

Example 2.12 *For an infinite GaAs QW of width 8 nm, and effective mass 0.067 m_0, the values of subband energies are $E_n = 87.7, \ 350, \ 789.2 \dots (meV)$*

2.10.2.2 *Density of states in two dimensions*

The DOS in a QW structure can be calculated following the similar approach as followed for the bulk (3D) case. Here we assume that the lengths of the sample are L_x and L_y, respectively along the x and y directions. Assuming periodic boundary conditions along x and y, the allowed values of wave vectors are,

$$k_x = \frac{2\pi n_x}{L_x}, \quad k_y = \frac{2\pi n_y}{L_y} \qquad (2.78)$$

where n_x and n_y are integers and $k = \sqrt{k_x^2 + k_y^2}$

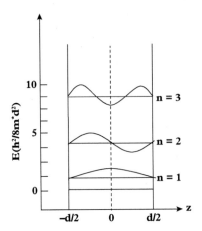

Figure 2.9 *Energy levels and envelope functions for the infinite square well.*

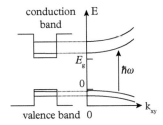

Figure 2.10 *E-k diagram of 2D electron and hole gas.*

The area in the (k_x, k_y) plane that contains one allowed value of k is $4\pi^2/L_xL_y$. The area between two circles corresponding to E and $E+ dE$ is $2\pi kdk$. The number of states associated with this area is

$$dN = \frac{2\pi kdk}{4\pi^2}L_xL_y = \frac{kdk}{2\pi}L_xL_y \tag{2.79}$$

The change in energy is as the one obtained from Eq. (2.77)

$$dE = \frac{\hbar^2 kdk}{m_{||}} \tag{2.80}$$

Therefore, the DOS will be

$$\frac{dN}{dE} = \frac{m_{||}L_xL_y}{2\pi\hbar^2} \tag{2.81}$$

Introducing a factor 2 for spin and normalizing to unit area, one obtains for the 2DDOS function as

$$\rho_{2D} = \frac{m_{\parallel}}{\pi \hbar^2} \qquad (2.82)$$

Eq. (2.82) is independent of energy. In a QW, each subband contributes this amount to the total DOS. The profile is like a stair case as shown in Fig. 2.11.

Example 2.13 *For bulk GaAs, the DoS is $1.75 \times 10^{36} m^{-2} J^{-1}$. If energy interval is considered as 15 meV, the total number of states will be $4.2 \times 10^{15} m^{-2}$.*

2.10.2.3 Finite quantum well

A real finite QW is characterized by a potential at the heterointerface as shown in Fig. 2.12.

$$V(z) = 0 \quad |z| < d/2 \qquad (2.83a)$$

$$= V_0 \, |z| > d/2 \qquad (2.83b)$$

The envelope functions in the barriers penetrate into the two barrier layers due to finiteness of the barrier height V_0. The solutions of the effective mass equation take the

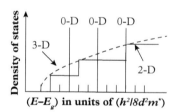

Figure 2.11 *The staircase-like DOS function for 2D electrons.*

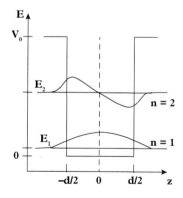

Figure 2.12 *Potential profile for a finite QW structure.*

form

$$\phi_n(z) = A_1 cos(k_w z), \qquad |z| < d/2 \qquad (2.84a)$$

$$= A_2 exp\left[-k_b\left(z - \tfrac{d}{2}\right)\right] \quad z > d/2 \qquad (2.84b)$$

$$= A_3 exp\left[+k_b\left(z + \tfrac{d}{2}\right)\right] \quad z < -d/2 \qquad (2.84c)$$

The solutions given previously are even functions. As are constants and the subscripts w and b refer, respectively, to the well and barrier. The odd solutions can also be written similarly by replacing cos by sin in Eq. (2.84a). The exponential functions in the other two are unchanged. The energy eigenvalues for both the cases may be expressed as

$$E_n = \frac{\hbar^2 k_w^2}{2m_w} \quad or \quad E_n = V_0 - \frac{\hbar^2 k_b^2}{2m_b} \qquad (2.85)$$

The values of k_w and k_b are determined by the boundary conditions at $\pm d/2$ that $\phi_n(z)$ and $\frac{1}{m}\frac{d\phi_n}{dz}$ are continuous. For even solutions one obtains

$$A_1 cos\left(k_w \frac{d}{2}\right) = A_2 \qquad (2.86a)$$

$$\frac{k_w}{m_w} A_1 sin\left(k_w \frac{d}{2}\right) = A_2 \frac{k_b}{m_b} \qquad (2.86b)$$

The non-trivial solution of Eq. (2.86a and b) is

$$\frac{k_w}{m_w} tan\left(k_w \frac{d}{2}\right) = \frac{k_b}{m_b} \qquad (2.87)$$

From Eq. (2.85) one obtains by eliminating E_n

$$\frac{\hbar^2 k_w^2}{2m_w} + \frac{\hbar^2 k_b^2}{2m_b} = V_0 \qquad (2.88)$$

Eqs. (2.87) and (2.88) may be solved graphically or numerically for k_w and k_b and their values are then substituted in Eq. (2.85) to give the energy eigenvalues.

Example 2.14 *From the graphical solution of the energy eigen value of finite QW, it can be proved that the number of bound states N can be determined by the condition $(N-1)\frac{\pi}{2} \le \sqrt{2m_w V_0}\left(\frac{d}{2\hbar}\right) \le \frac{N\pi}{2}$. If $V_0 = 0.3\,eV$, and $d = 8$ nm, then for finite GaAs QW, the number of bound states for electron will be 2.*

2.10.3   Quantum wires and quantum dots

In Section 2.10.2, it was discussed that the confinement of electrons or holes in a narrow 1D potential well leads to 2DEG. If further confinement has been introduced to the motion of electrons, the subband structure and electronic properties will be different.

Let us assume the confinement is extended into two dimensions, that is, along the y- and z-direction. In this case, the only direction the electrons are free to move along is the x direction. So, we encounter a problem of a 1DEG, and the corresponding structure is called a QWR.

The envelope function of electron confining in a 2D potential $V(y, z)$ should satisfy the following effective mass equation:

$$\left[-\frac{\hbar^2}{2m_e}\left(\frac{d^2}{dy^2} + \frac{d^2}{dz^2}\right) + V(y, z)\right]\phi(y, z) = E_{mn}\phi(y, z) \tag{2.89}$$

Here the effective mass has been assumed to be isotropic. In the simplest situation of infinite barrier model, we can write $\phi(y, z) = \phi(y)\phi(z)$ and both of them are sinusoidal functions. In that case, the subband energies will be given by

$$E_{mn} = \frac{\hbar^2}{2m_e}\left[\left(\frac{n\pi}{d_y}\right)^2 + \left(\frac{m\pi}{d_z}\right)^2\right] \tag{2.90}$$

where ds are the widths of the well in the respective directions and m and n are integers including zero but both cannot be zero.

Therefore, the dispersion relation expressing the free motion of electrons along the x-direction will be given as

$$E(k_x) = E_{mn} + \frac{\hbar^2 k_x^2}{2m_e} \tag{2.91}$$

The expression for DoS function will be obtained as

$$\rho_{1D} = \frac{2d_x}{\pi\hbar}\left[\frac{m_e}{2(E - E_{mn})}\right]^{1/2} \tag{2.92}$$

d_x is the length of the sample and it follows therefore that the DOS for 1D shows singularities at the subband edges.

Now, suppose that the barriers are created along all three directions, that is, there exists a 3D potential well for both electrons and holes. In this case, the motion of the carriers is confined in all three dimensions leading to the 0D carrier system. The corresponding structure is called a QD. If the shape of the QD is spherical; it is referred to as a Superatom.

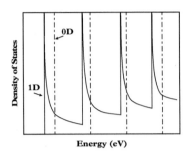

Figure 2.13 *DOS function of a QWR (1D) and a QD (0D) structure*

If infinite barrier approximation has been applied, the energy levels can be expressed as

$$E_{n_x,n_y,n_z} = \frac{\hbar^2\pi^2}{2m_e}\left[\left(\frac{n_x}{d_x}\right)^2 + \left(\frac{n_y}{d_y}\right)^2 + \left(\frac{n_z}{d_z}\right)^2\right] \qquad (2.93)$$

where ds are the dimensions in the respective directions.

The DOS for QDs consists of delta functions centered on each discrete level. The DOS functions for QWRs and QDs are shown in Fig. 2.13.

Example 2.15 *The shift of ground state energy of a GaAs QW due to additional confinement along another direction (y-direction) with d_y = 80 nm can be calculated by the relation*

$$\Delta E_e = \left(\frac{\hbar\pi}{d_y}\right)^2 \Big/ 2m_e = 0.88\,meV$$

2.10.4   Multiple quantum wells and superlattices

The repetition of single QWs leads to Multiple Quantum Wells (MQWs) as shown in Fig. 2.14. The individual QWs in MQW structure are uncoupled. As a result, the energy levels remain discrete.

The coupling between the QWs occurs due to the small thickness of the barrier layer or low barrier height. In that case, the exponential tails of the envelope functions in the adjacent two wells overlap, which in turn give rise to coupling. For two coupled wells, the coupling leads to two closely spaced energy levels centered about the original degenerate energy level. The presence of N number of coupled wells gives rise to N number of closely spaced energy levels. If the spacing between the levels is small enough, the levels are distributed continuously and an energy band, called Miniband will be formed.

The main difference between an MQW and the Superlattice (SL) structure arises due to well-to-well coupling. In an SL, the width of the barrier layer is also small. In this case, the periodicity $a = d + b$, where d and b are, respectively, the well and barrier width, must be much greater than the interatomic distance. As the lattice spacing $a \gg a_0$, the structure is rightly called the SL. The SL behaves as an anisotropic bulk medium having a different E-k relationship as shown in Fig. 2.15. The energy band exist in the

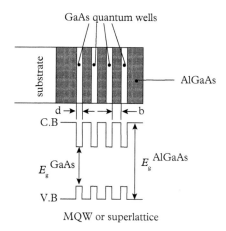

Figure 2.14 *Schematic diagram of a semiconductor multiple QW (MQW) or a superlattice*

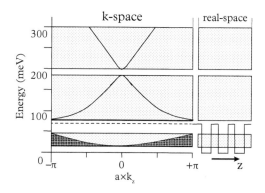

Figure 2.15 *Miniband (shaded region) and minigap (clear region) in a superlattice, shown in both real and k-space.*

minizone ranging from $-\frac{\pi}{a}$ to $+\frac{\pi}{a}$. As $a \gg a_0$, the bands are termed as minibands. The band diagram shows gaps, called minigaps, at the edges of the minizone (see Mukherji and Nag 1975).

Example 2.16 *We consider a GaAs/Al$_{0.3}$Ga$_{0.7}$As superlattice where the barrier height is 250.6 meV. The penetration length of the electron wavefunction into the barrier is given by the relation $\delta = \sqrt{(\hbar^2/2m_b V_0)}$. Taking m_b=0.0914 m_0, δ can be estimated as 1.29 nm. The wavefunction in the barrier is assumed to be zero at a distance of 4δ, a barrier width of 8δ = 10.32 nm will not lead to any coupling between adjacent wells.*

Example 2.17 *Let us assume a superlattice structure of QW width d = 5 nm, barrier width b = 8 nm. So, the period is 13 nm which is quite large compared to the lattice constant.*

So, the minizone will be confined in the range $\pm 2.42 \times 10^8 m^{-1}$ to $6.28 \times 10^9 m^{-1}$, where the lattice constant is assumed to be 0.5 nm.

2.10.5   Electron subband structure in multivalley semiconductor

For multivalley semiconductors, like Si, Ge, the effective mass is expressed as a tensor. The kinetic energy operator for the Schrodinger equation is

$$T = -\frac{\hbar^2}{2} \sum_{i,j} w_{ij} \frac{\partial^2}{\partial x_i \partial x_j} \tag{2.94}$$

where w_{ij}'s are the elements of the reciprocal effective mass tensor for the particular conduction band minimum. Since the potential energy $V(z)$ is a function of z only, the solution of the Schrodinger equation will be

$$\phi(x, y, z) = \beta(z) exp(ik_1 x + ik_2 y) \tag{2.95}$$

where 1...3 denote the principal axes of the constant energy ellipsoid. Substituting this into the effective mass Schrodinger equation, one obtains

$$-\frac{1}{2} w_{33} \hbar^2 \frac{d^2\beta}{dz^2} - \hbar^2 (w_{13}k_1 + w_{33}k_2) \frac{d\beta}{dz} + \left[qV(z) + E' \right] \beta(z) = 0 \tag{2.96}$$

where

$$E = E' + \frac{\hbar^2}{2} \left[w_{11}k_1^2 + 2w_{12}k_1 k_2 + w_{22}k_2^2 \right]$$

Now making the substitution

$$\beta(z) = \alpha(z) exp \left[-iz \frac{(w_{13}k_1 + w_{23}k_2)}{w_{33}} \right] \tag{2.97}$$

to eliminate the first derivative with respect to z, we find

$$\frac{d^2\phi_i}{dz^2} + \frac{2m_3}{\hbar^2} \left[E''_i + qV(z) \right] \phi_i(z) = 0 \tag{2.98}$$

Where $m_3 = \frac{1}{w_{33}}$ and a subscript i have been introduced to the label of solutions. Thus, the total energy becomes

$$E_i(k_1, k_2) = E''_i + \frac{\hbar^2}{2}\left[\left(w_{11} - \frac{w_{13}^2}{w_{33}}\right)k_1^2 + 2\left(w_{12} - \frac{w_{13}w_{23}}{w_{33}}\right)k_1 k_2 + \left(w_{22} - \frac{w_{23}^2}{w_{33}}\right)k_2^2\right]$$

(2.99)

In the previous equation, E''_i denotes the subband energy and the other term is the kinetic energy in the xy plane.

2.11 Strained layers

Application of strain to a semiconductor brings important changes in its physical properties, particularly in the band gap and effective mass. These changes are utilized in making better electronic and photonic devices. In this section, the growth of strained structures, and some changes in physical properties are discussed briefly.

2.11.1 Growth of strained layers

The usual method of applying strain is to grow an epitaxial layer, having a small degree of lattice mismatch with the thick substrate. For a sufficiently thin epi-layer, having thickness less than a critical thickness, the grown overlayer maintains the same crystal structure as that of the substrate and such a growth is termed as pseudomorphic.

For growth of an epitaxial layer having a lattice mismatch with the substrate, misfit dislocation or voids occur at the interface between the two as well as in a few monolayers. After that, the epitaxial layer grows freely with its own lattice constant. However, for a sufficiently small lattice mismatch, and provided that the thickness of the epitaxial layer is below a critical value, the growth may allow the epilayer to grow with the same lattice constant as the substrate. There is no misfit dislocation, but the epilayer is under strain. The growth is called pseudomorphic. The critical layer thickness is determined by the competition between the strain energy and the energy of the formation of the misfit dislocation. The strain energy increases linearly with increasing thickness, but the formation energy for the misfit dislocation increases superlinearly with increasing layer thickness, and then saturates. The two energies are equal at critical thickness. Below the thickness pseudomorphic growth is preferred as the strain energy is lower, but for thickness larger than the critical value, the growth occurs with a misfit dislocation as the energy of its formation is lower.

2.11.2 Expression for critical thickness

Different expressions for the critical layer thickness are given in the literature. We will now quote the expression given by Voisin (1988)

$$h_c = \frac{1 - v/4}{4\pi(1 + v)} b\varepsilon^{-1}\left[\ln\left(\frac{h_c}{b}\right) + \theta\right]$$

(2.100)

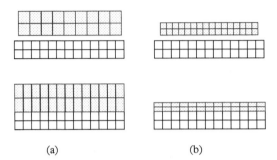

(a)   (b)

Figure 2.16 *Schematic diagram depicting growth of lattice-mismatched epilayer on a substrate. Top: epilayer and substrate shown isolated; bottom: growth of strained layer with in-plane lattice matching. (a) Compressively strained and (b) tensile strained growth. The effect of perpendicular strain is also illustrated.*

where, ν is the Poisson ratio, b is the dislocation Burgers vector ($\approx 4\text{Å}$), ε is the in-plane strain (to be defined later), and θ is a constant (~1). The calculated value of h_c is 9 nm for $\varepsilon = 1\%$.

Epitaxial layers grown pseudomorphically may be compressively strained when its lattice constant is higher than that of the substrate. Similarly, an epilayer with lower lattice constant will be tensile strained. These situations are illustrated in Fig. 2.16. The top part shows isolated layers, and the bottom part shows the pseudomorphically grown structure.

2.11.3   Strain symmetric structures and virtual substrates

Pseudomorphically grown strained layers are used both in electronic and photonic devices. In the latter applications, the altered band gap due to strain is exploited for tailoring the optoelectronic properties of the semiconductors. Epitaxial layers exceeding critical layer thickness also find use as they provide virtual substrates for different layers in a multilayered structure. In this case, the dislocations are confined near the interface between the substrate and the thick epilayer and do not affect the properties of the epi-layers. The virtual substrates allow growth of strain-symmetric superlattice structures, in which two layers A and B are tensile and compressively strained alternately. The buffer or virtual substrate has a lattice constant intermediate of the values for the constituent materials A and B.

Example 2.18 *Consider growth of alternate layers of $In_{0.4}Ga_{0.6}As$ (A) and $In_{0.65}Ga_{0.35}As$ (B) on InP substrate (S). The lattice constants are $a(InP) = 5.8686$ Å, $a(InAs) = 6.058$ Å, and $a(GaAs) = 5.653$ Å. Using Vegard's law, $a(A) = 5.815$ Å and $\varepsilon(A) = +0.0092$, $a(B) = 5.9163$ Å, and $\varepsilon(B) = -0.0081$. For strain compensation the thickness should be $d_A/d_B = \varepsilon(B)/\varepsilon(A) = 1.136$. Thus, if $d_A = 5$ nm, then $d_B = 5.68$ nm. The strain is defined as $\varepsilon_{A,B} = a_S - a_{A,B}/a_S$.*

2.11.4 Strain, stress, and elastic constants

Stress is the force per unit area on a face of a crystal and in most cases it is anisotropic. We need to introduce a stress tensor, having elements, σ_{ij}, $i, j = x, y, z$. The non-diagonal elements, σ_{ij}, $i \neq j$, represent shear components and cause the crystal to rotate. In the presence of an equal and opposite component, the shear stress deforms the lattice. A cubic lattice thus gets a non-rectangular shape, and consequently the crystal axes become non-orthogonal. However, this type of deformation is rarely found in typical semiconductors, so we safely put $\sigma_{ij} = 0$, for $i \neq j$. The normal components of stress expanding or contracting are the crystals along the crystal axes, but the cubic lattice has its shape changed into a rectangular one. The six faces of the cubic crystal are now acted upon by three normal forces, σ_{11}, σ_{22}, and σ_{33} which from now are denoted by σ_1, σ_2, and σ_3. The forces are treated as positive if they are directed outwards. The crystal is deformed or strained due to stress. A strain tensor is introduced which is diagonal and has three elements $\varepsilon_1, \varepsilon_2$, and ε_3. If $\varepsilon_i > 0$, the crystal is elongated along the ith axis and if $\varepsilon_i < 0$, the crystal is compressed along the ith axis.

Example 2.19 *Consider growth of $In_{0.2}Ga_{0.8}As$ on GaAs. Using previous parameters, a (InGaAs) = 0.5734 nm and a(GaAs) = 0.5653 nm. Therefore $\varepsilon_i < 0$ and the crystal is compressed.*

For an isotropic medium one may relate stress and strain by $\sigma = C\varepsilon$ following Hooke's law, where C is the Young's (or rigidity) modulus. This scalar relation is however modified in a crystal due to the tensor nature of both stress and strain, and Hooke's law is written as

$$[\sigma] = [C] \cdot [\varepsilon] \tag{2.101}$$

The components C_{ij} of the tensor **C**, the total number of which is 36 in general, as there are six independent components of stress tensor, are known as the elastic stiffness coefficients or elastic moduli. For cubic crystals the numbers reduce drastically so that one may write

$$\begin{bmatrix} \sigma_1 \\ \sigma_2 \\ \sigma_3 \end{bmatrix} = \begin{bmatrix} C_{11} & C_{12} & C_{12} \\ C_{12} & C_{11} & C_{12} \\ C_{12} & C_{12} & C_{11} \end{bmatrix} \begin{bmatrix} \varepsilon_1 \\ \varepsilon_2 \\ \varepsilon_3 \end{bmatrix} \tag{2.102}$$

Note that only two elastic moduli, C_{11} and C_{12}, need be considered now. For crystals with less symmetry, there are more C_{ij} components. Another component C_{44} is to be included in the elastic moduli tensor even for cubic crystals if shear is to be considered. In common semiconductors $C_{11} > C_{12}$, and both the moduli are positive and are usually described in units of 10^{11} dynes/cm^2.

Consider now that a film with natural lattice constant a_f is grown pseudomorphically on a substrate of lattice constant a_s along the z-direction. Let the directions 1,2,3 coincide with the x-, y-, and z-directions. As the lattice constants of the film and the substrate match, the film experiences biaxial stress, with stresses along all the four x and y faces, but no stress is applied to the z faces. When $a_f < a_s$, the film or epilayer will be under biaxial tensile strain making the stress components directed outward from the four x and y faces ($\sigma_1 = \sigma_2 > 0$). Similarly, for $a_f > a_s$, the film will under biaxial compressive strain and the stress components will be directed inwardly through x and y faces ($\sigma_1 = \sigma_2 < 0$). The stress and strain components along x and y directions must be equal by symmetry. This sets $\sigma_1 = \sigma_2$ and $\varepsilon_1 = \varepsilon_2$, and further with $\sigma_3 = 0$, one obtains from the first and third equations in (2.102)

$$\sigma_1 = C_{11}\varepsilon_1 + C_{12}\varepsilon_1 + C_{12}\varepsilon_3 \qquad (2.103a)$$

$$0 = C_{11}\varepsilon_1 + C_{12}\varepsilon_1 + C_{12}\varepsilon_3 \qquad (2.103b)$$

The strain along z-direction may now be expressed as

$$\varepsilon_3 = -\left(2C_{12}/C_{11}\right)\varepsilon_1 \qquad (2.104)$$

Since both C_{11} and C_{12} are positive quantities, the deformation along z will be opposite to that along x and y directions. From Eq. (2.103a), one may write

$$\sigma_1 = C_{11}\varepsilon_1[1 + C_{12}/C_{11} - 2(C_{12}/C_{11})^2] \qquad (2.105)$$

When the crystal under uniform stress is applied equally to all the six faces of the cubic crystal, a uniform pressure change, (dP) directed inward develops, and then $\sigma_1 = \sigma_2 = \sigma_3 = -dP$. Adding up all the three equations resulting from Eq. (2.102), we find

$$-3dP = (C_{11} + 2C_{12})(\varepsilon_1 + \varepsilon_2 + \varepsilon_3) \qquad (2.106)$$

The volume of the crystal is now $V + dV = a^3(1 + \varepsilon_1)(1 + \varepsilon_2)(1 + \varepsilon_3)$, where a is the lattice constant and the unstrained volume is $V = a^3$. The fractional change in the volume of the crystal, neglecting second-order product terms like $\varepsilon_1\varepsilon_2$, etc., is given by

$$dV/V \approx \varepsilon_1 + \varepsilon_2 + \varepsilon_3 \qquad (2.107)$$

2.11.5   Band structure modification by strain

The presence of biaxial strain shifts HH and LH bands by different amounts, so that the degeneracy of the two at $\mathbf{k} = 0$ is lifted. The shifts are given as follows, in terms of the deformation potential constants a and b:

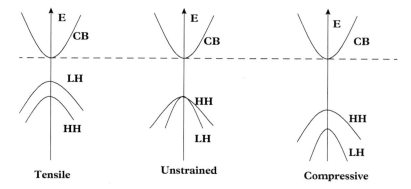

Figure 2.17 *Effect of compressive and tensile strains on the HH and LH band edges.*

$$\Delta E_{hh}(\varepsilon) = \left[2a\left(\frac{C_{11} - C_{12}}{C_{11}}\right) + b\left(\frac{C_{11} + 2C_{12}}{C_{11}}\right)\right]\varepsilon \qquad (2.108\text{a})$$

$$\Delta E_{lh}(\varepsilon) = \left[2a\left(\frac{C_{11} - C_{12}}{C_{11}}\right) - b\left(\frac{C_{11} + 2C_{12}}{C_{11}}\right)\right]\varepsilon \qquad (2.108\text{b})$$

The shifts are measured with respect to the conduction band edge which is assumed to have a fixed energy. It follows easily from the previous equations, and as shown in Fig. 2.17, the LH band is at the top for tensile strain, whereas the HH band is on top for the compressively strained layer. The sign of strain thus determines the position of the two bands now.

The splitting between the HH and LH bands causes a dramatic decrease of the DOS mass as a function of stress. This has an impact on the reduction of the threshold current density in the laser.

Example 2.20 *The splitting between $HH - LH$ is calculated for $In_{0.2}Ga_{0.8}As$ layer grown on a (001) GaAs substrate, taking $a = 0.xx$ eV, $b = -2.0$ eV, $C_{11} = 11.5 \times 10^{11}$ dynes/cm^2 and $C_{12}=5.5 \times 10^{11}$ dynes/cm^2. The splitting produced is $\Delta E_{hh}(\varepsilon) - \Delta E_{lh}(\varepsilon) = 2b[(C_{11} + 2C_{12})/C_{11}]\varepsilon = 5.91\varepsilon$. The strain is $a(GaAs)-a(InGaAs)/a(GaAs)= 5.653 - 5.734/5.653 = -0.014$. The splitting is $\Delta E_{hh}(\varepsilon) - \Delta E_{lh}(\varepsilon) = 0.085$ eV.*

Problems

2.1 *The electron effective mass values for Si are $m_l = 0.92\ m_0$ and $m_t = 0.19\ m_0$. Calculate the DOS effective mass.*

2.2 *Calculate the DOS effective mass for holes in GaAs, given $m_{hh} = 0.5\ m_0$ and $m_{lh} = 0.087\ m_0$.*

2.3 *The hole effective mass in semiconductors is defined as* $m_h = \dfrac{m_0}{A \mp \{B^2 + (C^2/5)\}^{1/2}}$, *where* − (+) *signs refer to HH (LH). The A, B, C parameters are related to the Luttinger parameters as*

$$\gamma_1 = -2m_0(A + 2B)/3\hbar^2$$

$$\gamma_2 = -m_0(A - B)/3\hbar^2$$

$$\gamma_3 = -m_0 C/3\hbar^2$$

Calculate the HH and LH effective masses in GaAs for which $\gamma_1 = 6.85$, $\gamma_2 = 2.1$, *and* $\gamma_3 = 2.9$.

2.4 *Obtain an expression for the position of the Fermi level with respect to the CB edge in terms of the electron density.*

2.5 *Show that the expression for electron density in a semiconductor obeying Fermi statistics is*
   $n = N_c F_{1/2}(\eta)$, *where* $F_{1/2}(\eta) = (2/\sqrt{\pi}) \int_0^\infty \dfrac{x^{1/2}\, dx}{1 + exp(x - \eta)}$ *and*

$$\eta = (E_F - E_c)/k_B T$$

2.6 *Prove that the position of Fermi level in an intrinsic semiconductor is in the middle of the band gap when the effective masses of electrons and holes are equal.*

2.7 *Calculate the energy separation between 1s and 2s states of P impurity in Si. Take effective mass of electron as* $0.26\, m_0$ *and* $\varepsilon = 12\varepsilon_0$.

2.8 *Prove from Eq. (2.36) that for extremely degenerate semiconductors* $\mu = e\tau(E_F)/m_e$, *where* E_F *is the Fermi energy.*

2.9 *Using Eq. (2.36) obtain the expression for the mobility for electrons obeying Boltzmann statistics.*

2.10 *The electron concentration in an n-type semiconductor is* 10^{17} *cm*$^{-3}$ *and the electrons are shared by two conduction band valleys separated by 0.1 eV. The electron mobilities at 300 K in lower (upper) valleys are 10,000 (500) cm*^{2}/*V.s. Calculate the overall mobility.*

2.11 *Mobilities limited by impurity and phonon scatterings are expressed respectively as* $AT^{3/2}$ *and* $BT^{-1/2}$, *where A and B are constants, and T is the temperature. Show that the overall mobility attains a maximum at a certain T.*

2.12 *Using Eq. (2.37) and the parameters given for Si obtain the values of low field mobility and saturation drift velocity for electrons.*

2.13 *The mean energy of electrons is* $(1/2)\, k_B T$ *per degree of freedom. The diffusion length is* $L = v_{th}\tau$, *where* v_{th} *is the thermal velocity, and* τ *is the mean free time. Using these show that* $D/\mu = k_B T/e$.

2.14 *Solve continuity equation for diffusion and recombination to show that the excess electron density follows Eq. (2.54) for a semi-infinite semiconductor.*

2.15 *Prove that the diffusion length* L_n *is the mean length an excess electron diffuses in a p-type semiconductor before being lost by recombination.*

2.16 *The lattice constants for different semiconductors are given in parentheses: GaAs (5.6533), AlAs (5.66), InAs (6.0584), and InP (5.8688), where all the values are given in Å.*
Calculate the composition of $In_{1-x}Ga_xAs$ *lattice matched to InP. Find also the composition of* $In_{1-z}Al_zAs$ *lattice matched to both InGaAs and InP.*

2.17 *The band gap of* $In_{1-x}Ga_xAs$ *is expressed as 0.36+ 1.064x eV and that of InP is 1.36 eV. The band offsets in lattice matched InGaAs/InP heterojunction are given by* $\Delta E_c/\Delta E_v = 40/60$ *Calculate the values of* ΔE_c *and* ΔE_v.

2.18 *Prove that the Fermi-wavevector for a degenerate electron gas is* $k_F = \left(3\pi^2 n\right)^{1/3}$, *where n is the electron density. From this prove that the thickness d of a QW should be less than* $2(\pi/3n)^{1/3}$ *to exhibit quantization.*

2.19 *Obtain the maximum value of* d *when the electron density is* 10^{19} cm^{-3} *and extreme degeneracy condition is valid.*

2.20 *For quantization of electron gas another characteristic length is the mean free path. Establish the limiting condition for well width d in terms of the mobility of the electron gas.*

2.21 *Calculate the limiting value of d in the previous problem when the carrier density is* 10^{18} cm^{-3}, *and the mobility is 8000 cm²/V.s.*

2.22 *A GaAs QW is grown along the z-direction and has a thickness* d = 5 nm. *The effective masses of HH and LH along the z-direction are* $m_{hhz} = \frac{m_0}{(\gamma_1 - 2\gamma_2)}$ *and* $m_{lhz} = \frac{m_0}{(\gamma_1 + 2\gamma_2)}$, *and the in-plane masses are* $m_{hh||} = \frac{m_0}{(\gamma_1 + \gamma_2)}$ *and* $m_{lh||} = \frac{m_0}{(\gamma_1 - \gamma_2)}$.
Using $\gamma_1 = 6.85$ *and* $\gamma_2 = 2.1$, *calculate the mass values and lowest subband energies for HH and LH. Which subband will have higher population? Which subband will have higher 2DEG mobility?*

2.23 *Prove that the electron density per unit area of a 2DEG can be expressed as* $n_{2D} = (m_e k_B T/\pi \hbar^2) \ln\{1 + \exp(E_F - E_1)/k_B T\}$, *assuming that only the lowest subband* E_1 *is occupied.*

2.24 *Modify the previous expression for 2DEG density for extreme degeneracy. Calculate the difference between the Fermi level and subband energy for* $n_{2D} = 10^{12}$ cm^{-2}. *Assume* $m_e = 0.07 m_0$.

2.25 *The inversion layer in an Si metal-oxide semiconductor field-effect transistor (MOS-FET) is quantized and only the lowest subband is occupied. Prove that the 2DEG density may be expressed as* $en_{2D} = C(V_g - V_{th})$, *where C is the gate capacitance per unit area,* V_g *is the gate voltage, and* V_{th} *is the threshold voltage.*
Calculate the position of the Fermi level with respect to lowest subband energy for $V_g = 1$, *1.5 and 2 V. Assume* $V_{th} = 0.1$ V, $C = 10^{-7}$ F/m² *and* $m_e = 0.19 m_0$ *and extreme degeneracy for 2DEG.*

2.26 *Obtain the expression for DOS function of 1DEG.*

2.27 *Assuming non-degenerate statistics for electrons, prove that the expression for electron density of 1DEG is $n_{1D} = (2m_e k_B T / \pi \hbar^2)^{1/2} exp[(E_F - E_1)/k_B T]$, where E_1 is the lowest subband energy.*

2.28 *Prove that, at low temperature, when the Fermi function has a box-like nature $n_{1D} \propto (E_F - E_1)$. Also find the expression for the proportionality constant.*

2.29 *Solve Schrodinger equation for a quantum box of dimension $d_x \times d_y \times d_z$ by separation of variables and show that the quantized energy levels may be expressed by Eq. (2.93).*

2.30 *Using $\gamma_1 = 4.95$ and $\gamma_2 = 1.65$, calculate the binding energies for 2D HH and LH excitons in InP. The electron effective mass is 0.077 m_0.*

2.31 *The Luttinger parameters for GaAs (InAs) are $\gamma_1 = 6.8(20.4)$ and $\gamma_2 = 1.9(8.3)$. The electron effective mass for GaAs (InAs) are 0.067 (0.023). Calculate the binding energies of 1s HH and LH excitons in $In_{0.53}Ga_{0.47}As$.*

2.32 *Assume that the reduced mass equals the electron-effective mass and the binding energy of quasi-2D exciton is determined by an average electron effective mass defined as $m_{av} = m_w d_w + m_b d_b/(d_w + d_b)$, where d is the effective thickness and w and b refer respectively to the well and barrier.*

   Using these approximations, prove that the exciton binding energy in a QW increases first with d_w, reaches a maximum, and then decreases.

Reading List

Ando T, Fowler A B, and Stern F (1982) Electronic properties of two-dimensional systems. *Review of Modern Physics* 54: 437–672.

Balkanski M and Wallis R F (2000) *Semiconductor Physics and Applications.* Oxford UK: Oxford University Press.

Bastard G (1988) *Wave mechanics Applied to Semiconductor Heterostructure.* Les Ulis: Les Editions de Physique.

Basu P K (2003) *Theory of Optical Processes in Semiconductors: Bulk and Microstructures.* Oxford, UK: Clarendon Press.

Basu P K, Mukhopadhyay B, and Basu R (2015) *Semiconductor Laser Theory.* Boca Raton, FL: CRC Press.

Cohen M L and Chelikowsky J R (1989) *Electronic structure and optical properties of semiconductors.* In: *Springer Series in Solid-State Science 75,* M Cardona (ed.). Berlin: Springer.

Datta S (1989) *Quantum Phenomena.* Reading, Mass: Addison-Wesley.

Davies John H (1998) *The Physics of Low-Dimensional Semiconductors: An Introduction.* Cambridge, UK: Cambridge University Press.

Harrison P (2000) *Quantum Wells, Wires and Dots: Theoretical and Computational Physics.* Chichester: John Wiley & Sons, Ltd.

Manasreh Omar (2005) *Semiconductor Heterojunctions and Nanostructures.* N2Y: McGraw Hill.

Mitin V, Strocio M A, and Kochelap (1999) *Quantum Heterostructures*. NY: John Wiley.

Nag B R (1980) *Electron Transport in Compound Semiconductors*. Berlin: Springer-Verlag.

Neamen D A (1992) *Semiconductor Physics and Devices: Basic Principles*. NY: McGraw-Hill.

Ridley B K (2000) *Quantum Processes in Semiconductors*, 5th edn. Oxford: Clarendon Press.

Roblin P and Rohdin H (2002) *High-Speed Heterostructure Devices*. Cambridge, UK: Cambridge University Press.

Shur M (2003) *Physics of Semiconductor Devices*. Englewood Cliffs, NY: Prentice-Hall.

Singh J (2003) *Electronic and Optoelectronic Properties of Semiconductor Structures*. Cambridge, New York: Cambridge University Press.

Smith R A (1964) *Semiconductors*. Cambridge, UK: Cambridge University Press.

Streetmann B G and Banerjee S (2006) *Solid State Electronic Devices*, 6th edn. Englewood Cliffs, NY: Prentice Hall International.

Sze S M (1969) *Physics of Semiconductor Devices*. New York: John Wiley.

Weisbuch C and Vinter B (1991) *Quantum Semiconductor Structures*. San Diego: Academic Press.

Yu P and Cardona M (2010) *Fundamentals of Semiconductors*, 4th edn. Berlin: Springer.

References

Hall R N (1952) Electron-hole recombination in Germanium. *Physical Review* 152: 387.

Hu J, Xu G, Stotz J A H, Watkins S P, Curzon A E, and Thewalt M L W M L W (1998) Type II photoluminescence and conduction band offsets of GaAsSb/InGaAs and GaAsSb/InP heterostructures grown by metalorganic vapor phase epitaxy. *Applied Physics Letters* 73(19): 2799–2801.

Joyce W B and Dixon R W (1977) Analytic approximations for the Fermi energy of an ideal Fermi gas. *Applied Physics Letters* 31: 354–356.

Luttinger J M and Kohn W (1955) Motion of electrons and holes in perturbed periodic fields. *Physical Review* 97: 869–883.

Mukherji D and Nag B R (1975) Band structure of semiconductor superlattices. *Physical Review B* 12: 4338–4342.

Shockley W and Read W T (1952) Statistics of recombination of electrons and holes. *Physical Review* 87: 835–842.

Vande Walle C G (1989) Band lineups and deformation potentials in the model-solid theory. *Physical Review B* 39: 1871–1883.

Voisin P (1988) in *Quantum Wells and Superlattices in Optoelectronic Devices and Integrated Optics*, Vol. 861, p. 88. Bellingham, WA: SPIE.

3

Macroscopic theory of optical processes

3.1 Introduction

Optical processes in gases, liquids, and solids including nanostructures can best be explained with the help of quantum theory. The present book aims at developing the basic models relying on quantum mechanics, which make use mainly of perturbation theory, and in some cases, theories beyond perturbation. However, many experimental results related to optical processes can be qualitatively explained by using classical theories developed using Maxwell's equations. The quantum mechanical theory for light matter interaction requires the quantization of electro-magnetic (EM) waves and the concept of photons. For this purpose, also, the solution of wave equation and the exact form of electric and magnetic fields need be known. In addition, different waveguides, resonators, and microcavities are essential components to observe novel physical phenomena and to make nanophotonic devices. This also necessitates the study of the behaviour of EM waves in dielectrics, semiconductors, and metals.

This chapter outlines the theory of optical processes in bulk solid materials from the previously mentioned view-point, emphasizing the macroscopic theories. The starting points are the four Maxwell's equations and the constitutive equations. The wave equations developed from Maxwell's equations and the relationships between the absorption coefficient and the refractive index with real and imaginary parts of the susceptibility are established. The concept of phase and group velocities of EM waves is introduced to serve as the background for understanding slow and fast waves treated in a later chapter. The well-known damped harmonic oscillator model is included to introduce dispersion of EM waves and establish its parallelism with quantum oscillators treated in later chapters. This chapter then introduces the three basic processes of light matter interaction, namely, absorption, and spontaneous and stimulated emissions, by considering the well-known two-level atomic system. The respective rates of the processes are expressed in terms of Einstein's A and B coefficients. The conditions for net absorption and stimulated emission are pointed out. The concepts introduced here are utilized in later chapters to develop the theory of optical processes in real materials, like semiconductors and their nanostructures.

Semiconductor Nanophotonics. Prasanta Kumar Basu, Bratati Mukhopadhyay, and Rikmantra Basu, Oxford University Press.
© P.K. Basu, B. Mukhopadhyay, R. Basu (2022). DOI: 10.1093/oso/9780198784692.003.0003

Metals support plasma waves due to presence of a very large concentration of free electrons. The starting point in understanding plasma in conductors is again Maxwell's equation, as will be discussed in a later chapter.

3.2 Optical constants

The optical processes in semiconductors and nanostructures can be described by using the quantum theory of radiation, which will be introduced in a later section. However, the behaviour of many of the optical constants in a conductor or a dielectric, like the absorption coefficient or permittivity, and their interrelationship can be understood by using the classical wave equation.

3.2.1 Wave equation

The following four Maxwell's equations related to the behaviour of electromagnetic waves form the starting point.

$$\nabla \times F + \frac{\partial B}{\partial t} = 0 \tag{3.1}$$

$$\nabla \times H = \frac{\partial D}{\partial t} + J = \varepsilon \left(\partial F / \partial t \right) + \sigma F \tag{3.2}$$

$$\nabla \cdot D = \rho \tag{3.3}$$

$$\nabla \cdot B = 0 \tag{3.4}$$

where F and H are the electric and magnetic fields, $D = \varepsilon F, B = \mu H$, J and ρ are the current and charge densities, respectively. The permittivity of the medium is $\varepsilon = \varepsilon_r \varepsilon_0$, where ε_r and ε_0 are the relative and free space permittivities, respectively, and $\mu = \mu_r \mu_0$, where μ, μ_r and μ_0 are, respectively, the permeability, relative permeability, and free-space permeability. The conductivity of the medium is denoted by σ.

Let us first consider the propagation of EM waves in free space without any free charge. Then $\varepsilon_r = \mu_r = 1, \rho = J = 0$ and we may rewrite Eqs. (3.1) and (3.3) as

$$\nabla \times H = \varepsilon_0 \left(\partial F / \partial t \right) \tag{3.5}$$

and

$$\varepsilon_0 \nabla \cdot F = 0 \tag{3.6}$$

Taking the curl of both sides of Eq. (3.1) and using Eq. (3.5), one obtains

$$\nabla \times \nabla \times F = -\mu_0 \frac{\partial}{\partial t} \nabla \times H = \varepsilon_0 \mu_0 \frac{\partial^2}{\partial t^2} F = \frac{1}{c^2} \frac{\partial^2}{\partial t^2} F \tag{3.7}$$

where $c = (\varepsilon_0 \mu_0)^{-1/2}$ is the velocity of light in free space. Using the vector identity $\nabla \times \nabla \times F = \nabla (\nabla \cdot F) - \nabla^2 F$ and noting from Eq. (3.6) that $\nabla \cdot F = 0$, we may convert Eq. (3.7) to arrive at the wave equation in the following form:

$$\nabla^2 F = \frac{1}{c^2} \frac{\partial^2 F}{\partial t^2} \tag{3.8}$$

The wave equation for the magnetic field may be expressed similarly. Possible solutions for the wave equations are plane propagating waves expressed in the following forms:

$$F(r, t) = F_0 \{exp[i(q \cdot r - \omega t)] + c.c\} \tag{3.9}$$

$$H(r, t) = H_0 \{exp[i(q \cdot r - \omega t)] + c.c\} \tag{3.10}$$

where **q** is the wave vector for EM waves and *c.c.* stands for *complex conjugate*. Using Eq. (3.9) in Eq. (3.8) we express the dispersion relation as

$$q = \omega/c \tag{3.11}$$

Eqs. (3.1), (3.5), and (3.6) lead to the following interrelationships between the electric and magnetic field vectors:

$$q \times F = \omega \mu_0 H \tag{3.12a}$$

$$q \times H = \omega \mu_0 F \tag{3.12b}$$

It appears therefore from Eq. (3.12a and b.) that **q, F** and **H** are mutually orthogonal and the field magnitudes are related by

$$H = (\varepsilon_0/\mu_0)^{1/2} F \tag{3.13}$$

For each direction of q there are two independent polarization directions of the electric field vector.

Example 3.1 *The ratio of F/H in free space is called the intrinsic impedance. Putting the values of ε_0 and μ_0 in Eq. (3.13) the value of intrinsic impedance is 377 ohms.*

3.2.2   Light absorption in a medium: Macroscopic theory

Let us now consider a dielectric medium at the surface of which light of intensity I_0 is incident. If $z = 0$ denotes the coordinate of the surface of the medium, then the variation of light intensity within the medium is expressed by the well-known expression

$$I(z) = I_0 exp(-\alpha z) \tag{3.14}$$

where α is called the absorption coefficient of the material, which is a function of frequency or wavelength. We now aim at relating the absorption coefficient with the dielectric properties of the medium, such as the refractive index and extinction coefficient, by using Maxwell's equations.

Taking the divergence of both sides of Eq. (3.2), one obtains

$$\nabla \cdot \nabla \times H = \varepsilon_0 \frac{\partial}{\partial t}(\nabla \cdot F) + \nabla \cdot J \tag{3.15}$$

Now $\nabla \cdot (\nabla \times H) = 0$. Using Eq. (3.3) we rewrite the previous equation as

$$\nabla \cdot J = -(\partial \rho / \partial t) \tag{3.16}$$

which is the continuity equation. Let us assume that the medium contains only bound charges and therefore $\sigma = 0$. We may then express both ρ and J in terms of the polarization P as

$$\rho = -\nabla \cdot P \tag{3.17}$$

$$J = \frac{\partial P}{\partial t} \tag{3.18}$$

The polarization is proportional to the electric field, so that

$$P = \varepsilon_0 \chi F \tag{3.19}$$

where χ, the susceptibility is a function of the frequency ω.
    In presence of polarization Eqs. (3.2) and (3.3) take the following forms

$$\nabla \times H = \varepsilon_0 (1 + \chi)\frac{\partial F}{\partial t} \tag{3.20}$$

$$\nabla \cdot (\varepsilon_0 F + P) = 0 \tag{3.21}$$

Putting Eq. (3.19) in Eq. (3.21) one obtains

$$\varepsilon_0 (1 + \chi)\nabla \cdot F = 0 \tag{3.22}$$

Now taking the curl of both sides of Eq. (3.1), yields the following results:

$$\nabla \times \nabla \times F = -\mu_0 \frac{\partial}{\partial t} \nabla \times H = \varepsilon_0 \mu_0 (1 + \chi) \frac{\partial^2}{\partial t^2} F \tag{3.23}$$

Following the same steps as indicated after Eq. (3.7), one obtains the modified wave equation that reads

$$\nabla^2 F = \frac{1 + \chi}{c^2} \frac{\partial^2 F}{\partial t^2} \tag{3.24}$$

and the modified dispersion relation becomes

$$\left(\frac{qc}{\omega}\right)^2 = 1 + \chi = \varepsilon_r = \varepsilon_1 + i\varepsilon_2 \tag{3.25}$$

In Eq. (3.25) ε_r is the relative permittivity and is a function of frequency ω, but independent of electric field as long as the field is low, and is in general a complex quantity having its real and imaginary parts denoted by ε_1 and ε_2, respectively.

The susceptibility is a complex quantity, so that $\chi = \chi_r + i\chi_i$. The dispersion relation is expressed in terms of real and imaginary parts of the refractive index η as $(kc/\omega) = \eta_r + i\eta_i$. The subscripts r and i in χ and η refer to the real and imaginary parts of the respective quantities. It is easy then to arrive at the following relations:

$$\varepsilon_1 = 1 + \chi_r = \eta_r^2 - \eta_i^2 \tag{3.26}$$

$$\varepsilon_2 = \chi_i = 2\eta_r\eta_i \tag{3.27}$$

The expression for the electric field propagating along the z-direction becomes

$$F = F_0 exp[i(qz - \omega t)] = F_0 exp\left\{i\omega\left(\frac{\eta_r z}{c} - t\right)\right\} exp\left(\frac{-\eta_i \omega z}{c}\right) \tag{3.28}$$

A similar equation may be written for the magnetic field. The magnitudes of the two fields are now related by

$$H = (\varepsilon_0/\mu_0)^{1/2} (\eta_r + i\eta_i) F \tag{3.29}$$

As in the previous case, **F**, **H**, and **q** are orthogonal to each other.

The intensity of the lightwave I defined as the energy crossing per unit area per unit time, is given by the Poynting vector $I = S = (F \times H)$. The energy flow is averaged over one cycle of oscillation. Using the vector theorem $Re(A) \times Re(B) = (1/2)Re(A \cdot B^*)$, where Re stands for real part and $\star$ means conjugation, the cycle-averaged intensity may be written as

$$I = \frac{1}{2}\varepsilon_0 c\eta |F(r, t)|^2 \tag{3.30}$$

The absorption coefficient introduced in Eq. (3.14) may be related to the extinction coefficient η_i given by Eq. (3.27) by using Eqs. (3.30) and (3.28) as follows:

$$\alpha = \frac{2\omega\eta_i}{c} = \frac{\omega \, \varepsilon_2}{c \, \eta_r} \tag{3.31}$$

3.2.3   Electromagnetic waves in a conductor

The wave equation gets modified for a conductor, in which the free charge ρ_f and free current density due to the flow of the free charge cannot be controlled. The current density is related to the electric field by Ohm's law as

$$J_f = \sigma F \tag{3.32}$$

where σ is the conductivity. The modified forms of Maxwell's equations are as follows:

$$\nabla\times F = -\frac{\partial B}{\partial t} \quad (i) \quad \nabla\times B = \mu\sigma F + \mu\varepsilon\frac{\partial F}{\partial t} \; (ii) \quad \nabla\cdot F = \frac{\rho_f}{\varepsilon} \; (iii) \quad \nabla\cdot B = 0 \; (iv) \tag{3.33}$$

The continuity equation for the free current reads

$$\nabla\cdot J_f = -\frac{\partial\rho_f}{\partial t} \tag{3.34}$$

Using Ohm's law and Gauss law (iii) in Eq. (3.33) one obtains

$$\frac{\partial\rho_f}{\partial t} = -\sigma\left(\nabla\cdot F\right) = -\frac{\sigma}{\varepsilon}\rho_f \tag{3.35}$$

The time variation of the free charge density may be expressed as

$$\rho_f(t) = \rho_f(0)\exp\left[-\left(\sigma/\varepsilon\right)t\right]$$

Where $\rho_f(0)$ is the initial free charge density. The initial free charge density will flow out to the edges. Neglecting this transient and putting the free charge density to zero, we obtain from Eq. (3.33)

$$\nabla \times F = -\frac{\partial B}{\partial t} \quad (i) \quad \nabla \times B = \mu\sigma F + \mu\varepsilon \frac{\partial F}{\partial t} \quad (ii) \quad \nabla \cdot F = 0 \quad (iii) \quad \nabla \cdot B = 0 \quad (iv) \qquad (3.36)$$

We now apply the curl to (i) and (ii) in Eq. (3.36) as before, and obtain the modified wave equations for **F** and **B**

$$\nabla^2 F = \varepsilon\mu_0 \frac{\partial^2 F}{\partial t^2} + \sigma\mu_0 \frac{\partial F}{\partial t} \quad (i) \qquad \nabla^2 B = \varepsilon\mu_0 \frac{\partial^2 B}{\partial t^2} + \sigma\mu_0 \frac{\partial B}{\partial t} \quad (ii) \qquad (3.37)$$

The solutions are again plane wave in nature and are given by

$$F = F_0 exp\{j(q \cdot r - \omega t)\} \quad (i) \qquad B = B_0 exp\{j(q \cdot r - \omega t)\} \quad (ii) \qquad (3.38)$$

It is easy to find by substituting Eq. (3.38) in Eq. (3.37) that the wave vector is a complex quantity expressed as

$$q^2 = \mu\varepsilon\omega^2 + j\mu\sigma\omega \qquad (3.39)$$

Writing $q = q_r + jq_i$, where r and i denote, respectively, the real and imaginary terms, and by taking the square root, we obtain

$$q_{r(i)} = \omega \sqrt{\frac{\varepsilon\mu}{2}} \left[\sqrt{1 + \left(\frac{\sigma}{\varepsilon\mu}\right)^2} + (-)1 \right]^{1/2} \qquad (3.40)$$

The solution for the electric field propagating along the z-direction takes the form

$$F(z, t) = F_0 exp(-q_i z)exp\{j(q_r z - \omega t)\} \quad (i);$$
$$B(z, t) = B_0 exp(-q_i z)exp\{j(q_r z - \omega t)\} \quad (ii) \qquad (3.41)$$

The wave is then attenuated along the z-direction because of q_i. The amplitude reduces by a factor $1/e$ at a depth, known as the skin depth, expressed by

$$d = \left(\frac{1}{q_i}\right) \qquad (3.42)$$

The wavelength, the propagation speed and the index of refraction are determined by the real part q_r of the wave vector.

For a poor conductor $\sigma \ll \omega\varepsilon$, $q_r \cong \omega\sqrt{\varepsilon\mu}$, $q_i \cong (\sigma/2)\sqrt{\mu/\varepsilon}$ and the skin depth is independent of frequency. On the other hand, $\sigma \gg \omega\varepsilon$ for a good conductor, and

$$q_r \cong q_i \cong \sqrt{\omega\sigma\mu/2} \qquad (3.43)$$

The skin depth therefore decreases with increasing frequency.

Example 3.2 *The conductivity of copper at low frequency is 5.9×10^7 mhos.m^{-1}. Assuming that the relative permeability is 1, the skin depth at 10 MHz is $d \cong [\omega\sigma\mu/2]^{-1/2}$*
$= 20.7 \ \mu m$.

3.3 Phase and group velocities

In earlier subsections, the EM radiation has been assumed to be monochromatic, characterized by a single angular frequency ω, as indicated in Eqs. (3.9) and (3.10). In practice, however, EM radiation has a finite linewidth and is thus made of a number of planar waves having different frequency components. The phase velocity refers to the velocity with which a constant phase moves along a medium, whereas the group velocity determines how the energy is transported through the medium. These two velocities are treated in almost all textbooks in Electromagnetic Theory. We briefly introduce these two velocities in order to prepare the background for studies on slow and fast light. Our presentation follows essentially the treatment provided by Vornehm and Boyd (2009).

We write the plane wave solution for the electric field, given by Eq. (3.12), considering propagation along the z-direction only, as follows:

$$F(z, t) = F_0 exp[j(qz - \omega t)] = F_0 exp(j\phi); \phi = qz - \omega t \tag{3.44}$$

The phase of the wave is denoted by ϕ; constant values of ϕ define the phase front. The motion of the phase front is governed by the equation

$$qz - \omega t = 0 \tag{3.45}$$

The phase front moves by a distance z over time t, and we have

$$z = \frac{\omega}{q(\omega)} t = \frac{c}{n(\omega)} t \equiv v_p(\omega) t \tag{3.46}$$

Eq. (3.11) has been used to arrive at the previous result and absorption has been neglected. The phase velocity, or the speed of propagation of the phase front, is expressed as

$$v_p(\omega) = \frac{\omega}{q(\omega)} = \frac{c}{n(\omega)} \tag{3.47}$$

In general, the wave is not monochromatic and contains many Fourier components. We consider the propagation of a pulse and examine how it retains its original shape under propagation. The complex wave vector $q_c(\omega)$ is defined as

$$q_c(\omega) = q(\omega) + j\frac{\alpha(\omega)}{2} = \frac{n(\omega)\omega}{c} + j\frac{\alpha(\omega)}{2} \tag{3.48}$$

This assumes that the Fourier components are within a narrow bandwidth $\Delta\omega$ around a centre frequency ω_0. Writing $\omega = \omega_0 + \Delta\omega$, we may expand the real and imaginary parts of q in a Taylor series of about ω_0 and write

$$q(\omega) = q_0 + q_1\Delta\omega + (1/2)q_2\Delta\omega^2 + \cdots\cdots\cdots = \sum_{m=0}^{\infty} \frac{q_m}{m!}(\Delta\omega)^m \qquad (3.49\text{-i})$$

$$\alpha(\omega) = \alpha_0 + \alpha_1\Delta\omega + (1/2)\alpha_2\Delta\omega^2 + \cdots\cdots\cdots = \sum_{n=0}^{\infty} \frac{\alpha_n}{n!}(\Delta\omega)^n \qquad (3.49\text{-ii})$$

Different derivatives are defined as

$$q_m = \left.\frac{d^m q}{d\omega^m}\right|_{\omega=\omega_0} (i); \quad \alpha_n = \left.\frac{d^n q}{d\omega^n}\right|_{\omega=\omega_0} (ii) \qquad (3.50)$$

We now consider a pulse and write the electric field in terms of its Fourier components as

$$
\begin{aligned}
F(z,t) &= F_1 exp\{i[q_c(\omega_1)z - \omega_1 t]\} + F_2 exp\{i[q_c(\omega_2)z - \omega_2 t]\} + \cdots \\
&= \sum_j F_j exp(-j\omega_j t)exp[iq(\omega_j)z]exp[-\alpha(\omega_j)z/2]
\end{aligned}
\qquad (3.51)
$$

We now define $\Delta\omega_j = \omega_j - \omega_0$, substitute Eqs. (3.49(i)) and (3.49(ii)) in (3.51) and obtain the following modified form for (3.51):

$$F(z,t) = \sum_j F_j exp[-i(\omega_0 + \Delta\omega_m)t]exp\left[iz\sum_{m=0}^{\infty} \frac{q_m}{m!}(\Delta\omega_j)^m\right] exp\left[-\frac{z}{2}\sum_{n=0}^{\infty} \frac{\alpha_n}{n!}(\Delta\omega_j)^n\right]$$
$$\qquad (3.52)$$

Taking the ω_0, q_0 and α_0 terms outside the sums, and also taking the q_1 term outside the sum over m, we obtain

$$F(z.t) = F_0 exp(-\alpha_0 z/2)exp\left[-i(q_0 z - \omega_0 t)\sum_j F_j exp\left[-i\left(\Delta\omega_j t - q_1 z\right)t\right]\right]$$
$$exp\left[iz\sum_{m=2}^{\infty} \frac{q_m}{m!}(\Delta\omega_j)^m\right] exp\left[-\frac{z}{2}\sum_{n=1}^{\infty} \frac{\alpha_n}{n!}(\Delta\omega_j)^n\right]$$
$$\qquad (3.53)$$

In order of the shape of the pulse remains unchanged after propagating through the medium, it is required that the relative amplitudes and phases of the plane wave Fourier components should be constant. If $F(0,0)$ and $F(z,t)$ are, respectively, the amplitudes at

$z = 0$ and $t = 0$, and at z in time t,the two can at most be related by a simple complex constant, so that

$$F(z, t) = re^{i\phi}F(0, 0) \qquad (3.54)$$

Here r and ϕ are, respectively, the changes in amplitude and phase during propagations and both of them are free variables. Using Eq. (3.52) the right-hand side of Eq. (3.54) may be expanded as

$$re^{i\phi}F(0, 0) = re^{i\phi}\sum_j E_j e^{-i\omega_j 0} exp\left[ik(\omega_j) \cdot 0\right] exp[-\alpha(\omega) \cdot 0/2] = re^{i\phi}\sum_j F_j \qquad (3.55)$$

In order to maintain pulse shape Eq. (3.54) must be satisfied; this necessitates that Eqs. (3.53) and (3.55) must be equal. To satisfy Eq. (3.54), the following conditions must be satisfied:

$$
\begin{aligned}
r &= exp[-\alpha_0 z/2] & (i) \\
exp(i\phi) &= exp[i(k_0 z - \omega_0 t)] & (ii) \\
t &= k_1 z & (iii) \\
k_2 &= k_3 = \cdots\cdots\cdots = 0 & (iv) \quad and \\
\alpha_1 &= \alpha_2 = \alpha_3 = \ldots = 0 & (v)
\end{aligned}
\qquad (3.56)
$$

Eqs. (3.56)(i) and (ii) are automatically satisfied as r and ϕ are free variables. Eq. (3.56) (iii), as will be shown, defines group delay and group velocity. Eqs. (3.56)(iv) and (v) determine pulse propagation and pulse distortion.

The time t taken by the pulse to propagate a distance z, as given by Eq. (3.56)(iii) represents group delay expressed as

$$\tau_g = t = k_1 z = z\left.\frac{dk}{dz}\right|_{\omega=\omega_0} \qquad (3.57)$$

The group velocity v_g is the velocity with which the pulse envelope propagates, and can be defined as

$$v_g = \frac{z}{\tau_g} = \left(\frac{dk}{dz}\right)^{-1} = \left[\frac{d}{d\omega}\left(\frac{n\omega}{c}\right)\right]^{-1} = c\left(n + \omega\frac{dn}{d\omega}\right)^{-1} = \frac{c}{n_g} \qquad (3.58)$$

In analogy with Eq. (3.47) defining the phase index, the group index can be defined as previously calculated

$$n_g = n + \omega\frac{dn}{d\omega} \qquad (3.59)$$

Example 3.3 *The values of the refractive index (RI) of most semiconductors lie in the range 3 < n < 4. Assuming a RI = 3, the phase velocity of light becomes 10^8 m/s. The group index, on the other hand, can be an increased or decreased manifold in some special situations, leading to slow or fast light.*

Example 3.4 *The phase index and group index of silica at 1300 nm are, respectively, 1.477 and 1.462. The phase velocity is 2.073 × 10^8 m/s, and the group velocity is 2.052 × 10^8 m/s.*

3.4   Susceptibility of a material: A classical model

As pointed out already, both the refractive index $n(\omega)$ and the absorption coefficient $\alpha(\omega)$ of a medium are related to the susceptibility $\chi(\omega)$ of the material. An applied electric field shifts the positions of the charged particles like electrons and ions from their equilibrium positions, thereby inducing polarization. The electric dipole moment per unit volume is called the polarization density P, which is related to the applied electric field F by the relation

$$P = \varepsilon \chi F \tag{3.60}$$

As noted already, the polarization changes the permittivity of the medium which is expressed as

$$\varepsilon = \varepsilon_0 \varepsilon_r = \varepsilon_0 (1 + \chi) \tag{3.61}$$

A classical model, known as the Lorentz model, is generally used to describe the frequency variation of $\chi(\omega)$, $n(\omega)$, and $\alpha(\omega)$. In this model, an electron bound to a nucleus is displaced by a time-varying electric field of the form $F(t) = F_0 exp(-i\omega t)$ from its position of equilibrium and executes damped harmonic oscillation, just like a mass attached to a spring. The equation of motion of the electron may be written as

$$m_0 \frac{d^2 x}{dt^2} = -\Gamma \frac{dx}{dt} - \beta x - eF_0 exp(-i\omega t) \tag{3.62}$$

where m_0 is the mass of the electron, Γ is the damping constant, β is related to the restoring force, and x is the displacement. The resonant frequency of the oscillator is given by $\omega_0^2 = \beta/m_0$. Using the damping coefficient $\gamma = \Gamma/m_0$, the solution for displacement becomes

$$x = \frac{eF_0}{m_0(\omega_0^2 - \omega^2 - i\omega\gamma)} \tag{3.63}$$

If we assume that there are N such oscillators and a fraction f of them has the resonant frequency ω_0, the polarization of the materials arising out of the displacement of fN oscillators is

$$P = \frac{fNex}{V} = \frac{fNe^2 F_0 exp(-i\omega t)}{m_0 V(\omega_0^2 - \omega^2 - i\omega\gamma)} \tag{3.64}$$

The susceptibility is then expressed as

$$\chi = \frac{fNe^2}{m_0 \varepsilon_0 V} \frac{1}{(\omega_0^2 - \omega^2 - i\omega\gamma)} \tag{3.65}$$

We separate the real and imaginary parts of the susceptibility and use Eqs. (3.26) and (3.27) and obtain

$$1 + \chi_r = \eta_r^2 - \eta_i^2 = 1 + \frac{\omega_0^2 S_0(\omega_0^2 - \omega^2)}{(\omega_0^2 - \omega^2) + \omega^2\gamma^2} \tag{3.66}$$

and

$$\chi_i = 2\eta_r\eta_i = \frac{\omega_0^2 S_0 \omega\gamma}{(\omega_0^2 - \omega^2) + \omega^2\gamma^2} \tag{3.67}$$

The dimensionless quantity S_0 appearing in Eqs. (3.66) and (3.67) denotes the strength of interaction between the oscillator and the EM wave and is expressed as

$$S_0 = \frac{fNe^2}{\varepsilon_0 m_0 V\omega_0^2} \tag{3.68}$$

The variation of normalized χ_r and χ_i with normalized frequency ω/ω_0 is shown in Fig. 3.1. As noted already, the absorption is related to χ_i and shows a peak at resonance. For smaller damping, the curve for χ_i is sharper. Using the approximation $\omega \approx \omega_0$, and $\gamma \ll \omega_0$ one obtains from Eq. (3.67):

$$\chi_i = 2\eta_r\eta_i = \frac{\omega_0^2 S_0 \omega\gamma}{(\omega_0 - \omega)^2 + (\gamma/2)^2} \tag{3.69}$$

Thus, the variation may be described by a Lorentzian lineshape function such that

$$L(\omega) = \frac{1}{2\pi} \frac{\gamma}{(\omega_0 - \omega)^2 + (\gamma/2)^2} \tag{3.70}$$

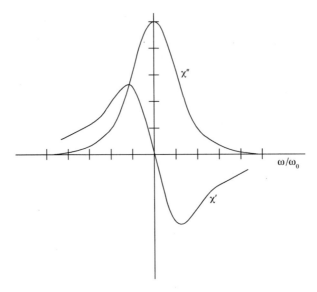

Figure 3.1 *Variation of real and imaginary parts of the susceptibility with normalized frequency.*

The absorption coefficient is $\alpha = 2\omega\eta_i/c$, and assuming a unity refractive index, can be expressed as

$$\alpha = \frac{\omega_0^2 S_0 (\gamma/4c)}{(\omega_0 - \omega)^2 + (\gamma/2)^2} \tag{3.71}$$

It also has a Lorentzian lineshape and the full-width at half maxima (FWHM) is γ.

The extinction coefficient becomes maximum at resonance frequency and decreases as $|\omega - \omega_0|^{-2}$ as the frequency deviation increases. The real part of the RI, on the other hand, decreases like $|\omega - \omega_0|^{-1}$ with increasing $|\omega - \omega_0|$. Eq. (3.67) also indicates that the RI > 1 at low frequency, ω increases up to the resonant frequency ω_0, decreases with a negative slope up to a minimum, and then increases again. The material exhibits normal dispersion in the frequency ranges where the slope $dn/d\omega$ is positive, and there is an anomalous dispersion in the small frequency interval where $dn/d\omega$ is negative.

If the material has a number of resonant frequencies ω_j, the general expression for the susceptibility is

$$\chi(\omega) = \frac{Ne^2}{\varepsilon_0 m_0 V} \sum_j \frac{f_j}{\omega_j^2 - \omega^2 - i\omega\gamma_j} \tag{3.72}$$

and $\sum_j f_j = 1$

In the presence of a number of resonant frequencies having wavelengths λ_i, the RI may be expressed by Sellmeier equation as follows:

$$n^2 = 1 + \sum_j \frac{A_i \lambda^2}{\lambda^2 - \lambda_i^2} \tag{3.73}$$

Where A_i and λ_i are constants called Sellmeier coefficients.

Example 3.5 *The RI in GaAs is expressed as $n^2 = 7.10 + \left[3.78\lambda^2 / \lambda^2 - 0.2767 \right]$, $0.89 \leq \lambda \leq 4.1$ μm, where the wavelength is in μm. The RI at a photon energy of 1 eV ($\lambda = 1.24$ μm) is 3.42.*

The quantum mechanical theory gives an almost similar equation and in addition gives the expressions for f_j and γ_j.

To complete this section, one should note that the real and imaginary parts of the dielectric function are not independent, but are related to each other by the Kramers–Kronig relation:

$$\varepsilon_1(\omega) = \frac{1}{\pi} P \int\limits_{-\infty}^{+\infty} \frac{\varepsilon_2(\omega')d\omega'}{\omega' - \omega}$$

$$\tag{3.74}$$

$$\varepsilon_2(\omega) = -\frac{1}{\pi} P \int\limits_{-\infty}^{+\infty} \frac{\varepsilon_1(\omega')d\omega'}{\omega' - \omega}$$

where P stands for the Cauchy principal value of the integral that follows.

Similar relationships exist between the real and imaginary parts of the susceptibility χ.

3.5   Einstein's model for light-matter interaction

As mentioned in the Introduction, optical processes in semiconductors can best be understood by using quantum theory based on perturbation theory and beyond. The following chapters will develop the relevant theories step by step. However, it is instructive to know the basic processes of light-matter interactions from a phenomenological theory developed by Einstein in his classic paper published as early as in 1917 (Einstein 1917). Einstein pointed out that EM waves can interact with matter in three different ways causing transitions in the energy levels of the matter. These processes are (1) absorption, (2) spontaneous emission, and (3) stimulated emission. In the following we discuss these processes and express the transition rates for each process.

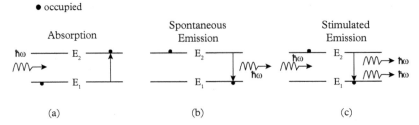

Figure 3.2 *Three basic processes of interaction of radiation with a two-level atomic system;*
(a) absorption, (b) spontaneous emission, and (c) stimulated emission. In each of (a), (b), and (c), the left
half depicts the initial state of occupancy of a level, while the right half shows the state of occupancy
after the process.

3.5.1 Absorption and emission

Figure. 3.2 illustrates the processes by considering a hypothetical atom, which has only two electronic energy levels E_1 (lower level) and E_2 (upper level). The population densities in the lower and upper levels are N_1 and N_2, respectively. A collimated monochromatic beam of light with frequency $f = (E_2 - E_1)/h$ may interact with the atoms in the following three ways.

3.5.1.1 Absorption

Under normal conditions, all materials absorb light. The absorption process is illustrated by Fig. 3.2(a). The left part of it shows the initial state of the atom; the lower state is occupied by an electron and the upper state is empty. When a photon of energy $hf=E_2-E_1$ is absorbed, an electron moves from the lower level to the upper level, as shown by the right part of Fig. 3.2(b). The intensity of the incident light gets attenuated due to absorption, but the direction of propagation and polarization of light remains unaltered. The rate of absorption is given by

$$\frac{dN_1(t)}{dt} = -B_{12}N_1\rho(f) = -\frac{dN_2(t)}{dt} \qquad (3.75)$$

where B_{12} is the Einstein B- coefficient for absorption and ρ is the energy density of the incident photon flux.

3.5.1.2 Spontaneous emission

In this process, as shown in Fig. 3.2(b), an atom initially in the upper level jumps down spontaneously to the lower level and a photon of energy hf is emitted. The left and right parts of Fig. 3.2(b) indicates the initial and final occupancy of the levels, as before. The direction of propagation and polarization of emitted radiation is arbitrary. The corresponding rate equation is

$$\frac{dN_1(t)}{dt} = -A_{21}N_2(t) = -\frac{dN_2(t)}{dt} \qquad (3.76)$$

where A_{21} is called the Einstein A-coefficient and is related to the spontaneous emission lifetime by $\tau = (A_{21})^{-1}$

3.5.1.3   *Stimulated emission*

In this case, as illustrated in Fig. 3.2(c) in which the left and right parts depict the initial and final conditions of the atomic system, the incident photon induces the atom to make a downward transition from the upper to the lower level. The energy is released in the form of a photon which has the same frequency, phase, polarization, and direction of propagation as the incident photon stimulating the transition. The corresponding rate equation is

$$\frac{dN_2(t)}{dt} = -B_{21}N_2\rho(f) = -\frac{dN_1(t)}{dt} \tag{3.77}$$

where B_{21} is the Einstein B-coefficient for stimulated emission.

3.5.2   Absorption and emission rates

We may write the rates of spontaneous emission, stimulated emission, and absorption from Eqs. (3.75)–(3.77) as

$$R_{spon} = A_{21}N_2, \quad R_{stim} = B_{21}N_2\rho(f), \quad and \quad R_{abs} = B_{12}N_1\rho(f) \tag{3.78}$$

In thermal equilibrium, atomic densities obey Boltzmann statistics and accordingly

$$\frac{N_2}{N_1} = \exp\left[-(E_2 - E_1)/k_B T\right] = \exp\left(-hf/k_B T\right) \tag{3.79}$$

where k_B is the Boltzmann constant and T is the absolute temperature. Under steady state, the rates for upward and downward transitions should be equal leading to

$$A_{21}N_2 + B_{21}N_2\rho(f) = B_{12}N_1\rho(f) \tag{3.80}$$

Using Eq. (3.79) in Eq. (3.80), the spectral density may be expressed as

$$\rho(f) = \frac{A_{21}/B_{21}}{(B_{12}/B_{21})\exp(\hbar\omega/k_B T) - 1} \tag{3.81}$$

In thermal equilibrium, the radiation spectral density must be identical with that for blackbody radiation given by Planck formula

$$\rho_{BB}(f) = \frac{8\pi h f^3}{c^3} \frac{1}{\exp(hf/k_B T/hf/k_B T) - 1} \tag{3.82}$$

Comparing Eq. (3.81) and (3.82) gives the relations

$$A_{21} = \left(8\pi h f^3 / c^3\right) B_{21} = \left(2\hbar\omega^3 / \pi c^3\right) \left(2\hbar\omega^3 / \pi c\right) B_{21} \ and \ B_{12} = B_{21} \tag{3.83}$$

The ratio of stimulated to spontaneous emission rates is therefore

$$\frac{R_{stim}}{R_{spom}} = \left[\exp\left(\hbar\omega/k_B T\right) - 1\right]^{-1} \tag{3.84}$$

Example 3.6 *The relative importance of the rates of stimulated emission and spontaneous emission rate may be assessed from Eq. (3.84). If the wavelength is 0.8 μm, the corresponding photon energy is $\hbar\omega = 1.55$ eV. At room temperature, $k_B T \approx 25$ meV and the ratio is 8.43×10^{-26}.*

It appears from Eq. (3.84) and Example 3.3 that in the visible and infrared region ($\hbar\omega \sim 1$ eV) spontaneous emission always dominates over stimulated emission around room temperature ($k_B T \approx 25$ meV).

It also appears from Eqs. (3.78) and (3.84) that the stimulated emission rate will exceed the absorption rate, which is also a stimulated process, that is, $R_{stim} > R_{abs}$, only when $N_2 > N_1$. This condition is referred to as population inversion. This situation can be achieved only when the system deviates from its thermal equilibrium condition and it may be created by a mechanism known as *pumping*. The population inversion is a necessary condition for the operation of a laser, as will be now established.

In the previous discussion, the atomic energy levels are assumed to be sharp, but in real situations, the levels are broadened due to interatomic collisions and several other processes. The spectra of the emitted radiation, no longer monochromatic, are characterized by a line shape function, $L(\omega)$, which may be either a Lorentzian or a Gaussian in nature. The total number of stimulated emission per unit volume per second, in the presence of broadening, may be written as

$$W_{21} = N_2 \int B_{21}\rho(\omega)L(\omega)d\omega = N_2 \frac{\pi^2 c^3}{\hbar\tau_{sp}n_r^3} \int \frac{\rho(\omega)}{\omega^3} L(\omega)d\omega \tag{3.85}$$

3.5.3   Absorption and amplification of light in a medium

The model described in Section 3.5.2 and the rates of transitions may be used to obtain an expression for the absorption or amplification coefficient of EM waves. Though the system considered is a hypothetical two-level atom, the concepts can readily be applied to semiconductors and nanostructures, with suitable ramifications.

We now concentrate on a small volume element containing atoms having two energy levels as before and consider an EM wave with small linewidth to be incident on this element and to propagate through it along the z-direction. The volume element we consider is defined by two parallel planes P_1 and P_2, perpendicular to the z-direction and located at z and $z + dz$, respectively, as shown in Fig. 3.3. The volume of the element

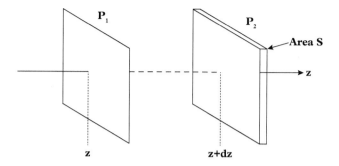

Figure 3.3 *Propagation of radiation along a z-direction.*

is Sdz, where S is the area of both P_1 and P_2. The number of stimulated emissions per unit time in the volume element is $W_{21}Sdz$ and the energy added to the beam due to this process is $W_{21}\hbar\omega Sdz$. Similarly, the energy given out by the beam due to absorption is $W_{12}\hbar\omega Sdz$. The net rate of absorption of energy in the volume element in the frequency interval between ω and $\omega + d\omega$ is the difference between the two. We do not consider spontaneous emission, as it contributes negligibly to the energy of the directional beam.

The intensity of light at planes P_1 and P_2 are respectively $I_\omega(z)$ and $I_\omega(z + dz)$. Therefore, the total energy leaving the volume element per unit time is

$$[I_\omega(z + dz) - I_\omega(z)]\, S = \frac{\partial I_\omega}{\partial z} Sdz \qquad (3.86)$$

Equating this with the net absorption rate within the volume element, one obtains

$$\frac{\partial I_\omega}{\partial z} Sdz = -(W_{12} - W_{21})\hbar\omega Sdz \qquad (3.87)$$

Using now the expression for W_{21} as in Eq. (3.85), and considering a small frequency interval $d\omega$, one obtains

$$\frac{\partial I_\omega}{\partial z} = -\frac{\pi^2 c^3}{n_r^2 \tau_{sp}\omega^2} L(\omega)\rho(\omega)(N_1 - N_2) \qquad (3.88)$$

The energy density and the intensity are related by $I_\omega = (c/n_r)\rho(\omega)$, where c is the velocity of light in free space and n_r is the refractive index. Eq. (3.88) may therefore be expressed as

$$\frac{\partial I_\omega}{\partial z} = -\frac{\pi^2 c^2}{n_r^2 \tau_{sp}\omega^2}(N_1 - N_2)L(\omega)I_\omega = -\alpha_\omega I_\omega \qquad (3.89)$$

Integrating Eq. (3.89), and assuming that the populations in two states do not depend on the intensity, one arrives at the well-known expression,

$$I_\omega(z) = I_\omega(0)exp(-\alpha_\omega z) \qquad (3.90)$$

where $I_\omega(0)$ is the intensity at the surface of the medium and α_ω is the absorption coefficient of the medium at an angular frequency ω. Since $N_1 > N_2$, under usual circumstances, there is always an absorption of EM waves in a medium and the intensity decays. However, if a situation is created such that $N_2 > N_1$, that is, a state of population inversion is created, then the EM waves will get amplified. In that case the stimulated emission rate will exceed the rate of absorption. The expression for intensity, instead of the absorption coefficient α_ω, uses the gain coefficient $k_\omega = -\alpha_\omega$. The amplification of the E = M waves is governed by the expression

$$I_\omega(z) = I_\omega(0)exp(k_\omega z) \qquad (3.91)$$

Where k_ω is a positive quantity as may be found from Eq. (3.89) since $N_2 > N_1$.

Example 3.7 *Let the intensity of light falling at the surface of a semiconductor be 1.0 W/cm²*
and the absorption coefficient is $\alpha = 5$ cm⁻¹. The intensity is reduced to 0.607 W/cm² as
the beam traverses 1 mm. The reduction is thus 2.17 dB.

3.5.4   Self-sustained oscillation in a gain medium

The photonic device exploiting the amplification of light is called a laser. Although a laser may be used to amplify weak light signal, in most cases the Laser refers to a photonic device which emits light supporting self-sustained oscillation. It is well-known that in order to convert an amplifying medium into an oscillator, a positive feedback is necessary. Three essential parts then constitute a laser structure: (1) *a gain medium*, in which the condition for population inversion is established by (2) a *pump*, and (3)*a mechanism to provide feedback*. Usually, a Fabry–Perot (FP) resonator consisting of a pair of plane parallel mirrors as shown in Fig. 3.4 provides the positive feedback.

The plane-parallel mirrors form a resonant cavity which selectively supports different modes. Modes depict patterns of EM field or power distribution within the resonant cavity and each mode is characterized by a set of integers (i, j, k). Simply put, the EM

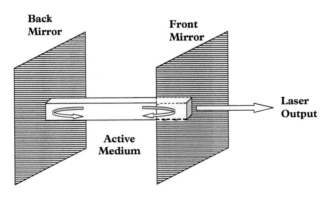

Figure 3.4 *FP resonator.*

waves suffer reflections from the end mirrors and form a standing wave pattern. The separation between the mirrors, L, is such that it supports an integral multiple of half wavelengths, i, of the EM field. The reflectivity of the back mirror is nearly 100% while that of the front mirror is purposely made lower ($\sim$ 90%) in order that light comes out of this mirror.

The gain medium has a well-defined frequency or energy range over which positive gain is obtained. Let the medium enclosed by the mirrors have a gain coefficient g and an absorption (loss) coefficient α. Let R_1 and R_2 denote the reflectivities of the two mirrors. If I_0 be the initial intensity then after a round trip of length $2L$ within the cavity, the intensity becomes

$$I(2L) = I_0 R_1 R_2 exp[(g - \alpha)2L] \tag{3.92}$$

When $g > \alpha$, the wave is amplified. Self-sustained oscillation results when $I(2L) = I_0$ and the corresponding gain reaches its threshold value g_{th}, which may be expressed as follows:

$$g_{th} = \alpha + \frac{1}{2L} ln \left(\frac{1}{R_1 R_2} \right) \tag{3.93}$$

Example 3.8 *A semiconductor laser has a cavity length L = 200 μm, the absorption coefficient of the material is 500 m^{-1} and the mirror reflectivities are 0.3 and 1. The threshold gain has a value of 6.07 $\times$ 10^3 m^{-1}.*

Eq. (3.93) is the general expression for threshold gain which ensures that at threshold the gain must equal the absorption and other loss in the gain medium and the mirror loss represented by the second term. The threshold gain is related to threshold current density in a semiconductor laser, the expression for which will be developed in Chapter 5. A simple method of evaluating the threshold current in a semiconductor laser by treating the conduction band (CB) and valence band (VB) states as two sharp atomic levels is assigned as a problem.

Problems

3.1 *Show that the intensity is related to the amplitude of the electric field by the relation $I = (1/2)\varepsilon v_p F_0^2$.*

3.2 *Calculate the electric field in a medium when the intensity is 1.0 W/m^2. The RI of the medium is 3.5.*

3.3 *An EM wave falls normally at the interface of two dielectrics of RI s n_1 and n_2. Applying proper boundary conditions at the interface prove that the intensity reflection coefficient from the interface is $R = \left(\frac{n_1 - n_3}{n_1 + n_2} \right)^2$*

3.4 *Calculate the skin depth of a copper at 1GHz. A cylindrical copper wire has a radius of 1 mm. Calculate the change in ac resistance of the wire from its value at dc resistance at 1 GHz and 100 GHz.*

3.5 *For a good conductor $\sigma/\varepsilon \gg \omega$, the frequency of operation. Using the values given for copper in Example 3.2, estimate the maximum frequency over which copper can behave as a good conductor.*

3.6 *Work out the steps to arrive at Eq. (3.40) from Eq. (3.39).*

3.7 *Free electrons in a conductor are not bound to any particular atom or molecule, so that the term $\beta x = 0$ may be put in Eq. (3.62). From this obtain the expression for conductivity of a conductor. Show that the conductivity is purely imaginary when damping is negligible.*

3.8 *When the conductivity is purely imaginary, show that the RI can be expressed as $n = \left[1 - (\omega_p/\omega)^2\right]^{1/2}$, where ω_p is the plasma frequency. Use Eq. (3.39) to derive the result.*

3.9 *The absorption spectra of a medium are described by a delta function. Indicate how the susceptibility will vary with frequency.*

3.10 *The resistance of a network in which R and C are connected in parallel is given by $R(\omega) = R/(1 + \omega^2\tau^2)$, where $\tau = RC$. Using the Kramers–Kronig relation, obtain the expression for the frequency variation of the reactive part $X(\omega)$.*

3.11 *Prove that the maximum of $\chi'(\omega)$ occurs at a frequency at which $\chi''(\omega)$ reaches half the maximum value.*

3.12 *The spontaneous emission lifetime in a two-level atomic system is 0.1 ps. Calculate the value of the corresponding Einstein coefficient.*

3.13 *At what temperature the rates of spontaneous and stimulated emission will be equal for a photon of wavelength 1 µm?*

3.14 *Calculate the emission frequency due to transition from 2P level to 1S level in a hydrogen atom. Calculate also the value of B coefficient if the spontaneous emission lifetime is 1.6 ns.*

3.15 *The first (1) and second (2) excited states in an atom are 1.1 eV and 3.4 eV, above the ground (0), respectively. The Einstein coefficients are $A_{21} = 5 \times 10^7$, $A_{20} = 2 \times 10^7$ and $A_{10} = 10^8$ sec^{-1}. Calculate the corresponding emission wavelengths, the lifetime in state 2 and the relative probability of spontaneous transition from state 2 to states 1 and 0.*

3.16 *The lineshapes for broadening of the two levels are Lorentzian with widths ΔE_1 and ΔE_2. Show that the absorption spectra for transition between the two states will have a Lorentzian lineshape of width $\Delta E_1 + \Delta E_2$.*

3.17 *The two mirrors in a FP cavity have field reflection(transmission) coefficients $r_1(t_1)$ and $r_2(t_2)$. The electric field incident on mirror 1 is E_i and by entering into the cavity is reflected and transmitted by the mirrors by making several round trips.*

Show by summing all the fields transmitted outside by mirror 2, the total field is

$$E_t = \frac{E_i t_1 t_2 exp[-(g-\alpha)L]}{1-r_1 r_2 exp[-2(g-\alpha)L]}$$

where $g(\alpha)$ is the gain (absorption) coefficient of the medium. From the previous expression, derive the condition for threshold given by Eq. (3.93).

3.18 *Use Eq. (3.89) for the gain coefficient and assume that threshold population inversion $N_{th} = (N_2 - N_1)_{th} \approx (\tau_r/ed) j_{th}$, where J is the current density, d is the thickness of the active layer, and τ_r is the recombination lifetime. Assume also that lasing condition is achieved at frequency ω_0 when the lineshape function is maximum. Obtain the expression for J_{th} in terms of absorption and mirror losses.*

3.19 *Calculate the threshold current density of a GaAs laser operating at 0.84 μm having length = 400 μm, $\Delta\omega_0 = 10^{14}$ Hz, $\alpha = 10^3$ m^{-1}, d = 2 μm, $\eta_r = 3.6$ and injection efficiency $\eta_{inj} = \tau_r/\tau_{=sp} = 0.8$. Use the reflectivity for air-GaAs interface for the front mirror, but $R_2=0.95$ for the rear mirror.*

3.20 *The gain in a laser at 1.0 μm wavelength is 1 m^{-1}. Assume $\Delta\omega = 10^{14}$ $\eta_r = 3$ and $\tau_{sp} = 1.0$ ps. Calculate the difference of population. Laser action takes place between two atomic levels.*

Reading List

Agrawal G P and Dutta N K (1986) *Long Wavelength Semiconductor Lasers*. New York: van Nostrand Reinhold.

Basu P K (1997) *Theory of Optical Processes in Semiconductors: Bulk and Microstructures*. Oxford: Oxford University Press.

Basu P K, Mukhopadhyay B, and Basu R (2015) *Semiconductor Laser Theory*. Boca Raton, FL: CRC Press (Taylor & Francis).

Griffiths D J (1989) *Introduction to Electrodynamics*. Englewood Cliffs, NJ: Prentice Hall.

Loudon R (1973) *The Quantum Theory of Light*. Oxford: Oxford University Press.

Singh J (2003) *Electronic and Optoelectronic Properties of Semiconductor Structures*. Cambridge: Cambridge University Press.

Verdeyen J T (1989) *Laser Electronics*, 2nd edn. Englewood Cliffs, NJ: Prentice Hall.

References

Einstein A (1917) Zur Quantentheorie der Strahlung. *Physikalische Zeitschrift* 18: 121.

Vornehm J E and Boyd R W (2009) Slow and fast light. In: *Tutorials in Complex Photonic Media*, M A Noginov, G Dewar, M W McCall, and N A Zheludev (eds.) Chapter 19, pp. 647–686. Bellingham, Washington, USA: SPIE Press.

4

Photons and electron-photon interactions

4.1 Introduction

The macroscopic theory for light-matter interaction developed in Chapter 3 is primarily based on the wave equation derived from a Maxwell equation. The theory develops the relationships between various optical constants. The three basic processes of light-matter interaction, as proposed by Einstein, are introduced by using the appropriate rate equations. These rate equations are used to establish the condition for amplification of electromagnetic (EM) waves in a hypothetical two-level atomic system.

The present chapter provides the basic quantum mechanical theory for light-matter interaction using time-dependent perturbation theory. It starts with the classical derivation of electric field in a rectangular cavity and evaluates the density-of-modes (DOM). The EM field is then quantized, the concept of photons is introduced, and the modes are described in terms of creation and annihilation operators. The form of perturbation Hamiltonian is then developed and with this the rates of transition for the basic processes in a two-level atomic system are obtained. The expression for Einstein's A-coefficient is then derived.

Though the main emphasis in this book is on processes in semiconductors, the theory developed in this chapter may be extended for systems of lower dimensionality or systems replicating atomic systems, as for example, Quantum Dot (QD) structures having sharp atom-like energy levels.

4.2 Wave equation in a rectangular cavity

According to wave-particle duality, EM waves can be considered to be made up of particles termed photons. To understand quantization of EM waves, it is useful to consider the behaviour of EM waves within a rectangular cavity having finite dimensions and to study the electric field distribution within the cavity. An EM mode follows a specific pattern of field variation. We discuss the mode patterns and the DOM within the cavity by using classical Maxwell's equations.

Semiconductor Nanophotonics. Prasanta Kumar Basu, Bratati Mukhopadhyay, and Rikmantra Basu, Oxford University Press.
© P.K. Basu, B. Mukhopadhyay, R. Basu (2022). DOI: 10.1093/oso/9780198784692.003.0004

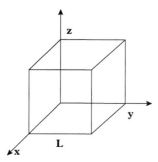

Figure 4.1 *Cubic optical cavity and its geometry. L denotes the length of each side.*

4.2.1 Modes

Consider a cubic structure with length L on each side and with the edges aligned parallel to the coordinate axes, x, y, and z as shown in Fig. 4.1. The walls of the cavity are, assumed to be perfectly conducting and therefore the tangential components of the electric field, vanish at the cavity boundaries.

Let us consider one Cartesian component: the x-component for example, of the electric field vector. The component F_x satisfies the following wave equation:

$$\frac{\partial^2}{\partial x^2}F_x + \frac{\partial^2}{\partial y^2}F_x + \frac{\partial^2}{\partial z^2}F_x = \frac{1}{c^2}\frac{\partial^2}{\partial t^2}F_x \tag{4.1}$$

Now, by separation of variables, we can write

$$F_x = X(x)\,Y(y)Z(z)\,T(t) \tag{4.2}$$

Substituting Eq. (4.2) in Eq. (4.1) and dividing by F_x, one obtains

$$\frac{1}{X}\frac{d^2 X}{dx^2} + \frac{1}{Y}\frac{d^2 Y}{dy^2} + \frac{1}{Z}\frac{d^2 Z}{dz^2} = \frac{1}{c^2 T}\frac{d^2 T}{dt^2} \tag{4.3}$$

Let us assume

$$\frac{1}{\mathcal{J}}\frac{d^2 \mathcal{J}}{dj^2} = -q_j^2 \quad \mathcal{J} = X, Y, Z \quad and\, j = x, y, z \tag{4.4}$$

$$\frac{1}{c^2 T}\frac{d^2 T}{dt^2} = -q^2 \tag{4.5}$$

Therefore

$$q^2 = q_x^2 + q_y^2 + q_z^2$$

Eq. (4.5) gives

$$T(t) = Cexp(-i\omega t), \quad \omega = qc \tag{4.6}$$

where ω is the angular frequency and C is a constant. Now F_x is tangential to the planes $y = 0$ and $y = L$ and $z = 0$ and $z = L$. To have vanishing fields at these planes, the solutions for y and z directions should be of the form $\sin q_y y$ and $\sin q_z z$ respectively, where

$$q_y = \frac{n\pi}{L}, \quad q_z = \frac{p\pi}{L}, \quad n, p = 0, 1, 2 \ldots \tag{4.7}$$

Similarly, x and z dependencies of F_y should be $\sin q_x x$ and $\sin q_z z$, respectively with

$$q_x = \frac{m\pi}{L}, \quad m = 0, 1, 2, \ldots \tag{4.8}$$

In a similar way, x and y dependencies of F_z should be $\sin q_x x$ and $\sin q_y y$. It may also be noted that $\partial F_z/\partial z$ should vanish at the planes $x=0$ and $x=L$ as x dependencies of F_y and F_z are given by sine functions. Thus, $\nabla \cdot F = 0$ on these planes leads to $\partial F_x/\partial x = 0$, so that F_x should vary as $\cos q_x x$ with q_x given by Eq. (4.8). Similarly, the y dependence of F_y and z dependence of F_z will also be cosine functions.

Therefore, the complete solution of the field components will be

$$F_x = F_{0x} \cos q_x x \, \sin q_y y \, \sin q_z z \, e^{-j\omega t}$$
$$F_y = F_{0y} \sin q_x x \, \cos q_y y \, \sin q_z z \, e^{-j\omega t}$$
$$F_z = F_{0z} \sin q_x x \, \sin q_y y \, \cos q_z z \, e^{-j\omega t} \tag{4.9}$$

$$\text{So } \nabla \cdot F = 0 \quad \text{gives } F_0 \cdot q = 0 \tag{4.10}$$

and $q = x q_x + y q_y + z q_z$ where x, y, z are unit vectors in the respective directions. The dispersion relation, that is, the $\omega - q$ relation, by using Eqs. (4.5) and (4.6), can be expressed as

$$\omega^2 = c^2 q^2 = c^2 \left(q_x^2 + q_y^2 + q_z^2\right) = \frac{c^2 \pi^2}{L^2}(m^2 + n^2 + p^2) \tag{4.11}$$

From Eq. (4.10), it is clear that **F** is normal to **k** and there are again two independent directions of **F** for each **q**. The field pattern obtained from Eq. (4.9) represents standing wave in the cavity and is called modes of oscillation of the cavity.

Example 4.1 *The frequency of oscillation in a cubic cavity of each side = 1 μm and for*
m=n=p=1 is 2.6 × 10¹⁴ Hz.

4.2.2 Density-of-modes

To calculate the number of modes per unit volume per unit frequency interval, we consider first the number of modes whose x-component of the wave vector **q** lies between q_x and $q_x + dq_x$. This will be the number of modes lying between $(L/\pi)q_x$ and $(L/\pi)(q_x + dq_x)$, and is approximately $(L/\pi)dq_x$. Similar expressions may be obtained for y and z component of **q**. If $V = L^3$ is the physical volume of the cavity, there will be $(L/\pi)dq_x$, $(L/\pi)dq_y$ and $(L/\pi)dq_z$ modes or $(V/\pi^3)dq_xdq_ydq_z$ modes in the range $dq_xdq_ydq_z$ of **q**. Therefore, the number of modes per unit volume in **q**- space is (V/π^3). If $P(q)dq$ is considered to be the number of modes with **q** vector lying between **q** and **q**+d**q**, then

$$P(q)dq = 2.\frac{V}{8\pi^3}4\pi q^2 dq \tag{4.12}$$

Here, the term $4\pi q^2 dq$ is the volume element enclosed by spheres of radii **q** and **q** + d**q** in **q** space. The factor 2 arises from the fact that for each **q**, two independent modes of polarization are possible and the factor 1/8 takes into account the fact that in counting the modes, only the positive values of q_x, q_y, and q_z are considered.

If $p(q)dq$ is the number of modes per unit volume, then

$$p(q)dq = \frac{1}{\pi^2}q^2 dq \tag{4.13}$$

If $p(\omega)d\omega$ is the number of modes per unit volume in the frequency interval $d\omega$, then

$$p(\omega)d\omega = p(q)dq = \frac{1}{\pi^2}q^2 dq = \frac{1}{\pi^2}\left(\frac{\omega}{c}\right)^2\frac{d\omega}{c}$$

and therefore

$$p(\omega)d\omega = \frac{\omega^2 d\omega}{\pi^2 c^3} \tag{4.14}$$

For a dielectric medium of refractive index η, the velocity c is replaced by c/η. The previous equation can be written in terms of the DOM in the energy intervals $\hbar\omega$ and $\hbar(\omega + d\omega)$, and is given by

$$G(\hbar\omega) = \frac{\eta^3\omega^2}{\pi^2 c^3 \hbar} \tag{4.15}$$

The expression for radiation energy density $p(\omega)d\omega$ between ω and $\omega + d\omega$ can be given by using Eq. (4.14) and Bose–Einstein distribution law as

$$p(\omega)d\omega = \frac{\hbar\omega^3\eta^3}{\pi^2 c^3}\frac{d\omega}{exp(\hbar\omega/k_BT) - 1} \tag{4.16}$$

Example 4.2 *A cavity of volume $V = 1mm^3$ is formed by using a material of index 3.6. The total number of modes N, over a frequency interval $dv = 3 \times 10^{10}$ Hz around a frequency $v = 3 \times 1014$ Hz is obtained from $N = 8\pi v^2 \eta^3 V dv / c^3$ and is 1.17×10^5.*

4.3   Quantization of the radiation field

EM radiation is equivalent to an assembly of particles termed as photons. The photons themselves are independent harmonic oscillators corresponding to the modes. We now develop the theory of quantization of EM waves, by considering the harmonic oscillator model, properties of operators associated, and then applying these properties to an assembly of harmonic oscillators.

4.3.1   One-dimensional harmonic oscillator

The classical Hamiltonian for a harmonic oscillator with unit mass can be expressed as

$$H = \frac{(p^2 + \omega^2 q^2)}{2} \tag{4.17}$$

where p is the momentum, ω is the angular frequency, and q is the displacement of the oscillator. In Quantum Mechanics, p and q are operators and follow the commutation relation as

$$[q \cdot p] = i\hbar \tag{4.18}$$

Example 4.3 *The commutation relation given by Eq. (4.18) is proved by using a one-dimensional (1D) case, x, and p_x being the position and momentum operators, respectively. One may write*

$$[q \cdot p] = i[\hat{x}(-i\hbar\partial/\partial x) - (-i\hbar\partial/\partial x)\hat{x}]\psi = x(-i\hbar\partial/\partial x)\psi - (-i\hbar\partial/\partial x)x\psi$$
$$= -i\hbar\{x\partial\psi/\partial x - (x\partial\psi/\partial x + \psi)\} = i\hbar\psi$$

Now we define one destruction operator a and one creation operator a^+ of the oscillator such that

$$a = \frac{(\omega q + ip)}{\sqrt{2\hbar\omega}} \tag{4.19a}$$

$$a^+ = \frac{(\omega q) - ip}{\sqrt{2\hbar\omega}} \tag{4.19b}$$

Using Eqs. (4.18) and (4.19a and b), it can be shown that

$$[a, a^+] = aa^+ - a^+a = 1 \tag{4.20}$$

$$H = \hbar\omega \left(a^+a + \frac{1}{2} \right) \tag{4.21}$$

Let us assume that $|n\rangle$ is an eigen state of the harmonic oscillator with eigen value E_n. Therefore, the eigen value equation using Eq. (4.21) will be

$$H|n\rangle = \hbar\omega \left(a^+a + \frac{1}{2} \right) |n\rangle = E_n|n\rangle \tag{4.22}$$

Using commutation relation given in Eq. (4.20), we can write

$$\hbar\omega \left(a^+a + \frac{1}{2} \right) a^+|n\rangle = Ha^+|n\rangle = (E_n + \hbar\omega)a^+|n\rangle \tag{4.23}$$

From this equation, it is clear that the state $a^+|n\rangle$ is also an eigen state with eigen value $E_n + \hbar\omega$. With these new values of eigen state and eigen value as

$$|n + 1\rangle = a^+|n\rangle \quad E_{n+1} = E_n + \hbar\omega \tag{4.24}$$

one may obtain

$$H|n + 1\rangle = E_{n+1}|n + 1\rangle \tag{4.25}$$

It may therefore be concluded that, if there is a harmonic oscillator energy level E_n, there exists another higher level with energy separation $\hbar\omega$. With this argument, the energy levels will be an equally spaced ladder as shown in Fig. 4.2(a) and the eigen values are expressed as

$$E_n = \left(n + \frac{1}{2} \right) \hbar\omega \quad n = 0, 1, 2, \tag{4.26}$$

As shown in Fig. 4.2(b), the destruction operator a destroys a quantum of energy $\hbar\omega$ and as a result the energy of the oscillator goes from E_n to E_{n-1}. On the other hand, the creation operator a^+ raises the energy of the nth state to the next higher $(n +1)$th state. The states $|n\rangle$ are simultaneous eigen states of H and the number operator $n = a^+a$, and thus

$$n|n\rangle = a^+a|n\rangle = n|n\rangle \tag{4.27}$$

The eigen value n indicates the number of quanta $\hbar\omega$ excited above the oscillator ground state. Using normalization value, it follows that

$$a|n\rangle = \sqrt{n}|n - 1\rangle \tag{4.28a}$$

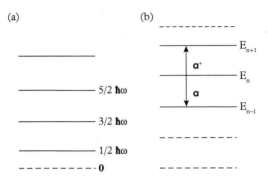

Figure 4.2 *(a) Energy level of quantum mechanical harmonic oscillator; (b) property of creation and destruction operator.*

$$a^+| n\rangle = \sqrt{n+1}| n+1\rangle \tag{4.28b}$$

The non-vanishing matrix elements of a and a^+ are given by

$$\langle n-1|a|n\rangle = \sqrt{n} \tag{4.29a}$$

$$\langle n+1|a^+|n\rangle = \sqrt{n+1} \tag{4.29b}$$

4.3.2   Classical field equation for radiation

Let us start with the Maxwell equation in the absence of current and free charge carriers, $\nabla \cdot B = 0$ and assume $B = \nabla \times A$ where A is the vector potential. Working under the Coulomb gauge $\nabla \cdot A = 0$, one can write

$$B = \mu_0 H = \nabla \times A \tag{4.30a}$$

$$F = -\frac{\partial A}{\partial t} \tag{4.30b}$$

Using the previous equations, the wave equation for A in free space can be written as

$$\nabla^2 A = \frac{1}{c^2}\frac{\partial^2 A}{\partial t^2} \tag{4.31}$$

where $c = 1/\sqrt{\mu_0\varepsilon_0}$. We solve the previous Eq. (4.31) by the method of separation of variables, by putting $A(r,t) = A(t)X(r)$, and thus we can write

$$A(t)\nabla^2 X(r) = X(r)\frac{1}{c^2}\frac{d^2 A}{dt^2} \tag{4.32}$$

From Eq. (4.32), the z-component of $X(r)$ can be written as

$$\frac{c^2}{X_z(r)}\nabla^2 X_z(r) = \frac{1}{A(t)}\frac{d^2A}{dt^2} = -\omega^2 \ (say) \tag{4.33}$$

Thus

$$\nabla^2 X_z(r) = -\frac{\omega^2}{c^2}X_z(r) = -q^2 X_z(r) \tag{4.34}$$

So, it can easily be concluded that $X_z(r) \sim exp(iq.r)$

The condition $\nabla \cdot A = 0$ gives $q \cdot A = 0$. The allowed values of q and ω are determined from the boundary conditions. Let us consider that the radiation is confined within a cubical cavity of length L and the cavity is regarded as a region of space without any real boundaries. We here take running wave solutions and subject them to periodic boundary conditions.

Therefore $A(x = 0, y, z) = A(x = L, y, z)$ etc.
or $exp(iq_xL) = 1 = exp(iq_yL) = exp(iq_zL)$
Thus,

$$q_x = 2\pi\nu_x/L, \quad q_y = 2\pi\nu_y/L, \quad q_z = 2\pi\nu_z/L, \quad \nu_x, \nu_y, \nu_z = 0, \pm1, \pm2, \ldots$$

Therefore, the equation for $A(t)$ can be written as

$$\frac{d^2A_q(t)}{dt^2} + \omega_q^2 A_q(t) = 0 \tag{4.35}$$

The solution of the previous equation is given as

$$A_q(t) = A_q exp\left(-i\omega_q t\right) \tag{4.36}$$

Including the space dependence, the complete expression for the vector potential becomes

$$A = \sum_q \left\{ A_q exp\left(-i\omega_q t + iq.r\right) + A_q^* exp\left(i\omega_q t - iq.r\right) \right\} \tag{4.37}$$

which follows the expressions for the electric and magnetic fields as

$$F = -\frac{\partial A}{\partial t} = \sum_q F_q \tag{4.38a}$$

$$H = \frac{1}{\mu_0}\nabla \times A = \sum_q H_q \tag{4.38b}$$

where

$$F_q = i\omega_q \left\{ A_q exp\left(-i\omega_q t + iq.r\right) - A_q^* \left(i\omega_q t - iq.r\right) \right\} \tag{4.39a}$$

$$H_q = \frac{i}{\mu_0}q \times \left\{ A_q exp\left(-i\omega_q t + iq.r\right) - A_q^* \left(i\omega_q t - iq.r\right) \right\} \tag{4.39b}$$

The total energy of the radiation field is given by

$$E = \frac{1}{2} \int (\varepsilon_0 \mathbf{F} \cdot \mathbf{F} + \mu_0 \mathbf{H} \cdot \mathbf{H}) dV \tag{4.40}$$

Using Eqs. (4.39a and b) and evaluating the time average, the energy of the qth mode is given by

$$\langle E_q \rangle = 2\varepsilon_0 V \omega_q^2 A_q A_q^* \tag{4.41}$$

Now, replacing the mode variables A_q and A_q^* by position coordinate Q_q and momentum P_q, we get

$$A_q = \frac{1}{\sqrt{4\varepsilon_0 V \omega_q^2}} \left(\omega_q Q_q + i P_q \right) \varepsilon_q \tag{4.42a}$$

$$A_q^* = \frac{1}{\sqrt{4\varepsilon_0 V \omega_q^2}} \left(\omega_q Q_q - i P_q \right) \varepsilon_q \tag{4.42b}$$

In the previous equations, as Q_q and P_q are scalar quantities, a unit polarization vector ε_q has been introduced to indicate the directional property of a particular mode. Using the transformations given in Eq. (4.42a and b), the single mode energy can be written from the Eq. (4.41) as

$$\langle E_q \rangle = \frac{P_q^2 + \omega_q^2 Q_q^2}{2} \tag{4.43}$$

Comparing the previous equation with the Eq. (4.17), it may be therefore concluded that the energy $\langle E_q \rangle$ is in the form of a classical harmonic oscillator of unit mass.

4.3.3   Quantization of the field

In complete analogy with the 1D harmonic oscillator, the classical vector potentials A_q and A_q^* for the cavity mode q expressed in terms of P_q and Q_q can now be converted into quantum mechanical operators expressed in terms of p_q and q_q as

$$A_q = \frac{1}{\sqrt{4\varepsilon_0 V \omega_q^2}} \left(\omega_q Q_q + i P_q \right) \varepsilon_q \rightarrow \frac{1}{\sqrt{4\varepsilon_0 V \omega_q^2}} \left(\omega_q q_q + i p_q \right) \varepsilon_q = \sqrt{\frac{\hbar}{2\varepsilon_0 V \omega_q}} a_q \varepsilon_q$$

$$\tag{4.44a}$$

$$A_q^* = \frac{1}{\sqrt{4\varepsilon_0 V \omega_q^2}} \left(\omega_q Q_q - i P_q \right) \varepsilon_q \rightarrow \frac{1}{\sqrt{4\varepsilon_0 V \omega_q^2}} \left(\omega_q q_q - i p_q \right) \varepsilon_q = \sqrt{\frac{\hbar}{2\varepsilon_0 V \omega_q}} a_q^\dagger \varepsilon_q$$

$$\tag{4.44b}$$

Here, the classical Fourier components A_q and A_q^* of the vector potential have been replaced by the destruction and creation operators respectively and a proper multiplication factor and a unit vector have been used. Therefore, the quantum mechanical expression for the total vector potential can be written as

$$A = \sum_q \sqrt{\frac{\hbar}{2\varepsilon_0 V \omega_q}} \varepsilon_q \left\{ a_q exp\left(-i\omega_q t + i\mathbf{q}.\mathbf{r}\right) + a_q^+ exp\left(i\omega_q t - i\mathbf{q}.\mathbf{r}\right) \right\} \tag{4.45}$$

The vector potential is now an operator and the electric field operator $\mathbf{F_q}$ and $\mathbf{H_q}$ are now to be obtained as

$$F_q = i\sqrt{\frac{\hbar\omega_q}{2\varepsilon_0 V}} \varepsilon_q \left\{ a_q exp\left(-i\omega_q t + i\mathbf{q}.\mathbf{r}\right) - a_q^+ exp\left(i\omega_q t - i\mathbf{q}.\mathbf{r}\right) \right\} \tag{4.46a}$$

$$H_q = i\sqrt{\frac{\hbar c^2}{2\mu_0 V \omega_q}} \mathbf{q} \times \varepsilon_q \left\{ a_q exp\left(-i\omega_q t + i\mathbf{q}.\mathbf{r}\right) - a_q^+ exp\left(i\omega_q t - i\mathbf{q}.\mathbf{r}\right) \right\} \tag{4.46b}$$

a_q^+ and a_q are now the creation and destruction operators for the qth mode. A quantum of energy $\hbar\omega_q$ is therefore either added to or removed from the cavity electromagnetic mode of wave vector $\mathbf{q}$. The number operator for the qth mode is $n_q = a_q^+ a_q$ and its eigen value n_q has the possible values 0, 1, 2,..... For the operators, the following two relations are valid:

$$a_q \left| n_q \right\rangle = \sqrt{n_q} \left| n_q - 1 \right\rangle \tag{4.47a}$$

$$a_q^+ \left| n_q \right\rangle = \sqrt{n_q + 1} \left| n_q + 1 \right\rangle \tag{4.47b}$$

A state of the total radiation field may be expressed as $\left| n_{q1}, n_{q2}, n_{q3}, ... \right\rangle$, where n_{q1}, n_{q2}, ... are the number of photons excited. As different cavity modes are independent, the total field can be written as

$$\left| n_{q1}, n_{q2}, n_{q3}, ... \right\rangle = \left| n_{q1} \right\rangle \left| n_{q2} \right\rangle \left| n_{q3} \right\rangle ..... \tag{4.48}$$

The total state of the field is normalized as the states of the individual cavity modes are assumed to be normalized. An operator for a particular mode q_i affects only the photons in that particular mode. Therefore

$$a_{qi}^+ \left| n_{q1}, n_{q2}, n_{q3}, ... \right\rangle = \sqrt{n_{qi} + 1} \left| n_{q1}, n_{q2}, n_{q3}, ... \right\rangle \tag{4.49}$$

4.4 Time-dependent perturbation theory

The rate of transition from a discrete state $\left| i \right\rangle$ to a group of states $\left| f \right\rangle$ can be expressed as

$$W_{i \rightarrow f} = \frac{2\pi}{\hbar} \left| H_{fi} \right|^2 \rho(E = E_i \pm \hbar\omega) \tag{4.50}$$

where H_{fi} is the perturbation and $\rho(E)$ is the density of final states expressed as a function of energy.

The transition rate from $|i\rangle$ to $|f\rangle$ is given by

$$W_{i \to f} = \frac{2\pi}{\hbar}|H_{fi}|^2 \delta(E_i - E_f \pm \hbar\omega) \tag{4.51}$$

Consider now a transition from state $|i\rangle$ to a single state $|f\rangle$ where $E_f > E_i$. Let the energy E_f be smeared so that the probability function $f(E_f)\,dE_f$ gives the probability of the states between the energy E_f and $E_f + dE_f$. By normalizing $f(E_f)$, we can write

$$\int_{-\infty}^{+\infty} f(E_f)\,dE_f = 1 \tag{4.52}$$

The transition rate will then be

$$W_{i \to f} = \frac{2\pi}{\hbar}|H_{fi}|^2 \int_{-\infty}^{+\infty} \delta\left(E_i - E_f + \hbar\omega\right) f(E_f)\,dE_f = \frac{2\pi}{\hbar}|H_{fi}|^2 f(E_f = E_i + \hbar\omega) \tag{4.53}$$

Eq. (4.53) is referred to the Fermi Golden Rule in the literature.

Now we define another function $S_f(\nu)$ in the frequency domain such that $S_f(\nu)\,(\omega_f) = 2\pi\hbar f(E_f)$ and

$$\int_{-\infty}^{+\infty} S_f(\nu)\,d\nu = 1$$

Therefore, Eq. (4.53) can be written as

$$W_{i \to f} = \frac{1}{\hbar^2}|H_{fi}|^2 S_f(\nu) \tag{4.54}$$

4.5   Interaction of an electron with the electromagnetic field

We now aim to develop the theory of electron-photon interaction by using the transition rate given by Eq. (4.53): the Fermi Golden Rule, obtained from time-dependent perturbation theory.

4.5.1 Perturbation Hamiltonian

Consider an electron in an arbitrary static potential $V(\mathbf{r})$ subjected to an electromagnetic field of vector potential $A(r, t)$. We noted earlier that $\nabla \cdot A = 0$. The Hamiltonian of the electron is then written as

$$H = \frac{1}{2m_0}(\mathbf{p} - eA)^2 + V(\mathbf{r}) \quad \mathbf{p} = i\hbar\nabla \tag{4.55}$$

or

$$H = \frac{p^2}{2m_0} - \frac{e}{2m_0}(\mathbf{p} \cdot A + A \cdot \mathbf{p}) + \frac{e^2 A^2}{2m_0} + V(\mathbf{r}) \tag{4.56}$$

Considering p and q being canonically conjugate momentum and position operators, which satisfy the commutation relations, one may obtain using $\nabla \cdot A = 0$.

$$\mathbf{p} \cdot A = A \cdot \mathbf{p} - i\hbar\nabla \cdot A = A \cdot \mathbf{p} \tag{4.57}$$

Using Eq. (4.57), Eq. (4.56) reduces to

$$H = \frac{p^2}{2m_0} - \frac{e}{m_0}(A \cdot \mathbf{p}) + \frac{e^2 A^2}{2m_0} + V(\mathbf{r}) \tag{4.58}$$

In the previous equation, the ratio of the 3rd term to the 2nd term is

$$\frac{e|A|}{2p} = \frac{e|F_0|}{2\omega p} \approx \frac{e}{\omega p}\sqrt{\frac{2I}{\varepsilon_0 c\eta}} \tag{4.59}$$

where I is the average intensity and

$$F = -\frac{\partial A}{\partial t}.$$

Example 4.4 *We make an estimate of the ratio given by Eq. (4.59), by assuming* $I = 10^4 \ Wm^{-2}$, $\omega = 10^{14} \ s^{-1}$, $p = (3kBT/m_0)^{1/2} = 1 \times 10^{-25} kg.m.s^{-1}$ *and RI* $\eta = 3.6$. *The previous ratio is* $\sim 1.4 \times 10^{-5}$.

The previous example shows that the ratio is quite low for usual values of optical intensity. In the usual linear theory, the third term can be neglected and the total Hamiltonian including the electromagnetic field is given by

$$H = V(\mathbf{r}) + \frac{p^2}{2m_0} + \sum_{q \cdot \lambda} \hbar\omega_q \left(a_q^+ a_q + \frac{1}{2}\right) - \frac{eA \cdot \mathbf{p}}{m_0} = H_0 + H_\lambda \tag{4.60}$$

The first three terms represent the unperturbed Hamiltonian, and the last term is the perturbation given by H_λ. Let $|i\rangle$ is the eigen function of the whole system and is given by the product of an electronic eigen function and a radiation field eigen function. Thus, $| i \rangle = | m \rangle | \{n_q\} \rangle$. So, the matrix element for transition between two states $| i \rangle$ and $| j \rangle$ is as follows

$$\langle j | H_\lambda | i \rangle = -\frac{e}{m_0} \sum_q \sqrt{\frac{\hbar}{2\varepsilon_0 V \omega_q}} \langle j | a_q^+ \varepsilon_q exp\,(i q \cdot r) \cdot p + a_q \varepsilon_q exp\,(-i q \cdot r).p | i \rangle \quad (4.61)$$

Now we apply dipole approximation $exp(i q \cdot r) \approx 1$ and consider one electromagnetic mode, so that the total eigen function will be the product of an electronic part $| \alpha \rangle$ and radiation part $| n \rangle$.
Therefore, we obtain

$$\langle j | H_\lambda | i \rangle = -\frac{e}{m_0} \sqrt{\frac{\hbar}{2\varepsilon_0 V \omega_q}} \langle \alpha, n | a_q^+ \varepsilon_k \cdot p + a_q \varepsilon_q \cdot p | \beta, 1 \rangle \quad (4.62)$$

For the electronic part $\langle \alpha | p | \beta \rangle = i \omega_{\alpha\beta} m_0 \langle \alpha | r | \beta \rangle \quad \omega_{\alpha\beta} = \left(E_\alpha - E_\beta \right) / \hbar$

For emission $\omega_{\alpha\beta} = -\omega_k$ and for absorption $\omega_{\alpha\beta} = \omega_k$. Therefore,

$$\langle j | H_\lambda | i \rangle = i e \sqrt{\frac{\hbar \omega_q}{2\varepsilon_0 V}} \langle j | a_q^+ \varepsilon_q \cdot r - a_q \varepsilon_q \cdot r | i \rangle \quad (4.63)$$

Using Eq. (4.46a) for F_q, the previous equation can be rewritten as

$$\langle j | H_\lambda | i \rangle = \langle j | -e r \cdot F | i \rangle \quad (4.64)$$

Thus, the interaction Hamiltonian is of the form $H_\lambda = \mu \cdot F$, where $\mu = -e r$ is the dipole moment operator.

This result can be used to derive an expression for the spontaneous transition rate of an atom from an excited state $|2\rangle$ to a lower state $|1\rangle$. If the atom is assumed to interact with a single radiation mode, the quantum mechanical system will be made up of one atom and one radiation mode $\mathbf{q}$ as shown in Fig. 4.3. When the atom is in the initial excited state $| 2 \rangle$, the radiation mode $\mathbf{q}$ is in the state $| n_q \rangle$. Thus, the initial state can be designated as

$$| 2, n_q \rangle = | 2 \rangle | n_q \rangle \quad (4.65)$$

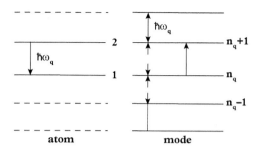

Figure 4.3 *The quantum mechanical system consisting of two-level atom and photons.*

After transition, if the atom is in the lower state $|1\rangle$ with the state of radiation mode $|n_k + 1\rangle$, the final state will be

$$|1, n_q + 1\rangle = |1\rangle |n_q + 1\rangle \qquad (4.66)$$

The initial and final energies are given by

$$E_i = E_2 + \hbar\omega_q \left(n_q + \frac{1}{2}\right) \qquad (4.67)$$

$$E_f = E_1 + \hbar\omega_q \left(n_q + \frac{3}{2}\right) \qquad (4.68)$$

$$\text{and } E_i - E_f = E_2 - E_1 - \hbar\omega_q \qquad (4.69)$$

Using Eqs. (4.46) and (4.64), we get the interaction Hamiltonian as

$$H_\lambda = -ie\sqrt{\frac{\hbar\omega_k}{2\varepsilon V}}[a_k^+ exp(-i\boldsymbol{k} \cdot \boldsymbol{r}) - a_k exp(i\boldsymbol{k}.\boldsymbol{r})]\varepsilon_k \cdot \boldsymbol{r} \qquad (4.70)$$

Using the property of the creation operator $\langle n_k + 1 | a_k^+ | n_k \rangle = \sqrt{n_k + 1}$, the transition rate will be

$$W = \frac{2\pi}{\hbar}|H_{fi}|^2 \delta (E_i - E_f)$$

$$= \frac{2\pi e^2}{\hbar}\frac{\hbar\omega_q}{2\varepsilon V}|\langle 1, n_q + 1 | a_q^+\varepsilon_q \cdot \boldsymbol{r} | 2, n_q\rangle|^2 \delta (E_2 - E_1 - \hbar\omega_q) \qquad (4.71)$$

$$= \frac{\pi e^2 \omega_q}{\varepsilon V}|\langle 1 | \varepsilon_q \cdot \boldsymbol{r} | 2\rangle|^2 (n_q + 1) \delta (E_2 - E_1 - \hbar\omega_q)$$

The part of W which is proportional to n_q is the stimulated emission rate and the part independent of n_q is the spontaneous emission rate. Thus,

$$W_{stim}(q) = \frac{\pi e^2 \omega_q}{\varepsilon V} |\varepsilon_q \cdot r_{12}|^2 \delta(E_2 - E_1 - \hbar\omega_q) n_q \qquad (4.72a)$$

$$W_{spon}(q) = \frac{\pi e^2 \omega_q}{\varepsilon V} |\varepsilon_q \cdot r_{12}|^2 \delta(E_2 - E_1 - \hbar\omega_q) \qquad (4.72b)$$

where $r_{12} = \langle 1\,|r|\,2\rangle$

The total spontaneous emission rate can be obtained by summing Eq. (4.72b) over all modes and is given by

$$W_{spon}(q) = \sum_q \frac{\pi e^2 \omega_q}{\varepsilon V} |\varepsilon_q \cdot r_{12}|^2 \delta(E_2 - E_1 - \hbar\omega_q) \qquad (4.73)$$

If the modes are closely spaced, the summation can be replaced by integration as follows:

$$\sum F(q) = \iiint F(q)p(q)dq \qquad (4.74)$$

where $F(q)$ is a function of q and $p(q)dq$ is the number of electromagnetic modes with q-vectors within a differential volume dq in q space and is given by

$$p(q)dq = \frac{V dq}{8\pi^3} = \frac{q^2\, dq\, sin\theta\, d\theta\, d\phi\, V}{8\pi^3} \qquad (4.75)$$

Thus

$$W_{spon} = \frac{e^2}{8\pi^2 \varepsilon \hbar} \sum_{\lambda=1,2} \int_0^\infty \int_0^\pi \int_0^{2\pi} \omega_q |\varepsilon_q \cdot r_{12}|^2 \delta(\omega_0 - \omega_k) q^2\, dq\, sin\theta\, d\theta\, d\phi \qquad (4.76)$$

Here λ indicates two polarization directions of the radiation. Using $\omega_q = qc/\eta$, Eq. (4.76) takes the form

$$W_{spon} = \frac{e^2 \omega_0^3 \eta^3}{8\pi^2 \varepsilon_0 c^3} \sum_{\lambda=1,2} \int_0^\pi \int_0^{2\pi} |\varepsilon_q \cdot r_{12}|^2 sin\theta\, d\theta\, d\phi \qquad (4.77)$$

Let us assume that r_{12} is parallel to q_z in q-space as shown in Fig. 4.4. The polarization direction ε_{q1} is assumed to lie in the plane formed by r_{12} and q. As ε_{q2} is normal to both q and ε_{q1}, it is also perpendicular to r_{12}. Therefore

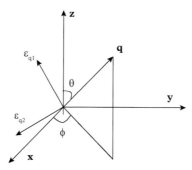

Figure 4.4 *Coordinate system used to evaluate the integral of Eqn. (4.77).*

$$W_{spon} = \frac{e^2\omega_0^3\eta^3\,|r_{12}|^2}{8\pi^2\varepsilon_0 c^3\hbar} \int_0^\pi \int_0^{2\pi} \sin^3\theta\, d\theta d\phi = \frac{e^2\omega_0^3\eta^3}{3\pi\varepsilon_0 c^3\hbar}|r_{12}|^2 \tag{4.78}$$

This equation gives the value of Einstein's A coefficient which is equal to τ_{spon}^{-1}.

Example 4.5 *Calculation of r_{12} for two hydrogenic states. Consider $2p \rightarrow 1$'s transition between hydrogenic states. As $|r_{12}|^2 = |x_{12}|^2 + |y_{12}|^2 + |z_{12}|^2$, where $|x_{12}| = \langle 1\,|x|\,2\rangle$, etc., use of hydrogenic wavefunctions give $|x_{12}| = 0.75\,a_B$, where a_B is the Bohr radius. Using the result, one obtains $\tau_{sp} = 1.6$ ns, see Problem 4.11.*

4.5.2   Oscillator strength

According to classical theory, a fraction f_0 of a number, N, of oscillators contribute to the susceptibility. By the Quantum Theory of absorption for a two level atomic system, an expression for f_0 can be obtained which is known as oscillator strength.

We know that the absorption coefficient α is related to the imaginary part of susceptibility χ'' and χ is in turn related to the dipole matrix element $|r_{12}|^2$. A comparison between the two points out that the two-level system may be thought of as an oscillator having oscillation frequency $\omega_0 = (E_2 - E_1)/\hbar$.

Using the relations $\alpha = 2\omega\eta_i/c$ and $2\eta_r\eta_i = \varepsilon_2 = \chi''$, the absorption coefficient can be expressed as

$$\alpha = \frac{\omega}{c}\chi'' = \frac{f_0 N e^2 \pi}{2\varepsilon_0 m_0 c} S_f(\omega) \tag{4.79}$$

with

$$S_f(\omega) = \frac{\gamma/2\pi}{(\omega_0 - \omega^2) + (\gamma/2)^2} \tag{4.80}$$

In terms of Einstein's A coefficient, the absorption (gain) coefficient of a two-level system can be expressed as

$$\alpha = A_{21} \frac{\pi^2 c^2}{\omega_0^2} S_f(\omega) N \tag{4.81}$$

Equating the expressions obtained, the oscillator strength for $\eta = 1$ is given as

$$f_0 = \frac{2\omega_0 m_0}{3\hbar} |r_{12}|^2 \tag{4.82}$$

Replacing the dipole matrix element by momentum matrix element, the oscillator strength has been obtained as

$$f_0 = \frac{2|p_{12}|^2}{3\hbar \omega_0 m_0} \tag{4.83}$$

4.6   Second-order perturbation theory

There are many situations in which we should follow the second-order perturbation theory. The examples of such processes are two-phonon absorption involving simultaneously one photon and one phonon. The second-order perturbation theory is utilized to calculate the transition probability in interband transitions in indirect band gap semiconductors and free carrier absorption where both photon and phonon are involved.

The expression for the transition probability for a second-order process represented by a transition from state $|i\rangle$ to an intermediate state $|m\rangle$ with the help of a quantum energy $\hbar\omega_{k'}$ and finally to the state $|f\rangle$ via another quantum of energy $\hbar\omega_k$ is given as

$$W_{if} = \frac{2\pi}{\hbar} \sum_{k,k'} \sum_m \frac{\left|H_{fm}^k\right|^2 \left|H_{mi}^{k'}\right|^2}{\left(E_m - E_i - \hbar\omega_{k'}\right)^2} \delta\left(E_f - E_i - \hbar\omega_k - \hbar\omega_{k'}\right) \tag{4.84}$$

Problems

4.1 *Obtain an expression for the frequency of oscillation of mode (m, n, p) in a rectangular cavity of size $L_x \times L_y \times L_z$.*

4.2 *The dimensions L_x and L_y in a rectangular cavity are small to support only the fundamental mode in these directions. Prove that the frequency separation between two adjacent longitudinal modes is given by $\Delta f_p = c/2\eta L_z$. Calculate the frequency separation for a long cavity formed by a dielectric of RI = 3.5 having a length = 500 μm.*

4.3 *A cubic cavity is 1μm long on each side. Calculate the frequency of the lowest order mode in this cavity. Assume RI = 3.2.*

4.4 *Obtain the expressions for the DOM in a planar cavity of negligible thickness.*

4.5 *A wire-like cavity supports only longitudinal modes. Obtain the expression for the mode density in the ideal 1D cavity.*

4.6 *Verify from Eq. (4.9) that only one integer in the mode number (m, n, p) can be zero in a cubic cavity, putting a limit to the smallest frequency in a cubic cavity. Show that the smallest frequency can be expressed as $\omega_0 = \sqrt{2}\pi c/L$, where L is the length of each side.*

4.7 *Derive the commutation relation Eq. (4.20) satisfied by the creation and destruction operators.*

4.8 *Prove that the Hamiltonian for harmonic oscillator can be expressed by Eq. (4.20).*

4.9 *Use Bohr–Sommerfeld quantization rule $\int pdq = nh$ to obtain the expression $E = n\hbar\omega$ for a harmonic oscillator.*

4.10 *The wavefunction of a harmonic oscillator is expressed as $\phi(x) = C\exp(-\alpha x^2)$, where C is the normalization constant and α is the variational parameter. The expectation value of the ground state energy is given by $E' = \sum\langle\phi|H|\phi\rangle/\langle\phi|\phi\rangle$. Show from $dE'/d\phi = 0$, that the ground state energy id $E_0 = (1/2)\hbar\omega$. H is the Hamiltonian for harmonic oscillator.*

4.11 *Show that $B_{12} = \pi e^2|D_{12}|^2/3\varepsilon_0\hbar^2$, where $D_{12} = \langle 1|d|2\rangle$, and d is the electric dipole moment.*

4.12 *Show that the B-coefficient for absorption in the two-level system can be expressed as $B_{12} = \left[(\pi e^2\hbar)/(m_0^2\varepsilon_0^2\eta^2\hbar\omega)\right]|\langle 1|\varepsilon_\lambda \cdot p|2\rangle|^2$.*

4.13 *The stimulated emission rate in a two-level atomic system equals the spontaneous emission rate with a light beam intensity I_0. If the wavelength of the beam is 1 μm and the spectral width is $\Delta f = 10^7$ Hz, calculate the value of I_0.*

4.14 *Show that the average thermal energy per mode under equilibrium is $\langle E_k\rangle = (1/2)\hbar\omega_k + \hbar\omega_k/\left[\exp(\hbar\omega_k/k_BT) - 1\right]$.*

4.15 *Prove that $A = B\hbar\omega\sum p(k)$, where p(k) is the photon DOS.*

Reading List

Anselm A (1981) *Introduction to Semiconductor Theory*, M M Samokhvalov (trans.) Moscow: Mir Publishers.

Basu P K (1997) *Theory of Optical Processes in Semiconductors: Bulk and Microstructures*. Oxford: Oxford University Press.

Heitler W (1954) *The Quantum Theory of Radiation*, 3rd edn. Oxford: Oxford University Press.

Loudon R (1983) *The Quantum Theory of Light*, 2nd edn. Oxford: Oxford University Press.

Yariv A (1982) *An Introduction to the Theory and Applications of Quantum Mechanics*. New York: Wiley.

Yariv A (1989) *Quantum Electronics*, 3rd edn. New York: Wiley.

5

Electron-photon interactions in bulk semiconductors

5.1 Introduction

An electromagnetic (EM) wave entering into any medium suffers absorption and its propagation is governed by the refractive index (RI) of the medium. The absorption and RI in matter are interrelated. The classical EM theory was employed in Chapter 3 to find this interrelationship and the phase and group velocity of waves. The three basic processes in light-matter interactions were also introduced in that chapter and the rates for transition in two atomic levels were expressed in terms of Einstein's coefficient. The requirements for achieving gain, that is negative absorption coefficient, were also established there.

Chapter 4 dealt with quantization of EM waves. After introducing the concept of photons, the interaction between an atomic system with photons are described in terms of the interaction Hamiltonian. In Chapter 4, the theory to calculate the transition rates for the three basic processes related to light-matter interaction was developed by considering just two energy levels. The transition rate, calculated by using Fermi Golden Rule, depends on the matrix element of transition between the two states. The initial and final state wavefunctions are assumed as a product of a one electron wavefunction and photon wavefunction, which itself is a product of wavefunctions of all the photon modes.

With the background developed in earlier chapters, we are now in a position to begin our desired journey into the domain of semiconductor Nanophotonics. This will be accomplished in several phases, the first phase of which results in the present chapter. Here, we develop the microscopic theory of optical processes in bulk semiconductors using the concepts developed in earlier chapters. The theory presented in this chapter serves as the guideline to develop and understand processes occurring in nanostructures step-by-step. Basically, we introduce the band picture, in place of atomic levels considered in earlier chapters, and consider the density-of-states (DOS) functions and occupational probabilities of the carriers. First of all, a classification of different absorption processes in semiconductors is given. The theory of absorption is then developed by using a semiclassical formalism, in which the EM radiation is considered classically,

Semiconductor Nanophotonics. Prasanta Kumar Basu, Bratati Mukhopadhyay, and Rikmantra Basu, Oxford University Press.
© P.K. Basu, B. Mukhopadhyay, R. Basu (2022). DOI: 10.1093/oso/9780198784692.003.0005

and the behaviour of carriers is described by using Bloch functions. The theory outlines first the method to obtain absorption coefficient for the fundamental band-to-band transition in a direct-bandgap semiconductor valid for 0 K. The theory is then modified for finite temperatures taking into account band picture, DOS of electrons and holes, and the occupational probabilities. Next developed is the intervalence band absorption. A simple classical Drude model is employed to treat the freecarrier absorption. The opposite process of absorption, that is, recombination and luminescence due to direct-interband transition, is then introduced. The theory of Auger recombination is developed for the most important transition, while other recombination processes, namely, the trap assisted recombination and surface recombination, are discussed briefly. A brief description of carrier-induced changes in the bandgap, and the RI then follows.

We end this chapter by deriving expression for a gain coefficient in semiconductors, which will be needed to understand laser action in bulk semiconductors.

In the present chapter, we discuss optical processes in direct-gap semiconductors only, as most of the studies related to Nanophotonics consider nanostructures made of such semiconductors. Also, this chapter considers single particle transitions. Excitonic processes will be discussed in Chapter 7.

As already pointed out, the theory presented in this chapter for bulk semiconductors will be modified for quantum nanostructures taking into account the modified form of Bloch functions for carriers, altered DOS, and occupational probabilities, and most importantly the selection rules for transitions.

5.2   Absorption processes in semiconductors

The absorption processes in semiconductors can best be understood with the help of quantum mechanical theory. This theory treats the radiations as a bunch of photons. This bunch, when incident on a semiconductor, induces transitions amongst different energy levels in different bands in the material. A few representative transitions occurring in a semiconductor are illustrated in Fig. 5.1 by using the $E-\mathbf{k}$ diagram.

The process indicated by A in Fig. 5.1 represents direct valence to conduction band transitions (constant $\mathbf{k}$ vector), the interband absorption, that occurs in a direct-band gap semiconductor, like GaAs. Process B illustrates transition from the valence band to the indirect-conduction band involving a photon (vertical line (1) and a scattering agent like a phonon (wavy line (2)). Such transitions occur in indirect-gap semiconductors like germanium (Ge) and silicon (Si). Process C shows inter-valence band transitions; in the figure transition between the HH and the SO bands has been illustrated, though other types like heavy-hole (HH) to light-hole (LH) transitions are also possible. Process D depicts free-carrier transitions in a valence band involving scattering by impurities, phonons, or other mechanisms. Process E is for free-carrier transitions in conduction band aided by impurities or photon-phonon interactions. Depending on the photon energy, a single type or a combination of a few types of all the absorption processes indicated in the figure may be important. We shall discuss the origin and characteristics of these processes one by one. Indirect band-to-band transition (Process B) is not of interest in the present context.

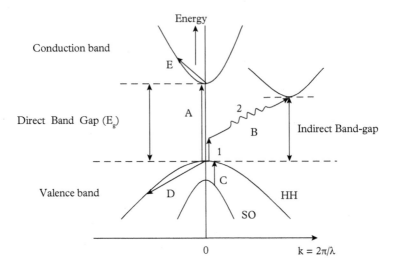

Figure 5.1 *Illustration of the different absorption processes in semiconductors.*

5.3   Fundamental absorption in direct-gap

Many, if not almost all, photonic devices are made of direct-gap semiconductors. Optical processes in nanostructures and nanophotonic devices studied so far are also concerned with this type of semiconductor. Studies of absorption or gain and emission in a direct-gap semiconductor are therefore of extreme importance. In the following, we discuss first the basic conservation laws for optical transitions in direct-gap semiconductors and then proceed to obtain the expression for the absorption coefficient.

5.3.1   Conservation laws

When photons of energy greater than the fundamental band gap are incident on a semi-conductor sample, electrons from the valence band absorb the photons and move to the conduction band. As a consequence, an empty state or a hole is created in the valence band. The absorption of photons thus leads to the generation of excess electron-hole pairs. The absorption coefficient rises rapidly as the photon energy $\hbar\omega \geq E_g$, the bandgap energy. This process is shown schematically in Fig. 5.2.

Two conditions must be fulfilled for this band-to-band transition. The first one, the energy conservation condition, states that the energy of the electron should equal the sum of energies of the hole and photon. This leads to the expression

$$E_c(\mathbf{k}_c) = E_v(\mathbf{k}_v) + \hbar\omega \tag{5.1}$$

where $\mathbf{k}_c$ and $\mathbf{k}_v$ are, respectively, the electron and the hole wavevectors.

The absorption threshold occurs when the photon energy equals the band gap energy. We may define a threshold or cut-off wavelength from the previous equation as follows:

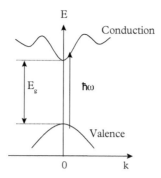

Figure 5.2 *Fundamental absorption process in a direct gap semiconductor.*

$$\lambda_c = \frac{hc}{E_g} = \frac{1.24}{E_g(eV)}(\mu m).$$

The last ratio indicates that the value of the cut-off wavelength is in μm, when the energy is in eV.

The second condition to be satisfied is the momentum conservation, stated as follows:

$$k_c = k_v + q \tag{5.2}$$

However, in most of the situations the photon wave vector $\mathbf{q} \ll \mathbf{k}_c$ and $\mathbf{k}_v$. Therefore, we obtain

$$k_c = k_v \tag{5.3}$$

This **k**-conservation is indicated by a vertical line in the $E-\mathbf{k}$ diagrams in Figs. 5.1 and 5.2.

Example 5.1 *If the wavelength of the photon is 1 μm, the wavevector is $q = 2\pi/\lambda = 2\pi \times 10^6 m^{-1}$. For an electron with effective mass $m_e= 0.1m_0$ and energy of 25 meV, the wavevector $k_c = (2m_eE/\hbar^2)^{1/2} = 2.56 \times 10^8 m^{-1}$f. Thus $q \ll k_c$.*

Since the wavevector remains essentially unchanged after photon absorption, it is easy to conclude that scattering by a momentum conserving agency, like phonons, is needed to bring the electron in the conduction band in indirect-gap semiconductor to complete the absorption process. This is illustrated by process *B* in Fig. 5.1 (vertical line satisfying **k** conservation + wavy line corresponding to phonon participation). The probability of such a transition is extremely low.

5.3.2 Calculation of absorption coefficient

Calculation of the absorption coefficient involves finding the probability of transition from a valence band state to a conduction band state. Here, we shall use the semiclassical approach by treating electromagnetic waves classically but describing the electrons as quantum mechanical (Bloch) waves (*see list of books in Suggested Readings*). An alternative way is to consider the radiation as collection of particles (photons) and to use the photon creation, annihilation, and number operators to describe the absorption, stimulated, and spontaneous emissions.

The unperturbed one electron Hamiltonian in the semiclassical treatment is given by

$$H_0 = p^2/2m_0 + V(r) \tag{5.4}$$

The electric field and magnetic fields of the EM wave are expressed as

$$F = \frac{\partial A}{\partial t} \text{ and } B = \nabla \times A \tag{5.5}$$

$$H = \frac{1}{2m_0}(p + eA)^2 + V(r) \tag{5.6}$$

Writing the term
$(p + eA)^2 = (eA.p) + (ep.A)$ and noting that $p = (\hbar/i)\nabla$, we may write

$$(p.A)f(r) = A \cdot \left(\frac{\hbar}{i}\nabla f\right) + \left(\frac{\hbar}{i}\nabla \cdot A\right)f \tag{5.7}$$

It is safe to neglect the term involving A^2 in Eq. (5.6) (see Example 4.4), so that first-order perturbation theory is valid.

Since, $\nabla \cdot A = 0$, the Hamiltonian given by Eq. (5.6) can be approximated by using Eq. (5.4) as

$$H = H_0 + \frac{e}{m_0}A.p \tag{5.8}$$

The second term in the right-hand side of Eq. (5.8) represents the interaction between radiation and Bloch electrons and is referred to as the electron-radiation interaction Hamiltonian H_{eR}. It is written in two different forms as

$$H_{eR} = \frac{e}{m_0}A.p \text{ or } H_{eR} = (-e)r.F \tag{5.9}$$

The equivalence of the two can be established when the wave vector of the EM field is small, the so-called electric dipole approximation. The first form of H_{eR} is more frequently used.

We now consider transition from a valence band state $|v\rangle$ with energy E_v and wave vector $\mathbf{k}_v$ to a conduction band state $|c\rangle$ with energy E_c and wave vector $\mathbf{k}_c$. The rate of such a transition depends on the following squared matrix element:

$$|\langle c|H_{eR}|v\rangle|^2 = (e/m_0)^2|\langle c|A \cdot p|v\rangle|^2 \tag{5.10}$$

The matrix element involves integration in both space and time. The time integration involves the time dependency of the vector potential, as given in Eq. (4.37), and time dependencies of the Bloch functions, and takes the following form:

$$\int exp(iE_c t/\hbar) exp(-i\omega t) exp(-iE_v t/\hbar) dt \propto \delta[E_c(k_c) - E_v(k_v) - \hbar\omega] \tag{5.11}$$

The δ-function indicates energy conservation in the absorption process as expressed by Eq. (5.1). Similarly, the matrix element involving the complex conjugate term $exp(-i\omega t)$ of the vector potential in Eq. (4.37), gives rise to $\delta[E_c(k_c) - E_v(k_v) + \hbar\omega]$, and describes stimulated emission of a photon due to transition from state E_c to state E_v. Writing the Bloch functions of electrons and holes, respectively, as

$$|c\rangle = U_{c,k_c}(r) exp(ik_c \cdot r) \tag{5.12}$$

and

$$|v\rangle = U_{v,k_v}(r) exp(ik_v \cdot r) \tag{5.13}$$

and using the form of vector potential as given by Eq. (4.37), we obtain for the space-dependent factor

$$|\langle c|A \cdot p|v\rangle|^2 = |A_0|^2 \left| \int U^*_{c,k_c} exp[i(q - k_c) \cdot r](\hat{q} \cdot p) U_{v,k_v} exp(ik_v \cdot r) dr \right|^2 \tag{5.14}$$

As **p** is a differential operator, we may write

$$pU_{v,k_v} exp(ik_v \cdot r) = exp(ik_v \cdot r)pU_{v,k_v} + \hbar k_v U_{v,k_v} exp(ik_v \cdot r) \tag{5.15}$$

The previous second term, when multiplied by U^*_{c,k_c} and integrated over space vanishes due to orthogonality of U^*_{c,k_c} and U_{v,k_v}. We are now left with an integral containing the first term, which can be broken into two parts as follows, in view of the periodicity of functions U_c and U_v,

$$\int U^*_{c,k_c} exp[i(q - k_c + k_v) \cdot r]p \, U_{v,k_v} dr$$

$$= \left(\sum_l exp[i(q - k_c + k_v) \cdot R_l] \right) \int_{unit\,cell} U^*_{c,k_c} exp[i(q - k_c + k_v) \cdot r']p \, U_{v,k_v} dr' \tag{5.16}$$

In the previous example, we have written $r = R_l + r'$, where R_l is a lattice vector and r' lies within one unit cell. One notes that

$$\sum_l exp[i(q - k_c + k_v) \cdot R_l] - \delta(q - k_c + k_v)$$

meaning thereby that the wave vector (momentum) is conserved in the absorption process. Use of the conservation condition in the integral over unit cell in Eq. (5.16) yields the following simplified result:

$$\int_{unit\ cell} U^*_{c,k_c} exp[i(\boldsymbol{q} - \boldsymbol{k}_c + \boldsymbol{k}_v) \cdot \boldsymbol{r}'] \boldsymbol{p} U_{v,k_v} d\boldsymbol{r}' = \int_{unit\ cell} U^*_{c,k_v+q} \boldsymbol{p} U_{v,k_v} d\boldsymbol{r}' \qquad (5.17)$$

This equation is further simplified by assuming $\boldsymbol{q} << \boldsymbol{k}_v$. Writing $U_{k+q} \approx U_{k+q} \cdot \nabla_k U_k + \dots$, and keeping only the first term, the matrix element may be expressed as

$$|\langle c|\hat{e} \cdot \boldsymbol{p}|v\rangle|^2 = \left(\int_{unit\ cell} U^*_{c,k}(\hat{e} \cdot \boldsymbol{p}) U_{v,k_v} d\boldsymbol{r}' \right)^2 \qquad (5.18)$$

This approximation is known as dipole approximation. In most cases, the momentum matrix element defined previously does not depend on $\boldsymbol{k}$. We therefore write

$$|\langle c|H_{eR}|v\rangle|^2 = (e/m_0)^2 |A_0|^2 |(\hat{e} \cdot \boldsymbol{p})_{cv}|^2 \qquad (5.19)$$

where $|(\hat{e} \cdot \boldsymbol{p})_{cv}|$ denotes the integral over the unit cell. The transition probability R for photon absorption per unit time by using Fermi Golden rule is expressed as

$$R = (2\pi/\hbar) \sum_{k_c,k_v} |\langle c|H_{eR}|v\rangle|^2 \delta[E_c(\boldsymbol{k}_c) - E_v(\boldsymbol{k}_v) - \hbar\omega] \qquad (5.20)$$

Eq. (5.20) gives the rate of transition per unit volume and $R\hbar\omega$ gives the rate of loss of power per unit volume due to absorption of photons. The rate of power loss, or of intensity I, may also be related to the absorption coefficient α as follows:

$$-\frac{dI}{dt} = -\left(\frac{dI}{dz}\right)\left(\frac{dz}{dt}\right) = \frac{c}{\eta}\alpha I \qquad (5.21)$$

Thus,

$$\alpha = \frac{1}{I}\frac{\eta}{c}\frac{dI}{dt} = \frac{1}{I}\frac{\eta}{c}R\hbar\omega \qquad (5.22)$$

Both R and I contain $|A_0|^2$ and therefore α is independent of intensity of light.

The summation over both $\boldsymbol{k}_c$ and $\boldsymbol{k}_v$ in Eq. (5.20) for the transition rate R is reduced to a single summation by writing $\boldsymbol{k}_c = \boldsymbol{k}_v = \boldsymbol{k}$ and then the summation is converted into integration by noting

$$\sum_k \rightarrow \frac{2}{(2\pi)^3} \int 4\pi k^2 dk$$

This leads to

$$I_k = \sum_k \delta[E_c(\boldsymbol{k}) - E_v(\boldsymbol{k}) - \hbar\omega] = \frac{2}{8\pi^3} \int 4\pi k^2 dk \delta[E_c(\boldsymbol{k}) - E_v(\boldsymbol{k}) - \hbar\omega] \qquad (5.23)$$

Assuming parabolic *E-k* relationships,

$$E_c(\boldsymbol{k}) = E_g + \frac{\hbar^2 k^2}{2m_e} \text{ and } E_v(\boldsymbol{k}) = -\frac{\hbar^2 k^2}{2m_h} \qquad (5.24)$$

where m_e and m_h are the electron and hole effective masses, respectively, the argument of the δ-function in Eq. (5.23) becomes $E_g + (\hbar^2 k^2 / 2m_r) - \hbar\omega$, where the reduced mass is given by

$$m_r^{-1} = m_e^{-1} + m_h^{-1} \qquad (5.25)$$

The integration is performed by transforming k to $E(k)$ and then using the δ-function. Using the expressions for R and I, the final expression for the absorption coefficient becomes

$$\alpha(\hbar\omega) = \frac{e^2 (2m_r)^{3/2}}{2\pi\varepsilon_0 cnm_0^2 \hbar^3 \omega} (\hbar\omega - E_g)^{1/2} \langle |p_{cv}^2| \rangle, \quad \hbar\omega \geq E_g \qquad (5.26)$$

The term $\langle |p_{cv}^2| \rangle$ is the average of squared momentum matrix element for unpolarized light. The average is obtained by considering the polarization dependence of the term $|(\hat{e} \cdot p)_{cv}|$.

For transitions involving states near the band edges, both $U_{c, k}$ and $U_{v, k}$ are given by their zone centre values. In this approximation

$$\text{Conduction band } U_{C,0} = | s \rangle \qquad (5.27)$$

$$\text{Valence band : heavy hole states } |3/2, 3/2\rangle$$
$$= \left(-1/\sqrt{2}\right) (| p_x \rangle + i| p_y \rangle) |\alpha\rangle$$
$$|3/2, -3/2\rangle = \left(1/\sqrt{2}\right) (|p_x\rangle - i|p_y\rangle) |\beta\rangle \qquad (5.28a)$$

$$\text{Light hole states } |3/2, 1/2\rangle = (-1/6)(| p_x \rangle + i| p_y \rangle)|\beta\rangle - 2| p_z \rangle|\alpha\rangle \qquad (5.28b)$$
$$|3/2, -1/2\rangle = (1/6)(| p_x \rangle - i| p_y \rangle)|\alpha\rangle + 2| p_z \rangle|\beta\rangle$$

Symmetry allows only the following matrix elements to be nonzero:

$$\langle p_x|p_x|s \rangle = \langle p_y|p_y|s \rangle = \langle p_z|p_z|s \rangle = p_{cv} \qquad (5.29)$$

The following transitions are then allowed, and the corresponding matrix elements are given by

$$\langle HH|p_x|s \rangle = \langle HH|p_y|s \rangle = \left(1/\sqrt{2}\right) \langle p_x|p_x|s \rangle \qquad (5.30a)$$

$$\langle LH|p_x|s \rangle = \langle LH|p_y|s \rangle = \left(1/\sqrt{6}\right) \langle p_x |p_x| s \rangle \qquad (5.30b)$$

It is also to be noted that

$$\langle HH| \, p_z| \, s\rangle = 0 \tag{5.31}$$

The squared matrix element for light polarized along different directions may now be listed. This dependence on orientation is useful in Quantum Wells (QWs) because heavy-hole(HH) and light-hole(LH) states become no longer degenerate there.

z-polarized light: HH→CB: No coupling

$$\text{LH} \rightarrow \text{CB} : |p_{cv}|^2 = (2/3)|\langle p_x|p_x|s\rangle|^2 \tag{5.32}$$

$$\text{x} - \text{polarized light} : \text{HH} \rightarrow \text{CB} : |p_{cv}|^2 = (1/2)|\langle p_x|p_x|s\rangle|^2$$

$$\text{LH} \rightarrow \text{CB} : |p_{cv}|^2 = (1/6)|\langle p_x|p_x|s\rangle|^2 \tag{5.33}$$

$$\text{y} - \text{polarized light} : \text{HH} \rightarrow \text{CB} : |p_{cv}|^2 = (1/2)|\langle p_x|p_x|s\rangle|^2$$

$$\text{LH} \rightarrow \text{CB} : |p_{cv}|^2 = (1/6)|\langle p_x|p_x|s\rangle|^2 \tag{5.34}$$

For finite values of k, there is some mixing of HH and LH states. As may be seen for x-y polarized light HH-CB coupling is three times stronger than LH-CB coupling.

In the present context, we are considering unpolarized light. Therefore, using the previous examples, we may write

$$\langle|p_{cv}|^2\rangle = \frac{2m_0^2 E_g (E_g + \Delta)}{3m_e (E_g + 2\Delta/3)} \tag{5.35}$$

In the literature, a quantity E_P defined as follows is used to describe the momentum matrix element

$$E_p = \frac{2}{m_0} |\langle p_x|p_x|s\rangle|^2 \tag{5.36}$$

The values of E_p for most semiconductors lie in the range 20–25 eV.

The final expression for the absorption coefficient is therefore

$$\alpha(\hbar\omega) = \frac{e^2 (2m_r)^{3/2}}{2\pi\varepsilon_0 cnm_0^2 \hbar^3 \omega} (\hbar\omega - E_g)^{1/2} \langle|p_{cv}|^2\rangle \tag{5.37}$$

The absorption coefficients for a few common semiconductors are presented in Fig. 5.3 as a function of wavelength. The presence of a cut-off wavelength can easily be identified

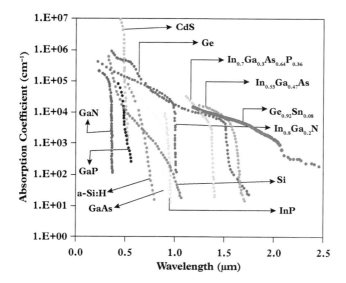

Figure 5.3 *Absorption coefficients of a few common semiconductors as a function of wavelength.*

in the diagram for direct-gap materials like GaAs, InP, and the ternary and quaternary alloys. The rise in absorption below λ_c (above E_g) is also apparent.

The absorption coefficient is sometimes expressed in terms of a dimensionless quantity called the *oscillator strength*, denoted as f_{vc} and defined as

$$f_{vc} = \frac{2P^2}{m_0(E_{kc} - E_{kv})} \tag{5.38}$$

The absorption coefficient may be expressed as follows in terms of the oscillator strength

$$\alpha(\omega) = \frac{e^2(2m_r)^{3/2}}{4\pi\varepsilon_0 cm_0 n(\omega)\hbar^2} f_{vc}(\hbar\omega - E_g)^{1/2} \tag{5.39}$$

Example 5.2 *We use Eq. (5.37) to calculate the absorption coefficient in GaAs. We take $m_r = 0.065m_0$, $n = 3.6$, a factor of 2/3 for unpolarized light and $2|p_{cv}|^2/m_0 = E_p = 23 eV$. The absorption coefficient is $\alpha = 5.25 \times 10^4 \left[(\hbar\omega - E_g)^{1/2}/\hbar\omega\right] cm^{-1}$ and the energies are in eV. Take $\hbar\omega = 1.5 eV$ and $E_g = 1.43 eV$. The value is $0.97 \times 10^4 cm^{-1}$ at 0 K. The value is reduced at finite temperatures.*

5.3.3   Absorption coefficient at finite temperature

The absorption coefficient expressed by Eq. (5.37) applies to the ideal situation for 0 K, when all states in the valence band are occupied by electrons and all states in the conduction band are empty. The expression becomes modified at finite temperature by including the Fermi functions for both electrons and holes. One must include a factor $f_v(\mathbf{k}_v)[1-f_c(\mathbf{k}_c)]$ in the integrand in Eq. (5.20), where $f_v(\mathbf{k}_v)$ denotes that a state of wave vector $\mathbf{k}_v$ in the VB is occupied by an electron and $[1-f_c(\mathbf{k}_c)]$ indicates that a state $\mathbf{k}_c$ in the CB is unoccupied by an electron. One should not forget that the calculation of net absorption coefficient, or gain coefficient under suitable condition, needs to take into account opposite processes, both spontaneous and stimulated emissions. The detailed treatment will be given in Section 5.8.

5.4   Intervalence band absorption

As noted already, the valence bands in typical semiconductors comprise LH, HH, and split-off (SO) bands. At elevated temperatures or in heavily doped p-type materials having Fermi energy below the valence band edge, transitions are possible from LH band to HH band or from SO band to HH or LH band.

An example of such transitions has been given in Fig. 5.1 as process (C) indicating a transition from SO to HH band. The different transitions are further illustrated in Fig. 5.4, in which processes A, B, and C denote, respectively the LH-HH, SO-LH, and SO-HH transitions. The intervalence band transitions are forbidden at $\mathbf{k} = 0$, due to quantum mechanical selection rules. Direct (vertical) transitions become possible at $\mathbf{k} \neq 0$. The momentum matrix elements for this transitions are proportional to k. The absorption bands are broad.

In the following, we shall outline the method of obtaining the absorption coefficient for inter-valence band absorption (IVBA) by considering transition between SO and HH bands.

The absorption coefficient for transition from a state of energy $E(\mathbf{k_s})$ in SO band to a state of energy $E(\mathbf{k_h})$ in HH band is given as

$$\alpha(\hbar\omega) = \frac{2\pi}{\hbar}\frac{1}{(c/n_r)}\left(\frac{eA_0}{m_0}\right)^2 \sum_{k_s,k_h}\langle|M_{k_s,k_h}|^2\rangle(f_h - f_s)\delta|E(\mathbf{k}_s - \mathbf{k}_h) + \hbar\omega|\delta_{k_s,k_h} \qquad (5.40)$$

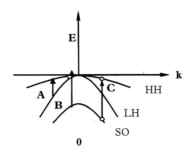

Figure 5.4 *Schematic of different types of intervalence band absorption.*

The energies are expressed as

$$E(k_h) = -\frac{\hbar^2 k_h^2}{2m_h} \quad \text{and} \quad E(k_s) = -\frac{\hbar^2 k_s^2}{2m_s} \tag{5.41}$$

The argument of the energy conserving δ-function becomes

$$E(k_s) - E(k_h) + \hbar\omega = -\Delta_{so} - \frac{\hbar^2 k_s^2}{2m_s} + \frac{\hbar^2 k_h^2}{2m_h} + \hbar\omega \tag{5.42}$$

The double summation in Eq. (5.40) can be converted to a single integral by using the Kronecker delta specifying k-conservation, i.e.$k_s = k_h$, so that

$$E(k_s) - E(k_h) + \hbar\omega = \hbar\omega - \Delta_{so} + \frac{\hbar^2 k_h^2}{2m_r} \tag{5.43}$$

$$\frac{1}{m_r} = \frac{1}{m_h} - \frac{1}{m_s} \tag{5.44}$$

m_r being the reduced mass.

The total summation in Eq. (5.40) reduces to

$$\sum_{k_h} \rightarrow \frac{2V}{(2\pi)^3} \int k_h^2 \sin\theta \, d\theta \, d\phi \, dk_h = \frac{2V}{(2\pi)^3} \int 4\pi k_h^2 \, dk_h \tag{5.45}$$

The square of the absolute value of the matrix element averaged over all directions is

$$|M(k_s, k_h)|^2 = \hbar^2 A_{sh} k_h^2 \tag{5.46}$$

where A_{sh} is a dimensionless constant, not to be confused with the Einstein A-coefficient.

Putting Eqs. (5.45) and (5.46) in Eq. (5.40), the integration may be easily performed by using the energy conserving δ-function. Using the expression for A_0 the expression for intervalence band absorption coefficient becomes

$$\alpha(\hbar\omega) = \frac{e^2 A_{sh}(2m_r)^{5/2}}{\pi \varepsilon_0 n_r c \hbar^2 m_0^2} \frac{(\hbar\omega - \Delta)^{3/2}}{\hbar\omega} (f_h - f_s) \tag{5.47}$$

5.5   Free carrier absorption

Semiconductor materials are usually doped to obtain either p or n region as well as to have required level of conductivity. Doping gives a substantial number of free electrons or holes in different regions. By absorbing photons, these free carriers make a transition from a state of wave vector $\mathbf{k}$ to another state of wave vector $\mathbf{k'}$ in the same band. In Fig. 5.1, process D and process E illustrate respectively such transitions in the valence band and the conduction band, termed respectively as free-hole (D) and free-electron (E) absorption. The general term 'freecarrier absorption' is used to describe this type of absorption.

In freecarrier absorption, a change in wave vector occurs, and the wave vector is conserved satisfying the relation $|\mathbf{k'} - \mathbf{k}| = \kappa$. The photon wave vector $\kappa = \omega \eta(\omega)/c$. In order to satisfy both energy and momentum conservations, the energy of the photon should be prohibitively large, larger than the typical band gap even. Therefore, the theory for band-to-band absorption discussed earlier cannot be applied to intraband free-carrier absorption.

The intraband processes occur in two steps. First, a photon of energy, quite small compared to the bandgap and hence of negligible momentum, is absorbed and then a scattering, by either phonons, or defects like impurities, alloy disorder, or other imperfections, occurs to place the electron in state $\mathbf{k'}$. The main contribution to wavevector change comes from the scattering agencies. Quantum mechanical calculation of the free-carrier absorption has been described in a number of textbooks and monograms (Basu 1997; Ridley 2000; Nag 1980).

In the present section, we use classical electromagnetic theory and Drude model for conduction (Basu 1997; Basu et al 2015) to describe free carrier absorption. The theory is simple and it also gives the expression for the carrier induced change in the RI. Furthermore, there is not much difference between the expressions obtained by classical and quantum methods.

The equation of motion of a free electron in the conduction band of a semiconductor under the influence of a sinusoidal electric field takes the following form:

$$m_e \frac{d^2x}{dt^2} + m_e g \frac{dx}{dt} = -eF_0 exp(i\omega t) \tag{5.48}$$

where x is the displacement of the electron, m_e is its effective mass, g is a damping coefficient, and F_0 is the amplitude of the impressed electric field varying with angular frequency ω. The first term is the force term, the second term represents the damping of electron motion by scattering with lattice vibrations (phonons) and impurities, etc., and the right-hand side represents the applied force. The steady state solution for displacement is

$$x = \frac{(eF_0)/m_e}{\omega^2 - i\omega g} exp(i\omega t) \tag{5.49}$$

The displacement of free carriers will produce additional polarization, P_1 given by

$$P_1 = -nex \tag{5.50}$$

where n denotes the free carrier concentration per unit volume. The total polarization is now $P = P_0 + P_1$, where P_0 is the polarization present in the material without free carriers. The relative permittivity is now given as

$$\varepsilon_r = \frac{\varepsilon}{\varepsilon_0} = 1 + \frac{P}{\varepsilon_0 F} = 1 + \frac{P_0}{\varepsilon_0 F} + \frac{P_1}{\varepsilon_0 F} = \eta_0^2 + \frac{P_1}{\varepsilon_0 F} \tag{5.51}$$

where η_0 is the index of refraction without free carriers. Using Eqs. (5.49), and (5.50) in Eq. (5.51), we may obtain

$$\varepsilon_r = \eta_0^2 - \frac{(ne^2)/(m_e\varepsilon_0)}{\omega^2 - i\omega g} \tag{5.52}$$

We separate the real and imaginary parts of the relative permittivity and write

$$\varepsilon_{rr} = \eta_0^2 - \frac{(ne^2)/(m_e\varepsilon_0)}{\omega^2 + g^2} \tag{5.53}$$

$$\varepsilon_{ri} = \frac{(ne^2 g)/(m_e\omega\varepsilon_0)}{\omega^2 + g^2} \tag{5.54}$$

At steady state $d^2x/dt^2 = 0$, and from Eq. (5.48)

$$m_e g \frac{dx}{dt} = eF \tag{5.55}$$

Using the following relation between the drift velocity and the mobility, μ_e:

$$\frac{dx}{dt} = \mu_e F \tag{5.56}$$

Eqs. (5.55) and (5.56) yield

$$g = \frac{e}{\mu_e m_e} = \frac{1}{\tau} \tag{5.57}$$

In Eq. (5.57), the well-known expression $\mu_e = e\tau/m_e$ relating the electron mobility and the momentum relaxation time τ has been used.

Example 5.3 *Assume an electron mobility = 8000 cm²/V.s for GaAs; the effective mass is 0.067 m_0, wherem_0 is the free electron mass. The value of g is 3.26 × 10¹² s⁻¹ and of τ is 0.3 ps.*

Since typical values for $\omega \approx 10^{15}$ s⁻¹, one may neglect g in the denominators of Eqs. (5.53) and (5.54), and obtain the following modified equations by using Eq. (5.57):

$$\varepsilon_{rr} = \eta_0^2 - \frac{ne^2}{m_e\varepsilon_0\omega^2} \tag{5.58}$$

$$\varepsilon_{ri} = \frac{ne^3}{m_e^2\varepsilon_0\omega^2\mu_e} \tag{5.59}$$

The exponential loss coefficient α is related to ε_{ri} by $\alpha = k\varepsilon_{ri}/\eta_r$, where η_r is the RI and k is the light wave vector given by $k = \omega/c$. We may write therefore

$$\alpha_{fc} = \frac{ne^3\lambda_0^2}{4\pi^2\eta_r m_e^2 \mu_e \varepsilon_0 c^3} \qquad (5.60)$$

Eq. (5.58) indicates that there is a change of refractive index also due to free carriers. Writing the small change as $\Delta\eta_e$, the corresponding expression may be written as

$$\Delta\eta_e = \frac{ne^2\lambda_0^2}{8\pi^2 c^2 \eta_r \varepsilon_0 m_e} \qquad (5.61)$$

The change in the RI due to free carriers in GaAs, InP, and InGaAsP is given by Bennett et al (1990) in the form

$$\Delta\eta_e = -\frac{6.9\times10^{-22}}{\eta_r(\hbar\omega)^2}\left\{\frac{n}{m_e} + p\left[\frac{m_{hh}^{1/2} + m_{lh}^{1/2}}{m_{hh}^{3/2} + m_{lh}^{3/2}}\right]\right\} \qquad (5.62)$$

where the photon energy is in eV, effective masses are ratios, and the carrier densities are in cm^{-3}. The prefactor may easily be calculated by using Eq. (5.61)

Example 5.4 *The free carrier absorption in GaAs is calculated by using Eq. (5.60) with the following values of parameters: $\lambda_0 = 0.85 \ \mu m$, $\mu_e = 7000 \ cm^2 V^{-1} s^{-1}$, $\eta_r = 3.6$, $m_e = 0.067 \ m_0$, and $n = 10^{18} \ cm^{-3}$. We find $\alpha_{fc} = 0.86 \ cm^{-1}$.*

Example 5.5 *The quaternary $In_{0.82}Ga_{0.18}As_{0.6}P_{04}$ is lattice matched to InP and has a band gap of 0.954 eV corresponding to $\lambda_0 = 1.3 \ \mu m$. Here $y = 0.6$. The mass values may be obtained from $m_e = 0.080 - 0.039y = 0.0566$, $m_{hh} = 0.46$, and $m_{lh} = 0.12 - 0.099y + 0.030y^2 = 0.0714$. Taking $\eta_r = 3.4$, $n = p = 10^{18} \ cm^{-3}$, $\Delta\eta_e = -0.42 \times 10^{-2}$. The change in RI due to free carriers is small.*

5.6   Recombination and luminescence

Emission is the process opposite to absorption. In semiconductors, excess electrons and holes are created by photoexcitation, energetic particle bombardment or by injection in a p-n junction. When these excess carriers recombine, energy is given up in the form of photons. The general name luminescence is given to the phenomenon of light emission. In this section, the recombination and luminescence processes will be discussed.

5.6.1   Luminescence lifetime

The excess electron-hole pairs created in a semiconductor have relatively short lifetime after which they recombine to emit a photon. The recombination process for a direct-gap semiconductors is shown in Fig. 5.5.

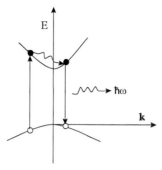

Figure 5.5 *Recombination process in a direct gap semiconductor.*

It has been shown in Chapter 2 that the excess carriers decay exponentially with time as

$$\Delta n(t) = \Delta n(0)e^{-t/\tau} \tag{5.63}$$

where $\Delta n(0)$ is the initial excess electron density. It is convenient to introduce a recombination rate defined as follows

$$R = -\frac{d\Delta n}{dt} = \frac{\Delta n}{\tau} \tag{5.64}$$

The recombination lifetime is a measure of the average time an excess carrier pair spends in the sample before being lost by recombination. The recombination in a semiconductor may be both intrinsic and extrinsic.

In addition to radiative recombination, there occurs also non-radiative recombination caused by defects, impurities, traps and surface states, in which the excess energy is given up in the form of phonons or heat waves. The ratio of radiative to total recombination rate is the internal quantum efficiency (IQE), which is expressed as

$$\eta_i = \frac{R_r}{R_r + R_{nr}} \tag{5.65}$$

The symbols R_r and R_{nr} denote, respectively, the rates for radiative and non-radiative recombination in a material per unit volume. The total recombination rate for spontaneous process becomes

$$R_{sr} = R_r + R_{nr} \tag{5.66}$$

In terms of the lifetimes of radiative (τ_r) and non-radiative (τ_{nr}) processes given by Eq. (5.64) the IQE may be written as

$$\eta_i = \frac{\tau_r^{-1}}{\tau_r^{-1} + \tau_{nr}^{-1}} = \left(1 + \frac{\tau_r}{\tau_{nr}}\right)^{-1} \tag{5.67}$$

Example 5.6 *In an indirect-gap semiconductor $\tau_r = 10^{-3}$ s and $\tau_{nr} = 10^{-7}$ s. Therefore IQE $= 10^{-4}$. On the other hand, in a direct gap semiconductor $\tau_r = 10^{-8}$ s and $\tau_{nr} = 10^{-7}$ s. The IQE = 0.9.*

5.6.2   Carrier lifetime: Dependence on carrier density

The spontaneous recombination rate in a non-degenerate semiconductor is expressed as

$$R_{sr} = B_r np \tag{5.68}$$

where, n and p, are respectively, the electron and hole concentrations. The recombination coefficient B_r in Eq. (5.68) may be derived quantum mechanically.

The thermal equilibrium electron and hole concentrations, denoted by n_0 and p_0, respectively, are related by $n_0 p_0 = n_i^2$, where n_i is the intrinsic carrier concentration. In the presence of excess carriers the spontaneous recombination rate becomes,

$$R_{sr} = B_r(n_0 + \Delta n)(p_0 + \Delta p) \tag{5.69}$$

Since $\Delta n = \Delta p$ and the spontaneous recombination rate $R_{sr}^0 = B_r n_0 p_0$ in absence of excess carriers, the recombination rate for the excited carriers R_{sr}^{exc} may be written as

$$R_{sr}^{exc} = R_{sr} - R_{sr}^0 = B_r \Delta n(n_0 + p_0 + \Delta n) \tag{5.70}$$

Therefore, the excess carrier lifetime, as related to recombination rate by Eq. (5.64), becomes

$$\tau_r = [B_r(n_0 + p_0 + \Delta n)]^{-1} \tag{5.71}$$

In the high injection case, $\Delta n \gg n_0$ or p_0, and $\tau_r = [B_r(\Delta n)]^{-1}$, while in the low injection case, $\Delta n \ll n_0$ or p_0 and $\tau_r = [B_r(n_0+p_0)]^{-1}$.

Example 5.7 *The carrier lifetime in GaAs is calculated for $\Delta n = 10^{18}$ cm^{-3}. Using $B_r = 10^{-10}$ cm^{3}s^{-1} the lifetime is 10^{-8} sec.*

5.6.3   Absorption and recombination

The coefficient B_r depends on the band structure of the material. Its value is large in direct-gap material and hence the recombination lifetime is small. On the other hand, for an indirect gap the value of B_r is extremely low, and hence the recombination lifetime is extremely large. The value of the recombination coefficient may be obtained from the absorption data or from microscopic calculation. A relationship between the lifetime and the absorption coefficient has been developed by van Roosebroeck and Shockley (1954). The reader is referred to that work.

5.6.4   Microscopic theory of recombination

The microscopic theories of band-to-band recombination and of absorption in a direct-gap semiconductor are developed in almost the same manner. Two states $|2\rangle$ in the conduction band and $|1\rangle$ in the valence band are considered and the spontaneous emission rate due to transition $|2\rangle \rightarrow |1\rangle$ may be expressed as

$$R_{sp}(\hbar\omega) = \frac{2\pi}{\hbar}\left(\frac{eA_0}{m_0}\right)^2 \sum_{1,2} \langle|p_{12}|^2\rangle G(\hbar\omega)f_2(1-f_1)\delta(E_{12}-\hbar\omega) \tag{5.72}$$

where p_{12} is the momentum matrix element between the two states, and G is the photon DOS. Using the following expressions for $|2\rangle$ and $|1\rangle$:

$$|2\rangle = |c, \mathbf{k}_c\rangle = U_c(\mathbf{r})exp(i\mathbf{k}_c \cdot \mathbf{r}) \text{and } |1\rangle = |v.\mathbf{k}_v\rangle = U_v(\mathbf{r})exp(i\mathbf{k}_v \cdot \mathbf{r}) \tag{5.73}$$

and noting that

$$\langle c, \mathbf{k}_c|e^{i\mathbf{q}\cdot\mathbf{r}}\hat{q}\cdot\mathbf{p}|v, \mathbf{k}_v\rangle = p_{cv} \tag{5.74}$$

the spontaneous emission rate as given in Eq. (5.72) may be written as

$$[R_{sp}(\hbar\omega)] = \frac{2\pi}{\hbar}\left(\frac{eA_0}{m_0}\right)^2 \sum_{\mathbf{k}_c,\mathbf{k}_v} \langle|p_{cv}|^2\rangle\delta_{\mathbf{k}_c,\mathbf{k}_v}G(\hbar\omega)f_c(\mathbf{k}_c)f_v{}'(\mathbf{k}_v)\delta[E(\mathbf{k}_c)-E(\mathbf{k}_v)-\hbar\omega]$$

$$\tag{5.75}$$

Assuming for the moment that the probability factors are unity, we put the expressions for A_0, G, and the squared matrix element averaged over all polarizations in Eq. (5.75) to obtain,

$$R_{sp} = \frac{e^2 n_r}{3\pi\varepsilon_0 m_0 c^3 \hbar^2}\frac{\langle|p_{cv}|^2\rangle}{m_0}\hbar\omega = \frac{1}{\tau_{sp}} \tag{5.76}$$

The spontaneous emission lifetime is related to the Einstein's A-coefficient by $\tau_{sp} = A_{21}^{-1}$, as two discrete states are involved.

Example 5.8 *For GaAs, $2\langle|p_{cv}|^2\rangle/m_0 = 23eV$ and $\eta_r = 3.6$. Let us take $\hbar\omega = 1.43eV$. We obtain $R_{sp} = 1.14 \times 10^9 \, s^{-1}$ and $\tau_{sp} = 0.58ns$.*

In the general case, when f_e and $f_h{}'$ are not a unity, a summation of the following form appears from Eq. (5.75):

$$I = \sum_{\mathbf{k}_c,\mathbf{k}_v} \delta_{\mathbf{k}_c,\mathbf{k}_v}f_c(\mathbf{k}_c)f_h{}'(\mathbf{k}_v)\delta[E(\mathbf{k}_c)-E(\mathbf{k}_v)-\hbar\omega] \tag{5.77}$$

The argument of the energy-conserving δ-function may be written in terms of the reduced mass using k-conservation, as in Eqs. (5.23–5.25), and proceeding in a similar manner, we arrive at the following expression for the recombination rate:

$$R_{sp}(\hbar\omega) = \frac{e^2\eta_r\hbar\omega\langle|p_{cv}|^2\rangle}{2\pi^3c^3m_0^2\varepsilon_0\hbar^2}\left(\frac{2m_r}{\hbar^2}\right)^{3/2}(\hbar\omega - E_g)^{1/2}f_c(k_v)f_v'(k_v) \tag{5.78}$$

When quasi-Fermi levels for both the electrons and holes are away from the respective band edges by a few k_BT, the distribution functions may be approximated as Boltzmann distributions expressed as

$$f_c(k_v) = exp\{-[E_c(k_v) - F_e]/k_BT\} \tag{5.79a}$$

$$f_v'(k_v) = exp\{[E_v(k_v) - F_h/k_BT]\} \tag{5.79b}$$

Writing
$\Delta F = F_e - F_h$ and noting that $E_c(k_v) - E_v(k_v) = \hbar\omega$, the product may be written as

$$f_c(k_v)f_v'(k_v) = exp\{-[E_g - \Delta F]/k_BT\}\times exp[-(\hbar\omega - E_g)/k_BT] \tag{5.80}$$

Putting this in Eq. (5.78), the recombination rate may be expressed as

$$R_{sp}(\hbar\omega) = C'(\hbar\omega, T)(m_r)^{3/2}(\hbar\omega - E_g)^{1/2}exp\{-[E_g - \Delta F]/k_BT\}$$
$$\times exp[-(\hbar\omega - E_g)/k_BT]^{1/2} \tag{5.81}$$

where the prefactor C' contains the material parameters and the fundamental constants. The quasi-Fermi levels are related to the electron density, n, and the hole density, p, by

$$n = 2(m_ek_BT/2\pi\hbar^2)^{3/2}exp[-(E_c - F_e)/k_BT] \tag{5.82a}$$

$$p = 2(m_hk_BT/2\pi\hbar^2)^{3/2}exp[(E_v - F_h)/k_BT] \tag{5.82b}$$

This allows the recombination rate to be expressed in terms of the np product as follows

$$R_{sp}(\hbar\omega) = [2\pi/(\pi k_BT)^{3/2}]C(\hbar\omega, T)np(m_r/m_em_h)^{3/2}(\hbar\omega - E_g)^{1/2}exp[-(\hbar\omega - E_g)/k_BT] \tag{5.83}$$

where

$$C(\hbar\omega, T) = \frac{e^2\eta_r\hbar\omega}{2\pi\varepsilon_0\hbar^2m_0^2c^3}\left(\frac{2\pi\hbar^2}{k_BT}\right)^{3/2}\langle|p_{cv}|^2\rangle \tag{5.84}$$

In order to obtain the total spontaneous emission rate $R_{sp}(\hbar\omega)$ is to be integrated over all photon energies. Assume that $C(\hbar\omega, T)$ is slowly varying so that it may be treated as constant in the range over which $(\hbar\omega - E_g)^{1/2}exp[-(\hbar\omega - E_g)/k_BT]$ is appreciable; the integral takes the form $\int_0^\infty x^{1/2}exp(-x)dx = \sqrt{\pi/2}$

and thus

$$R_{sp} = \int\limits_0^\infty R_{sp}(\hbar\omega)d(\hbar\omega) = npC(\hbar\omega_{max}, T)\left(\frac{1}{m_e + m_h}\right)^{3/2} \qquad (5.85)$$

The photon energy at which the integrand is maximum is given by $\hbar\omega_{max} = E_g + k_BT/2$. Using $B = R_{sp}/np$ (Eq. (5.68)), the recombination lifetime is now expressed as

$$\frac{1}{\tau_r} = (n_0 + p_0 + \Delta n)\frac{(2\pi)^{1/2}e^2\hbar}{\varepsilon_0 c^3 m_0^2 (k_BT)^{3/2}}(\eta_r\hbar\omega)_{\hbar\omega_{max}}\frac{\langle|p_{cv}|^2\rangle}{(m_e + m_h)^{3/2}} \qquad (5.86)$$

In arriving at the previous equation, only one type of hole is considered. It is straight-forward to extend the calculation by including both the light and heavy holes and the expression is

$$\frac{1}{\tau_r} = (n_0 + p_0 + \Delta n)\frac{(2\pi)^{1/2}e^2\hbar}{\varepsilon_0 c^3 m_0^2 (k_BT)^{3/2}}(\eta_r\hbar\omega)_{\hbar\omega_{max}}\langle|p_{cv}|^2\rangle\sum_{v=h,l}\frac{[m_v/(m_e + m_v)]^{3/2}}{m_{hh}^{3/2} + m_{lh}^{3/2}} \qquad (5.87)$$

The excess carrier lifetime thus varies as $T^{3/2}$ and is inversely proportional to the band gap as for most III–V semiconductors $(2/m_0)\langle|p_{cv}|^2\rangle \approx 20eV$.

Example 5.9 *The recombination lifetime for GaAs is calculated by using $m_e = 0.067\,m_0$, $m_{hh} = 0.48\,m_0$, $m_{lh} = 0.09\,m_0$, $\eta_r = 3.6$, $E_p = 23\,eV$, $E_g = 1.424\,eV$, $\hbar\omega_{max} = 1.437eV$ at 300 K and injected carrier density. The calculated values are $B_r = 4.62 \times 10^{-10}\,cm^3 s^{-1}$ and $\tau_r = (B_r\Delta p)^{-1} = 2.2\,ns$.*

5.7 Non-radiative recombination

The non-radiative recombination in direct-gap semiconductors involves defects or occurs by Auger process. The excess carriers in indirect-gap group IV semiconductors like silicon decay mostly by nonradiative processes. In the following, these two processes are discussed briefly.

5.7.1 Recombination via traps

The theory of this process was developed by Shockley and Read (1952) and modified by Hall (1952). The model is referred to as the SRH model. The mechanism involves four electron and hole transitions as shown in Fig. 5.6. A trap level E_t first captures an electron (process 1) and then a hole is captured by the trap filled by the electron (process 3):

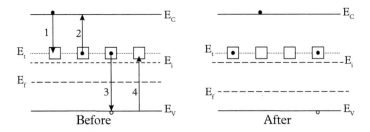

Figure 5.6 *Four basic steps in trap assisted recombination: (1) electron capture, (2) electron emission, (3) hole capture, and (4) hole emission.*

the direction of the arrow indicates the transition of an electron, and the hole moves oppositely. The combined result of electron and hole capture is a recombination of the pair. As shown in the right part of the diagram, the trap is free to capture another EHP after this. The inverse processes are the emission of an electron from the filled trap into the conduction band (process 2) and the emission of a hole from an empty trap into the valence band (process 4).

The rate of electron capture by the traps, R_{nc}, is proportional to the number of electrons and to the number of empty traps, so that

$$R_{nc} = C_n n (1 - f_t) N_t \tag{5.88}$$

Here N_t is the trap density, f_t is the occupancy function of the trap level, and C_n is the capture coefficient of the electrons expressed in terms of a capture cross section, σ_n, for electrons and the electron thermal velocity, $v_{thn} = (3k_B T/m_e)^{1/2}$, as

$$C_n = \sigma_n v_{thn} \tag{5.89}$$

The rate of electron emission from the traps, R_{ne}, is proportional to the number of filled traps, $f_t N_t$, and is given by

$$R_{ne} = e_n f_t N_t \tag{5.90}$$

where e_n is the emission probability of the electron. Under thermal equilibrium $R_{nc} = R_{ne}$, and hence from (5.88) and (5.90),

$$C_n n_0 = e_n f_{t0} / (1 - f_{t0}) \tag{5.91}$$

The equilibrium electron concentration, n_0, is expressed as

$$n_0 = N_c exp[(E_F - E_c)/k_B T] \tag{5.92}$$

and the ratio $f_{t0}/(1-f_{t0})$, when f_{t0} is the equilibrium occupancy of the trap level is expressed by using Fermi Dirac occupation function as:

$$f_{t0}/(1 - f_{t0}) = exp[-(E_t - E_F)/k_B T]$$ (5.93)

Here, E_F, E_c, and E_t, denote, respectively, the Fermi level, conduction band edge and the energy of the trap level. Using Eqs. (5.91) and (5.93), we may write

$$e_n = N_c exp[(E_t - E_c)/k_B T] = n_t C_n$$ (5.94)

The difference between electron capture and electron emission rates is given by

$$R_n = R_{nc} - R_{ne} = C_n N_t[(1 - f_t)n - f_t n_t]$$ (5.95)

Proceeding in a similar fashion, the difference between the hole capture and hole emission rates may be written as

$$R_p = R_{pc} - R_{pe} = C_p N_t[f_t p - (1 - f_t)p_t]$$ (5.96)

The capture coefficient for holes may be defined in terms of capture cross section and thermal velocity for holes by replacing σ_n in Eq. (5.89) by σ_p, and by expressing the thermal velocity in terms of hole effective mass, m_h. Under steady state, there is no accumulation of charge and hence, electrons and holes must recombine in pairs. Thus

$$R_p = R_n = R$$ (5.97)

where R is the recombination rate. Equating Eqs. (5.95) and (5.96), the occupation function may be expressed as

$$f_t = \frac{nC_n + p_t C_p}{C_n(n + n_t) + C_p(p + p_t)}$$ (5.98)

Substituting this in Eq. (5.95), one finds

$$R = \frac{pn - n_i^2}{\tau_{pl}(n + n_t) + \tau_{nl}(p + p_t)}$$ (5.99)

The electron and hole lifetimes are expressed as:

$$\tau_{nl} = 1/(v_{thn}\sigma_n N_t)$$ (5.100a)

$$\tau_{pl} = 1/(v_{thp}\sigma_p N_t)$$ (5.100b)

The recombination rates under special conditions may easily be obtained from Eq. (5.99). Consider p type materials, in which electrons are minority carriers. Then

$n << p \approx N_A$: the acceptor density. In this case, $p >> p_t$ and also $p >> n_t$. Eq. (5.99) now reduces to

$$R = \frac{n - (n_i^2/N_A)}{\tau_{nl}} = \frac{n - n_0}{\tau_{nl}} \tag{5.101a}$$

Similarly for n type materials

$$R = \frac{p - (n_i^2/N_D)}{\tau_{pl}} = \frac{p - p_0}{\tau_{pl}} \tag{5.101b}$$

It is noted that the SRH model presented in this section is applicable for describing the nonradiative recombination process via a single deep-level recombination center in the forbidden gap of a semiconductor. Treatment of the nonradiative recombination process via multiple deep-level centers in the forbidden gap of the semiconductor can be found in a classical paper by Sah and Shockley (1958).

5.7.2   Auger recombination

The Auger band-to-band recombination process is important in small bandgap semiconductors such as indium antimonide (InSb) and mercury cadmium telluride (HgCdTe) materials and in the presence of high carrier density as in degenerate semiconductors or under heavy injection. Auger recombination may occur in five different ways: (1) direct band-to-band, (2) phonon assisted, (3) trap assisted electronic transition, (4) trap-assisted hole transition, and (5) donor-acceptor pair related. The Auger recombination rate R_a is approximately written as

$$R_a = Cn^3 \tag{5.102}$$

where C is called the Auger coefficient, and n is the injected carrier density. The corresponding recombination lifetime is then expressed as

$$\tau_A = n/R_a = (Cn^2)^{-1} \tag{5.103}$$

In most cases, the effect of Auger recombination is taken into account by using the experimentally obtained values of C. In the following, we shall give a simplified picture of band-to-band Auger process.

Example 5.10 *The Auger coefficient $C = 10^{-28}$ cm^{6}s^{-1}. Taking the injected electron density $n = 10^{18}$ cm^{-3}, the Auger recombination lifetime becomes 10 ns. Using $B_r = 0.12 \times 10^{-10}$ cm^{3}s^{-1}, the radiative lifetime becomes 8.3×10^{-8} s.*

5.7.2.1   Band-to-band Auger effect

Three different types of band-to-band Auger process in a direct-gap semiconductor, namely (CCCH), (CHHS), and (CHHL)processes are shown in Fig. 5.7 left, middle,

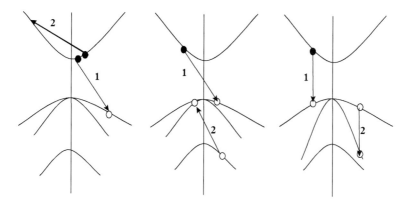

Figure 5.7 *(Left) Auger recombination (CHCC) process in n-type sample; (middle) CHHS process and (Right) CHHS process occurring in p-type sample.*

and right parts, respectively . The symbols C, H, L, and S, have their usual meanings. In the CCCH process, two electrons 1 and 2 in the conduction (C) band and one heavy hole (H) 1/ are involved. Here, electron 1 recombines with hole 1/; electron 2 absorbs the emitted photon and moves up to a higher state in the conduction band. The excited electron loses energy by phonon emission and eventually comes to a lower state to be in equilibrium with other electrons. It is easy to understand the other processes from Fig. 5.7, middle and right (b) and (c). The recombination is an inverse process of impact ionization, in which an energetic carrier generates an electron-hole pair.

CCCH Process   We outline a simple theory for the Auger rate of the CCCH process in which the states k_1 and k_2 are occupied, and states k_1', and k_2' are empty. The Auger rate should be proportional to the following occupation factor

$$P(k_1, k_2, k'_1) = f(k_1) f(k_2)[1 - f(k'_1)] \qquad (5.104)$$

The probability functions, assuming non-degenerate statistics, are

$$f(k_1) = (n/N_c)exp(-E_{ck_1}/k_B T); \ f(k_2) = (n/N_c)exp(-E_{ck_2}/k_B T); \ \text{and}$$
$$1 - f(k'_1) = (p/N_v)exp(-E_{vk'_1}/k_B T)$$

Here, n and p are, respectively, the electron and hole densities, and N_c and N_v are the effective DOS for the conduction and valence bands. The probability factor may now be written as

$$P(k_1, k_2, k'_1) = \frac{np}{N_c N_v} \frac{n}{N_c} \exp\left(-\frac{E_{ck_2} + E_{vk'_1} + E_{ck_1}}{k_B T}\right)$$

$$\approx \frac{n}{N_c} \exp\left(-\frac{E_g + E_{ck_2} + E_{vk'_1} + E_{ck_1}}{k_B T}\right) \tag{5.105}$$

Our aim is now to find the energy for which the exponent in Eq. (5.105) is maximum. Writing the energies in the exponent of the first term on the right-hand side in terms of wavevectors, we may write

$$\frac{k_1^2}{2m_e} + \frac{k_2^2}{2m_e} + \frac{{k'_1}^2}{2m_h} = \frac{1}{2m_e}[k_1^2 + k_2^2 + \mu {k'_1}^2]; \mu = \frac{m_e}{m_h} \tag{5.106}$$

It is easy to conclude that the probability factor will be maximum for the lowest energy values of the wavevectors k_1, k'_1 and k_2. Since k'_2 is the largest wavevector, we choose

$$k_1 + k'_1 + k_2 = -k'_2 \tag{5.107}$$

We also write $k_1 = ak'_1$ and $k_2 = bk'_1$ and obtain from Eq. (5.106)

$$k'^2_2 = (a^2 + b^2 + \mu)k'^2_1 + k_g^2; \quad \frac{\hbar^2 k_g^2}{2m_e} = E_g \tag{5.108}$$

Eq. (5.108) leads to the expression

$$k'_2 = (a + b + 1)k'_1 \tag{5.109}$$

Squaring this equation and eliminating k'^2_2 using Eqs. (5.107) and (5.108), we obtain

$$k'^2_1(1 + 2ab + 2a + 2b - \mu) = k_g^2 \tag{5.110}$$

The Auger rate is maximum when the following is minimized

$$k_1^2 + k_2^2 + \mu {k'_1}^2 = {k'_1}^2(a^2 + b^2 + \mu) \tag{5.111}$$

Using Eq. (5.108), this reduces to

$${k'_1}^2(a^2 + b^2 + \mu) = \frac{(a^2 + b^2 + \mu)}{(1 + 2ab + 2a + 2b - \mu)} k_g^2 \tag{5.112}$$

The quantity minimizes when $a = b = \mu$. The energy value for the initial state electron is then expressed as

$$E_{ck_1} = E_{ck_2} = \mu E_{vk'_1} = \left(\frac{\mu^2}{1 + 3\mu + 2\mu^2}\right) E_g \tag{5.113}$$

The maximum probability function is now written as

$$P(k_1, k_2, k'_1) = \frac{n}{N_c} exp\left(-\frac{1+2\mu}{1+\mu}\frac{E_g}{k_B T}\right) \tag{5.114}$$

The final state energy of the electron is

$$E_{ck'_2} = \frac{1+2\mu}{1+\mu}E_g \tag{5.115}$$

When $\mu \ll 1$, the previous energy may be approximately expressed as

$$E_{ck'_2} \approx (1+\mu)E_g \tag{5.116}$$

There is a decrease in the Auger rate once this threshold energy is exceeded, since the occupation factor decreases with increasing energy. The threshold depends on the band gap and the ratio of electron and hole masses. The Auger rate is changed due to strain as it changes the hole mass.

The general expression for the Auger rate for a fixed initial state, k_i is now given:

$$W_{Auger}(k_1) = 2\left(\frac{2\pi}{\hbar}\right)\left(\frac{e^2}{\varepsilon}\right)\left(\frac{1}{2\pi}\right)^6 \times \int dk_2 \int dk'_1 \int dk'_2 |M|^2 P(k_1, k_2, k'_1)\delta(E_{ck_1}$$
$$+ E_{ck_2} - E_{vk'_1} - E_{ck'_2}) \tag{5.117}$$

The matrix element M involves the screened Coulomb potential. The evaluation of the multiple integral is rather complicated asthe bandstructure must be known accurately. The other processes like direct CHHS, CHHL processes and phonon-assisted processes may be calculated by using proper band structures and matrix elements.

In most of the situation the cubic dependence on the carrier density of the Auger rate is employed and the value of the proportionality constant is obtained from the experimental data.

5.7.3   Other recombination

In this subsection recombination processes via surface states and through complexes are briefly described.

5.7.3.1   *Surface recombination*

The crystal structure is abruptly terminated at the surfaces and interfaces giving rise to a large number of electrically active states at the surface. In addition, a large number of impurities may be present at the surfaces and interfaces, since they are exposed during the device fabrication process. The result is an increase in recombination and the net recombination rate due to trap assisted recombination and generation is given by

$$R_s = \frac{pn - n_i^2}{p + n + 2n_i \cosh\left(\frac{E_i - E_{st}}{k_B T}\right)} N_{st} v_{th} \sigma_s \tag{5.118}$$

This expression is almost identical to that for SRH recombination. However, as the traps exist only at the surface, N_{st}, represents the two-dimensional (2D) density of traps.

Further simplification of Eq. (5.118) is possible. For example, for electrons in the quasi-neutral p-type region, $p>>n$ and $p>>n_i$ so that for $E_i = E_{st}$, it can be simplified to

$$R_{s,n} = R_{s,nc} - R_{s,ne} = v_s(n - n_0) \tag{5.119}$$

where the surface recombination velocity is expressed as

$$v_s = N_{st} v_{th} \sigma_s \tag{5.120}$$

5.7.3.2 Recombination of complexes

Bound EHPs, or excitons, play important role in the recombination processes. These processes will be discussed in Chapter 7.

5.8 Carrier effect on absorption and refractive index

Heavy doping in bulk material or a high level of injection by forward bias in a p-n junction create a large number of carriers in a semiconductor. As a result, changes in the band structure, absorption coefficient, and the refractive index occur in semiconductors. We shall discuss briefly the reasons for the changes and present expressions for the relevant parameters.

5.8.1 Band filling

Large carrier density decreases the fundamental absorption coefficient, most prominently in lower gap materials like InSb. The effect is known as Burstein–Moss effect. The electrons either injected or provided by the donors occupy the lower energy states in the conduction band, to block the transition of the electrons from the valence band. The same argument applies to absorption in p-type material, but due to larger DOS, more number of holes are needed to block the absorption. The effect is alternatively called as the band filling effect or sometimes Pauli blocking effect.

Assuming parabolic bands and **k**-conservation, the change in absorption at photon energy E due to band filling may be expressed as

$$\Delta\alpha(n, p, E) = \alpha(n.p.E) - \alpha_0(E) = \sum_{i=hh,lh} (C_i/E)(E - E_g)^{1/2}[f_v(E_{ai}) - f_c(E_{bi})] \quad (5.121)$$

where α_0 is the absorption coefficient without injection, C is the prefactor in Eq. (5.121), E_a is the energy level in the valence band, E_b is the energy level in the conduction band satisfying energy and momentum conservation, and n and p are, respectively, the injected electron and hole density.

The change in refractive index is similarly defined and is expressed in terms of $\Delta\alpha$ by using the following Kramers–Kronig relation

$$\Delta n_r(n.p, E) = \frac{2c\hbar}{e^2} P \int_0^\infty \frac{\Delta\alpha(n, p, E')}{E'^2 - E^2} dE' \quad (5.122)$$

where P stands for the principal value. To calculate the change in absorption for $n = p$, the Fermi levels are obtained first by numerical methods or by using approximate formula, and then the Fermi factors in Eq. (5.121) are evaluated. The calculated change in RI may be both negative and positive around the band gap. For InP below the band gap, the linear relation $\Delta n_r = -An$ holds good, where the density is in cm^{-3}. The values of A for different energies (given in parentheses) are 1.4×10^{-20} (1.2 eV), 7.7×10^{-21} (1.0 eV) and 5.6×10^{-21} (0.8 eV) (Bennet et al 1990).

5.8.2 Bandgap shrinkage

Large carrier density screens the lattice potential which an electron encounters.In a dense electron gas, the wavefunctions of adjacent electrons overlap, forming a gas of interacting particles. The electrons not only repel one another by Coulomb forces, but the electrons with the same spin also avoid each other. Thus the behaviour of an electron is controlled by other electrons: an effect known as correlation. These two effects lower the conduction band edge. A similar correlation effect for holes raises the valence band edge. The net effect is a band gap shrinkage, also known as the band gap renormalization.

The band gap shrinkage is expressed by the following relation:

$$\Delta E_g = - \left(\frac{e^2}{2\pi\varepsilon_0\varepsilon_s}\right) \left(\frac{3n}{\pi}\right)^{1/3} \quad (5.123)$$

where ε_s is the relative static permittivity of the semiconductor. The cube root dependence suggests that the shrinkage is proportional to the average interparticle spacing. The following modification by Bennett et al (1990) has been found to give good agreement with experiment

$$\Delta E_g(n) = \frac{\kappa}{\varepsilon_s}\left(1 - \frac{n}{n_{cr}}\right)^{1/3}; n_{cr} = 1.6\times10^{24}\left(\frac{m_e}{1.4\varepsilon_s}\right)^3 \qquad (5.124)$$

The critical carrier density n_{cr} is in cm^{-3}. The fitting parameter κ is 0.11 for holes, 0.125 for electrons, and 0.14 when both electrons and holes are present in equal numbers.

Example 5.11 *Taking $\varepsilon_s = 13\varepsilon_0$ and $n = (\pi/3) \times 10^{24}\ m^{-3}$ for GaAs, the band gap shrinkage is 22.2 meV from Eq. (5.124). $n_{cr} = 5.36 \times 10^{16}\ cm^{-3}$ for n-GaAs (see Bennett et al 1990).*

By assuming that the bandgap shrinkage causes a rigid shift of the absorption curve, the change in absorption is given by

$$\Delta\alpha(n, E) = (C/E)\left[(E - E_g - \Delta E_g(n))^{1/2} - (E - E_g)^{1/2}\right] \qquad (5.125)$$

The change in RI may then be calculated from Kramers–Kronig relation.

5.8.3   Free carrier absorption

The effect of free carrier on absorption and RI change has already been discussed in Section 6.6. The expression derived there should be modified to include the contributions from the heavy and light holes. Since the concentrations of heavy and light holes are proportional to their effective masses to the 3/2 power, the following expression is obtained by inserting the values of the fundamental constants (Bandyopadhyay and Basu 1992; Bennet et al 1990; Bottledooren and Baets 1989):

$$\Delta n_r = \frac{-6.9\times10^{-22}}{n_r E^2}\left[\frac{n}{m_e} + p\left(\frac{m_{hh}^{1/2} + m_{lh}^{1/2}}{m_{hh}^{3/2} + m_{lh}^{3/2}}\right)\right] \qquad (5.126)$$

5.9   Gain in semiconductors

In the macroscopic theory of optical processes presented in Chapter 3, it has been stated that a light beam falling on any material is usually absorbed and its intensity decays exponentially with distance, as expressed by Eq. (3.90). However, if a condition for population inversion is created, then instead of absorption, the electromagnetic radiation is amplified (see Eq. (3.91)). In arriving at this conclusion, a two-level atomic system is considered.

In this section, we develop the theory of absorption and gain in a real semiconductor at a finite temperature. Instead of sharp atomic levels, we take into account band picture, DOS functions, and Fermi occupation factors, and develop the expressions for absorption and gain. By doing so, we establish the condition for population inversion in a semiconductor. We express the gain coefficient in terms of the B coefficient. It is straightforward to relate the gain (absorption) coefficient with the absorption coefficient given by Eq. (5.37) in Section 5.3.2.

5.9.1   Absorption and gain

Consider an energy level E_2 in the conduction band and an energy level E_1 in the valence band of a direct gap semiconductor. For spontaneous emission from E_2 to E_1 to occur, the upper state is to be occupied and the lower state is to be empty (i.e. occupied by a hole). The Fermi occupational probabilities for the electrons in the conduction and valence bands are expressed by

$$f_c(E_2) = \{1 + exp[(E_2 - F_e)/k_B T]\}^{-1} \tag{5.127}$$

and

$$f_v(E_1) = \{1 + exp[(E_1 - F_h)/k_B T]\}^{-1} \tag{5.128}$$

where F_e and F_h are the respective quasi-Fermi levels (under non-equilibrium situation) for electrons and holes. The number of states in a small energy interval dE_2 in the CB is $S_c(E_2)dE_2$, where S_c is the DOS in the CB. Similarly $S_v(E_1)dE_1$ is the number of states in the small energy interval dE_1 around level E_1 in the VB, where S_v is the DOS in the VB. Assuming parabolic DOS functions, one may write:

$$S_c(E_2) = \frac{(2m_e)^{3/2}}{2\pi^2 \hbar^3}(E_2 - E_c)^{1/2} \tag{5.129a}$$

$$S_v(E_1) = \frac{(2m_h)^{3/2}}{2\pi^2 \hbar^3}(E_v - E_1)^{1/2} \tag{5.129b}$$

E_c and E_v are the respective band edge energies.

First consider stimulated emission from state E_2 to a state E_1. The rate depends on: (1) the transition probability per unit time expressed in terms of Einstein coefficient B_{21}, (2) the probability that E_2 is occupied, (3) the probability that E_1 is unoccupied, and (4) the photon density, $n_{ph}(\hbar\omega)$. Thus the rate of stimulated emission is

$$r_{21}(st) = B_{21}n_{ph}(\hbar\omega)S_c(E_2)S_v(E_1)\{f_c(E_2)[1 - f_v(E_1)]\} \tag{5.130}$$

Similar arguments give the rate of absorption from E_1 to E_2 as

$$r_{12}(abs) = B_{12}n_{ph}(\hbar\omega)S_c(E_2)S_v(E_1)\{f_v(E_1)[1 - f_c(E_2)]\} \tag{5.131}$$

The spontaneous emission rate involves Einstein's A coefficient and is written as

$$r_{21}(sp) = A_{21} S_c(E_2) S_v(E_1) \{f_c(E_2)[1 - f_v(E_1)]\} \tag{5.132}$$

In thermal equilibrium, the upward transition rate $[r_{12}(abs)]$ equals the total downward transition rate $[r_{21}(sp)+r_{21}(st)]$ and both the quasi-Fermi levels merge with equilibrium Fermi level. Following a similar calculation as given in Section 1.2.2, we obtain for the photon density:

$$n_{ph}(\hbar\omega) = \frac{n_r^3(\hbar\omega)^2}{\pi^2 \hbar^3 c^3 [exp(\hbar\omega/k_B T) - 1]} \tag{5.133}$$

It is assumed that $B_{12} = B_{21} = B$ and A_{21} is related to B by

$$A_{21} = \frac{n_r^3(\hbar\omega)^2}{\pi^2 \hbar^3 c^3} B \tag{5.134}$$

As already proved, the ratio A_{21}/B is equal to the number of EM modes per unit volume or mode density $m(E)dE$, and therefore

$$m(E)dE = \frac{n_r^3(E)^2}{\pi^2 \hbar^3 c^3} dE \tag{5.135}$$

It should be noted that the rates of radiative transitions equal the rate of absorption only under thermal equilibrium. In the non-equilibrium condition, the net stimulated emission rate is expressed as

$$r^0(st) = r_{21}(st) - r_{12}(abs) = B n_{ph}(E_{21}) S_c(E_2) S_v(E_1) \{f_c(E_2) - f_v(E_1)\} \tag{5.136}$$

It follows easily from Eq. (5.136) that the net stimulated emission rate becomes positive, i.e. $r^0(st) > 0$, when $f_c(E_2) > f_v(E_1)$. This leads to the condition

$$F_e - F_h \geq E_2 - E_1 > E_g \tag{5.137}$$

The previous condition, known as Bernard–Duraffourg condition (1961), states that the separation of the quasi-Fermi levels must exceed the band gap in order to make the rate of stimulated emission larger than the rate for absorption. A pumping scheme is needed to separate the quasi-Fermi levels in the p and n layers from the equilibrium condition, $F_e = F_h$. The most convenient way to accomplish this is to use a p-n junction, and to inject both electrons and holes in sufficient numbers by applying a forward bias, thereby creating a condition of population inversion in the junction.

The optical power gain coefficient per unit length, g, is defined as

$$\frac{dI}{dz} = gI \tag{5.138}$$

where I is the light intensity and z is the direction of propagation of the EM radiation. The intensity, the energy crossing per unit area per unit time, is expressed as

$$I = v_g \hbar \omega n_{ph}(\hbar \omega) \tag{5.139a}$$

where $v_g = d\omega/dk = c/\eta_r$ is the group velocity of light in the material. Using the time derivatives of both sides of Eq. (5.139a), we obtain

$$\frac{dI}{dt} = v_g \hbar \omega \frac{dn_{ph}(\hbar \omega)}{dt} = v_g \hbar \omega r^0(st) \tag{5.139b}$$

Using the relation $dI/dz = (dI/dt)(dz/dt)^{-1} = (1/v_g)(dI/dt)$, we may rewrite Eq. (5.139b) by using Eqs. (5.138) and (5.139a) as

$$\frac{dI}{dz} = gI = gv_g \hbar \omega n_{ph}(\hbar \omega) = \frac{1}{v_g}\frac{dI}{dt} \tag{5.140}$$

Comparison of Eq. (5.139b) with Eq. (5.140) gives

$$r^0(st) = v_g g n_{ph}(\hbar \omega) \tag{5.141}$$

The gain coefficient therefore is expressed by using Eq. (5.136) as

$$g = \frac{r^0(st)}{n_{ph}(\hbar \omega)}\frac{n_r}{c} = \frac{n_r}{c}BS_c(E_2)S_v(E_1)\{f_c(E_2) - f_v(E_1)\} \tag{5.142}$$

Time dependent perturbation theory may be used to express the Einstein coefficient B, the transition rate due to the stimulated emission of photons from state 2 to state 1, in the following form:

$$B = \frac{\pi e^2}{m_0^2 \varepsilon_0 n_r^2}\langle 1|\mathbf{p}|2 \rangle^2 \tag{5.143}$$

The momentum matrix element, denoted previously by $\langle 1|\mathbf{p}|2 \rangle$, connects states $|1>$ and $|2>$ in the valence and conduction bands, respectively, and $\mathbf{p}$ is the momentum operator, as already defined in Section 5.2.

Using the relation between the momentum operator and the position operator $\mathbf{r}$, given by $\mathbf{p} = m_0(d\mathbf{r}/dt)$, and assuming $r \propto exp(i\omega t)$, one obtains $\langle 1|\mathbf{p}|2 \rangle = m_0 \frac{d}{dt}\langle 1|\mathbf{r}|2 \rangle =$

$i\omega m_0 \langle 1|r|2 \rangle$. The B coefficient may therefore be written as

$$B = \frac{\pi e^2 \omega}{\varepsilon_0 \eta_r^2} \langle 1|r|2 \rangle^2 = \frac{\pi \omega}{\varepsilon_0 \eta_r^2} \mu^2, \; \mu^2 = \langle 1|er|2 \rangle^2 \qquad (5.144)$$

where μ is called the dipole moment. It is straightforward to show that

$$A_{21} = \frac{e^2 \eta_r \omega}{\pi m_0^2 \varepsilon_0 \hbar c^3} \langle 1|p|2 \rangle^2 = \frac{\eta_r \omega^3}{\pi \varepsilon_0 \hbar c^3} \mu^2 \qquad (5.145)$$

Eq. (5.136) has been derived by considering a pair of states E_2 and E_1, such that $E_2 - E_1 = \hbar\omega = E_{ph}$. There are a number of such pairs separated by the photon energy E_{ph}, in the two bands. The net stimulated emission rate is to be calculated by taking the sum over such pair of states. In the following, the conduction band state is denoted by E and the energy of the matching valence band state is given by $E_1 = E - E_{ph}$. The net stimulated emission rate is therefore expressed, by using Eqs. (5.136) and (5.143), as

$$R_{st}(E_{ph}) = \frac{\pi e^2 \hbar n_{ph}(E_{ph})}{m_0^2 \varepsilon_0 n_r^2 E_{ph}} \int_0^\infty \langle 1|p|2 \rangle^2 S_c(E) S_v(E - E_{ph})[f_c(E) - f_v(E - E_{ph})]dE \qquad (5.146)$$

The optical power gain coefficient can be derived similar methods described in this section and is given by

$$g(E_{ph}) = \frac{\pi e^2 \hbar}{m_0^2 \varepsilon_0 \eta_r c E_{ph}} \int_0^\infty \langle 1|p|2 \rangle^2 S_c(E) S_v(E - E_{ph})[f_c(E) - f_v(E - E_{ph})]dE \qquad (5.147)$$

Using Eq. (5.132), (5.134), and (5.143), the spontaneous recombination rate is expressed as

$$R_{sp}(E_{ph}) = \frac{e^2 n_r E_{ph}}{\pi m_0^2 \varepsilon_0 n_r \hbar^2 c^3} \int_0^\infty \langle 1|p|2 \rangle^2 S_c(E) S_v(E - E_{ph}) f_c(E)[1 - f_v(E - E_{ph})] dE \qquad (5.148)$$

It is interesting to note that if we put $f_v = 1$ and $f_c = 0$ in Eq. (5.147) and apply k-conservation condition, we may recover Eq. (5.37) for the absorption coefficient $\alpha(\hbar\omega)$.

Although the gain coefficient is calculated in general numerically, the qualitative nature of its variation with photon energy may be understood by using the following assumption for the quasi-Fermi levels:

$$F_e - E_c < 4k_BT \, and \, E_v - F_h < 4k_BT \tag{5.149}$$

The Fermi functions may then be expressed by the following linear relationships:

$$f_v(E) = \frac{1}{2} + \frac{F_h - E}{4k_BT} \, and \, f_c(E) = \frac{1}{2} - \frac{E - F_e}{4k_BT} \tag{5.150}$$

The integration in Eq. (5.148) may be performed analytically by considering parabolic DOS functions, leading to the following expression for the gain:

$$g(\hbar\omega) = K(\Delta F - \hbar\omega)(\hbar\omega - E_g)^2 \tag{5.151}$$

where all the constants are lumped into the prefact or K and $\Delta F = F_e$ -F_h.

Example 5.12 *Let us assume that the electron and hole densities injected into the active GaAs layer in a laser are n = p = 10^{24} m^{-3}. With m_e = $0.067m_0$ and m_h = $0.5 \, m_0$, the effective densities of states at T = 300 K are N_c=4.35×10^{23} m^{-3} and N_v = 8.87 × 10^{24} m^{-3}. The quasi-Fermi levels are F_e-E_c = 41.8 meV and E_v-F_h = 55.4 meV. Thus the range of photon energies for which laser action takes place is from 1.424 eV to 1.521 eV.*

Fig 5.8(a) shows how the band states are filled under heavy injection, and Fig. 5.8(b) shows how the gain spectra vary qualitatively for different injected carrier densities as parameter. The gain curve covers the range from E_g to quasi-Fermi level separation ΔF for a particular injection level, shows a maxima at $\hbar\omega$ = $(1/3)(E_g + 2\Delta F)$. The gain becomes negative (absorption) when $\hbar\omega$> ΔF. For increasing level of injection, the gain curve covers a wider range of photon energy, as the value of ΔF increases.

The maximum value of gain, g_{max}, for different injected carrier densities, as obtained from Fig. 5.8(b) is plotted in Fig. 5.8(c) against the corresponding carrier density.

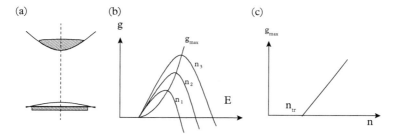

Figure 5.8 *(a) Band state filling under heavy injection. Shaded regions represent occupied states. (b) Variation of gain with photon energy with injected electron density as parameter. (c) Plot of maximum gain with injected carrier density; n_{tr} is the transparency carrier density.*

The maximum gain is zero below the transparency carrier density, n_{tr}, and then rises approximately linearly. However the actual variation is non-linear.

The position of the Fermi levels for a given carrier density is usually calculated numerically. Various approximate formulas are however available in the literature (Joyce and Dixon 1977; Nilsson 1973). Calculation of the quasi-Fermi level using one such formula is assigned as a problem in this chapter.

Thus far the states E_1 and E_2 are assumed to be sharp levels and the expression for gain, Eq. (5.147), has been derived accordingly. Scattering and other effects cause the levels to broaden. We modify now the expression for gain under level broadening. It is assumed that the energy states related to the transitions have an energy width $\hbar/\tau_{in}$, and that the spectral shape is Lorentzian and is given by

$$L(E) = \frac{1}{\pi} \frac{\hbar/\tau_{in}}{(E - E_{ph})^2 + (\hbar/\tau_{in})^2} \tag{5.152}$$

where τ_{in} is the relaxation time due to scattering and transitions. The gain coefficient is now modified to the following form:

$$g(E_{ph}) = \frac{\pi e^2 \hbar}{m_0^2 \varepsilon_0 \eta_r c E_{ph}} \int_0^\infty \langle 1|p|2\rangle^2 S_{red}(E)[f_c(E) - f_v(E - E_{ph})]L(E)dE \tag{5.153}$$

where S_{red} stands for reduced DOS which may be defined by replacing m_e by the reduced mass m_r in Eq. (5.129(a,b)).

The calculated gain spectra for an $In_{0.53}Ga_{0.47}As$ layer clad between two InP layers are plotted in Fig. 5.9 for different values of injected carrier densities. The peak gain

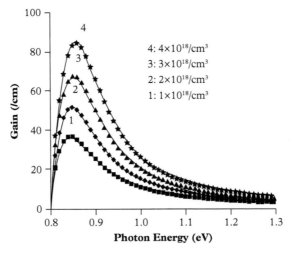

Figure 5.9 *Calculated gain spectra for InGaAs/InP for different injection carrier densities.*

Table 5.1 *Values of parameters for InAs and GaAs.*

Material	m_e/m_0	m_{hh}/m_0	m_{lh}/m_0	ε_s(F/m)	E_g(eV)	Δ(eV)
InAs	0.023	0.4	0.026	$15.15\,\varepsilon_0$	0.354	0.38
GaAs	0.0663	0.5	0.087	$13.1\,\varepsilon_0$	1.424	0.34

occurs at photon energies away from the gap energy and the gain peak shifts to higher photon energies with increased injection level. The values of the different parameters will be given. The values for the alloy are calculated by linear interpolation.

Example 5.13 *The maximum value of gain in an InP/In$_{0.53}$Ga$_{0.47}$As/InP DH laser for an injected carrier density of 4×10^{18} cm^{-3} is about 85 m^{-1}. The values of parameters used in the calculation are given in Table 5.1. The value is calculated with $\tau_{in} = 0.1$ ps.*

Problems

5.1 *Determine the energies of phonons involved in indirect band-to-band absorption in Si. Use the phonon spectra given in the literature.*

5.2 *Calculate the absorption coefficient of Si at 300 K using the empirical expression given by Bucher et al and quoted by Deen and Basu (2012).*

5.3 *Establish the equivalence of the two forms of perturbation Hamiltonian as given in Eq. (5.9).*

5.4 *Using energy and momentum conservations, prove that absorption from valence band to the degenerate conduction band starts at absolute zero at a photon energy*

$$\hbar\omega = E_g + (E_F - E_c)(1 + m_e/m_h).$$

5.5 *Calculate the value of squared momentum matrix element for GaAs using Eq. (5.35). Hence find the value of E_p.*

5.6 *Derive the general expression for the absorption coefficient taking into account the DOS functions and occupational probabilities of both the bands.*

5.7 *Assume that the occupational probabilities in Eq. (5.47) are governed by Maxwellian distribution. Show that the absorption coefficient is proportional to $exp[-(\hbar\omega - \Delta/k_B T)(m_s/m_s - m_h)]$ for a HH-SO transition.*

5.8 *Give reasons why the absorption spectra for IVBA is peaked. Prove that the peak occurs at a photon energy $(\hbar\omega)_{max} = \Delta + (3/2)k_B T[(m_s - m_h)/m_h]$.*

5.9 *Prove that, in an intraband free carrier absorption process, electrons must absorb or emit phonons in order to conserve momentum*

5.10   *Consider Eq. (5.60) for free carrier absorption. Assuming that the electron mobility is limited by deformation potential acoustic phonon scattering, comment how the free carrier absorption depends on temperature*

5.11   *Consider intraband free carrier absorption. Assume that the angle between the initial wave vector **k** and the light wave vector **κ** is θ. Show that the energy of the photon to satisfy energy and momentum conservation conditions may be expressed as*

$$\hbar\omega = 2\left[m^*\left(\frac{c}{n(\omega)}\right)^2 - \hbar k\left(\frac{c}{n(\omega)}\right)\cos\theta\right]$$

Take $m^* = 0.25\ m_0$, $n(\omega) = 3.5$ *and* $k = 0$. *Calculate the value of photon energy needed.*

5.12   *Starting from Eq. (5.60), valid for free electrons, obtain Eq. (5.62), the complete equation for RI change due to the free carrier (FCA) considering both electrons and holes.*

5.13   *Assume that the electron and hole densities at energies E_2 in CB and at E_1 in VB are given as $n(E_2) = A\exp[-(E_2 - E_c/k_BT)]$ and $p(E_1) = B\exp[-(E_v - E_1/k_BT)]$. Show that the spectral distribution of photon energy $\hbar\omega = E_2 - E_1$ is expressed as $P(\hbar\omega) = K\exp(\hbar\omega - E_g)\exp[-(\hbar\omega - E_g)/k_BT]$. A, B, and K are constants of proportionality.*

5.14   *Prove that the maximum recombination lifetime may be expressed as $\tau_r(max) = (2B_r n_i)^{-1}$ appropriate for an intrinsic semiconductor and that the radiative lifetime is reduced if the material is made n- or p-type.*

5.15   *Prove that for a p-type material the lifetime under low injection is given by $\tau_r = [B_r N_A]^{-1}$ where N_A is the acceptor concentration.*

5.16   *There is an upward shift of absorption threshold for high level of injection or due to heavy doping. Prove that the shift is proportional to $n^{1/3}$ at 0 K, where n is the electron density.*

5.17   *Derive the expression (5.123) assuming that the Fermi function is a box like that valid at T = 0 K.*

5.18   *Establish Eq. (7.11): the Bernard–Durrafourg condition.*

5.19   *Put $f_c = 1$ and $f_v = 0$ in Eq. (5.147) and apply **k**-conservation condition; show that it is possible to recover Eq. (5.37) for the absorption coefficient $\alpha(\hbar\omega)$.*

5.20   *Calculate $S_{c(v)}$ for $In_{0.53}Ga_{0.47}As$ using $m_e = 0.042m_0$, $m_{lh} = 0.0503m_0$, and $m_{hh} = 0.465m_0$. The injected electron and hole densities are $n = p = 10^{18}\ cm^{-3}$. Calculate the range of photon energies over which positive gain may be obtained at 0 K.*

5.21   *Derive the expression for the B coefficient as given by Eq. (5.144).*

5.22   *At low temperature, the Fermi functions are box like. Using this in Eq. (5.147) obtain the expression for gain.*

5.23 *Derive Eq. (5.151). Show that the maximum gain occurs at photon energy = (1/3)* $(E_g + 2\Delta F)$

5.24 *There is absorption instead of gain when the injection level is below the transparency carrier density. Using Eq. (5.151) give plots of absorption spectra for different injected carrier densities.*

5.25 *Work out the steps to show that the reduced DOS appears in Eq. (5.153) in place of $S_c \times S_v$ product in Eq. (5.147). Use energy and momentum conservation conditions to arrive at this result.*

5.26 *The maximum gain in a semiconductor is expressed as $g_{max} = \beta(\mathcal{J}/d - \mathcal{J}_0)$, where $\mathcal{J}$ is the current density, $\mathcal{J}_0$ is transparency current density, d is the thickness of active layer. Prove that the threshold condition in a double heterojunction laser.*

Reading List

Agrawal G P and Dutta N K (1986) *Long Wavelength Semiconductor Lasers*. NY: van Nostrand.

Anselm A I (1981) *Introduction to Semiconductor Theory*, M Samokhvalov (trans.). Moscow: Mir Publishers.

Basu P K (1997) *Theory of Optical Processes in Semiconductors: Bulk and Microstructures*. Oxford, UK: Clarendon Press.

Basu P K, Mukhopadhyay B, and Basu R (2015) *Semiconductor Laser Theory*. Boca Raton, FL: CRC Press (Taylor & Francis).

Bebb H B and Williams E W (1971) Photoluminescence I: Theory, Chapter 4 in *Semiconductors and Semimetals*. vol. 8. R K Willardson and A C Beer (eds.) pp. 181–320. New York: Academic Press.

Casey, H C Jr., and Panish M B (1978) *Heterostructure Lasers, part A, Fundamental Principles*. NY: Academic Press.

Chuang S L (2009) *Physics of Photonic Devices*. New York: John Wiley.

Coldren L A, Corzine S W, and Masanovi'c M L (2012) *Diode Lasers and Photonic Integrated Circuits*, 2nd edn. New York, NY, USA: Wiley.

Deen M J and Basu P K (2012) *Silicon Photonics: Fundamentals and Devices*. Chichester, UK: Wiley.

Nag B R (1980) *Electron Transport in Compound Semiconductors*. Berlin: Springer.

Numai T (2004) *Fundamentals of Semiconductor Lasers*. New York: Springer Verlag.

Ridley B K (2000) *Quantum Processes in Semiconductors*, 5th edn. Oxford: Clarendon Press.

Shur M (1990) *Physics of Semiconductor Devices*. Englewood Cliffs, NJ: Prentice Hall.

Singh Jasprit (2003) *Electronic and Optoelectronic Properties of Semiconductor Structures*. Cambridge, New York: Cambridge University Press.

Wang S (1989) *Fundamentals of Semiconductor Theory and Device Physics*. Englewood Cliffs, NJ: Prentice Hall.

Yariv A (1989) *Quantum Electronics*, 3rd edn. New York: Wiley.

Yu P and Cardona M (1995) *Fundamentals of Semiconductors*. Berlin: Springer.

References

Bandyopadhyay A and Basu P K (1992) A comparative study of phase modulation in In-GaAsP/InP and GaAs/AlGaAs based P-i-N and P-p-n-N structures. *Journal of Lightwave Technology* 10(10): 1438–1442.

Bennett B R, Soref R A, and Del Alamo J A (1990) Carrier induced change in refractive index of InP, GaAs, and InGaAsP. *IEEE Journal of Quantum Electronics* 26: 113–122.

Bernard M G A and Duraffourg G (1961) Laser conditions in semiconductors. *Physics Statistical Solutions* 1: 699–703.

Bottledooren D and Baets R (1989) Influence of bandgap shrinkage on the carrier induced refractive index change in InGaAsP. *Applied Physics Letters* 54: 1989–1991.

Hall R N (1952) Electron-hole recombination in Germanium. *Physical Reviews* 152: 387–387.

Joyce W B and Dixon R W (1977) Analytic approximations for the Fermi energy of an ideal Fermi gas. *Applied Physics Letters* 31: 354–356.

Nilsson N G (1973) An accurate approximation of the generalized Einstein relation for degenerate semiconductors. *Physica Status Solidi* A 19: K75–K78.

Sah C T and Schokley W (1958) Electron-hole recombination statistics in semiconductors through flaws with many charge conditions. Physical Review 109: 1103–1115.

Shockley W and Read W T (1952) Statistics of recombination of electrons and holes. *Physical Review* 87: 835–842.

Van Roosebroeck W and Shockley W (1954) Photon radiative recombination of electrons and holes in Ge. *Physical Review* 94: 1558–1560.

6

Optical processes in quantum wells

6.1 Introduction

The simplest semiconductor nanostructure is a Quantum Well (QW), in which the thickness along one direction is comparable to the de Broglie wavelength. In usual semiconductors, this is of the order of a few tens of a nanometre. In the order of increasing complexity, a Quantum Wire (QWR) comes next and then follows a Quantum Dot (QD)or a nanoparticle.

The understanding of optical processes in semiconductor QW forms the starting point in the subject Nanophotonics. The aim of the present chapter is to develop this background. The relevant theory follows the perturbative approach.

The basic approach followed in this chapter is to modify the theory of optical processes for bulk semiconductors developed in Chapter 5. In the analyses presented in Chapter 5, electrons and holes are described by Bloch functions having quasi-free motion along all three dimensions (3Ds). In addition, the proper density-of-states (DOS) and occupational probabilities of carriers in different bands are considered. In the present chapter, electrons and holes are assumed to be confined along a particular direction. This leads to quantized energy levels and modification of wave functions. It is then natural to extend the bulk model to accommodate the two-dimensional (2D) nature of the carriers, changes in E-**k** relationship, and of the DOS functions. The Fermi occupational probabilities are also modified due to the 2D nature of the electron (hole) gas. It is also necessary to consider the polarization dependence of the matrix element for transitions.

Various optical processes described in this chapter begin with the theory of fundamental band-to-band transitions from the valence band (VB) subbands to conduction band (CB) subbands, followed by intersubband transitions within the same band. The recombination processes are discussed next. Following this, the theory of gain coefficient in QW structure is developed. Some important mechanisms reducing gain in QWs, like intervalence band absorption, Auger recombination, etc., are then introduced. Strained QWs find their places in Photonics. The effects of strain on a few important optical processes are discussed.

In all the theoretical treatments developed in the present chapter only analytical methods are described. For more realistic and accurate analysis, more sophisticated methods,

Semiconductor Nanophotonics. Prasanta Kumar Basu, Bratati Mukhopadhyay, and Rikmantra Basu, Oxford University Press.
© P.K. Basu, B. Mukhopadhyay, R. Basu (2022). DOI: 10.1093/oso/9780198784692.003.0006

for example the multiband **k.p** method, and numerical methods have been and are being employed by different workers. Suitable references for numerical methods are included in the chapter, without giving details. Excitonic effects are discussed in Chapter 7.

6.2   Optical processes in quantum wells

The three basic optical processes, namely, absorption, spontaneous and stimulated emission, also occur in QWs. Taking the case of absorption, various processes may still be understood with the help of Fig. 5.1. The only difference is that a QW transitions takes place between two quantized energy levels. Fig. 6.1 illustrates some of the representative transitions occurring in a QW, both in real space and in the E-k diagram. The processes shown are (1) absorption from a heavy-hole (HH) subband to CB subband, (2) absorption from a light-hole(LH) subband to the CB subband, (3) intersubband absorption from a lower to a higher subband in the CB, and (4) absorption from a lower HH subband to a higher HH subband. Because of the 2D nature of the carriers, the theory developed for bulk semiconductors should be modified by considering 2D DOS functions, altered selection rules, and modified transition matrix elements. In the following subsections, we develop the theory of interband and intersubband absorption, recombination, and gain in QWs.

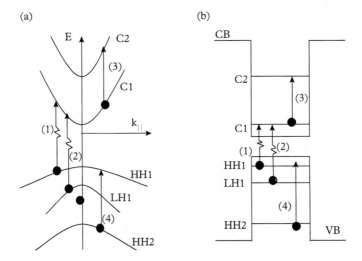

Figure 6.1 *Different optical processes occurring in a QW; (a) in E–k diagram; (b) in real space.*

6.3   Interband absorption

The theory of absorption in QWs is developed for a rectangular QW having type I band alignment. A common example is a gallium arsenide (GaAs) QW sandwiched between two $Ga_{1-x}Al_xAs$ layers. The simplified band diagram shown in Fig. 6.2 has abrupt band discontinuities ΔE_c and ΔE_v at the heterointerfaces, and the carrier energies are quantized due to restricted motion along the z-direction. Two subbands marked by E_{c1} and E_{c2} and two subbands E_{h1} and E_{h2} are shown in Fig. 6.2.

6.3.1   Absorption coefficient

The expression for absorption coefficient in a QW is developed following the method outlined in Section 5.2.2.2. For this purpose, the transition rate is calculated by using Eq. (5.13) after incorporating proper modifications. Here we use $r_{3D} = (r, z)$ and $k_{3D} = (k, k_z)$ to denote position and wavevectors, respectively, where the subscript 3D denotes the 3D vectors, r and k denote in-plane (x-y plane) position and wave vectors, respectively, and subscript z refers to the quantities along the z-direction, the direction of quantization. The absorption of photons occurs by transfer of an electron in the m-th HH subband with 2D wave vector k_h to a state of 2D wavevector k_e in the n-th subband in the conduction band. The wavefunctions are then written as

$$|1\rangle = |h, m, k_h\rangle = U_h(r, z)exp(i k_h \cdot r)\phi_{hm}(z) \qquad (6.1a)$$

$$|2\rangle = |c, n, k_e\rangle = U_c(r, z)exp(i k_e \cdot r)\phi_{cn}(z) \qquad (6.1b)$$

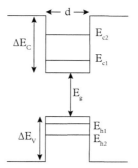

Figure 6.2 *Simplified band diagram of a rectangular QW showing a few subbands.*

where Us are the cell-periodic parts of the Bloch function and ϕ' s are envelope functions. The photon wave vector is written as $q_{3D} = (q, q_z)$, and using this the momentum matrix element may be written as

$$p_{12} = |\langle c, n, k_e | \exp(iq \cdot r + iq_z z) \hat{e} \cdot p | h, m, k_h \rangle| \tag{6.2}$$

Considering the r-integration and using the arguments presented after Eq. (5.8), we note that

$$k_e = k_h + q \approx k_h \tag{6.3}$$

If the momentum of the photon is neglected, which is justified, we note that the momentum is conserved in the QW layer plane. However, as the free motion of the particles is inhibited along the z-direction, there is no momentum conservation (selection rule) for that direction. The matrix element may therefore be written as

$$|p_{12}|^2 = \langle |p_{cv}|^2 \rangle_{QW} \delta_{k_e, k_h} C_{mn} \tag{6.4}$$

where C_{mn} is written as

$$C_{mn} = |\langle \phi_{hm} | \phi_{cn} \rangle|^2 = \left| \int \phi_{hm} \phi_{cn} dz \right|^2 \tag{6.5}$$

We write the momentum matrix element for transition between conduction and valence subbands in a QW as $\langle |p_{cv}|^2 \rangle_{QW}$, since it is different from the momentum matrix element for bulk semiconductors, as will now be shown.

The term C_{mn} occurs due to overlap between the envelope functions for electrons and holes in the respective subbands. In the ideal case where the potential barriers are infinite, both the envelope functions are either sin or cos functions. It is easy to prove then that $C_{mn} = \delta_{mn}$, where δ is the Kronecker δ-function. This simple selection rule becomes invalid in practical situations, since ΔE_c and ΔE_v have different finite values, and the effective masses are different in the well and the barrier. The envelope functions for electrons and holes decay in the barrier differently. In general, the value of C_{mn} is obtained by a numerical method, even for a rectangular QW. Approximate formulas for C_{mn} for narrow QWs are given by some workers (Asryan et al 2000)

We introduce a multiplying factor A_{mn} to relate squared momentum matrix elements for QW and bulk. Thus

$$\langle |p_{cv}|^2 \rangle_{QW} = A_{mn} \langle |p_{cv}|^2 \rangle_{bulk} \tag{6.6}$$

Using this and modifying Eqs. (5.13) and (5.15), the absorption coefficient for photon energy $\hbar\omega$ may be expressed as

$$\alpha(\hbar\omega) = \frac{2\pi\eta}{\hbar c}\left(\frac{eA_0}{m_0}\right)^2 A_{mn}C_{mn}\sum_{k_e,k_h} \langle|p_{cv}|^2\rangle_{bulk}\delta_{k_e,k_h}(f_e - f_h)\delta[E_e(k_e) - E_h(k_h) - \hbar\omega]$$

(6.7)

The **E-k** relationships for the final and initial states are written as

$$E(k_e) = E_g + E_{cn} + \frac{\hbar^2 k_e^2}{2m_e}$$

(6.8a)

$$E(k_h) = -E_{hm} - \frac{\hbar^2 k_h^2}{2m_h}$$

(6.8b)

The occupational probabilities are expressed as

$$f_e(E_e) = \{1 + exp[(E_{cn} + \hbar^2 k_e^2/2m_e - F_e)/k_B T]\}^{-1}$$

(6.9a)

$$f_h(E_h) = \{1 + exp[(E_{hm} + \hbar^2 k_h^2/2m_h - F_h)/k_B T]\}^{-1}$$

(6.9b)

where F_e and F_h are the respective quasi-Fermi levels. The double summation in Eq. (6.7) is now reduced to a single summation over k_h due to the **k**-conservation. Furthermore, the argument of the energy conserving δ-function is,

$$E(k_e) - E(k_h) - \hbar\omega = (E_g + E_{cn} + E_{hm}) + (\hbar^2 k_h^2/2m_r) - \hbar\omega$$

(6.10)

where m_r is the reduced mass. The sum over k_h in Eq. (6.7) is now converted into an integral by using the usual method:

$$\sum_{k_h} \rightarrow \frac{2S}{(2\pi)^2}\int \delta(E_{gmn} + \hbar^2 k_h^2/2m_r - \hbar\omega)2\pi k_h dk_h[f_h(k_h) - f_e(k_e)]$$

(6.11)

where S is the surface area and $E_{gmn} = E_g + E_{cn} + E_{hm}$ is the effective band gap in the QW. The integration in Eq. (6.11) is facilitated by the presence of the δ-function and the final expression takes the form:

$$\alpha(\hbar\omega) = \frac{e^2 \langle|p_{cv}|^2\rangle_{bulk}}{\varepsilon_0 m_0^2 c\hbar n_r \hbar\omega d}m_r C_{mn}A_{mn}(f_h - f_e)H(\hbar\omega - E_{gmn})$$

(6.12)

where $H(x)$ is the Heaviside step function.

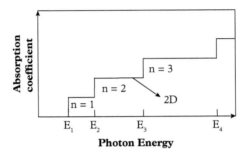

Figure 6.3 *Ideal step-like absorption spectra of QW.*

It is to be noted from Eq. (6.12) that the absorption coefficient is proportional to the joint DOS function. It also applies to the bulk semiconductor, as seen in Eq. (5.19). Therefore, while in the bulk $\alpha(\hbar\omega) \propto (2m_r/\hbar^2)^{3/2}(\hbar\omega - E_g)^{1/2}$, in QWs α is constant for one pair of subbands. As more and more pairs of subband are involved, α changes in a step-like fashion and its expected variation is shown in Fig. 6.3.

6.3.2 Polarization dependent momentum matrix elements

The momentum matrix element for optical transitions in a QW, $\langle|p_{cv}|^2\rangle_{QW}$ depends on the orientation between the electric field vector of the electromagnetic (EM) wave with respect to z, the direction of quantization. This may be understood by considering first that the electron wave vector **k** is parallel to the z-axis. We shall consider only HH and LH bands. If α and β denote, respectively, the spin-up and spin-down states, then the wavefunctions for the conduction band are written as

$$|s\alpha\rangle \text{ and } |s\beta\rangle \tag{6.13a}$$

The corresponding wavefunctions for the heavy hole are

$$|3/2, 3/2\rangle = \frac{1}{\sqrt{2}}|(x + iy)\alpha\rangle \tag{6.13b}$$

$$|3/2, -3/2\rangle = \frac{1}{\sqrt{2}}|(x - iy)\beta\rangle \tag{6.13c}$$

Similarly for light holes

$$|3/2, 1/2\rangle = \frac{1}{\sqrt{6}}|2z\alpha(x + iy)\beta\rangle \tag{6.13d}$$

$$|3/2, -1/2\rangle = \frac{1}{\sqrt{6}}|2z\beta(x - iy)\alpha\rangle \tag{6.13e}$$

Using these wavefunctions the momentum matrix elements between the conduction band and HH band for different axes may be expressed as

$$x\text{- axis}: \frac{1}{\sqrt{2}}\sqrt{3}M; \ y\text{- axis}: \pm i\frac{1}{\sqrt{2}}\sqrt{3}M; \ z\text{- axis}: 0 \qquad (6.14)$$

M is defined as

$$\sqrt{3}M = \langle s|p_x|x \rangle = \langle s|p_y|y \rangle = \langle s|p_z|z \rangle = \left(\frac{m_0}{m_e} - 1\right)^{1/2}\left[\frac{1}{2m_e}\frac{E_g(E_g + \Delta)}{E_g + (2/3)\Delta}\right]^{1/2} \qquad (6.15)$$

where m_0 is the free electron mass, m_e is the electron effective mass, E_g is the band gap energy and Δ is the split-off energy due to spin-orbit interaction. Due to the presence of the factor $\sqrt{3}$ the matrix element becomes M, when averaged over all directions

As already assumed, the z-axis is the axis of growth of the QW and xy is the layer plane. Assume also that the emitted light propagates along the $+x$ axis. The propagation is then with transverse electric (TE) transverse magnetic (TM) mode when the electric field vector E is along the $y(z)$ direction, as illustrated in Fig. 6.4(a). In the polar coordinate system shown in Fig. 6.4(b), the propagation vector $\mathbf{k}_{3D}$ makes an angle θ with the z-axis. For an infinite barrier QW, $\mathbf{k}_{3D}$ has discrete values of $k_z = (n\pi/L)$ and continuous values of k_x and k_y denoting the polar and azimuthal angles by θ and ϕ, respectively, as in Fig. 6.4(b), the momentum matrix elements between the conduction and HH bands are

$$x\text{-axis}: \frac{1}{\sqrt{2}}\sqrt{3}M(\cos\theta \ \cos\phi \mp i\sin\phi); \ y\text{-axis}: \frac{1}{\sqrt{2}}\sqrt{3}M(\cos\theta \ \sin\phi \pm i\cos\phi) \qquad (6.16a)$$

$$z\text{-axis}: -\frac{1}{\sqrt{2}}\sqrt{3}M \ \sin\theta \qquad (6.16b)$$

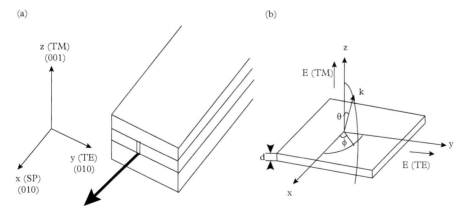

Figure 6.4 *(a) QW layer and electric field vector for TE polarization. (b) Polar coordinate system to obtain polarization dependence of matrix element.*

As the square of optical transition matrix element is proportional to $\langle 1|E \cdot p|2\rangle^2$, only the momentum matrix element with a component parallel to E contributes to the optical transition. We consider a wave vector k_n for a quantum number n, and average the square of the momentum matrix element over all the directions in the xy plane, by fixing the z-component k_{nz} of k_n. The following expressions are obtained for different transitions.

6.3.2.1   Conduction band-heavy-hole transition

For transverse electric(TE) mode, $E\|y$, and the average of the squared momentum matrix element is obtained by taking the second expression in Eq. (6.16a), and it is expressed as

$$
\begin{aligned}
\langle M^2 \rangle_{hh,TE} &= \frac{3M^2}{2} \frac{1}{2\pi} \int_0^{2\pi} (\cos^2\theta \sin^2\phi + \cos^2\phi)d\phi = \frac{3M^2}{4}(1 + \cos^2\theta) = \frac{3M^2}{4}\left(1 + \frac{k_z^2}{k^2}\right) \\
&= \frac{3M^2}{4}\left(1 + \frac{E_{z,n}}{E_n}\right)
\end{aligned}
$$

$$(6.17)$$

where $E_{z,n}$ and E_n, are respectively the quantized energy and the total energy of the n-th subband.

For TM mode $E \parallel z$, and using Eq. (6.16b), the squared average momentum matrix element is given by

$$
\langle M^2 \rangle_{hh,TM} = \frac{3M^2}{2}\sin^2\theta = \frac{3M^2}{2}(1 - \cos^2\theta) = \frac{3M^2}{2}\left(1 - \frac{k_z^2}{k^2}\right) = \frac{3M^2}{2}\left(1 - \frac{E_{z,n}}{E_n}\right)
$$

$$(6.18)$$

Since this vanishes at the subband edge where $E_{z,n} = E_n$, we conclude that optical transitions are allowed by TE modes only.

6.3.2.2   Conduction band-light-hole transition

The average of squared momentum matrix element for TE polarization is

$$
\langle M^2 \rangle_{lh,TE} = \frac{M^2}{4}(1 + \cos^2\theta) + M^2\sin^2\theta = \frac{M^2}{4}\left(1 + \frac{E_{z,n}}{E_n}\right) + M^2\left(1 - \frac{E_{z,n}}{E_n}\right) \quad (6.19)
$$

Similarly for TM mode

$$
\langle M^2 \rangle_{lh,TM} = \frac{M^2}{2}\sin^2\theta + 2M^2\cos^2\theta = \frac{M^2}{2}\left(1 - \frac{E_{z,n}}{E_n}\right) + 2M^2\frac{E_{z,n}}{E_n} \quad (6.20)
$$

Example 6.1 *Consider a GaAs QW with L = 10 nm having infinite barriers. Using standard parameters* $m_{ez} = 0.067\,m_0$, $m_{hhz} = 0.45\,m_0$, $m_{e\|} = 0.067m_0$, $m_{hh\|} = 0.112\,m_0$

$|p_{cv}|^2 = 2.7m_0E_g$, $C_{mn} = 1$, $A_{11} = 1.5$, $n = 3.6$, *the value of absorption becomes* $4.13{\times}10^5 m^{-1}$. *The effective band gap is 1.49 eV. It is assumed that* $f_e = 1$ *and* $f_h = 0$.

Example 6.2 *Let us assume that the kinetic energy of the electrons and holes are zero, so that* $\cos\theta = E_{cn}(\boldsymbol{k})/E_{cn} = 1$. *Putting this in Eq. (6.16b), we note that for the HH→CB transition* $A_{11} = 3/2$ *for TE polarization and* $A_{11} = 0$ *for TM polarization. The values agree with the values given in Eq. (6.16a).*

6.4 Intersubband absorption

Intersubband transitions involve two different subbands within the same band, either conduction or valence band. Fig. 6.1 shows two such processes: intersubband absorption in each of the CB and VB. Fig. 6.5 illustrates a few types of the intersubband absorption processes. In Fig. 6.5(a), the transition between the ground and the first excited subbands in the conduction band of the QW is shown. Another intersubband transition, from the lone subband in the CB of anarrow QWto the previous continuum states the hetero barrier is shown in Fig. 6.5(b). Fig. 6.5(c) shows a few intervalence subband transitions, like transitions between HH→HH, LH→LH subbands; HH→LH and HH→SO transitions are also possible. In Fig. 5.6(d), a transition from a HH subband to a conduction subband in type II QW is shown to provide an example of spatially indirect transition.

Intersubband transitions form the key operations in quantum cascade lasers (Basu et al 2015) and quantum well infrared photodetectors working at mid-infrared ranges.

Example 6.3 *Consider the values of subband energies given in Example 3.3. The difference is 169 meV and the wavelength of photon involved in the transition is 7.34 μm.*

Example 6.4 *the wavelength for intersubband transition between HH1 and HH2 in a Ge QW may be calculated by using Eqs. (3.37a) and (3.38a). The values of* $\gamma_1 = 13.38$ *and* $\gamma_2 = 4.24$ *for Ge. The effective mass perpendicular to the layer plane is* $= 0.204$

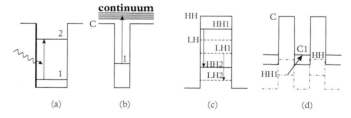

Figure 6.5 *Various intersubband transitions: (a) from subband 1 to subband 2 in the conduction band; (b) from subband 1 to continuum states; (c) between hole subbands; and (d) indirect transition in real space from hole subband to an electron subband in Type II QWs.*

m_0. With width of the well $d = 10$ nm, EHH1 $=18.46$ meV and EHH2 $= 4 \times 18.46$ meV. The wavelength of emission is $\lambda = 1.24.\ 103/3.18.46 = 22.3\ \mu m$.

6.4.1   Interconduction subband absorption: Isotropic mass

We outline the theory of interconduction subband absorption and the selection rules, assuming an ideal QW having isotropic electron mass, the common example being a GaAs QW. As usual, the QW is assumed to grow along the z-direction, the barrier height is infinite, and absorption takes place between two subbands, i and f, $f > i$ in the conduction band. The wavefunctions in these subbands may be written as

$$\psi_i(k, z) = (Sd)^{-1/2} U_{ci}(r) exp(ik \cdot r) \phi_i(z) \tag{6.21}$$

$$\psi_f(k, z) = (Sd)^{-1/2} U_{cf}(r) exp(ik \cdot r) \phi_f(z) \tag{6.22}$$

In the previous equations S is the area, d is the well width, $\mathbf{k}$ and $\mathbf{r}$ are 2D vectors, and ϕ' s are envelope functions. Note that ϕ' s vary slowly in comparison to the central cell function, U_c, assumed to remain unchanged for both the subbands, and the matrix element may be written as

$$\langle \psi_f | V_p | \psi_i \rangle = \langle U_{cf} | V_p | U_{ci} \rangle_{cell} \langle \phi_f | \phi_i \rangle + \langle U_{cf} | U_{ci} \rangle_{cell} \langle \phi_f | V_p | \phi_i \rangle \tag{6.23}$$

The matrix element and the scalar product denoted by the subscript *cell* in Eq. (6.23) are evaluated over a unit cell and V_p stands for the interaction potential.

Since the states denoted by i and f lie in the same band, the CB in this case, the first term is zero since ϕ_i and ϕ_f are the eigenfunctions of the same effective mass Hamiltonian and hence are orthogonal to each other.

Using **k.p** theory, the cell periodic functions Us are expressed as

$$U_{cnk} = U_{n0} + \frac{\hbar}{m_0} \sum_{m_0} \frac{k \cdot \langle m0 | p | n0 \rangle U_{m0}}{E_{n0} - E_{m0}} \tag{6.24}$$

These eigenfunctions are used to calculate the scalar product in the second term in Eq. (6.23). Noting that the effective mass in **k.p** theory is

$$\frac{1}{m_e} = \delta_{ij} + \frac{2}{m_0} \sum_{m_0} \frac{\langle m0 | p | n0 \rangle \langle n0 | p | m0 \rangle}{E_{n0} - E_{m0}} \tag{6.25}$$

the matrix element, as given by Eq. (6.23), is now expressed as

$$\langle \psi_f | V_p | \psi_i \rangle = -\frac{eA_0 \langle \phi_f | e \cdot p | \phi_i \rangle}{m_e} \tag{6.26}$$

We have used the standard form of V_p. Notice from Eq. (6.26) the appearance of the effective mass m_e. However, the matrix element for interband transitions involve free-electron mass m_0 as in Eqs. (5.19) and (6.6). The previous matrix element involving momentum may be converted into an expression involving position operator, and thus

$$\langle \psi_f | V_p | \psi_i \rangle = -\frac{eA_0(E_i - E_f)}{i\hbar} a \cdot \langle \phi_f | r | \phi_i \rangle \qquad (6.27)$$

Note that ϕ_i and ϕ_f are functions of z only and orthogonal to each other. Therefore, only the matrix element $\langle \phi_f | z | \phi_i \rangle$ becomes non-zero and $\langle \phi_f | x | \phi_i \rangle = \langle \phi_f | y | \phi_i \rangle = 0$.

Using the expression for A_0 and performing the usual mathematical work, the expression for absorption coefficient becomes

$$\alpha_{mn}(\hbar\omega) = \frac{\omega e^2}{dn_r \varepsilon_0 c} \sum_f (f_{ei} - f_{ef})^2 \langle z \rangle^2 \delta(E_f - E_i - \hbar\omega); \langle z \rangle = \langle \phi_f | z | \phi_i \rangle \qquad (6.28)$$

In Eq. (6.28), f_{ei} and f_{ef} denote, respectively, the Fermi occupational probabilities for electrons in the initial and final subbands, and the summation is over all the final states satisfying momentum and energy conservation. As usual, the summation is converted into integration over the 2D wave vector $\mathbf{k}$. The presence of the δ-function simplifies the integration, and the final expression for the absorption coefficient including a Lorentzian broadening with Γ as the linewidth reads

$$\alpha_{mn}(\hbar\omega) = \sum_{i,f} \frac{\omega e^2 m_e k_B T}{dn_r \varepsilon_0 c \hbar^2} \langle z \rangle^2 \ln \left\{ \frac{1 + \exp[(E_F - E_i)/k_B T]}{1 + \exp[(E_F - E_f)/k_B T]} \right\} \frac{(\Gamma/2)}{(\hbar\omega - E_{fi})^2 + (\Gamma/2)^2} \qquad (6.29)$$

In this expression, E_F is the Fermi level, $E_{fi} = E_f - E_i$. Now the ratio in the argument of the ln term is proportional to the difference in population in the two subbands $(N_{iu} - N_f)$. We rewrite the previous expression as

$$\alpha_{mn}(\hbar\omega) = \frac{\omega e^2}{\eta_r \varepsilon_0 c d} \langle z \rangle^2 (N_i - N_f) L(\hbar\omega, E_{fi}) \qquad (6.30)$$

where $L(\hbar\omega, E_{fi})$ stands for the Lorentzian lineshape given previously. The peak of the absorption occurs at E_{fi}.

The oscillator strength for transition from the ground subband to the first excited subband is expressed as $f_{osc} = (2m_0 E_{21}/\hbar^2)\langle z \rangle^2$. In a QW with an infinite barrier, $\langle z \rangle = 16d/9\pi^2$ (see Problem 6.12). The value of f_{osc} is $f_{osc} = (m_0/m_e)(256/\pi^2) = 14.34$, when $m_e = 0.067 \, m_0$ as in GaAs.

Example 6.5 *The oscillator strength for transition from the first excited subband (2) to the ground subband (1) is expressed as $f_{osc} = (2m_0 E_{21}/\hbar^2)\langle z_{21} \rangle^2$. The expression for $\langle z_{21} \rangle$ is given previously; we obtain $f_{osc} = (m_0/m_e)(256/\pi^2) = 14.34$, using $m_e = 0.067 \, m_0$ for GaAs.*

Example 6.6 *Consider a GaAs/Ga$_{0.7}$Al$_{0.3}$As QW with QW width of 6.5 nm. The values of subband energies are the same as in a GaAs QW of L = 10.13 nm with infinite barrier height. For an electron density of 1.6 × 10^{17} cm^{-3}, E$_1$ = 54.7 meV, E$_2$ = 218.8 meV, and E$_F$ = 6.49 meV. Taking half width Γ = 5 meV and expressing $\langle z \rangle^2$ in terms of f$_{osc}$, the peak absorption coefficient using Eq. (6.30) becomes 2.9×10^5 m^{-1}.*

6.4.2 Interconduction band absorption: Anisotropic mass

Although the fundamental interband absorption is very weak in indirect gap semiconductors, absorption from a lower subband to a higher subband in the CB of such semiconductors is quite appreciable. However, the theory in Section 6.4.1 needs some modification for Si, Ge, or SiGe alloys, because of the anisotropic nature of effective mass. We express the interaction potential between an electron and the radiation field as

$$V_p = (eA_0/m_0)e \cdot p \tag{6.31}$$

The matrix element for transition from a Bloch state ψ_k to another state $\psi_{k'}$ may be written as

$$M(k, k') = (ieA_0/\hbar)e \cdot \int \psi_{k'} (\nabla_k E_k) \psi_k d^3 r \tag{6.32}$$

where the wavefunctions may be written as in Eqs. (6.21) and (6.22). In a multivalley semiconductor, the E-k relationship is given by

$$E_k = \sum_{i,j} \frac{\hbar^2 k_i k_j}{2m_{ij}}, i,j = x, y, z \tag{6.33}$$

Using this, the matrix element defined by Eq. (6.32) may be written as

$$M(k, k') = (ie\hbar A_0)\delta_{k,k'} \left[\langle k_x \rangle \left\{ \frac{e \cdot i_x}{m_{xx}} + \frac{e \cdot i_y}{m_{xy}} + \frac{e \cdot i_z}{m_{xz}} \right\} + \langle k_y \rangle \left\{ \frac{e \cdot i_x}{m_{yx}} + \frac{e \cdot i_y}{m_{yy}} + \frac{e \cdot i_z}{m_{yz}} \right\} \right.$$
$$\left. + \langle k_z \rangle \left\{ \frac{e \cdot i_x}{m_{zx}} + \frac{e \cdot i_y}{m_{zy}} + \frac{e \cdot i_z}{m_{zz}} \right\} \right] \tag{6.34}$$

In Eq. (6.34), i_x and so on are the unit vectors along the respective directions, $\langle k_x \rangle = \langle \phi_f | k_x | \phi_i \rangle$, and $\langle k_y \rangle$, and $\langle k_z \rangle$ are similarly defined. As already noted, $\langle k_y \rangle = \langle k_z \rangle = 0$, since ϕ_f and ϕ_i are orthogonal. The matrix element thus reduces to

$$M(k, k') = -ie\hbar A_0 \langle k_z \rangle \delta_{k,k'} \left\{ \frac{e \cdot i_x}{m_{zx}} + \frac{e \cdot i_y}{m_{zy}} + \frac{e \cdot i_z}{m_{zz}} \right\} \tag{6.35}$$

For direct-gap materials, the nondiagonal elements in the effective mass tensor are zero. Therefore, the previous matrix element contains only the term involving m_{zz}, and introducing the isotropic mass m_e for m_{zz} in Eq. (6.35), we get back Eq. (6.26): the expression for the matrix element for isotropic effective mass. Eq. (6.35) reveals an interesting difference from the case of the isotropic mass. Here the xy polarized electric field of the radiation can also induce intersubband transitions through the terms involving m_{xz} and m_{yz}.

Writing $\langle k_z \rangle$ in terms of $\langle z \rangle$ and following the prescribed method, the expression for the absorption coefficient takes the form:

$$\alpha(\hbar\omega) = \frac{\omega e^2 \langle z \rangle^2 m_{zz}}{n_r \varepsilon_0 cL} \left\{ \frac{e \cdot i_x}{m_{zx}} + \frac{e \cdot i_y}{m_{zy}} + \frac{e \cdot i_z}{m_{zz}} \right\} (N_i - N_f) L(\hbar\omega, E_{fi}) \qquad (6.36)$$

An estimate of the absorption coefficient for different QW growth directions and for different polarization directions has been made by Yang et al (1989). It is mentioned that the value of α for a GaAs QW at 10 μm may be 4800 cm^{-1}, whereas for Si or Ge QWs the peak values may be 20,000 cm^{-1} for [100] growth and x-polarization and even 81,000 cm^{-1} for [111] growth and z-polarization. The experimental values obtained so far are indeed quite large ($\sim 4\times10^4$ cm^{-1}), at wavelengths near 5 μm. The transition energies depend on the strain generated due to lattice mismatch between the active layer and the substrate.

Example 6.7 *The intersubband absorption coefficient for conduction band electrons in Si oriented along the [100] plane will be calculated for L = 3 nm, for which the energy difference between two subbands E_{L2}–E_{L1} = 92 meV. This corresponds to a wavelength $\lambda = 1.24\times10^3/92 = 13.47$ μm. Putting these values in Eq. (6.36) and noting that m_{zz} = 0.92 m_0, m_{zx} = m_{zy} = 0, and taking Γ = 10 meV, N_i = 10^{16} m^{-2}, and N_f = 0, we obtain α = 4.69 × 10^5 m^{-1}.*

6.4.3   Intervalenceband absorption

Intervalenceband absorption (IVBA) is an important loss process in lasers. This may be understood from Fig. 6.6, in which a photon emitted by the transition from the conduction subband to the HH subband, (indicated by a) is reabsorbed by an electron in the split-off (SO) band to move into the HH band (*b* in the figure). In other words, a hole is transferred from the HH to the SO band. IVBA is more significant the less the bandgap is, for example, in InGaAsP alloy. It increases the threshold current of lasers.

The photon energy involved in the IVBA process may be expressed by using the energy and momentum conservations as

$$E_{IVBA} = m_s(\hbar\omega - \Delta_n)/(m_{hh} - m_s) \qquad (6.37)$$

where $\Delta_n = \Delta + E_{sn} - E_{hhn}$, Δ is the spin-orbit splitting energy in the bulk and E_{sn} and E_{hhn} are the n-th subband energies in the SO and HH bands, respectively, and m_s denotes the effective masses in the respective bands.

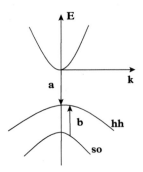

Figure 6.6 *Intervalence band absorption. The photon emitted through process a is reabsorbed in process b by the SO-HH transition.*

Calculation of the IVBA absorption coefficient, by assuming that the matrix element is independent of $\mathbf{k}$, follows the method outlined in Section 6.4.1 and the expression is

$$\alpha(\hbar\omega) = \frac{B}{d}\frac{m_s m_{hh}}{m_{hh} - m_s}\langle|p_{shh}|^2\rangle(f_s - f_{hh}) \tag{6.38}$$

where B is a constant of proportionality, p_{shh} is the momentum matrix element, and fs denote the Fermi occupation probabilities. To simplify assume that $f_s \sim 1$ and $(1-f_{hh})$ is described by the Boltzmann distribution. This allows us to write

$$\alpha_{shh}(QW) = \frac{e^2\omega}{c\eta_r\varepsilon_0 d\hbar^2}\frac{m_s m_{hh}}{m_{hh} - m_s}p\frac{\sum_n \langle z\rangle_n^2 exp\{[-E_{hhn} - E(\mathbf{k})]/k_B T\}}{\sum_n exp(-E_{hhn}/k_B T)} \tag{6.39}$$

In Eq. (6.39), $\langle z\rangle_n^2$ is the squared average dipole matrix element for a pair of subbands, expression terms represent the Boltzmann factor involving the energies of the pair of subbands, and p is the total hole density.

Calculations performed by Asada et al (1984) for InGaAs/InP QWs indicate that the IVBA is of the same order as in bulk.

The effect of reduction of in-plane hole mass on IVBA will be discussed in a later section of the present chapter in connection with strained QW lasers.

6.5   Recombination in quantum wells

The rate of spontaneous emission from a state of energy E_2 in CB to a state of energy E_1 in VB is given by

$$R_{sp}(\hbar\omega)d\hbar\omega = \frac{2\pi}{\hbar}\sum|p_{21}|^2 G(\hbar\omega)f_2(1 - f_1)\delta(E_{21} - \hbar\omega)d\hbar\omega \tag{6.40}$$

The wavefunctions for the n-th conduction subband and the m-th valence subband are given by Eq. (6.1a,b). The squared matrix element is given by Eq. (6.2). Using the expression for the optical DOS, $G(\hbar\omega)$, the spontaneous emission rate is given by

$$R_{sp}(\hbar\omega) = \frac{e^2 \eta_r m_r(\hbar\omega)}{\pi^2 m_0^2 \varepsilon_0 c^3 \hbar^4} \sum_{m,n} \langle |p_{cv}|^2 \rangle C_{mn} f_e(E_{cn} - F_e)[1 - f_h(E_{vm} - F_h)] \qquad (6.41)$$

where F_e and F_h are the quasi-Fermi levels. The emitted radiation is unpolarized and hence the factor $A_{mn} = 1$.

The total recombination rate is obtained by integrating Eq. (6.41) over all photon energies and thus

$$R_{sp} = B' \int f_e(E')[1 - f_h(E)] dE \qquad (6.42)$$

where B' is the prefactor. The occupational probabilities are expressed in terms of electron and hole concentrations, n and p respectively, in the following way:

$$f_e = \frac{n\pi\hbar^2}{m_e k_B T} \frac{1}{[exp\{(E - F_e)/k_B T\} + 1]} \qquad (6.43a)$$

$$f_h = \frac{p\pi\hbar^2}{m_h k_B T} \frac{1}{[exp\{(F_h - E)/k_B T\} + 1]} \qquad (6.43b)$$

The total recombination rate may be calculated from Eq. (6.42) using Eqs. (6.43a,b). However, a useful insight into the behaviour of recombination may be obtained by using non-degenerate statistics for the carriers. We may write

$$f_e(1 - f_h) \rightarrow exp[-(\hbar\omega - E_{gmn})/k_B T] exp[-(E_{gmn} - \Delta F)/k_B T]$$

where $\Delta F = F_e - F_h$. Inserting this in Eq. (6.41) and noting that the prefactor is a slowly varying function of $\hbar\omega$, the integral takes the form $\int_0^\infty exp(-x) dx = 1$. Noting also that the product $np \propto exp[-(E_{gmn} - \Delta F)/k_B T]$, the total spontaneous emission rate may be expressed as

$$R_{sp} = \frac{e^2 \eta_r m_r \hbar\omega}{\pi^2 m_0^2 \varepsilon_0 dc^3 \hbar^4} \sum_{m,n} \langle |p_{cv}|^2 \rangle C_{mn} \frac{(\pi\hbar^2)^2}{m_e m_h (k_B T)} np = Bnp \qquad (6.44))$$

The lifetime of excess carriers in the QW is defined as $\tau = (Bn_0)^{-1}$.

Example 6.8 *The recombination lifetime in a GaAs QW is estimated now. Take $m_e = 0.067$ m_0 and $m_{hh} = 0.112\ m_0$, so that $m_r = 0.042\ m_0$, $\eta_r = 3.6$ and $\langle |p_{cv}|^2 \rangle = 2.7 m_0 E_g$, a carrier density $n = 10^{12}$ cm^{-2}, and $C_{mn} = 1$. The calculated value of $B_{3D} = B_{2D}.d =$ 8.33×10^{-8} and $\tau = 1.2$ ns. The measured value is 6 ns, close to this estimate.*

The excess electron-hole pairs in a QWare usually created by injection or by photoexcitation. Let an intense photon pump of energy $\hbar\omega_p > E_{gmn}$ be incident on the sample to lift electrons from a valence subband to states well above the subband edge in the CB. The electrons and holes are essentially hot, and they give up their excess energy by longitudinal optical(LO) phonon emission to quickly relax to the bottom or top of the respective subbands. The excess electron-hole pairs then recombine to give the emission (photoluminescence in the case of photon absorption) spectra. From Eq. (6.41), the spectra depend on $f_e(1-f_h)$. The Fermi functions, as expressed by Eqs. (6.43(a) and (b)), use electron and hole temperatures T_e and T_h, in place of T. The expression for the recombination spectra takes the form

$$R_{sp}(y) \propto exp(-y/k_B T^*)H(y) \tag{6.45}$$

where

$$\frac{1}{k_B T^*} = \frac{1}{k_B T_e}\frac{m_h}{M} + \frac{1}{k_B T_h}\frac{m_e}{M}; \; y = \hbar\omega - E_{gmn}; \; M = m_e + m_h$$

and $H(y)$ denotes Heaviside step function. If $T_e = T_h = T^*$, the tail of the photoluminescence spectra gives the carrier temperature (Basu and Kundu 1985).

6.6   Loss processes in quantum wells

The threshold characteristics of a QW laser are determined by various loss processes within the well and in the barriers. Amongst them, the free carrier absorption, intervalence band absorption, and Auger processes play dominant roles. All these processes occurring in bulk semiconductors and double heterostructure (DH) lasers have been discussed in earlier chapters. In this section, we shall consider the IVBA and the Auger recombination and point out how the restricted motion in the QW modifies the relevant expressions for bulk semiconductors.

6.6.1   Intervalence band absorption

This process has been introduced in Section 6.4.3, and illustrated by SO→HH transitions. In addition, LH→HH and SO→LH processes are also possible. In all these processes, the emitted photon is reabsorbed by a carrier, thus contributing to an overall reduction of the number of photons.

6.6.2   Auger recombination

The Auger process plays the most detrimental factor in increasing the loss and thereby the threshold current density in InGaAsP-InP DH lasers. The same is true for InGaAs QW lasers.

Various band-to-band Auger processes in bulk semiconductors are shown in Fig. 5.7. The same figures can be used to illustrate the Auger processes in a QW. The differences are that the transitions occur between different subbands and that the electrons and holes have a 2D motion. In this subsection, only the theory for a CCCH process is developed to illustrate the method of calculation.

The modified expression for the Auger rate for a QW takes the following form:

$$R_a = \frac{2\pi}{\hbar} \left[\frac{1}{(2\pi)^2} \right]^4 \frac{1}{d} \iiint \int |M_{if}|^2 P(1, 1', 2, 2') \delta(E_i - E_f) dk_1 dk_2 dk_{1'} dk_{2'} \quad (6.46)$$

where R_a is the Auger recombination rate per unit volume, $|M_{if}|^2$ is the squared matrix element for transition from the initial (i) state to the final (f) state, and P denotes the difference of probabilities of the Auger process and the inverse process of impact ionization. The presence of well width d indicates that the rate is per unit volume. The in-plane wave vectors are k_1, k_2, $k_{1'}$ and $k_{2'}$

The squared matrix element is approximated as $|M_{if}|^2 \approx 4 \, |\bar{M}_{if}|$, where

$$M_{if} = \iint \psi_{1'}^*(r_1) \psi_{2'}^*(r_2) \frac{e^2}{4\pi\varepsilon|r_1 - r_2|} \psi_1(r_1)\psi_2(r_2) dr_1 dr_2 \quad (6.47)$$

and $\psi(r)$'s are the QW wavefunctions. We express the Fourier transform of the Coulomb interaction, which is assumed to be unscreened, as follows:

$$\frac{1}{|r_1 - r_2|} = \frac{1}{(2\pi)^3} \int \frac{4\pi}{q^2} exp[iq \cdot (r_1 - r_2)] dq \quad (6.48)$$

and obtain

$$\bar{M}_{if} = \frac{e^2}{\varepsilon} \frac{1}{(2\pi)^3} \int \frac{1}{q^2} I_{1'1}(q) I_{2'2}(q) dq \quad (6.49)$$

The general form of the integrals in Eq. (6.49) is

$$I_{mn}(q) = \int \psi_m^*(r)\psi_n(r) exp(iq \cdot r) dr \quad (6.50)$$

Note that the position coordinate $r = \rho, z$ and the wavefunctions in the QW have the form

$$\psi_m(r) = U_m(r)\phi_m(z) exp(ik_m \cdot \rho) \quad (6.51)$$

where U_m is the cell periodic part. The envelope function ϕ_m will be taken as sin function, as in an infinite barrier, and k_m and ρ are the 2D wave vector and 2D position vector, respectively. Eq. (6.50) is now

$$I_{mn}(q) = (2\pi)^2 \delta(-k_m + k_n + Q) F_{mn} G_{mn}(q_z) \quad (6.52)$$

where $q^2 = Q^2 + q_z^2$; $F_{mn} = \int U_m^*(r)U_n(r)dr$ and $G_{mn} = \int \phi_m^*(z)\phi_n(z)exp(iq_zz)dz$

The components of q in the plane and along the z-direction are denoted respectively by Q and q_z. The integral F_{mn} is over a unit cell of the crystal. The matrix element may now be written as

$$\bar{M}_{if} = \frac{e^2}{2\varepsilon}V_0\delta(k_1 + k_2 - k_{1'} - k_{2'}) \tag{6.53}$$

where

$$V_0 = 4\pi F_{11'}F_{22'}\int \frac{G_{11'}G_{22'}}{|k_1 - k_{1'}|^2 + q_z^2}dq_z$$

In writing the denominator of the previous equation, we have used

$$q^2 = Q^2 + q_z^2 = |k_1 - k_{1'}|^2 + q_z^2$$

Using Eqs.(6.46) and(6.53), the Auger rate is expressed as

$$R_a = \frac{8\pi}{\hbar}\left(\frac{1}{4\pi^2}\right)^4 \frac{1}{d}\left(\frac{e^2}{2\varepsilon}\right)^2 I \tag{6.54}$$

where

$$I = \iiint \int P(1,1',2,2')dk_1 dk_{1'}dk_2 dk_{2'}|V_0(k_1,k_{1'})|^2\delta(k_1 + k_2 - k_{1'} - k_{2'})\delta(E_i - E_f) \tag{6.55}$$

Using energy conservation condition we may write

$$\frac{\hbar^2}{2m_e}k_1^2 + \frac{\hbar^2}{2m_e}k_2^2 = \frac{\hbar^2}{2m_h}k_{1'}^2 - E_Q + \frac{\hbar^2}{2m_{et}}k_{2'}^2$$

where m_e and m_h are, respectively, the electron and hole effective masses at the band edges, m_{et} is the electron effective mass at energy $E_{2'}$, and E_Q is the effective gap in the QW.

To evaluate the integral I in Eq. (6.55), we introduce the following integration variables:

$$h = k_1 + k_2; \; j = k_1 - k_2$$

and obtain the following:

$$I = \frac{m_e}{2\hbar^2}\int\int dk_{1'}\, dk_{2'} \int_{-1}^{1} dt\, P(1,1',2,2')|V_0(k_1,k_{1'})|^2\,|j_0| \tag{6.56}$$

where $t = cos\theta$ and θ is the angle between $\mathbf{h}$ and $\mathbf{j}$. The integral is to be evaluated under the condition $j_0^2 \geq 0$, where

$$j_0^2 = 2\left(\beta_e k_{2'}^2 - \beta k_{1'}^2 - \frac{2m_e E_Q}{\hbar^2}\right) - |k_{1'} + k_{2'}|^2$$

$\beta_e = m_e/m_{et}$ and $\beta = m_e/m_h$. We further introduce two new variables

$$z_1 = k_{1'} + \frac{k_{2'}}{1 + 2\beta}; \; z_2 = k_{2'} \tag{6.57}$$

and obtain

$$f_0^2 = \alpha_e z_2^2 - z_1^2 (1 + 2\beta) - \frac{4m_e}{\hbar^2} E_Q \tag{6.58}$$

where $\alpha_e = \beta_e - [\beta/(1 + 2\beta)]$. To have $f_0^2 \geq 0$, one must have $\alpha_e z_2^2 > 2m_e E_Q/\hbar^2$. Using the relation $E_{2'} = \hbar^2 z_2^2/2m_e$, we obtain

$$E_{2'} > E_T = \beta_e E_Q/\alpha_e \tag{6.59}$$

Therefore, the Auger process takes place only when the energy $E_{2'}$ reachesa threshold energy E_T. When $E_Q \gg k_B T$, the inverse process of impact ionization is less proba- ble. Ignoring impact ionization and using non-degenerate distribution, the probability $P(1, 1', 2, 2')$ may be expressed as

$$P(1, 1', 2, 2') = \frac{n^2 p}{N_e^2 N_v} exp\left(\frac{-E_{2'} + E_Q}{k_B T}\right) \tag{6.60}$$

where n and p are the electron and hole concentrations, respectively, and N_c and N_v are the respective maximum DOS. The energy conservation relation is

$$E_1 + E_2 = E_{2'} - E_Q - E_{1'} \tag{6.61}$$

Eq. (6.60) indicates that the probability P decreases rapidly as $E_{2'}$ increases. The $|V_0|^2$ term in Eq. (6.56) is then replaced by its value at $E_{2'} = E_T$. Using Eqs. (6.57) and (6.60) in Eq. (6.56), we finally obtain

$$I = \frac{m_e}{\hbar^2} \frac{n^2 p}{N_e^2 N_v} |V_T|^2 exp\left(\frac{E_Q}{k_B T}\right) I_1 \tag{6.62}$$

where

$$I_1 = \iint dz_1 dz_2 exp\left(\frac{-\hbar^2 z_2^2}{2m_e k_B T}\right) = \frac{8\pi^2 \alpha_e}{1 + 2\beta}\left(\frac{2m_{et} k_B T}{\hbar^2}\right)^2 exp\left(-\frac{m_e}{m_{et}\alpha_e} \frac{E_Q}{k_B T}\right) \tag{6.63}$$

Here $|V_T|^2$ is the value of $|V_0|^2$ evaluated at E_T. Using Eqs. (6.55) and (6.62) we obtain

$$R_a \propto \frac{n^2 p}{N_e^2 N_v} exp\left(-\frac{\Delta E_c}{k_B T}\right) \tag{6.64}$$

where

$$\Delta E_c = E_T - E_Q = \frac{m_{et} E_Q}{m_h + 2m_e - m_{et}} \tag{6.65}$$

In the previous theory the carriers are assumed to obey non-degenerate distribution. Nevertheless, it is easy to conclude from Eqs. (6.64) and (6.65) that the Auger rate increases with decreasing band gap and with increasing temperature. Further, if the hole mass is reduced, ΔE_c will increase and the Auger rate will be lowered.

6.6.3   Free carrier absorption

As noted in Chapter 5, free electron or free hole absorption contribute to the loss process in a bulk material. The same is true for QWs also. Several authors have treated free carrier absorption in QWs and have given relevant expressions. The treatment becomes complicated as more than one subband is occupied. The reader is referred to a few publications related to free carrier absorption in QWs (see, e.g. Vurghaftman and Meyer 1999).

6.7   Gain in quantum wells

The most important application of QWs as photonic devices is in lasers. Today, in all practical systems including communication and networking, consumer products, illuminations, and many others, DH lasers are almost of no use. QW lasers, both strain free and strained, are invariably used in all these areas. In this section, the characteristics of QWs relevant for lasers and examples of gain spectra obtained analytically or otherwise will be presented for unstrained structures. Strained QWs will be treated in Section 6.8.

6.7.1   Gain characteristics

Under sufficient injection, a QW may be converted from an absorbing medium to a gain medium. It may be noted from Eq. (6.12) that the absorption coefficient becomes zero when $f_e = f_h$. This occurs for the transparency carrier density, n_{tr}, assuming charge neutrality condition, $n = p$. For carrier densities exceeding n_{tr}, $f_e > f_h$, and $F_e - F_h > \hbar\omega$, there is amplification, rather than absorption of the EM waves. It follows easily that the maximum gain occurs at $\hbar\omega = E_{gmn}$, and its expression is

$$g_{max} = g_0[f_e(E_{cn}) - f_h(E_{vm})] \tag{6.66}$$

where

$$g_0 = \frac{e^2 \langle |p_{cv}|^2 \rangle}{\varepsilon_0 m_0^2 c \hbar \eta_r \hbar \omega d} m_r C_{mn} A_{mn} \tag{6.67}$$

Example 6.9 *We calculate the value of maximum gain in a GaAs-AlGaAs QW, assuming $f_e-f_h = 1$ and $C_{mn} = 1$. The parameters chosen are $d = 5$ nm, $\hbar\omega = 1.45$ eV, in-plane masses: $m_e = 0.067$ m_0, $m_{hh} = 0.112$ m_0, $\eta_r = 3.6$, $A_{mn}\langle|p_{cv}|^2\rangle = m_0 E_p/4$, $E_p = 25$ eV (this corresponds to the TE polarized case). The reduced mass is $m_r = 0.042$ m_0. The calculated value from Eq. (6.67) is $g_0 = 9.2 \times 10^5$ m^{-1} or 9200 cm^{-1}.*

The quantities $f_e(E_{cn})$ and $f_h(E_{vm})$ are the Fermi-occupational probabilities of electrons at the subband edges. Under charge neutrality, both of these may be expressed in terms of injected carrier density, n. It is straightforward to write n in terms of f_e as follows:

$$n = \sum_{l=0}^{\infty} n_l \ln\left[1 + \frac{f_e}{1 - f_e}\exp(-\varepsilon_{cl})\right]; \; n_l = \frac{k_B T m_{el}}{\pi \hbar^2 d} \tag{6.68}$$

In Eq. (6.68), ε_{cl} denotes the energy of the lth subband and $\varepsilon_{c0} = 0$, and the energies are normalized by $k_B T$. For holes, a similar equation may be written replacing f_e by $[1-f_h]$, ε_{cl} by ε_{vl}, and n_l by p_l.

We may express f_e and f_h in terms of injected carrier density as (Vahala and Zao 1988)

$$f_e = 1 - \exp(-n/N_s) \tag{6.69a}$$

$$f_h = \exp(-n/P_s) \tag{6.69b}$$

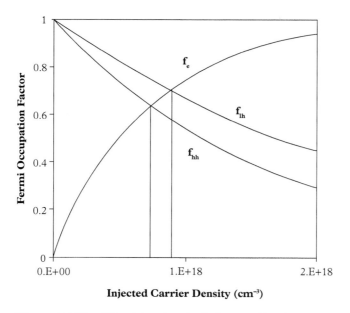

Figure 6.7 *Plot of Fermi functions for the lowest subbands.*

$$N_s = \sum_l n_l exp(-\varepsilon_{cl}) \qquad (6.69c)$$

$$P_s = \sum_l p_l exp(-\varepsilon_{hl}) \qquad (6.69d)$$

The validity criterion of the previous approximation is

$$\sum_{l=1} \frac{p_l exp(-\varepsilon_{vl})\{1 - exp(-\varepsilon_{vl})\}}{2P_s^2} p \ll 1 \qquad (6.70)$$

We may obtain a plot of f_e, f_{hh}, and f_{lh} as a function of injected electron density by using $m_e = 0.067\ m_0$, $m_{hh} = 0.15\ m_0$, $m_{lh} = 0.23\ m_0$, $d = 10$ nm, and $T = 300$ K, and assuming that a single subband each in CB and VB is occupied and the charge neutrality condition is obeyed. Fig. 6.7 shows the plots and the following conclusions may be drawn.

(1) Larger in-plane hole mass leads to $P_s > N_s$, even when a single subband is assumed to beoccupied. Since hole subbands are closely spaced, when more numbers of subbands are occupied, the actual value of P_s would be still larger than N_s. This explains why f_h decreases more slowly than the function f_e rises.

(2) At transparency $f_e = f_h$; the transparency occurs at a larger value of electron density.

(3) As in-plane mass of HH is lower, f_{hh} decreases more rapidly than f_{lh}. The transparency carrier density for HH subband is lower (7.4×10^{17} cm^{-3}) than the value for the LH subband ($9. \times 10^{17}$ cm^{-3}).

(4) A faster decrease of f_h with increasing injected carrier density may occur if the subband having lower in-plane effective mass is raised well above the other bands. In that case, a lower value of n_{tr} will result. This condition is achieved in strained QWs, that ensures alower threshold current in strained QW lasers.

(5) The effect of doping of the active layer is to reduce the value of n_{tr}. This may be understood by replacing n/N_s by $(n+N_D)/N_s$ and p/P_s by $(p+N_A)/P_s$, in Eqs. (6.69 (a–d)), where N_D and N_A are the donor and acceptor densities, respectively. The reduction is more pronounced for p-doping due to larger hole mass.

The effect of reduced in-plane mass on the position of quasi-Fermi level may be understood with Example 6.10.

Example 6.10 *Consider the expression fort 2D electron density at 0 K, $n_{2D} = (m_e/\pi\hbar^2)(F_e - E_0)$, Keeping n_{2D} unchanged, if the mass is decreased from m_{e1} to m_{e2}, the quasi-Fermi level is increased from F_{e1} to F_{e2}.*

We are now in a position to compare the ideal gain spectra for a DH and a QW laser. Fig. 6.8 shows the band diagram, gain spectra, and variation of maximum gain with injected carrier density for both DH and QW lasers. While the maximum gain for DH laser occurs at an energy exceeding the band gap, with its position shifting to higher

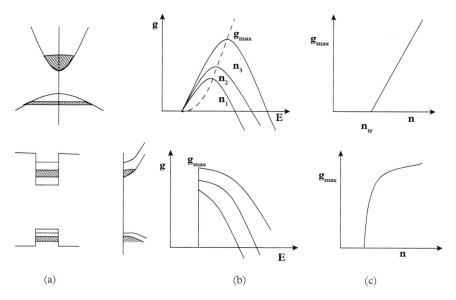

Figure 6.8 *(a) Band diagram, (b) gain spectra, and (c) g* $_{Max}$ *vs. n in a DH (top) and in a QW laser (bottom).*

energy with increasing n, the maximum gain for a QW always occurs at the effective gap energy. We may note the following differences as well.

(1) The emission starts at $E_{g,eff} = E_g(bulk) + E_{e1} + E_1$, that is at a higher energy in QW.

(2) In a QW, the peak gain g_{max}, always occurs at the effective band gap (Fig. 6.8(b)). The added injected carriers increase the value of peak gain. In a DH, added carriers shift the peak gain to higher energy and the carriers below g_{max} are wasted.

(3) In QWs, g_{max} saturates at some value of n, because all the available electron and hole states are fully inverted (see Fig. 6.8(c)). This is not the case in a DH, in which the peak gain increases monotonically with n.

(4) The transparency carrier density n_{tr} and the threshold current density are lower in QW lasers.

6.7.2 Model gain calculation: Analytical model

An analytical model for optical gain developed by Makino (1996) is presented first. It assumes parabolic E-**k** relation in all subbandsand in-plane momentum (**k**) conservation. The expression for the optical gain for transition from the n-th conduction subband to the n-th heavy hole subband is

$$g(\omega) = \omega\sqrt{\mu/\varepsilon}\sum_{n=0}^{\infty}(m_r/\pi\hbar^2 d)\int_{E_g+E_{cn}+E_{vn}}^{\infty}\langle R_{cv}^2\rangle(f_e - f_h)L(E_{cv})dE_{cv} \qquad (6.71)$$

In Eq. (6.71), ω is the angular frequency of light, μ and ε are the permeability and relative permittivity, m_r is the reduced mass, E_{cv} is the transition energy, and R_{cv}^2 is the dipole matrix element, $L(E_{cv})$ represents transition broadening, which is usually expressed by the following Lorentzian function

$$L(E_{cv}) = \frac{\hbar/\tau_{in}}{(E_{cv} - \hbar\omega)^2 + (\hbar/\tau_{in})^2} \tag{6.72}$$

where τ_{in} is the intraband relaxation time. The Fermi functions are given by

$$f_e = \frac{1}{[1 + exp(\varepsilon_{cn} - F_e)/k_B T]} \tag{6.73a}$$

$$f_h = \frac{1}{[1 + exp(\varepsilon_{vn} - F_h)/k_B T]} \tag{6.73b}$$

The symbols ε_{cn} and ε_{vn} represent the total energies of electrons and holes in the n-th subband. The densities of electrons and holes injected into the well may be related to the respective quasi-Fermi levels by the following expressions:

$$n = \frac{m_{ew} k_B T}{(\pi\hbar^2 d) \sum_n ln[1 + exp(F_e - E_{cn})/k_B T]} \tag{6.74a}$$

$$p = n = \frac{m_{hhw} k_B T}{(\pi\hbar^2 d) \sum_n ln[1 + exp(F_h - E_{vn})/k_B T]} \tag{6.74b}$$

The subscript w has been used to denote the effective masses in the well material.

6.7.2.1 *Approximate Fermi functions*

We rewrite the Fermi functions in terms of the ground state $(n = 0)$ energies as follows:

$$f_e = \frac{1}{[1 + exp\{(2\Delta E/\Gamma_e) - \chi_e\}]} \tag{6.75a}$$

$$f_h = \frac{1}{[1 + exp\{(2\Delta E/\Gamma_h) - \chi_h\}]} \tag{6.75b}$$

The symbols introduced previously are defined as

$$\Gamma_i = 2k_B T m_{iw}/m_r; \quad (i = e, hh) \tag{6.76a}$$

$$\chi_e = (F_e - E_{c0})/k_B T \tag{6.76b}$$

$$\chi_h = -(E_{v0} + E_g + F_h)/k_B T \tag{6.76c}$$

$$\Delta E = E_{cv} - E_{tr} \tag{6.76d}$$

$$E_{tr} = E_g + E_{c0} + E_{v0} \tag{6.76e}$$

The quantities χ_e and χ_h may be expressed approximately by

$$\chi_e = \ln[exp(n/N_s) - 1] \tag{6.77a}$$

$$\chi_h = \ln[exp(n/P_s) - 1] \tag{6.77b}$$

where N_s and P_s are given as

$$N_s = \frac{m_{ew} k_B T}{(\pi \hbar^2 d) \sum_n \ln[1 + exp(F_e - E_{cn})/k_B T]} \tag{6.78a}$$

$$P_s = \frac{m_{hhw} k_B T}{(\pi \hbar^2 d) \sum_n \ln[1 + exp(F_h - E_{vn})/k_B T]} \tag{6.78b}$$

The analytical expressions for the Fermi functions, given by Eqs. (6.75a) and (6.75b) depend on ΔE : the photon energy, and on carrier densities given by Eqs. (6.77a) and (6.77b).

6.7.2.2 Gain without broadening

If there is no broadening, the line shape function is given by

$$L(E_{cv}) = \pi \delta(E_{cv} - \hbar\omega) \tag{6.79}$$

In addition, if only the lowest subbands are taken into account, the gain expression given by Eq. (6.71) reduces to

$$g(\omega) = K(f_e - f_h) \tag{6.80}$$

The expression for K follows easily from Eq.(6.71) and ΔE in Eq. (6.76(d)) is given as

$$\Delta E = \hbar\omega - E_{tr}$$

Example 6.11 *The gain spectra of InGaAs QW with InGaAsP barrier for TE modes are shown in Fig. 6.9 for d = 8 nm. m_{ew} = 0.041 m_0 and m_{hhw} = 0.424 m_0. The subband energies are calculated by using the approximate formulas given by Makino (1996). The gain is plotted as a function of photon energy with injected carrier density as a parameter.*

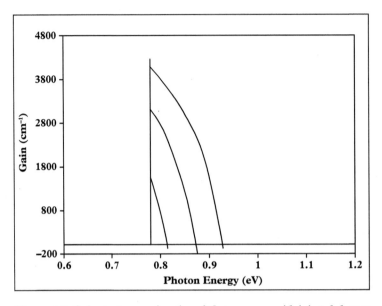

Figure 6.9 *Gain spectra as a function of photon energy with injected electron density as a parameter.*

Figs. 6.8 and 6.9 indicate that the maximum gain occurs at $\Delta E = 0$. The Fermi functions at the gain peak are obtained from Eq. (6.75(a,b)) by setting $\Delta E = 0$ and the expressions are;

$$f_{e0} = 1 - exp(-n/N_s) \qquad (6.81a)$$

$$f_{h0} = exp(-n/P_s) \qquad (6.81b)$$

These equations are originally developed by Vahala and Zah (1988).
  Here N_s may be expressed as

$$N_s = \sum_l n_l exp(-\varepsilon_{cl}) \qquad (6.82)$$

where the lth subband is located at ε_{cl} in units of $k_B T$ and $\varepsilon_{c0} = 0$. A similar equation may be written for P_s with f_e replaced by $1-f_h$ and ε_{cl} by ε_{vl} and n_l by p_l.
  Assume now that $f_{e0} - f_{h0} = Aln(n/N_0)$. The prefactor A is chosen such that at $n = eN_0$,
  $f_{e0} - f_{h0} = 1 - exp(-en/N_s) - exp(-en/P_s)$. With this we may write

$$g = g_0 ln(n/N_0) \qquad (6.83)$$

where

$$g_0 = K[1 - exp(-eN_0/N_s) - exp(-eN_0/P_s)] \qquad (6.84)$$

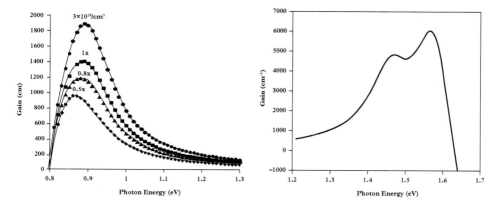

Figure 6.10 *Gain spectra of (a) InGaAs QW assuming one suband occupancy $\tau_{in} = 0.1$ Ps, and (b) of GaAs considering higher levels.*

The expression for gain Eq.(6.83) may be written in terms of the current density $\mathcal{J}$, by using the relation $\mathcal{J} = AN + BN^2 + CN^3$ and approximating this as $\mathcal{J} = B_{eff}N^2$, where B_{eff} is the effective recombination coefficient. It is now easy to show from Eq. (6.83) that

$$g = (g_0/2)\ln(\mathcal{J}/\mathcal{J}_0) \tag{6.85}$$

and

$$\Gamma g = \Gamma\xi\ln(\mathcal{J}/\mathcal{J}_0) \tag{6.86}$$

where Γ is the mode confinement factor, ξ is a proportionality constant and the transparency current density $\mathcal{J}_{tr} = B_{eff}N_0^2$. This form of modal gain has been used in a number of works.

6.7.2.3 *Effect of broadening*

The expression for gain considering Lorentzian broadening has also been derived analytically by Makino (1996). In Fig. 6.10, however, we present the gain spectra obtained for (a) InGaAs QW and (b) for a GaAs QW, by numerically integrating Eq. (6.71)and using complete form of the Fermi functions.

   Fig. 6.10(a) gives the plots of the gain spectra for InGaAs/InP system for different values of injection carrier densities assuming that all the carriers occupy the lowest subband. The calculated gain first increases with photon energy, reaches a peak and then decreases. With increased injected carrier densities, the peak gain increases and occurs at higher photon energies. Fig. 6.10(b) shows the gain spectra for a GaAs QW with $Al_{0.3}Ga_{0.7}As$ as the barrier. The 2D injected carrier density is 8×10^{12} cm^{-2}, which leads to a volume carrier density of 8×10^{18} cm^{-3} (d = 10 nm). In this case, since the second subband is occupied, the spectra contain two peaks corresponding to two subbands.

6.8   Strained quantum well lasers

The presence of strain in a layer induces a few important changes in the electronic properties. Use of strained QW as an active medium in lasers has some advantages over
unstrained QW lasers. In this section, some of the important features of strained layers will be discussed qualitatively. (See Agrawal and Dutta (1993); Arakawa and Yariv
(1985); Basu (1997); Basu, Mukhopadhyay, and Basu R (2015) in the Reading list; and
Nagarajan and Bowers (1999) for example.)

6.8.1   Gain spectra

The effect of strain is to change the band gap and in-plane effective masses. Considering in addition the effect of quantum confinement, the gain or absorption in a strained
QW may be calculated by using Eqs. (6.67) or (6.71), using proper in-plane masses. In
the following we give approximate expressions for the axial and in-plane masses under
quantization.

In the simplest situation at $\mathbf{k} = 0$, where $\mathbf{k}$ is the in-plane wavevector in the QW, the
HH and LH bands are decoupled. The subband energies for infinite barrier heights take
the form

$$E_n = \frac{\hbar^2}{2m_{iz}} \left(\frac{n\pi}{d}\right)^2, \quad i = hh, lh; n = 1, 2, 3, \dots. \tag{6.87}$$

The effective masses along the z-direction are given by

$$m_{hh(lh)z} = m_0/(\gamma_1 - (+)2\gamma_2) \tag{6.88}$$

The effective masses along the plane of the QW are expressed as

$$m_{hh(lh)\parallel} = m_0/(\gamma_1 + (-)\gamma_2) \tag{6.89}$$

Example 6.12 *Using the values of Luttinger parameters for GaAs:* $\gamma_1 = 6.85$ *and* $\gamma_2 = 2.1$,
we find $m_{hhz} = 0.38m_0$, $m_{lhz} = 0.09m_0$, $m_{hh\parallel} = 0.112m_0$ *and* $m_{lh\parallel} = 0.211m_0$.

Eqs. (6.87)–(6.89) and the previous example indicate that since $m_{hhz} > m_{lhz}$, the first
subband ($n = 1$) due to HH will be closest to the VB edge. However, the in-plane mass
for HH subbands will be lower than the in-plane mass for LH subbands. In spite of this,
the subbands will still be referred to as HH or LH subbands in conformity with the mass
value along the z-direction.

The in-plane dispersion relation for the hole subbands is, however determined by
considering the coupling between different bands, usually by employing multiband $\mathbf{k.p}$
perturbation theory.

The previous discussion suggests that proper calculation of gain in QWs, both strained and unstrained, must consider occupancy of different subbands and proper in plane dispersion relations.

6.8.2   Band gap tunability

The band gap of a semiconductor can be changed by applying either tensile or compressive strain. The active layers are usually grown on a thick buffer layer, which in turn is grown on a standard substrate. In this way, dislocations are confined in the vicinity of the interface between the substrate and the buffer. The lattice constant of the buffer layer is chosen in between those of the barrier B and active layer A. Accordingly, the layers A and B are either in tension or compression. A strain-symmetric multilayered structure can also be realized. In this way the band gap can be tuned over a wide range.

6.8.3   Reduced density-of-states

Fig. 6.11 shows the QW geometry we shall consider. The QW layer is grown along [001] axis, and the strain occurs along the same direction. This direction is denoted as the z-direction. The TE polarized wave has its electric field directed along y, while that for TM is along the z-direction.

Fig. 2.17 shows the band diagrams of a tensile strained layer, an unstrained layer, and a compressed layer. As may be noted, the degeneracy of HH and LH bands is removed by strain. The biaxial compression is a combination of a uniform hydrostatic and a uniaxial components. The hydrostatic component of stress increases the mean band gap, while the axial component splits the degeneracy and introduces an anisotropic band structure.

As noted already, in a QW laser the degeneracy of HH and LH bands is lifted, but due to close proximity of subbands the injected holes occupy a number of HH and LH subbands.As explained earlier in Section 6.7, the occupancy of more than one subbands results in a higher value of transparency carrier density. In the LH band the carrier density is lower, and the recombination lifetime is also shorter. Consequently, the threshold current density is due to equal contributions from the two bands, while in stimulated

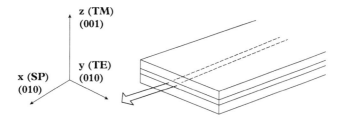

Figure 6.11 *Strained QW, electric field for TE and TM polarized waves is along y and z directions, respectively.*

emission only one band mainly contributes. Due to the isotropic nature of both polarization and gain in unstrained DH laser, both the threshold carrier and current densities increase. In contrast, when spontaneous emission and gain result out of contribution from only one polarization component, both the threshold carrier and current densities are reduced.

Following these arguments, the requirements for an ideal semiconductor laser are: (1) occupancy of carriers in only one valence band, (2) a low in-plane mass to achieve lower threshold current density, and (3) anisotropic polarization and gain above transparency; this means that the spontaneous emission and gain should be enhanced along the axis of the laser beam but suppressed in all other directions. All these requirements are satisfied by a strained QW laser, as will now be explained.

Consider first the effect of axial compressive strain. Its presence lifts the degeneracy of HH and LH bands. The highest valence subband has a lower in-plane effective mass due to which the transparency carrier density reduces (see also plots in Fig. 6.7 of f_e, f_h, and related discussion). The losses are also reduced as they depend on the number of injected carriers.

In tensile strained layers, the highest lying subband belongs to the LH band, but the LH and HH subbands shift in opposite directions due to quantum confinement. Hence, for some value of strain and well width, the energies of LH and HH subbands may be equal. The effective mass in this situation is large. With increasing tensile strain, however, the highest LH subbandis well above the HH subband, and again the DOS for LH states is reduced from the bulk value.

There is equal contribution from $|X\rangle$, $|Y\rangle$, $|Z\rangle$ like states in an unstrained bulk material. Carriers from two bands take part in spontaneous emission which has equal components of polarization along the three principal axes.Hence, the injected carriers make equal contribution to the TM gain (polarized along z), TE gain (polarized along y), and spontaneous emission (along x). This implies that only one third of all the holes at the correct energy are in the right polarization state to contribute to the laser mode.

Compressive strain pushes up the HH states, and the wavefunctions there have no $|Z\rangle$ character, but equal $|X\rangle$ and $|Y\rangle$ characters at $k_\parallel = 0$. Therefore, the TE gain is enhanced and TM gain is suppressed. Out of two injected carriers now, one contributes to the dominant polarization and in consequence, the threshold current decreases and the differential gain increases. With biaxial tension, the highest LH states have $(2/3) |Z\rangle$ character. In this case, two out of three injected carriers contribute to the TM gain. For lower strain, there is some coupling between the LH and HH states, but when the strain is large, the subbands separate out again.

6.8.4   Reduced intervalence band absorption

The IVBA process has been introduced in Section 6.4.3, and it represents an important loss process for low band gap materials used in long wavelength lasers. The energy of

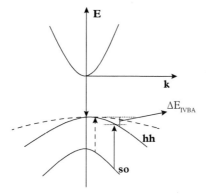

Figure 6.12 *Reduction of IVBA with reduced in-plane mass.*

the electron involved in IVBA is expressed as

$$E_{IVBA} = \frac{m_{so}(E_g - E_{so})}{(m_{hh\parallel} - m_{so})}$$ (6.90)

where the symbols have the usual meaning. Since the in-plane mass $m_{hh\parallel}$ is reduced under strain, the energy E_{IVBA} becomes larger, which in turn reduces the loss term, as the absorption coefficient depends on energy as $\exp(-E/k_B T)$. The net result is a decrease in the threshold current due to lower IVBA. The increase in energy ΔE_{IVBA} in the strained layer is illustrated in Fig. 6.12.

6.8.5   Auger recombination rate

Eq. (6.64) indicates that the Auger recombination rate depends on the densities of three types of carriers involved. The current density $\mathcal{J}_{aug}$ needed to compensate the Auger loss in a device for anundoped active region is written as

$$\mathcal{J}_{aug} = C_{AR}(T)n_{th}^3$$ (6.91)

where $C_{AR}(T)$ is the temperature dependent Auger coefficient and n_{th} is the threshold carrier density. It is usually assumed that the carrier density in the active layer is pinned at threshold. In the presence of strain, $\mathcal{J}_{aug}$ is reduced due to(1) decrease of n_{th} due to reduced hole mass and (2) strain-induced reduction of $C_{AR}(T)$. To understand the reduction, a simple parabolic band using Boltzmann statistics is considered to obtain the following expressions

$$C_{AR}(T) = C_0 \exp(-E_a/k_B T)$$ (6.92)

$$E_a(CHCC) = m_e E_g / (m_e + m_{hh}) \tag{6.93}$$

$$E_a(CHSH) = m_{so}(E_g - E_{so}) / (2m_{hh} + m_e - m_{so}) \tag{6.94}$$

It is evident that a lower value of m_{hh} increases E_a in both (CHCC) and (CHSH) processes and reduces the coefficient $C_{AR}(T)$.

To estimate phonon-assisted Auger recombination rate, a coefficient $C_p(T)$ replaces $C_0(T)$, where

$$C_p(T) = \frac{B}{e^x - 1} \left[\frac{1}{(E_1 + \hbar\omega_{LO})^2} + \frac{e^x}{(E_1 - \hbar\omega_{LO})^2} \right] ; \ x = \hbar\omega_{LO}/k_B T \tag{6.95}$$

B is proportional to the transition matrix element, $\hbar\omega_{LO}$ is the longitudinal optical (LO) phonon energy, and E_1, the kinetic energy associated with the forbidden intermediate state, is expressed as

$$E_1 = (m_{so}/m_{hh})(E_g - E_{so}), \quad \text{CHHS} - \text{ph} \tag{6.96a}$$

$$E_1 = (m_e/m_{hh})E_g, \quad \text{CHCC} - \text{ph} \tag{6.96b}$$

From Eq. (6.95), $C_p(T)$ is proportional to m_{hh}^2 if $\hbar\omega_{LO} \ll E_1$, and its value is reduced due to strain. However, this reduction is less significant than in the direct Auger process. The phonon assisted Auger process may not be a significant loss mechanism in long wavelength lasers.

Example 6.13 *We estimate the change in $C_{AR}(T)$ for the CHHS process, which is the most dominant Auger recombination process in strained InGaAsP QWlaser. Using $m_e =$ 0.037 m_0, m_{so} = 0.12 m_0, E_g = 0.79 eV, E_{so} = 0.39 eV, E_1 is 0.0364 eV for in-plane hole mass of 0.7 m_0. As the hole mass decreases to 0.15 m_0 due to strain, E_1 has a higher value of 0.2212 eV. The coefficient $C_{AR}(T)$ decreases from 2.5 × 10^{-1} to 2.02 × 10^{-4}: a decrease by three orders of magnitude (see Zhao B and Yariv A (1999) in Reading List).*

6.8.6 Temperature sensitivity

The threshold current density in a laser varies with temperature around room temperature as

$$\mathcal{J}_{th}(T) = \mathcal{J}_0 exp(T/T_0) \tag{6.97}$$

where $\mathcal{J}_0$ is an empirical constantand T_0,the characteristic temperature, may be expressed from Eq. (6.100) as

$$T_0 = \left[\frac{d}{dT}ln\{\mathcal{J}_{th}(T)\}\right]^{-1} = (T_2 - T_1)\left[ln\frac{\mathcal{J}_{th}(T_2)}{\mathcal{J}_{th}(T_1)}\right]^{-1} \tag{6.98}$$

A high value of T_0 is needed for stable operation of laser transmitter circuits. Typically $T_0 = 75$ K for long wavelength lasers, while for GaAs-GaAlAs systems, a higher value, ~ 300 K has been reported. The temperature dependence of threshold current is governed by the Auger process. A reduction of the Auger rate is highly desirable for increasing T_0.

The threshold current density is expressed as

$$\mathcal{J}_{th} = (A_{nr} + SA)n_{th} + B(T)n_{th}^2 + C_{AR}(T)n_{th}^3 \tag{6.99}$$

In Eq. (6.99) A_{nr} is the trap-assisted non-radiative recombination coefficient, S denotes surface recombination coefficient, A is the surface area and n_{th} the carrier density at threshold. $B(T)$ represents the temperature dependent radiative recombination coefficient. The last term is the non-radiative Auger recombination component in which C_{AR} denote the Auger coefficient. For an ideal QW structure, $B(T)$ decreases as n_{th} increases with temperature. We assume $n_{th}(T) \propto T^{1+x}$ ($x > 0$). The non-ideality factor, x, appearsdue to the occupancy of higher subbands, carrier spillover from the well into the barrier or optical confining layer, and intervalence band absorption. Considering the temperature dependence of C_{AR}, one may write

$$T_0 = \frac{T(1 + \mathcal{J}_{AR}/\mathcal{J}_R)}{1 + 2x + (3 + 3x + E_1/k_BT)(\mathcal{J}_{AR}/\mathcal{J}_R)} \tag{6.100}$$

Taking the ideal values, $\mathcal{J}_{AR}/\mathcal{J}_R = 0$, $E_1 = 0$, and $x = 0$, $T_0 = 300$ K, a value close to this has been observed in GaAs/GaAlAs QW system. However, presence of Auger recombination reduces the value considerably.

Example 6.14 *Putting $\mathcal{J}_{AR}/\mathcal{J} = 3$, $x = 0$ and $E_1 = 0$ in Eq. (6.100) $T_0 = 120$ K. Assuming $\mathcal{J}_{AR} >> \mathcal{J}_R$, the maximum value of T_0 is 100 K, even if $x = 0$ and $E_1 = 0$. Usually both E_1 and x are non-zero and therefore T_0 has a still lower value.*

The reported value for InGaAsP DH lasers is quite low, ~ 40 K. The small increase of the value for both compressive and tensile strained layers may be attributed to the reduction of Auger recombination. However, T_0 values exceeding 100 K are not reported.

Problems

6.1   *An approximate but accurate method of calculating subband energies in a GaAs QW has been presented by Mathieu et al (1992) in the Appendix A. Follow the method and use Eq. (A5) to calculate subband energies vs. well width for different composition x in $Ga_{1-x}Al_xAs$.*

6.2   *Assuming that momentum is conserved along the QW layer plane, work out the steps to prove that the absorption coefficient is given in terms of the reduced mass and follows a step-like variation.*

6.3   *Prove the equivalence of Eqs. (6.66) and (6.71): two expressions for gain expressed in terms of momentum and dipole matrix elements and the condition to be satisfied to get the equivalence.*

6.4   *The smaller in-plane mass leads to higher quasi-Fermi level. Prove this statement by using the expression for 2D carrier density at 0 K. Prove this statement for any finite temperature also.*

6.5   *The 2D electron density is 5×10^{11} cm^{-2}. Calculate the change in quasi-Fermi level if the in-plane mass is decreased from $0.1\ m_0$ to $0.05\ m_0$.*

6.6   *Using the linear approximation for the Fermi functions as given in Ch.05, show that the gain in QW may be expressed as $g(\hbar\omega) = K(\Delta F - \hbar\omega)H(\hbar\omega - E_{mn})$, where H is a step function and E_{mn} is the effective bandgap between the m-th conduction and the n-th valence subbands. Sketch the gain curves for different injected carrier densities.*

6.7   *Calculate the overlap integral for a parabolic QW in which the envelope functions are given by $\phi_n(z) = (\alpha/\sqrt{\pi 2^n n!})^{1/2} exp(-\alpha^2/2)zH_n[\alpha(z)]$.*

6.8   *Calculate the absorption coefficient in a GaAs QW with L = 10 nm and infinite barrier height at an energy E_{g11}. Take $C_{11} = 1$, $A_{11} = 1.5$ and n = 3.6, $f_h = 1$, and $f_e = 0$.*

6.9   *Repeat the exercise given in Problem 6.4 for an injected carrier density of 10^{11} cm^{-2}. Use Joyce–Dixon formula to calculate the quasi-Fermi levels.*

6.10 *Calculate the range of photon energy for amplification in a GaAs QW at 0K. Take $m_e = 0.067\ m_0$, $m_{hh\parallel} = 0.112 m_0$, $m_{hhz} = 0.042\ m_0$, d = 5 nm, $n = p = 10^{12}$ cm^{-2}.*

6.11 *A semiconductor A has a band gap of 1.3 eV. The first electron and hole subbands in a QW using A occur at 50 meV above and 20 meV below the respective band edges. The injected carrier density is 10^{12} cm^{-2} and $m_e = 0.04\ m_0$ and $m_h = 0.5\ m_0$. Assume that only the lowest subbands are occupied.*

   Determine the range of photon energies for which there will be positive gain in the QW.

6.12 *Obtain the expression for $\langle z \rangle$, for transition from subband m to subband n for a QW with infinite barrier. Discuss how the matrix element changes with difference (n−m). Show that $\langle z \rangle = 16L/9\pi^2$ for transition from lowest subband to the first excited subband.*

6.13 *Using the expression for $\langle z \rangle$ derived in previous problem, show that $f_{osc} = (m_0/m_e)(256/27\pi^2)$.*

6.14 *Prove that the recombination lifetime for a 2→1 intersubband transition is given by $\tau_r^{-1} = 2\pi n e^2 f_{12}/m_0\varepsilon_0\lambda^2 c$. Calculate the value assuming $f_{12} = 15$, n = 3.6, and $\lambda = 20\,\mu m$*

6.15 *Prove the relation $R_{sp}(y) \propto exp(-y/k_B T)H(y)$, given in Eq. (6.45).*

6.16 *Estimate the factor by which the IVBA changes if the HH mass is reduced from 0.5 m_0 to 0.16 m_0. Use $m_{so} = 0.12\,m_0$, $E_{so} = 320\,meV$, and $E_g = 0.8\,eV$.*

6.17 *The difference between Γ and L valleys in tensile strained Ge layer is 115 meV. The Ge layer is doped heavily so that the Fermi level touches the Γ valley. Find out the required donor density using parameter values for unstrained Ge.*

Reading List

Adams A R, Oreilly E P, and Silver M (1999) Strained layer quantum well lasers. In: *Semiconductor Laser I*. E Kapon (ed.) Chapter 2, pp. 123–176. San Diego: Academic.

Agrawal G P and Dutta N K (1993) *Semiconductor Lasers*, 2nd edn. New York: Van Nostrand Reinhold.

Arakawa A and Yariv A (1985) Theory of gain, modulation, response, and spectral linewidth in AlGaAs quantum-well lasers. *IEEE Journal of Quantum Electronics* 21: 1666–1674.

Basu P K (1997) *Theory of Optical Processes in Semiconductors*. Oxford, UK: Oxford University Press.

Basu P K, Mukhopadhyay B, and Basu R (2015) *Semiconductor Laser Theory*. Boca Raton, FL, USA: CRC Press,.

Chuang S L (2009) *Physics of Photonic Devices*. New York: John Wiley.

Coldren L A, Corzine S W, and Masanovi'c M L (2012) *Diode Lasers and Photonic Integrated Circuits*, 2nd ed. New York, NY, USA: Wiley.

Harrison P (2000) *Quantum Wells, Wires, and Dots, Theoretical and Computational Physics*. Chichester, UK: John-Wiley & Sons.

Manasreh M O (ed.) (1993) *Semiconductor Quantum Wells and Superlattices for Long Wavelength Infrared Detectors*. Boston: Artec.

Manasreh O (2005) *Semiconductor Heterojunctions and Nanostructures*. New York: McGraw-Hill.

Nagarajan R and Bowers J E (1999) High speed lasers. In: *Semiconductor Laser I*. E Kapon (ed.) Chapter 3, pp. 177–290. San Diego: Academic.

Numai T (2004) *Fundamentals of Semiconductor Lasers*. New York: Springer.

Singh J (2003) *Electronic and Optoelectronic Properties of Semiconductor Structures*. Cambridge: Cambridge University Press.

Zhao B and Yariv A (1999) Quantum well semiconductor laser. In: *Semiconductor Laser I*. E Kapon (ed.) Chapter 1, pp. 1–121. San Diego: Academic.

Zory P S (ed.) (1993) *Quantum Well Lasers*. San Diego: Academic.

References

Asada M, Kameyama A, and Suematsu Y (1984) Gain and intervalence band absorption in quantum-well lasers. *IEEE Journal of Quantum Electronics* QE-20: 745–753.

Asryan L V, Gun'ko N A, Polkovnikov A S, Zegrya G, Suris R A, Lau P-K, and Makino T (2000) Threshold characteristics of InGaAsP/InP multiple quantum well lasers. *Semiconductor Science and Technology* 15: 1131–1140.

Basu P K and Kundu S (1985) Energy loss of two-dimensional electron gas in GaAs-AlGaAs MQWs by screened electron-polar optic phonon interactions. *Applied Physics Letters* 47: 264–266.

MakinoT (1996) Analytical Formulas for the Optical Gain of Quantum Wells. *IEEE Journal of Quantum Electronics* 32(3): 493–501.

Mathieu H, Lefebvre P, and Christol P (1992) Simple analytical method for calculating exciton binding energies in semiconductor quantum wells. *Physical Review B* 46(7): 4092–4101.

VahalaK J and Zah C E (1988) Effect of doping on the optical gain and spontaneous noise enhancement factor in quantum well amplifiers by simple analytical expressions. *Applied Physics Letters* 52: 1945–1947.

Vurgaftman I and Meyer J R (1999)TE- and TM-polarized roughness-assisted free-carrier absorption in quantum wells at midinfrared and terahertz wavelengths. *Physical Review B* 60: 14294–14301.

Yang C L, Pan D S, and Somano R (1989) Advantages of indirect semiconductor quantum well system for infrared detection. *Journal of Applied Physics* 65: 3253–3258.

7

Excitons in semiconductors

7.1 Introduction

The earlier two chapters introduced various optical processes, first in bulk semiconductors, and then in Quantum Wells (QW)s. In the processes discussed, electrons and holes were not assumed to be correlated, that means that there existed no mutual Coulomb interaction between the particles.

There are many situations in which electron-hole pair bound by their Coulomb attraction dominates the optical processes. The bound pair is called an exciton. Excitons are characterized by a binding energy which may simply be calculated by using the elementary Bohr theory of hydrogen atom, with some modifications. It turns out that the value of binding energy is quite low in typical semiconductors. For gallium arsenide (GaAs), its value is only 4.2 meV. Excitonic processes in bulk semiconductors show their signatures at all but very low temperatures, therefore. The low-temperature absorption spectra show a single or at most a few peaks just below the fundamental band edge. It is natural that at higher temperatures, typically at room temperature, the absorption spectra show no such peaks, as the excitons totally dissociate due to the small binding energy. An excitonic signature is also observed in the recombination spectra. Application of an electric field modifies the excitonic absorption and recombination spectra.

The situation is somewhat different in QWs. The excitonic binding energy is much higher there and excitons survive even at room temperature, showing typical absorption peaks. Excitonic effects are also visible in luminescence spectra in QWs around room temperature. Furthermore, application of an electric field along the growth direction, or along the QW layer plane, influences the spectra in different ways.

In this chapter, excitonic processes in both bulk semiconductors and QWs will be discussed. First of all, the concept of exciton, already given in Chapter 2, will be elaborated for bulk semiconductors and the expressions for binding energy and absorption coefficient will be derived. Changes in the absorption spectra due to excitons from the known results given in Chapter 5 will be introduced. The recombination processes will then be discussed. Similar theories for QW will then be developed. The effect of electric

Semiconductor Nanophotonics. Prasanta Kumar Basu, Bratati Mukhopadhyay, and Rikmantra Basu, Oxford University Press.
© P.K. Basu, B. Mukhopadhyay, R. Basu (2022). DOI: 10.1093/oso/9780198784692.003.0007

field on absorption in bulk, known as Franz–Keldysh effect will be very briefly treated. More emphasis will be given on the effect of an electric field in the subband structure and excitonic processes in QWs. In this connection, a novel phenomenon in QWs, known as Quantum Confined Stark Effect (QCSE), will be given more attention.

Excitonic processes become more important as the dimension is reduced, that is, interesting phenomena are observed in Quantum Wires (QWRs) or nanowires and in Quantum Dots (QD)s or nanoparticles. The present chapter aims to prepare the background for understanding the processes in structures of reduced dimensionality, to be discussed in later chapters.

7.2   Excitons in bulk semiconductors

As already discussed, the ground state of a semiconductor material consists of a fully occupied valence band and a completely empty conduction band, well above the top of the valence band separated by a forbidden energy gap. When the material is excited, by absorbing a photon for example, an electron from the valence band is transferred to one of the unoccupied states in the conduction band. Thus electron-hole pairs are generated. These electron and holes are normally thought to be uncorrelated, as assumed in earlier chapters. However, in actual situation, particularly at low temperatures, an attractive Coulomb interaction is in play between the oppositely charged electron and hole. As a result, a hydrogen-like bound pair is created in place of a free electron-hole pair. These complex elementary excitations in solids that arise due to Coulomb interaction between electrons and holes are known as 'excitons'. It has its lowest energy level above the ground state energy of the unexcited crystal, the energy separation being $E_g - R_x$, slightly less than the band gap E_g. The symbol R_x denotes the exciton binding energy.

7.2.1   Classification

The excitons in solids are classified into one of two types: a Frenkel type or a Wannier type. The Frenkel-type is found in highly ionic or molecular crystals. The interaction between atoms is very weak in these solids, and any elementary excitation is more or less localized in space, covering at most a few atomic sites. The Frenkel excitons are however free to move in the crystal and thus provide a means for transfer of energy from one point to another. Wannier-type excitons, on the other hand, are created in semiconductors having covalent bonding, such as Gr. IV as well as in weakly ionic III–V compound semiconductors. In these materials, the electrons are shared by many atoms, and the excitons extend over many atomic sites in the crystal. Wannier excitons are also called as Mott–Wannier or simply Mott excitons. A conceptual picture of the two types is given in Fig. 7.1. Excitons have a very strong effect in absorption process in direct as well as indirect semiconductors.

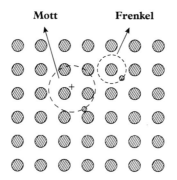

Figure 7.1 *A conceptual picture of Frenkel and Mott–Wannier type excitons.*

7.2.2   Exciton binding energy in bulk

Since two particles, an electron and a hole are involved, the problem is solved by using the two-particle Schrödinger equation

$$\left[-\frac{\hbar^2}{2m_e}\nabla_e^2 - \frac{\hbar^2}{2m_h}\nabla_h^2 - \frac{e^2}{4\pi\varepsilon\,|r_e - r_h|} \right]\psi_{ex} = E\psi_{ex} \qquad (7.1)$$

Here, m_e (m_h) are the electron (hole) effective mass and $|r_e - r_h|$ is the separation between the positions of the two particles. The standard two-body problem is transformed into a one-body problem by introducing the following transformations:

$$r = r_e - r_h;\; k = \frac{m_e k_e + m_h k_h}{m_e + m_h};\; R = \frac{m_e r_e + m_h r_h}{m_e + m_h};\; K = k_e - k_h \qquad (7.2)$$

The derivatives over r_e and r_h can be expressed in terms of derivatives with respect to r and R. Making use of the definition of total mass $M = m_e + m_h$ and reduced mass m_r, the two-particle Schrodinger Eq. (7.1) may be written as

$$\left[E_g - \frac{\hbar^2}{2m}\nabla_R^2 - \frac{\hbar^2}{2m_r}\nabla_r^2 - \frac{e^2}{4\pi\varepsilon r} \right]\Phi(R,r) = E\Phi(R,r) \qquad (7.3)$$

We note that centre-of-mass (CM) and relative co-ordinates are decoupled. Therefore, we use the following wavefunction in the form of a product wavefunction

$$\Phi(R,r) = g(R)\phi(r) \qquad (7.4)$$

Putting this in Eq. (7.3) we obtain two separate equations given as follows:

$$-\frac{\hbar^2}{2M}\nabla_R^2 g(R) = E_K g(R) \qquad (7.5)$$

and

$$\left[E_g - \frac{\hbar^2}{2m_r} \nabla_r^2 - \frac{e^2}{4\pi\varepsilon r} \right] \varphi_n(r) = E_n \varphi_n(r) \tag{7.6}$$

$$E = E_n + E_K \tag{7.7}$$

where n denotes different eigenstates.

Eq. (7.5) relates to a free particle indicating that the centre-of-mass(CM) motion is not affected by the Coulomb attraction, and its solution takes the form

$$g(R) = \frac{1}{V^{1/2}} exp(iK \cdot R) \tag{7.8}$$

The eigenvalues, representing the kinetic energy associated with the free motion of the CM, are therefore

$$E(K) = \frac{\hbar^2 K^2}{2M} \tag{7.9}$$

7.2.2.1 Hydrogenic states

The different eigenfunctions in Eq. (7.6) are expressed in terms of hydrogenic wave-functions in polar (r, θ, ϕ) coordinates as

$$\varphi_n(r) = R_{n,l}(r) Y_l^m(\theta, \varphi) \tag{7.10}$$

The radial part of the wavefunction is written as (Landau and Lifshitz 1975: p. 117)

$$R_{n,l}(r) = N_{np}l e^{-\frac{\rho}{2}} F(l+1-\eta; 2l+2; \rho) \tag{7.11}$$

where N_n is the normalization constant, F is the confluent hypergeometric function, and ρ is related to r. The symbol n denotes the excitonic levels; its negative values correspond to bound states lying below the band gap and positive values give rise to continuum states above the band gap. The following relations are valid for the parameter η in (7.11), the exciton Rydberg R_x and exciton Bohr radius a_B

$$\eta^2 = -\frac{R_x}{E}; \; R_x = \frac{\hbar^2}{2m_r a_B^2}; a_B = \frac{4\pi\varepsilon\hbar^2}{m_r e^2} \tag{7.12}$$

The expressions for various parameters appearing in Eqs. (7.11) and (7.12) and the s-state solution $|\varphi(r = 0)|^2$ for discrete and continuum states will now be quoted.

Discrete states (E< 0)   The parameter η is now the principal quantum number n and the other parameters may be expressed as

$$E_n = -\frac{R_x}{n^2} \tag{7.13a}$$

$$\rho = \frac{2r}{na_B} \tag{7.13b}$$

$$N = \frac{\frac{2}{na_B}}{(2l+1)!} \frac{(n+l)!}{(n-l-1)!2n} \tag{7.13c}$$

$$|\varphi_n(0)|^2 = (\pi a_B^3 n^3)^{-1} \tag{7.13d}$$

Continuum states ($E_n>0$)   In this case energy E is positive making the parameter η imaginary. This leads to

$$\eta = i\gamma = \frac{i}{ka_B} \tag{7.14a}$$

$$E(k) = -\frac{R_x}{(i\gamma)^2} = R_x(ka_B)^2 = \frac{\hbar^2 k^2}{2m_r} \tag{7.14b}$$

$$\rho = -2ikr \tag{7.14c}$$

$$N = [i^i/(2^l + 1)!][\Gamma(l + 1 - i\gamma)]exp(\pi\gamma/2) \tag{7.14d}$$

and

$$|F_n(0)|^2 = z\frac{exp(z)}{sinh\ z}, z = \pi\gamma \tag{7.14e}$$

From Eq. (7.14b) one obtains

$$E(k) = \frac{R_x}{\gamma^2} = \hbar\omega - E_g \tag{7.14f}$$

$$\gamma = \left[\frac{R_x}{\hbar\omega - E_g}\right]^{\frac{1}{2}} \tag{7.14g}$$

It is to be noted that k (or γ) is a quantum number and does not represent momentum.

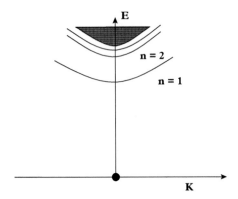

Figure 7.2 *E (energy) versus K (CM wave vector) relation for different discrete and continuum states (shaded region) of excitons. The ground state (E = 0, K = 0) is indicated by solid ball.*

The total energy of excitons is now expressed as

$$E = E_n + \frac{\hbar^2 K^2}{2M} \tag{7.15}$$

The energy level diagram of excitons is depicted in Fig. 7.2 in which the E-K relationship is shown for different excitonic levels. The ground state of the crystal is at the origin with $E = 0$ and $\mathbf{K} = 0$. The excited states correspond to two-particle energy. The minima of the lowest level excitonic E-K diagram lies slightly below the band gap energy at $E_g - R_x$. The E-K relationship is parabolic. The dispersion relations for higher lying discrete excitonic states (2s, 2p, etc.,) lies over the E-K diagram for 1s state; however, the spacing between the curves reduce as n^{-2}. The higher lying states are so closely spaced that they form a continuum starting from E_g and above.

The dispersion relation shows the usual periodicity in the reciprocal space. The E-K diagram shown in Fig. 7.2 applies to a region near the extrema in the first Brillouin zone.

Example 7.1 *The band gap in bulk GaAs at T = 0 K is Eg = 1:519 eV, m_e = 0.067 m_0, m_h = 0.51m_0, and ε = 12.9ε$_0$. Therefore, one obtains Bohr radius a_B = 11.6 nm and R_x = 4.8 meV. The values for bulk InAs at 0 K are: Eg = 0.415 eV, m_e = 0.023m_0, m_h = 0.41 m_0, ε = 15.15 ε$_0$, and the calculated values are a_B = 36.9 nm and R_x = 1.3 meV. Since the lattice constant of GaAs (InAs) is 0.56 nm (0.61 nm), the electron and hole are separated by many lattice constants. The excitons are Wannier-type.*

Example 7.2 *The excitonic Bohr radius and binding energy for Ge are calculated using m_e = 0.042 m_0 and m_{hh} = 0.284 m_0 so that the reduced mass is m_r = $m_e m_{hh} / (m_e + m_{hh})$ = 0.036 m_0. The effective Bohr radius is a_B = $\frac{4\pi\varepsilon\hbar^2}{m_r e^2}$ =23.5 nm, using ε = 16ε$_0$ for Ge. The exciton binding energy as expressed by Eq. (7.12) is E_0 = $\frac{4m_r e^4}{32\pi^2\hbar^2\varepsilon^2}$ =7.64 meV*

7.2.3   Calculation of absorption coefficient

We first introduce a simple method to calculate the absorption coefficient (Dutta 1989). Ignoring the cell-periodic part of the Bloch function, the two-particle wave function may be written as

$$\Phi(\boldsymbol{R},\boldsymbol{r}) = \frac{e^{i\boldsymbol{K}\cdot\boldsymbol{R}}}{\sqrt{V}} \cdot \frac{e^{i\boldsymbol{k}\cdot\boldsymbol{r}}}{\sqrt{V}} \tag{7.16}$$

In optical absorption $\boldsymbol{k}_e = \boldsymbol{k}_h$, so that $\boldsymbol{K} = 0$. As noted in Chapter 05, the absorption coefficient in band-to-band transition is calculated from

$$\alpha = \sum_k |M_{cv}(\boldsymbol{k})|^2 \, \delta(E - \hbar\omega) \tag{7.17}$$

In the present case the summation should be over the index n instead of $\boldsymbol{k}$. The new eigenstates can be expanded in terms of plane wave states as follows:

$$\varphi_n(\boldsymbol{r}) = \sum_k a_{n,k} \frac{e^{i\boldsymbol{k}\cdot\boldsymbol{r}}}{\sqrt{V}} \tag{7.18}$$

For normalization of $\varphi_n(\boldsymbol{r})$, one should have $\sum |a_{n,k}|^2 = 1$. The new matrix element is expressed as a linear combination of old matrix elements, such that

$$M_{cv}(n) = \sum_k a_{n,k} M_{cv}(\boldsymbol{k}) = M_{cv}(0) \sum_k a_{n,k} \tag{7.19}$$

Noting from Eq. (7.18) that $\sum_k a_{n,k} = \sqrt{V}\varphi_n(\boldsymbol{r} = 0)$, we may write

$$M_{cv}(n) = M_{cv}(0)\sqrt{V}\varphi_n(\boldsymbol{r} = 0) \tag{7.20}$$

Substituting this in Eq. (7.17) one obtains finally

$$\alpha(\hbar\omega) = \frac{2\pi\langle|p_{cv}|^2\rangle}{\eta\varepsilon_0\omega c} \left(\frac{e}{m_0}\right)^2 \sum_n |\varphi_n(\boldsymbol{r} = 0)|^2 \delta\left(\hbar\omega - E_g + \frac{R_x}{n^2}\right) \tag{7.21}$$

for the discrete excitonic states. Eq. (7.21) follows also from a more rigorous theory, to be presented in Section 7.2.4.

Using Eq. (7.13d), and assuming that excitonic lines are sharp, without being broadened, we may express Eq. (7.21) as

$$\alpha(\hbar\omega) = \frac{2\pi\langle|p_{cv}|^2\rangle}{\eta\varepsilon_0\omega c} \left(\frac{e}{m_0}\right)^2 \frac{1}{\pi a_B^3 n^3} \tag{7.22}$$

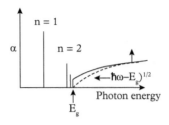

Figure 7.3 *Ideal excitonic absorption coefficient in a direct gap semiconductor. The sharp peaks occur due to excitonic absorption for hydrogenic levels 1 and 2. The solid line represent the absorption spectra in the continuum states; the dashed curve represents the expected band-to-band absorption starting from band gap energy E_g., which merges with the spectra due continuum states at higher photon energy.*

7.2.3.1 Absorption coefficient

The ideal absorption spectra consist of discrete lines corresponding to excitonic energies. The absorption coefficient decreases with increasing level n. The spectra are shown in Fig. 7.3.

For higher values of n, the excitonic lines are so close that they form a continuum. We may take n as a continuous variable for $\hbar\omega \to E_g$. Replacing sum in Eq. (7.21) by an integration over n, and using $x = \hbar\omega - E_g + R_x/n^2$ and Eq. (7.13d), we write

$$lim_{\hbar\omega \to E_g} \alpha(\hbar\omega) = lim_{\hbar\omega \to E_g} \frac{2(e/m_0)^2}{\eta\varepsilon_0 \omega c a_B^3} \langle|p_{cv}|^2\rangle \int_{\hbar\omega - E_g}^{\hbar\omega - E_g + R_x} dx \frac{\delta x}{R_x}$$

$$= \frac{(e/m_0)^2}{\eta\varepsilon_0 (E_g/\hbar) c a_B^3 R_x} \langle|p_{cv}|^2\rangle \tag{7.23}$$

For still higher photon energies, that is, when the continuum states are reached, the envelope function from Eq. (7.14e) reads

$$|F_n(0)|^2 = z\frac{exp(z)}{sinhz} = \frac{\pi\sqrt{R_x}}{\sqrt{\hbar\omega - E_g}} \frac{2}{1 - exp\left[-2\pi\sqrt{R_x/(\hbar\omega - E_g)}\right]} \tag{7.24}$$

where $z = \pi\gamma = \pi/a_B k$.

Proceeding as previously, the expression for absorption coefficient becomes

$$\alpha_{cont}(\hbar\omega) = A\frac{2\pi\sqrt{R_x}}{1 - exp\left[-2\pi\sqrt{R_x/(\hbar\omega - E_g)}\right]} \tag{7.25}$$

The band-to-band absorption coefficient is expressed by using the same prefactor A appearing in Eq. (7.25), such that

$$\alpha_{free}(\hbar\omega) = A(\hbar\omega - E_g)^{1/2} \tag{7.26}$$

The qualitative variation of $\alpha_{cont}(\hbar\omega)$ and $\alpha_{free}(\hbar\omega)$ are shown in Fig. 7.3 by a solid line and a dashed line, respectively. It is seen that the absorption coefficient with Coulomb interaction is enhanced from the values without the interaction for $\hbar\omega \approx E_g$ and above. The enhancement, termed as Sommerfeld factor, or Coulomb enhancement factor, is expressed as

$$C(\omega) = \frac{\alpha_{cont}}{\alpha_{free}} = \frac{2\pi\sqrt{R_x}}{1 - exp\left[-2\pi\sqrt{R_x/(\hbar\omega - E_g)}\right]} \cdot \frac{1}{\sqrt{\hbar\omega - E_g}} \qquad (7.27)$$

7.2.4 Excitonic absorption: Elliott formula

We now discuss a generalized formula for absorption coefficient derived by Elliott (1957) for an electron-hole pair bound by Coulomb attraction. The general formula for absorption coefficient for transition from an initial state $|i\rangle$ to a final state $|f\rangle$ by absorbing a photon of energy $\hbar\omega$ and wavevector $\mathbf{k}_{op}$ is given by

$$\alpha(\hbar\omega) = C_0 \left(\frac{2}{V}\right) \sum_{i,f} |\langle f|e^{i\mathbf{k}_{op}\cdot\mathbf{r}} \mathbf{e}_{op} \cdot \mathbf{p}|i\rangle|^2 \delta(E_f - E_i - \hbar\omega)[f(E_i) - f(E_f)] \qquad (7.28a)$$

where

$$C_0 = \frac{\pi e^2}{\eta_r c \varepsilon_0 m_0^2 \omega} \qquad (7.28b)$$

and other symbols are already defined. The two-particle wavefunction is expressed as a linear combination of the product of electron and hole wavefunctions and is written as

$$\Psi(\mathbf{r}_e, \mathbf{r}_h) = \sum_{\mathbf{k}_e} \sum_{\mathbf{k}_h} A(\mathbf{k}_e, \mathbf{k}_h) \psi_{c\mathbf{k}_e}(\mathbf{r}_e) \psi_{v-\mathbf{k}_h}(\mathbf{r}_h) \qquad (7.29)$$

where A is the amplitude function, $\psi_{c\mathbf{k}_e}(\mathbf{r}_e)$ and $\psi_{v-\mathbf{k}_h}(\mathbf{r}_h)$ are Bloch functions containing slowly varying plane wavefunctions and rapidly varying cell-periodic functions. We may introduce the following inverse Fourier transform of the amplitude function A under effective mass approximation for electron-hole pairs

$$\Phi(\mathbf{r}_e, \mathbf{r}_h) = \sum_{\mathbf{k}_e} \sum_{\mathbf{k}_h} A(\mathbf{k}_e, \mathbf{k}_h) \frac{exp(i\mathbf{k}_e \cdot \mathbf{r}_e)}{\sqrt{V}} \frac{exp(i\mathbf{k}_h \cdot \mathbf{r}_h)}{\sqrt{V}} \qquad (7.30)$$

The Fourier transform of the wavefunction $\Phi(\mathbf{r}_e, \mathbf{r}_h)$ is written as

$$A(\mathbf{k}_e, \mathbf{k}_h) = \int d^3\mathbf{r}_e \int d^3\mathbf{r}_h \Phi(\mathbf{r}_e, \mathbf{r}_h) \frac{exp(-i\mathbf{k}_e \cdot \mathbf{r}_e)}{\sqrt{V}} \frac{exp(-i\mathbf{k}_h \cdot \mathbf{r}_h)}{\sqrt{V}} \qquad (7.31)$$

It is to be noted that whereas $\Psi(r_e, r_h)$ contains the Bloch part, only plane wave parts are included in $\Phi(r_e, r_h)$, which satisfies the following effective mass equation:

$$[E_g + E_c(-i\nabla_e) - E_v(-i\nabla_e) + V(r_e, r_h)]\Phi(r_e, r_h) = E\Phi(r_e, r_h) \qquad (7.32)$$

The interaction potential $V(r_e, r_h) = e\mathbf{F} \cdot (r_e - r_h)$ in the presence of a uniform electric field $\mathbf{F}$ gives rise to Franz–Keldysh effect (Franz 1958; Keldysh 1958), whereas when it has the Coulomb interaction form, one encounters excitonic effect.

The two-particle effective mass in Eq. (7.32) can be written in centre-of-mass (CM) and relative coordinates and may be written as Eq. (7.2). The wavefunction is written in the product form as in Eq. (7.4). We define Fourier transform pair as

$$\varphi(r) = \sum_k a(k)\frac{exp(i\mathbf{k} \cdot r)}{\sqrt{V}} \qquad (7.33a)$$

$$a(k) = \int d^3r \varphi(r)\frac{exp(-i\mathbf{k} \cdot r)}{\sqrt{V}} \qquad (7.33b)$$

Therefore the two particle wavefunction is written as

$$\Phi(r_e, r_h) = \Phi(R, r) = \frac{exp(i\mathbf{k} \cdot R)}{\sqrt{V}} \sum_k a(k)\frac{exp[i\mathbf{k} \cdot (r_e - r_h)]}{\sqrt{V}} \qquad (7.34)$$

We now obtain the expression for optical matrix element for transition from the ground state in which all electrons lie in the valence band (VB) to the final excitonic state. This can be written as

$$|\langle f|e^{i\mathbf{k}_{op} \cdot r} e_{op} \cdot \mathbf{p}|i\rangle| = \sum_{k_e}\sum_{k_h} A*(k_e, k_h)\langle c, k_e|e^{i\mathbf{k}_{op} \cdot r} e_{op} \cdot \mathbf{p}|v, -k_h\rangle$$

$$\cong \sum_{k_e}\sum_{k_h} A*(k_e, k_h)e_{op} \cdot \mathbf{p}_{cv}(k_e)\delta_{k_e+k_h, k_{op}}$$

$$\approx \sum_k A*(k, -k)e_{op} \cdot \mathbf{p}_{cv}(k) \qquad (7.35)$$

In the previous derivation, the dipole approximation $\mathbf{k}_{op} = 0$ has been introduced, leading to the k-selection rule $\mathbf{k}_e + \mathbf{k}_h = \mathbf{k}_{op} = 0$. We also have $\mathbf{K} = \mathbf{k}_e + \mathbf{k}_h = 0$ and can write the matrix element as

$$\langle f|e^{i\mathbf{k}_{op} \cdot r} e_{op} \cdot \mathbf{p}|i\rangle = e_{op} \cdot \mathbf{p}_{cv} \sum_k A*(k, -k) = e_{op} \cdot \mathbf{p}_{cv} \sum_k a*(k) = e_{op} \cdot \mathbf{p}_{cv}\sqrt{V}\varphi*(0)$$

$$(7.36)$$

It is assumed that $e_{op} \cdot \mathbf{p}_{cv}$ is independent of k.

The absorption coefficient is obtained by substituting Eq. (7.36) in Eq. (7.28a) and the expression is

$$\alpha(\hbar\omega) = C_0 |e_{op} \cdot \boldsymbol{p}_{cv}|^2 2 \sum_n |\varphi_n(0)|^2 \delta(E_n + E_g - \hbar\omega) \tag{7.37}$$

where n corresponds to discrete and continuum states.

7.2.5 Excitonic recombination

As noted, the ground state in a semiconductor is formed by a fully occupied valance band (VB) and an empty conduction band (CB). The matrix element for transition from the ground state to the excited state having an exciton is expressed as

$$|\langle x|e \cdot \boldsymbol{p}|0\rangle|^2 = |\varphi(0)|^2 \langle |p_{cv}|^2\rangle \tag{7.38}$$

where the symbols have the usual meanings.

The emission probability is now proportional to the number of excitons instead of np product as in interband emission. The rate of spontaneous emission is now written as

$$R_{sp}(\hbar\omega) = \left(\frac{2\pi}{\hbar}\right) \sum_{E_x} |\langle x|H|0\rangle|^2 G(\hbar\omega) P(E_x)\delta(\hbar\omega - E_x)$$

$$= \eta \left(\frac{e^2 \hbar\omega}{\pi\varepsilon_0\varepsilon_r m_0^2 \hbar^2 c^3}\right) \langle |p_{cv}|^2\rangle \sum_{E_x} |\varphi(0)|^2 P(E_x)\delta(\hbar\omega - E_x)\delta_{K0} \tag{7.39}$$

where $\hbar\boldsymbol{K}$ is the momentum of the exciton. As photons have negligible momentum, only the excitons having zero kinetic energy $(\boldsymbol{K} = 0)$ can recombine. However, as phonons or defects may provide momentum, this restriction is relaxed. Moreover, the discrete excitonic states may have some linewidth due to homogeneous and inhomogeneous processes. Therefore, a proper line shape function $L_f(\hbar\omega - E_x)$ is to be included in previous expression replacing the δ function. Assuming that the lineshape function is Gaussian in nature and the exciton density is small, so that a Boltzmann distribution can describe the probability $P(E_x)$, the spontaneous emission rate may be expressed as

$$R_{sp}(\hbar\omega) \sim P(E_x = \hbar\omega) L_f(\hbar\omega - E_x) \approx L_f(\hbar\omega - E_x)exp(-\hbar\omega/k_B T) \tag{7.40}$$

Now using the following standard form for Gaussian shape function:

$$L_f(\hbar\omega - E_x) = \frac{1}{\sqrt{2\pi}\sigma} exp\left[\frac{-(\hbar\omega - E_x)^2}{2\sigma^2}\right] \tag{7.41}$$

where σ is the variance, the spontaneous emission rate takes the form

$$R_{sp}(\hbar\omega) \sim exp[-(E_x - \sigma^2/2k_B T)/k_B T]exp\left\{-[\hbar\omega - (E_x - \sigma^2/k_B T)/2\sigma^2]\right\} \tag{7.42}$$

It seems therefore that emission peak is downshifted from the absorption peak by an amount σ^2/k_BT. Note that the absorption coefficient in the quasi-continuum region $(\hbar\omega < E_g)$ is expressed as

$$\alpha(\hbar\omega) = 2\pi\sqrt{R_x}B(\hbar\omega) \tag{7.43}$$

On the other hand, in the true continuum region $(\hbar\omega > E_g)$

$$\alpha(\hbar\omega) = \frac{2\pi\sqrt{R_x}B(\hbar\omega)}{[1 - exp(-2\pi\gamma)]}; \ \gamma = [R_x/(\hbar\omega - E_g)]^{1/2} \tag{7.44}$$

The band-to-band absorption coefficient is expressed using the same prefactor $B(\hbar\omega)$ as

$$\alpha_{BB}(\hbar\omega) = B(\hbar\omega)(\hbar\omega - E_g)^{1/2} \tag{7.45}$$

It may easily be proved that when $\hbar\omega \gg E_g$, Eq. (7.44) reduces to Eq. (7.45).

The recombination rate may be expressed in terms of the absorption coefficient using the principle of detailed balance (van Roosbroeck and Shockley 1954). Thus

$$R_{sp}(\hbar\omega) = v_gG(\hbar\omega)\alpha(\hbar\omega)[exp\{(\hbar\omega - F)/k_BT\} - 1]^{-1} \tag{7.46}$$

where F_x is the quasi-Fermi level for excitons. Using Eq. (7.44) and Boltzmann distribution, Eq. (7.46) reduces to

$$R_{sp}(\hbar\omega) = v_gG(\hbar\omega)2\pi\sqrt{R_x}B(\hbar\omega)\ exp[-(\hbar\omega - F_x)/k_BT] \tag{7.47}$$

Eq. (7.47) indicates that excitonic recombination spectra vary exponentially with energy, see Fig. 7.4.

Example 7.3 *We make an order-of-magnitude calculation of excitonic absorption coefficient for 1s state in GaAs. The absorption spectra are assumed to have a Gaussian lineshape and the δ-function in Eq. (7.21) is replaced as follows:*

$$\delta(\hbar\omega - E) \to \frac{1}{\Gamma\sqrt{1.44\pi}}exp\left[-\frac{(\hbar\omega - E)^2}{1.44\Gamma^2}\right]$$

where Γ is the half width. Using $2p_{cv}^2 = 25$ eV and a factor 2/3 due to averaging, $a_B = 12$ nm, $\eta = 3.6$, and $\Gamma = 1$ meV, the value of peak absorption for the 1s state, according to Eq. (7.22), becomes 1.4×10^4 cm^{-1}, which comes close to the experimental value (see Fig. 7.5).

Excitons are Bosons and they obey Bose statistics. Thus there is no limit to the number of excitons in a single state. At a given temperature, T, excitons and free electron-hole

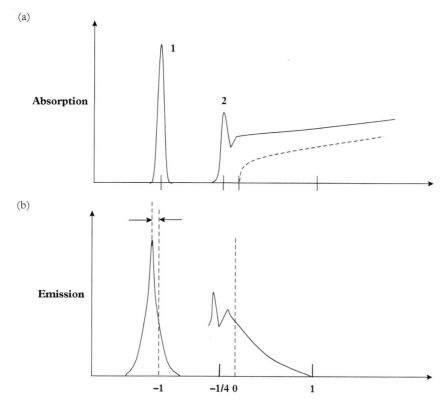

Figure 7.4 *Qualitative nature of excitonic absorption and emission spectra. Only n = 1 and 2 states are shown.*

pairs coexist. The number density of excitons, n_{ex}, and concentration of free electrons and holes, $n = n_e = n_h$, are related by the ionization equilibrium equation or Saha equation (Saha 1922)as

$$n_{ex} = n^2 \left(\frac{2\pi\hbar^2}{k_B T} \frac{1}{m_r} \right)^{3/2} exp \left(\frac{R_x}{k_B T} \right) \tag{7.48}$$

It follows easily that when $k_B T \gg R_x$, most of the excitons are ionized and free electrons and holes play their roles. On the other hand, for $R_x \geq k_B T$, a significant number of bound pairs exist in the crystal to influence the optical properties.

7.2.6 Line broadening mechanism

The ideal absorption spectra shown in Fig. 7.3 consist of a series of sharp lines due to discrete excitonic states and a continuous spectrum due to higher lying states. However, in typical experimental absorption spectra for both bulk material and QW, only the peaks

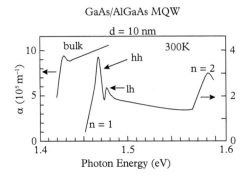

Figure 7.5 *Absorption spectra in bulk GaAs and in GaAs/AlGaAs MQW at 300 K.*

corresponding to the lowest excitonic states are observed at low temperature. In this section we shall present some typical experimental data to point out the deviation from the ideal theory developed so far and discuss the origin of the differences, first for bulk materials and then for QWs.

7.2.6.1   *Bulk semiconductor*

The absorption spectra for bulk GaAs are shown in Fig. 7.5. It is found that at very low temperature, only one sharp excitonic peak is visible, which may be attributed to the lowest exciton state. The absorption spectra due to higher lying discrete excitonic states merge with the continuum absorption spectra. Furthermore, with increasing temperature, the lowest peak broadens, and absorption diminishes. Finally around room temperature the excitonic peak is barely visible and the absorption spectra form a continuous curve.

The broadening of exciton absorption line is due to scattering of excitons by impurities, defects and phonons and is significant so that all but the lowest exciton peaks merge with one another as well as with the continuum. Therefore, only the peak, due to the lowest exciton state remains visible. The linewidths vary with temperature and for the lowest lying 1s state the linewidth may be expressed as

$$\Gamma(T) = \Gamma_0 + \sigma T + [\gamma / \{\exp(\hbar\omega_{LO}/k_B T) - 1\}] \qquad (7.49)$$

The first term in Eq. (7.49) represents the linewidth due to scattering of excitons by impurities and defects. The second term is the result of exciton-acoustic phonon scattering and the last term is the contribution by exciton-longitudinal optical (LO) phonon scattering.

The exciton-phonon scattering process is illustrated in Fig. 7.6 by using the E-K diagram of excitons. The dispersion relations of discrete excitonic states are shown by parabolic curves and of continuum states by the shaded region. The origin of the diagram represents the ground state of the crystal. The absorption of a photon of energy $\hbar\omega$ of negligible wavevector $\mathbf{k}_{op}$ is indicated by an almost vertical line by which the system comes to a discrete excitonic state. The exciton then absorbs an acoustic phonon,

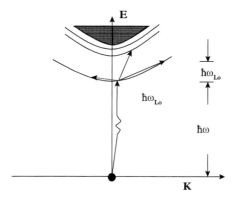

Figure 7.6 *Illustration of exciton-phonon scattering. An exciton with $K \approx 0$ is formed due to optical absorption. The exciton absorbs an acoustic phonon (dashed line) and goes to a state of wavevector K' in the same level. Absorption of an LO phonon brings the exciton in a state of higher energy in the same level or in a higher level (indicated by solid lines).*

thereby changing its wavevector from K to K', but remains in the same excitonic state due to small energy of the acoustic phonon. This process is denoted by a dashed line in the diagram. When the exciton absorbs a longitudinal optical (LO) phonon of energy $\hbar\omega_{LO}$, the final state with wavevector K' may be either in the same excitonic state or in a higher state, discrete or continuum. The LO phonon scattering processes are depicted by solid lines. In the diagram, only phonon absorption processes are shown, but phonon emission is also possible.

The lifetimes associated with the scattering processes have been calculated by Rudin et al (1990).

7.3   Excitonic processes in quantum wells

The excitonic phenomena in QWs show some similarities as well as many dissimilarities from what are observed in bulk materials. These features will be discussed in detail in later sections. In the following, we present first some preliminary concepts of excitonic phenomena in a QW made of a direct gap semiconductor. It is followed by an ideal theory of excitonic processes in a purely 2D system, that brings out the essential differences between the processes in bulk and QW.

7.3.1   Excitons in two dimensions: Preliminary concepts

The mutual coulomb interaction between electrons and holes in bulk semiconductors is more prominent at low temperatures. The exciton binding energy for GaAs has been found to be only 4.2 meV. Due to this small value and different line broadening mechanisms excitonic effects are barely visible at higher temperatures, as is found

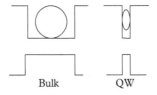

Figure 7.7 *Exciton orbits in bulk and QWs.*

in the representative experimental curves in Fig. 7.5. However, in a QW or other low-dimensional systems, the situation is altogether different, and excitons survive and possess larger oscillator strength even at room temperature. These features may be qualitatively understood from Fig. 7.7. As shown in the left part of Fig. 7.7, the excitonic orbit is spherical in bulk GaAs with an approximate Bohr radius of 15 nm. However, the orbit must be squeezed if it is to fit within a QW of width less than 15 nm, as shown in the right part of Fig. 7.7. Since the physical separation between the electron and hole along the z-direction reduces, the Coulombic attraction is enhanced, and consequently the binding energy increases. The hydrogenic levels and binding energies in a purely 2D system were first calculated by Shinada and Sugano (1966). They demonstrated that the binding energy for pure 2D systems increases four-fold from the corresponding value in bulk material. In the following, a brief outline of the theory developed by Shinada and Sugano will be presented.

7.3.2   Excitons in purely two-dimensional systems

The hydrogenic levels in a purely 2D system were calculated by Shinada and Sugano (1966). A real system is quite different from the ideal system they considered. However, the idealized theory is helpful in understanding the behaviour of excitons in 2D systems and we give here a brief sketch of their work. The equation describing the relative motion in a purely 2D system is

$$-\frac{\hbar^2}{2m_r}\left(\frac{\partial^2}{\partial x^2}+\frac{\partial^2}{\partial y^2}\right)\varphi - \frac{e^2}{4\pi\varepsilon(x^2+y^2)^{1/2}}\varphi = E\varphi \tag{7.50}$$

The wavefunction satisfying Eq. (7.50) is written in polar coordinate system (r,θ) as

$$\varphi = \left(2\pi\right)^{-\frac{1}{2}}R(r)exp(im\theta) \tag{7.51}$$

where m is an integer. The radial function $R(\mathbf{r})$ satisfies the following equation:

$$\left[-\frac{\hbar^2}{2m_r}\left\{\frac{1}{r}\frac{d}{dr}\left(r\frac{d}{dr}\right)-\frac{m^2}{r^2}\right\}-\frac{e^2}{4\pi\varepsilon r}\right]R(r) = ER(r) \tag{7.52}$$

The following changes of variables are introduced in Eq. (7.52):

$$R = \exp(-\rho/2)F; \quad \rho = \frac{r}{(a_0\lambda)\lambda^{-2}} = -4w; \quad w = \frac{E}{R_x} \tag{7.53}$$

where a_0 and R_x are, respectively, the Bohr radius and exciton binding energy for the bulk.

One now obtains

$$\rho\frac{d^2F}{d\rho^2} + (1-\rho)\frac{dF}{d\rho} + \left(2\lambda - \frac{1}{2} - \frac{m^2}{2}\right)F = 0 \tag{7.54}$$

As in the bulk case, we find bound and continuum excitonic states for negative and positive eigenvalues for Eq. (7.54).

Bound excitons (E <0): In this case, λ in Eq. (7.53) is real and is expressed as

$$\lambda = \left(\frac{1}{2}\right)(-w)^{-\frac{1}{2}} \tag{7.55}$$

Substituting $F = \rho^{|m|}L$ in Eq. (7.54) we get the following:

$$\rho\frac{d^2L}{d\rho^2} + (2|m| + 1 - \rho)\frac{dL}{d\rho} + \left(2\lambda - \frac{1}{2} - |m|\right)L = 0 \tag{7.56}$$

The solution of Eq. (7.56) is expressed in terms of the following power series:

$$L(\rho) = \sum_{\nu=0} \beta_\nu \rho^\nu \tag{7.57}$$

Inserting Eqs. (7.57) in (7.56) and comparing coefficients of different powers of ρ, one obtains the following recurrence relation:

$$\beta_{\nu+1} = \beta_\nu \frac{\nu - (2\lambda - 1/2 - |m|)}{(\nu+1)(\nu+|2m|+1)} \tag{7.58}$$

To ensure that $L(\rho)$ remains normalized, the series must terminate for $\nu = \nu_{max}$, so that for $\nu \geq \nu_{max}, \beta = 0$. Thus $\nu_{max} - (2\lambda - 1/2 - |m|) = 0$, as obtained from the numerator of (7.58), and thus

Table 7.1 *Normalized wavefunctions and binding energies of first few excitonic states in an ideal two-dimensional (2D) system.*

v_{max}	n	m	$R_{n,m}(r) = C\rho^{|m|} \exp(-\rho/2) \sum_v \beta_v \rho_v$	E_n
0	0	0	$R_{0,0}(r) = (4/a_0) \exp(-2r/a_0)$	$-4R_x$
1	1	0	$R_{1,0}(r) = \frac{4}{3\sqrt{3}a_0} \left\{1 - \left(\frac{4r}{3a_0}\right)\right\} \exp(-2r/3a_0)$	$-9R_x/4$
0	1	± 1	$R_{1,\pm 1}(r) = \frac{16}{9\sqrt{6}a_0} \left(\frac{r}{a_0}\right) \exp(-2r/3a_0)$	$-9R_x/4$

$$v_{max} + |m| + \frac{1}{2} = 2\lambda = n + \frac{1}{2} \tag{7.59}$$

The allowed values of the principal quantum numbers are $n = 0,1,2,\cdots$. From Eqs. (7.55) and (7.59) the bound state energies are given by

$$E_n = -\frac{R_x}{(n+1/2)^2}; \ n = 0, 1, 2, \ \ldots\ldots \tag{7.60}$$

The first few normalized wavefunctions and corresponding binding energies for a few lowest lying 2D excitonic states are shown in Table 7.1

Example 7.4 *The binding energies for the lowest heavy-hole (HH) and light-hole (LH) excitons in a GaAs QW is estimated by using the ideal theory. Take $m_e = 0.067\ m_0$, $m_{hh} = 0.112\ m_0$ and $m_{lh} = 0.0211\ m_0$, the hole masses being along the plane of the QW. The reduced masses are: $m_{rhh} = 0.042\ m_0$ and $m_{rlh} = 0.05\ m_0$. The binding energies for the HH and LH excitons are, respectively, 13.3 and 16 meV, the values of a_B are, 16.5 nm and 13.86 nm, respectively, and Bohr radii are 8.25 nm and 6.93 nm, respectively.*

Eq. (7.56) is Laguerre differential equations with solutions being associated Laguerre polynomials. The normalized wavefunctions are expressed as

$$\varphi_{n,m}(r) = \left[\frac{(n-|m|)!}{\pi a_0^2 (n+1/2)^3 \{(n+|m|)!\}^3}\right]^{1/2} exp(-\rho/2)\, \rho^{|m|} L_{n+|m|}^{2|m|}(\rho) exp(im\theta) \tag{7.61}$$

As may be noted from Table 7.1, the binding energy for lowest exciton is R_x in bulk, whereas it is $4R_x$ in 2D. The decrease in binding energy follows $1/n^2$ law in bulk but decreases as $1/n$ in 2D. Finally, the radius of Bohr orbit is $a_0/2$ in 2D, half of that in bulk.

  Unbound Excitonic (E > 0) The excitonic states with positive eigenvalues form a continuum of energy levels, which may be expressed in terms of a 2D wavevector **k** as follows:

$$E_k = \frac{\hbar^2 k^2}{2m_r} \tag{7.62}$$

where $w = a_0^2 k^2$. In the present case λ, given by Eq. (7.53) is a pure imaginary quantity. We write $\lambda = -i\gamma/2$ and $\gamma = (a_o k)^{-1}$, and rewrite Eq. (7.56) in the following form:

$$\rho\frac{d^2 L}{d\rho^2} + (2|m| + 1 - \rho)\frac{dL}{d\rho} - \left(|m| + \frac{1}{2} + i\gamma\right)L = 0 \tag{7.63}$$

The solutions are in terms of the confluent hypergeometric functions (Abramowitz and Stegun 1972) and are as follows (Landau and Lifshitz 1975):

$$L(\rho) = F(|m| + 1/2 + i\gamma; 2|m| + 1; \rho) \tag{7.64}$$

The confluent hypergeometric function $F(a; c; \rho)$ is a solution of the differential equation

$$\rho\frac{d^2 F}{d\rho^2} + (c - \rho)\frac{dF}{d\rho} - aF = 0 \tag{7.65}$$

The normalized wavefunctions are given by

$$\varphi_{km}(r) = \frac{1}{(2|m|)!} \left[\prod_{j=1}^m \frac{\left\{\left(j - \frac{1}{2}\right)^2 + \gamma^2\right\}}{S\cosh(\pi\gamma)}\right]^{1/2} exp\left[(\pi/2)\gamma\right] \times exp(-ikr)$$

$$\times (2kr)^{|m|} F(|m| + 1/2 + i\gamma; 2|m| + 1; 2ikr) exp(im\theta) \tag{7.66}$$

where S is the area of 2D crystal. When m = 0, the term $\prod_{j=1}^m \{(j - 1/2)^2 + \gamma^2\}$ is put to unity.

7.3.2.1　*Absorption intensities*

We shall use standard Elliott formula to calculate the absorption coefficients. The oscillator strength for discrete excitonic states is given by

$$f_{osc} = \frac{2}{Sm_0\hbar\omega} |\langle f| e^{i\eta k_{op} \cdot r} e_{op} \cdot p|i\rangle|^2 \tag{7.67}$$

For continuous or quasi-continuous states, the absorption coefficient is expressed as

$$\alpha(\hbar\omega) = \frac{\pi\varepsilon_r}{m_0 c\eta\omega a_0} |\langle f| e^{i\eta k_{op} \cdot r} e_{op} \cdot p|i\rangle|^2 D(\hbar\omega) \tag{7.68}$$

where D is the joint DOS function. As mentioned earlier, the matrix element for transition $|i\rangle \rightarrow |f\rangle$ is proportional to the value of the envelope function $\phi(\mathbf{r} = 0)$. We may therefore write

$$|\langle f| e^{i\eta k_{op}\cdot r} e_{op} \cdot p |i\rangle|^2 = S\varphi_{n(k),0}\langle |p_{cv}|^2\rangle \tag{7.69}$$

where $n\,(k)$ refers to discrete (continuous) states. It is to be recalled that $\phi_{n(k)}$ sare non-vanishing for $m = 0$ only and therefore only s-states are to be considered for calculating absorption. For discrete states

$$|\varphi_{n0}(0)|^2 = [\pi a_0^3(n + 1/2)^3]^{-1} \tag{7.70}$$

The oscillator strength for transition to the n-th excitonic state is therefore

$$f_n = (2/m_0\pi a_0^2\hbar\omega)(n + 1/2)^{-3}\langle |p_{cv}|^2\rangle \tag{7.71}$$

The spin degeneracy factor 2 has been included in Eq.(7.71). For large values of n, the states are distributed quasi continuously for which the DOS function may be defined as

$$D(E) = 2\left|\left(S\frac{\partial E}{\partial n}\right)^{-1}\right| = \frac{(n + 1/2)^3}{SR_x} \tag{7.72}$$

including the spin degeneracy factor. Using Eqs. (7.68) and (7.72) the absorption coefficient is written as

$$\alpha_{qc}(\hbar\omega) = [8\pi\varepsilon_s/(m_0 c\eta a_0\omega)]\langle |p_{cv}|^2\rangle \tag{7.73}$$

The envelope functions for the unbound states forming a continuum as obtained from Eq. (7.66) is

$$|\varphi_{k0}(0)|^2 = \frac{exp(\pi\gamma)}{S\,cosh(\pi\gamma)} \tag{7.74}$$

The joint DOS function is step like and is given by

$$D(E) = \frac{m_r}{\pi\hbar^2\,(E > 0)} \tag{7.75}$$

Using Eqs. (7.66), (7.74), and (7.75) the final expression for the absorption coefficient becomes

$$\alpha_{con}(\hbar\omega) = \frac{4\pi\varepsilon_s}{m_0^2 c\eta a_0\omega} \frac{exp(\pi\gamma)}{cosh(\pi\gamma)}\langle |p_{cv}|^2\rangle \tag{7.76}$$

The expression for band-to-band absorption coefficient in 2D systems is given by

$$\alpha_{bb}(\hbar\omega) = \frac{4\pi\varepsilon_s}{m_0^2 c\eta a_0\omega}\langle |p_{cv}|^2\rangle = \alpha_{con}(\hbar\omega)\frac{cosh(\pi\gamma)}{exp(\pi\gamma)} \tag{7.77}$$

In the limit $R_x \to 0$, or $E. >> R_x$, $\gamma = (R_x/E) \to 0$ and $exp(\pi\gamma)/cosh(\pi\gamma) \to 1$ and $\alpha_{con} \to \alpha_{bb}$, as expected. This conclusion holds true for bulk material also.

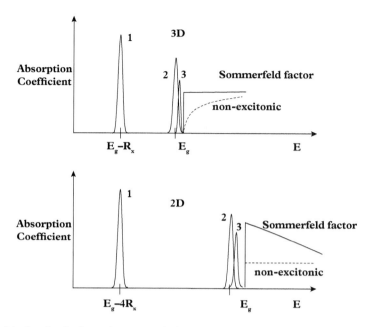

Figure 7.8 *Ideal excitonic absorption spectra in bulk and 2D for discrete and continuum states. The interband absorption spectra are also indicated by dashed lines.*

It may easily be concluded that the oscillator strength increases 16-fold in 2D from the value in bulk semiconductor.

Fig. 7.8 shows the ideal absorption spectra of discrete and continuum states in bulk and QWs. The discrete state absorption is shown broadened instead of using sharp lines.

7.3.3 Recombination of quantum well excitons

In the ideal situation, the exciton is discrete and hence both absorption coefficient and recombination lifetime are infinite. However, exciton states are broadened in actual practice. It is to be remembered that only the state with $\mathbf{K} = 0$ is involved in the recombination.

The radiative recombination rate may be calculated in the same manner as followed for free electron-hole pair recombination. The ground state means absence of electron and f_x denotes the probability of occupation of an excitonic state. With these facts into consideration, the spontaneous recombination rate is expressed as

$$R_{sp}(\hbar\omega) = \left(\frac{2\pi}{\hbar}\right)\left(\frac{eA_0}{m_0}\right)^2 \langle |p_{cv}|^2\rangle C_{mn} \sum_{E_x} |\varphi(0)|^2 G(\hbar\omega) f_x(E_x)\delta(\hbar\omega - E_x) \qquad (7.78)$$

Feldmann et al (1987) used the following arguments to calculate lifetime from Eq. (7.78). Instead of δ-function, a lineshape function, which is homogeneous due to acoustic phonon scattering at low temperature, is to be used. Taking the homogeneous linewidth to be $\Delta(T)$, the states within this width represent the recombination process, instead of a single state with $\mathbf{K} = 0$. Furthermore excitons are thermally distributed and only the fraction of excitons $r(T)$ within the width $\Delta(T)$ can participate in recombination. Assuming Boltzmann statistics, the probability function f_x is written as

$$f_x(E_x) = \frac{2\pi\hbar^2 N_x}{Mk_B T} exp(-E_x/k_B T) \tag{7.79}$$

The fraction $r(T)$ can be expressed now as

$$r(T) = \frac{\hbar^2 N_x}{M}[1 - exp\{-\Delta(T)/k_B T\}] \tag{7.80}$$

We replace δ-function in Eq. (7.78) by $1/\Delta(T)$, write $\varphi(0) = \sqrt{V}(2/\pi)^{1/2}(2/a_0)$, where $E_B = \hbar^2/2m_r a_0^2$, and use the expression for $G(\hbar\omega)$ to obtain the excitonic recombination rate as

$$R_{ex} = \frac{e\eta\hbar\omega^2\langle|p_{cv}|^2\rangle}{\pi m_0^2 \varepsilon c^3 \hbar^2} C_{mn} \frac{8\pi}{M} \frac{2}{\pi} \frac{E_x}{\Delta} \{1 - exp[-\Delta(T)/k_B T]\} N_x \tag{7.81}$$

The rate of change of exciton density is given by

$$\frac{dN_x}{dt} = -R_{ex} = -B_{ex}N_x = -\frac{N_x}{\tau_{ex}} \tag{7.82}$$

The excitonic recombination lifetime is $\tau_x = B_{ex}^{-1}$ and the full expression for B_{ex} appears before N_x in the right-hand side of Eq. (7.81).

Example 7.5 *We estimate the recombination lifetime in a GaAs QW. The mass values are $m_e = 0.067\ m_0$, $m_h = 0.15\ m_0$, so that $m_r = 0.046\ m_0$ and $M = 0.217\ m_0$. Taking the measured values, $E_B = 8\ meV$ and $\Delta = 0.22\ meV$ at $T = 1.85\ K$, $C_{mn} = 1$, standard value for p_{cv}, one obtains $B_{ex} = 5.6\times10^{10}$ and $\tau_x = 180\ ps$ (Kuhl et al 1989).*

Example 7.6 *We calculate the ratio of exciton density and free EHP density in bulk GaAs and GaAs QW at this stage, by using R_x (bulk) $= 4.7\ meV$ and $R_x(QW) = 10\ meV$, and $m_r = 0.057\ m_0$ and $n = 10^{16}\ cm^{-3}$ in both cases. The exciton densities, obtained from Eq. (7.48) are, respectively, $7.02\times10^{14}\ cm^{-3}$ and $8.59\times10^{14}\ cm^{-3}$.*

7.3.4   Binding energy of quasi two-dimensional systems

The QWs in real systems have a finite thickness and the barrier heights in both CB and VB are finite. The wavefunctions therefore leak into the barrier. The theoretical methods to calculate the waefunctions and binding energies include simple variational calculations, sophisticated numerical methods, and empirical methods based on the concept of fractional dimensional space (see, Pedersen 2017).

In this section, we shall present the variational method to illustrate the essential features involved in the quasi-2D nature of excitons in the QW. The infinite barrier model will be presented first for the sake of clarity and then modifications due to finite barrier will be introduced.

The effective mass Schrodinger equation for Coulomb interaction is written earlier in Eq. (7.1). Note that two confining potentials $V_e(z_e)$ and $V_h(z_h)$ are present at the interfaces, respectively at CB and VB. The z-coordinates of electrons and holes are denoted, respectively, by z_e and z_h. In the present case, the CM coordinate $\mathbf{R}$ and relative coordinate ρ are introduced for the QW layer plane, but the kinetic energy parts along z-direction is left unaltered. The effective mass equation for the QW case becomes

$$\left[\begin{array}{c} \frac{\hbar^2}{2M}(K_x^2 + K_y^2) + \frac{\hbar^2}{2m_r}(k_x^2 + k_y^2) + \frac{\hbar^2}{2m_e}k_{ze}^2 + \frac{\hbar^2}{2m_h}k_{zh}^2 \\ -\frac{e^2}{4\pi\varepsilon[\rho^2 + (z_e - z_h)^2]^{1/2}} + V_e(z_e) + V_h(z_h) \end{array} \right] \Psi = E\Psi \qquad (7.83)$$

where

$$M = m_e + m_h \text{ (a)}; \quad m_r = m_e^{-1} + m_h^{-1} \text{(b)}; \quad MR = m_e\rho_e + m_h\rho_h \text{(c)}; \quad \rho = \rho_e - \rho_h \text{ (d)}. \qquad (7.84)$$

The z-components of the wavevector of electrons and holes are denoted by k_{ze} and k_{zh}, respectively, $\mathbf{K}$ (k) and $\mathbf{R}$ (ρ) represent the 2D wavevector and position vector of the CM (relative) motion of the pair, respectively, m_r is the reduced mass and V_e (V_h) are the confining potentials having the values zero inside the well and $\Delta E_c(\Delta E_v)$ at the CB (VB) at the heterointerfaces. The wavefunction is expressed by following a similar one in bulk as:

$$\Psi = exp[i(K_x X + K_y Y)\phi_n(\rho, z_e, z_h)] \qquad (7.85)$$

where the exponential term indicates free motion of the CM along the layer plane.

Assuming first infinite barriers, the following simple variational wavefunction is chosen for 1s excitons:

$$\phi_{1s} = N\cos\left(\frac{\pi z_e}{L}\right) \cos\left(\frac{\pi z_h}{L}\right) exp\left\{-[\alpha_{1s}^2\rho^2 + \beta_{1s}^2(z_e - z_h)^2]^{1/2}\right\} \qquad (7.86)$$

where N is the normalization constant. The method of obtaining the variational parameters α and β and the binding energy has been discussed by Bastard et al (1982). Here we present a variational method using a single parameter, which ignores the electron-hole separation along the z-direction in Eq. (7.86). Thus the wavefunction is

$$\phi_{1s} = A\cos\left(\frac{\pi z_e}{L}\right)\cos\left(\frac{\pi z_h}{L}\right)\exp(-\alpha\rho) \tag{7.87}$$

Using the Hamiltonian for relative motion, $H = \left(\hbar^2 k^2/2m_r\right) - \left(e^2/4\pi\epsilon\rho\right)$, the expectation value is written as $E = \langle \phi_{1s} |H| \phi_{1s}\rangle = \left(\hbar^2\alpha^2/2m_r\right) - \left(e^2\alpha/2\pi\epsilon\right)$, and putting $\frac{\partial E}{\partial \alpha} = 0$, one obtains

$$\alpha = \left(\frac{e^2 m_r}{2\pi\epsilon\hbar^2}\right) \tag{7.88a}$$

$$E(\alpha) = -\frac{e^4 m_r}{8\pi^2\epsilon^2\hbar^2} \tag{7.88b}$$

The binding energy is four times the bulk value.

7.3.4.1 *Variational calculation for finite barriers*

The binding energy of excitons in real QWs has been calculated by a number of authors. Here we present a method due to Priester et al (1984) that effectively transforms the quasi 2D problem into a pure 2D problem. The effective mass equation is written in cylindrical coordinates as follows:

$$\left[E_e(k_0) - E_h(k_0) - \frac{\hbar^2}{2\mu}\left\{\frac{1}{r}\frac{\partial}{\partial r}r\frac{\partial}{\partial r} + \frac{1}{r^2}\frac{\partial^2}{\partial\theta^2} + U(r,\theta)\right\}\right]F(r,\theta) = EF(r,\theta) \tag{7.89}$$

In the previous example, k_0 is the extrema in the respective subbands, r is the 2D position vector, μ is the reduced mass involving the in-plane effective masses, and U is the effective Coulomb interaction and is written in the following way:

$$U(r,\theta) = -\frac{e^2}{4\pi\epsilon}\int\limits_{-\infty}^{\infty}\int\limits_{-\infty}^{\infty}\frac{f_e^{\,2}(z_e)f_h^{\,2}(z_h)}{|r_e - r_h|}dz_e dz_h \tag{7.90}$$

Here, f_e and f_h are the envelope functions for electrons and holes, respectively, for finite barriers and are cos functions for infinite barriers. The following one-parameter variational solution is now used.

$$F(r,\theta) = N\exp(-\alpha r) \tag{7.91}$$

where N is the normalization factor and α is the variational parameter. The expectation value of kinetic energy is $\hbar^2\alpha^2/2\mu$. The following integral is involved in evaluating the expectation value of the potential energy:

$$\langle F|U|F\rangle \rightarrow \int \frac{d\theta dr r exp(-2\alpha r)}{[(z_e - z_h)^2 + r^2]^{1/2}} \tag{7.92}$$

Integration over θ is trivial and integration over r yields

$$G(|z_e - z_h|) = \frac{|z_e - z_h|\pi}{2}[H_1(2\alpha|z_e - z_h|)] - |z_e - z_h| \tag{7.93}$$

(see Gradshteyn and Ryzhik 1980: p. 316). H_1 and N_1 are the first Struve and Neumann functions (Abramowitz and Stegun 1972). The variational energy which needs to be minimized with respect to α is now expressed as

$$E(\alpha) = \frac{\hbar^2\alpha^2}{2\mu} - \frac{e^2\alpha^2}{\pi\in}\int_{-\infty}^{\infty}\int_{-\infty}^{\infty}|z_e - z_h|\left\{\frac{\pi}{2}\left[\begin{array}{c}H_1(2\alpha|z_e - z_h|)\\-N_1(2\alpha|z_e - z_h|)\end{array}\right] - 1\right\}f_e^2(z_e)f_h^2(z_h)\,dz_e dz_h \tag{7.94}$$

The reduced mass μ used in all the previous equations is calculated by using the in-plane effective masses of electrons and holes suitably averaged over the well and barrier materials. As may be noted that the effective mass of holes along the z-direction are given by $1/m_{z\pm} = (1/m_0)(\gamma_1 \mp 2\gamma_2)$ and the in-plane effective masses for HH and LH are expressed as $(1/m_\pm) = (1/m_0)(\gamma_1 \pm \gamma_2)$, where +(−) signs apply to HH (LH) bands. To obtain the average in-plane effective masses, one needs to calculate the probabilities of finding electrons or holes in the well (w) and the barrier (b). The envelope functions for electrons are written as

$$f_e = A\cos(kz); |z| \leq L/2 \text{ and } B exp(-\beta z); |z| \geq L/2; k = \sqrt{2m_w E/\hbar^2};$$
$$\beta = \sqrt{2m_b(\Delta E_c - E)/\hbar^2}$$

A similar equation for holes is used. The coefficients A and B are obtained by matching f_e and $(1/f)(df/dz)$ at $z = L/2$. This yield $k\tan(kL/2) = (m_w/m_b)\beta$. The effective mass is the weighted average of $(1/m_w)$ and $(1/m_b)$, and the weight factors are the probability of finding the particle in respective regions. One finally obtains

$$\frac{1}{m} = \frac{1}{m_w}\left\{\frac{1 + (m_w/m_{zb})(m_{zw}/m_b - m_{zw}/m_w)\cos^2(kL/2)}{[1 + (m_{zw}/m_{zb} - 1)\cos^2(kL/2) + (kL/2)\tan((kL/2))]}\right\} \tag{7.95}$$

The values of binding energy for HH and LH excitons have been calculated. We show the general trend of variation of binding energies with well width in Fig. 7.15. The value is small for very narrow well width, as the envelope functions spread deeply into the barrier layers. The values increase with increasing width, as the envelope functions become more confined until it reaches a maximum. The binding energy then decreases with increasing well width, since the separation of electrons and holes increases to reduce the coulomb interaction. Finally, at large value of well width equal to the 3D Bohr radius, the binding energy reaches the bulk value.

7.3.4.2 *Absorption coefficient*

The method for calculating the absorption coefficient will now be illustrated by following the work by Miller et al (1985). It is assumed that the whole wavefunction is a product of (1) electron envelope function along z, (2) a hole envelope function along z, and (3) an excitonic wavefunction depending on the in-plane separation of electrons and holes. Eq. (7.91) may serve as an example. The ground state of the system has all the states occupied and the excited states are formed by removing an electron from the mth subband and placing it near the nth conduction subband. The hole wavevector has components (k_h, k_{zh}), (k_e, k_{ze}) corresponds to the electron, and for the excitons the in-plane wavevector has components $\mathbf{k} = (m_h \mathbf{k}_e - m_e \mathbf{k}_h)/M$ for relative motion and $\mathbf{K} = \mathbf{k}_e - \mathbf{k}_h$ for the CM motion.

The excited state wavefunction is a linear combination of all the excited state wavefunctions expressed as

$$|\Phi\rangle = \sum_{k_{ze}} \sum_{k_{zh}} \sum_{\mathbf{K}} A |\phi\rangle \tag{7.96}$$

where $|\phi\rangle$ is an excited state wavefunction formed by removing an electron from a VB state and placing it in CB state. The coefficient A is a function of many parameters as will be detailed

$$A = A(m, n, l, k_{ze}, k_{zh}, \mathbf{k}, \mathbf{K}, j)$$

where l and j denote, respectively, the excitonic state (1s, 2s, etc.,) and the VB index (HH, LH, etc.,). The Fourier transform of A is defined as

$$U = \frac{1}{\sqrt{P}} \sum_{\mathbf{k}} \frac{1}{\sqrt{Q}} \sum_{k_{zh}} A \exp[i(k_{ze} z_e + k_{zh} z_h + \mathbf{k} \cdot \mathbf{r})] \tag{7.97}$$

where P and Q are the number of unit cells in the plane, and along the z-direction, respectively, so that $N = PQ$. It is now assumed that U is a product of three terms, so that

$$U = f_{ne}(z_e) f_{mh}^*(z_h) \phi_{lk}(\mathbf{r}) \tag{7.98}$$

The wavefunctions are expressed as

$$f_{ne}(z_e) = \frac{1}{\sqrt{Q}} \sum_{k_{ze}} a_{nk_{ze}} \exp(ik_{ze} z_e); f_{mh}(z_h) = \frac{1}{\sqrt{Q}} \sum_{k_{zh}} b_{mk_{zh}} \exp(ik_{zh} z_h)$$

$$\phi_{lk}(\mathbf{r}) = \frac{1}{\sqrt{P}} \sum_{\mathbf{k}} c_{l,\mathbf{k},\mathbf{k}-\mathbf{K},m} \exp(i\mathbf{k} \cdot \mathbf{r}) \tag{7.99}$$

The absorption coefficient may be calculated from the following matrix element between the ground state $|0\rangle$ and the excited state $|\Phi\rangle$:

$$M_{mnlk} = \langle \Phi | e_\lambda \cdot \mathbf{p} | 0 \rangle \tag{7.100}$$

As noted in connection with bulk excitons, in the transitions $k_{ze} = k_{zh}$ and $\mathbf{K} = 0$. Therefore the matrix element may be written as

$$M_{mnl} = \langle |p_{cv}|^2 \rangle_{QW} \left[\left(\frac{1}{\sqrt{Q}} \right)^2 \sum_{k_z} a^*_{nk_z} b_{mk_z} \frac{1}{\sqrt{P}} \sum_k c_{lk\mathbf{K}} \right] \tag{7.101}$$

where $\langle |p_{cv}|^2 \rangle_{QW}$ is the polarization-dependent momentum matrix element as defined earlier. Using the Fourier expansions in Eq. (7.99), one may write

$$\langle f_{ne} | f_{mh} \rangle = \frac{1}{Q} \sum_z \sum_{k_{ze}} a^*_{nk_{ze}} b_{mk_{zh}} \exp\left[i(k_{zh} - k_{ze})z \right] = \frac{1}{Q} \sum_{k_z} a^*_{nk_z} b_{mk_z} \tag{7.102}$$

In Eq. (7.102) $\sum_z$ indicates sum over all unit cells; also using the usual approximation that a and b are slowly varying functions, $\exp[i(k_{zh} - k_{ze})z] = \delta_{k_{ze}k_{zh}}$. Furthermore

$$\frac{1}{\sqrt{P}} \sum_k c_{lk\mathbf{K}} = \phi_l(0) \tag{7.103}$$

The matrix element Eq. (7.101) is now rewritten by using (7.102) and (7.103), and then the multiplicative factors are included to give the following expression for the absorption coefficient:

$$\alpha(\hbar\omega) = B \sum_m \sum_n \sum_l |\langle f_{ne} | f_{mh} \rangle|^2 |\phi_l(0)|^2 \delta(E_m + E_n + E_l - \hbar\omega) \tag{7.104}$$

where

$$B = \frac{\pi e^2 \langle |p_{cv}|^2 \rangle_{QW}}{m_0^2 c \epsilon_0 \eta_r V \omega} \tag{7.105}$$

Note that $V = LS$, where S is the surface area of the QW. Since $|\phi_l(0)|^2 \propto S^{-1}$, the absorption coefficient given by Eq. (7.104) varies inversely as L, the effective width of the QW.

The calculation of the absorption coefficient requires the expressions for $\phi_l(0)$, f_e and f_h, all of which may be obtained in principle. The calculation should consider all the discrete and continuum states. However, in experimental results, only 1s or one or two more states show the usual peaks, and the other discrete states merge with the continuum due to broadening.

Example 7.7 *The absorption coefficient for 1s HH exciton for GaAs QW is calculated using*
$m_e = 0.067\,m_0$, $m_{hh} = 0.112\,m_0$, *giving a value* $\mu = 0.042\,m_0$ *for in-plane reduced mass,*
and $a_0 = 16.5\,nm$. *The* δ-*function in Eq. (7.104) is replaced by a Lorentzian with half*
width $\Gamma/2 = 4\,meV$. *This gives a multiplicative factor* $(\pi\Gamma/2)^{-1}$, *in the expression for*
peak absorption coefficient. Taking $L = 10\,nm$, $\hbar\omega = E_g$, $\langle|p_{cv}|^2\rangle_{QW} = 1.5 \times 2.7\,m_0 E_g$,
where 1.5 is the anisotropy factor, $\alpha_{peak} = 1.82 \times 10^4\,cm^{-1}$, *which is about 1.5 times*
the experimental value. The deviation is due to assumption of total confinement and the
value of peak energy chosen.

Experimental absorption spectra of high quality bulk GaAs sample, $\sim 1\mu m$ thick, and
50 periods of 10 nm GaAs/10 nm $Al_{0.3}Ga_{0.7}As$ multiple quantum waves (MQW)s of
total thickness 0.5 μm, are shown in Fig. 7.5 (see Schmitt-Rink et al 1989). The peak
for bulk sample is barely visible. For MQWs, the HH and LH peaks are resolved for
$n = 1$ (1s) state, but for peaks corresponding to $n = 2, 3$ states, the peaks merge. The
plateau between peaks is due to continuum states, where the 2D DOS is constant.

Experimental data near the 1s exciton peak for n = 1 subbands, which follow similar
variation as shown in Fig. 7.5, may be fitted by the following empirical relation proposed
by Chemla et al (1984):

$$\alpha(\hbar\omega) = \alpha_{hh}\exp[-(\hbar\omega - \hbar\omega_{xh})^2/2(\hbar\Gamma_{hh})^2] + \alpha_{lh}\exp[-(\hbar\omega - \hbar\omega_{xl})^2/2(\hbar\Gamma_{lh})^2]$$

$$+\frac{\alpha_c}{1 + \exp[(\hbar\omega_c - \hbar\omega)/\hbar\Gamma_c]} \times \frac{2}{1 + \exp[-2\pi\{(\hbar\omega_c - \hbar\omega)/R_x\}^{-1/2}]}$$

$$(7.106)$$

In Eq. (7.106), α_{hh}, α_{lh} and α_c refer respectively to the heavy hole, light hole and
continuum states, $\hbar\omega_{xh}$ and Γ_{hh}, are respectively the energy for peak absorption, and
broadening parameter for HH exciton and the other parameters may similarly be de-
fined. The empirical relation, along with the following parameters given in Example
7.8, may reproduce the experimental data.

Example 7.8 *The peak absorption coefficients obtained by fitting the experimental data are:*
$\alpha_{hh} = 11 \times 10^5$, $\alpha_{lh} = 6.3 \times 10^5$, $\alpha_c = 2.7 \times 10^5\,m^{-1}$. *The photon energies at which the*
peaks occur are $\hbar\omega_{xh} = 1.4575$, $\hbar\omega_{xl} = 1.466$ *and* $\hbar\omega_c = 1.4675\,eV$. *The broadening*
parameters are $\hbar\Gamma_{hh} = 3$, $\hbar\Gamma_{lh} = 3$, *and* $\hbar\Gamma_c = 5\,meV$, *and* $R_x = 10\,meV$. *The reader*
may reproduce the experimental curve obtained by Chemla et al (1984) by using these
parameters.

7.3.5 Line broadening mechanisms for two-dimensional excitons

Though the ideal absorption spectra in QWs contain discrete lines followed by a continuum, the experimental results show only one or two broadened peaks and then a
continuous variation. As in bulk, the broadening of lines occurs due to interaction of

excitons with various scattering agents, like phonons, impurities, defects and inhomogeneities. The observed linewidth for the lowest state may be expressed by the same Eq. (7.49). The first term represents the inhomogeneous linewidth, mainly arising out of fluctuations of well width and of alloy compositions and other defects. The second term is due to interactions between excitons and acoustic phonons: both deformation potential and piezoelectric. The interaction between excitons and LO phonons gives rise to the last term. Expressions for the transition rates for different phonon scattering processes are available (Basu and Ray 1991).

Example 7.9 *The expression for transition probability due to deformation potential acoustic phonon scattering is* $W_{dp}(\mathbf{K} = 0) = [(3k_B TM)/(4\hbar^3 \rho_m v_s^2 L)](D_e - D_h)^2$. *Take* $m_e = 0.067m_0$, $m_h = 0.34m_0$, $|D_e - D_h| = 7\ eV$, $v_s = 5.22 \times 10^5\ cm.s^{-1}$, $\rho_m = 5.36\ gm.cm^{-3}$ *and* $L = 10\ nm$. *The linewidth given by* $\Gamma = \hbar W/2$ *is* $0.927\ \mu eV \times T$, *and thus has a value* $0.28\ meV$ *at* $300\ K$.

7.3.5.1 *Inhomogeneous linewidth*

In all earlier discussions, the subband energies, exciton binding energies and linewidth are calculated by considering that the well width L is constant throughout the structure, meaning thereby that the heterointerfaces are atomically smooth. In practical QWs, however, there is always some roughness at the interfaces, so that there is a fluctuation in the well width.

The situation is illustrated in Fig. 7.9. The QW in the upper part of Fig. 7.9 has a mean well width L, but due to imperfections in the growth process, the width is extended in some regions and is reduced at some other regions. These changes occur along the *x-y* plane, but are small compared to the exciton Bohr radius, as depicted. The shift in the well width δL introduces a change in the electron subband energy, which may be expressed as $E_e = (\hbar^2/2m_e)(\pi^2/L^3)\delta L$, assuming the simple expression valid for infinite barrier height (Weisbuch et al 1981). There will be a similar change δE_h for the hole subband. In principle, the exciton binding energy should also change, but we might neglect the change in comparison to δE_e and δE_h. The exciton peak is still visible at $E_{ex} = E_g + E_e + E_h - E_B$, but is broadened having a linewidth given by

$$\Gamma = \delta E_e + \delta E_h = (\hbar^2/2)(\pi^2/L^3)\left(\frac{1}{m_e} + \frac{1}{m_h}\right)\delta L \tag{7.107}$$

Usually for good quality samples, $\delta L = \pm a/2$, where a is the lattice constant. The observed linewidth has been found to give reasonable agreement with the calculation.

When growth is interrupted for some time, the interfaces become more ordered as shown in the lower part of Fig. 7.9. Large islands of typically one monolayer height extending over few tens of a μm along the x-y plane are formed under this condition. The result is the formation of different QWs with widths L, $L + \delta L$, and $L - \delta L$. The well width fluctuations are of the order of a monolayer. Each island is long enough to accommodate one exciton as shown. The absorption and luminescence spectra in this situation contain three well resolved peaks at energies corresponding to L and $L \pm \delta L$.

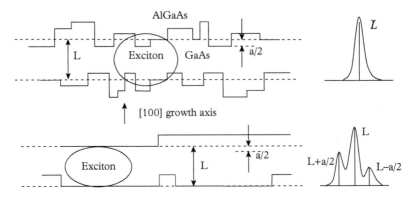

Figure 7.9 *Illustration of IFR in QW. (upper) random fluctuation of well width L due to growth; (lower) ordered interface arising out of interrupted growth process.*

The expressions for linewidth due to interface roughness scattering and alloy disorder scattering in QWs have been derived by several authors (see, for example, Basu 1991, Basu 1990; Singh and Bajaj 1985).

7.4   Effect of electric field in semiconductors

The effect of external electric field in the interband optical processes in bulk semiconductor has been studied long ago taking also into consideration the excitonic processes. Similar studies have also been conducted for QWs. In this section, the effect on bulk semiconductor will be discussed briefly, giving more emphasis in the processes occurring in QWs.

7.4.1   Effect of electric field in bulk semiconductors

The effect of external electric field on the absorption processes in bulk semiconductors was studied independently by Franz (1958) and by Keldysh (1958). The theory of electroabsorption considering or ignoring Coulomb interaction is elaborate and the reader is referred to the other sources (Basu 1997).

The absorption spectra in bulk semiconductor show a tail below the band gap energy and a decaying oscillatory nature for photon energies exceeding the band gap. The oscillation is known as the Franz–Keldysh oscillation. The phenomena can be explained qualitatively with the band diagram shown in Fig. 7.10. In (a) the field is small and there is little overlap between electron and hole wavefunctions. The absorption coefficient depends on the overlap and is low for $\hbar\omega < E_g$. In (b), the field is larger and absorption coefficient increases due to enhanced overlap. For $\hbar\omega > E_g$, as in (c) the relative phases of the envelope functions change and give rise to oscillation.

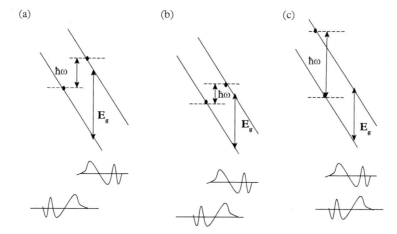

Figure 7.10 *Band diagrams to explain Franz Keldysh effect for different photon energies; (a) ħω<E$_g$, smaller field and smaller e-h overlap; (b) ħω<E$_g$, but larger field increases e-h overlap and hence absorption; (c) ħω>E$_g$, relative phases of wavefunctions produce oscillations.*

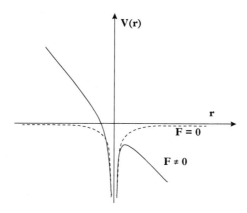

Figure 7.11 *Field ionization of excitons. Dashed lines correspond to profile of the Coulomb potential without field. Solid lines illustrate the asymmetry in the potential induced by the field. The lowering of potential in the right-hand side facilitates field ionization.*

The inclusion of excitonic effect has also been considered. The nature of spectra is the same. An additional point to note is that under sufficient field, the excitons are ionized, that is field tears apart the electron and hole. The effect is similar to the Stark effect in H-atom, see Fig. 7.11.

7.4.2   Effect of electric field on quantum wells

Application of an external electric field to a QW brings about interesting changes in optical properties. The electroabsorption and associated electrorefraction are orders of magnitude higher than in bulk semiconductors. Some novel photonic devices are already being used in practical systems exploiting these effects. In this section, we shall first discuss the effect of an electric field on the subband energies and envelope functions. Analytical results will be quoted mainly. The effect of field on the excitonic absorption and refraction will then be treated.

7.4.2.1   *Electric field perpendicular to the quantum well layer*

We assume that the electric field is applied along the z-direction, the direction along which the QW is grown. For simplicity, we assume infinite barrier, and take the origin of the z-axis at the centre of the well of width L. The Hamiltonian is then

$$H = -\frac{\hbar^2}{2m_e}\frac{\partial^2}{\partial z^2} + eFz = H_0 + eFz \tag{7.108}$$

where H_0 is the Hamiltonian without the field. For weak fields, the condition $eFL \ll (\hbar^2/2m_e)(\pi/L)^2$ is valid and using second-order perturbation theory, the change in energy for the n-th subband is written as

$$\Delta E_n = \sum_m{}' \frac{|\langle m0|eFz|n0\rangle|^2}{E_{m0} - E_{n0}} = \frac{2e^2 F^2 m_e L^2}{\hbar^2 \pi^2} \sum_m{}' \frac{|\langle m0|z|n0\rangle|^2}{n^2 - m^2} \tag{7.109}$$

The unperturbed subband energy for the n-th subband is given by

$$E_{n0} = \frac{\hbar^2}{2m_e}\left(\frac{n\pi}{L}\right)^2$$

The subscript 0 is used to denote zeroth order envelope functions and eigenvalues. Since $|n0\rangle$ and $|m0\rangle$ are sin functions, the matrix element in Eq. (7.109) may easily be expressed as

$$\langle m0|z|n0\rangle = \frac{8L}{\pi^2}\frac{nm}{n^2 - m^2} \tag{7.110}$$

The change in energy for the lowest subband (n = 1) may be calculated by summing over all possible m's starting from m = 2, and the result is

$$\Delta E_1 = -\frac{1}{24\pi^2}\left(\frac{15}{\pi^2} - 1\right)\frac{e^2 F^2 m_e L^4}{\hbar^2} \tag{7.111}$$

The hole subband energy also decreases in the same manner as the square of the electric field. There is therefore an overall decrease in the effective band gap due to the applied field.

Example 7.10 *Let* $F = 10^5$ *V/cm*, $L = 10$ *nm. The change in energy of the first electron subband is 1.93 meV.*

The previous perturbation calculation is not valid for large fields and thick wells, for which a variational calculation suits better. The applied field pushes the electron and hole wavefunctions in the opposite directions to the walls of the well, as depicted in Fig. 7.12. This brings asymmetry. Bastard et al (1983) chose the following form of the variational envelope function:

$$\phi(z) = N(\beta) \cos\left(\frac{\pi z}{L}\right) exp\left[-\beta\left(\frac{z}{L} + \frac{1}{2}\right)\right], \quad \frac{|z|}{L} \leq \frac{1}{2} \tag{7.112}$$

where β is the variational parameter and $N(\beta)$ is the normalization factor. As $F \to 0$, $\beta \to 0$, and the usual wavefunction for infinite barrier without field is recovered. On the other hand, for large F, the envelope functions are pushed to the barrier, as shown in Fig. 7.12 and the wavefunction is similar to Fang–Howard wavefunction (1966) valid for inversion layer in metal-oxide semiconductor field-effect transistors (MOSFET)s. The normalization factor is expressed as

$$|N(\beta)|^2 = \frac{2\beta(\beta^2 + \pi^2)exp(\beta)}{\pi^2 L \sinh \beta} \tag{7.113}$$

The expectation value of the subband energy, calculated in the usual fashion, may be expressed as

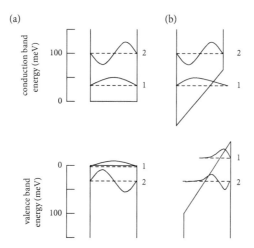

Figure 7.12 *Illustration of the effect of electric field in QW. (a) Subband and envelope function in conduction and valence bands without field; (b) Tilting of band edges in presence of field and the nature of envelope functions with electric field. The envelope functions become asymmetric and electron and hole envelope functions are pushed to opposite directions.*

$$E(\beta) = E_{10} \left[1 + \frac{\beta^2}{\pi^2} + \frac{\xi}{2} \left\{ \frac{1}{\beta} + \frac{2\beta}{\beta^2 + \pi^2} - \frac{1}{2} \coth \beta \right\} \right] \tag{7.114}$$

where $\xi = eFL/E_{10}$.

As usual, the energy is obtained by minimizing $E(\beta)$ with respect to β. For high values of the field β is large and Eq. (7.114) may be written as

$$E(\beta) = E_{10} \left[1 + \frac{\beta^2}{\pi^2} + \frac{\xi}{2} \left(\frac{3}{\beta} - \frac{1}{2} \right) \right] \tag{7.115}$$

Putting $\frac{\partial E}{\partial \beta} = 0$ gives

$$\beta_{min} = \left(\frac{3\xi\pi^2}{4} \right)^{1/3} \tag{7.116}$$

From this the change in lowest subband energy (n = 1) is obtained as

$$\Delta E_1 = -\frac{eFL}{2} + \left(\frac{3}{2} \right)^{5/3} \left[\frac{e^2 F^2 \hbar^2}{m_e} \right]^{1/3} \tag{7.117}$$

On the other hand, for small values of the field, Eq. (7.114) takes the following form:

$$\Delta E_1 = -\frac{1}{8} \left[\frac{1}{3} - \frac{2}{\pi^2} \right]^2 \frac{m_e e^2 F^2 L^4}{\hbar^2} \tag{7.118}$$

A comparison of the prefactors in Eqs. (7.111) and (7.118) points out that there is only 3% difference between the variational and perturbational calculations. The following expression for the energy is obtained from a semiclassical analysis assuming infinite barriers and a linearly varying potential eFz:

$$E_{SC} = -\frac{eFL}{2} + \left(\frac{\pi^2}{12} \right)^{\frac{1}{3}} \left(\frac{3}{2} \right)^{\frac{5}{3}} \left(\frac{e^2 F^2 \hbar^2}{m_e} \right)^{\frac{1}{3}} \tag{7.119}$$

The variational calculation has been extended with the inclusion of higher subbands (Ahn and Chuang 1987).

As noted earlier for infinite well the absorption coefficient vanishes when the electron and hole subband indices satisfies the inequality $m \neq n$, due to orthogonality condition. However in the presence of an electric field this is no longer valid even for $m \neq n$.

We now present a more general approach. The electron and hole envelope functions obey the following equation:

$$\left\{ -\frac{\hbar^2}{2m_i}\frac{d^2}{dz_i^2} \pm eFz_i \right\} \phi_{in}(z_i) = E_{in}\phi_{in}(z_i), \ i = e, h \tag{7.120}$$

In Eq. (7.120) + and − signs apply to electrons and holes, respectively. The envelope functions are made of two independent solutions of the Airy equation $Ai(x)$ and $Bi(x)$ (Abramowitz and Stegun 1972). Therefore, one may write

$$\phi_{in}(z_i) = aAi(x_{in}) + bBi(x_{in}) \tag{7.121}$$

The arguments of the Airy functions may be written as

$$x_{in} = -\left\{ \frac{2m_i}{(e\hbar F)^2} \right\}^{1/3} (E_{in} \mp eFz_i) \tag{7.122}$$

We obtain the following equations from the boundary conditions:

$$Ai(x_+)Bi(x_-) = Ai(x_-)Bi(x_+) \text{ and } \frac{b}{a} = \frac{Ai(x_+)}{Bi(x_-)} = \frac{Ai(x_-)}{Bi(x_+)}$$

where $x_\pm$ correspond to $z = \pm L/2$.

After obtaining the solutions, they can be plugged into the following expression for the absorption coefficient:

$$\alpha(E) = \frac{c}{L}\sum_{k_{\parallel},n,r} \delta\left(E - E_g - \frac{\hbar^2 k_{\parallel}^2}{m_r} - E_{en} - E_{hm} \right) I_{mn} \tag{7.123}$$

The symbols used in Eq. (7.123) are defined as

$$c = \frac{e^2\hbar^2\langle|p_{cv}^2|\rangle}{2\varepsilon_r\epsilon_0 A m_0^2 (E)^2}, E = \hbar\omega$$

$$I_{mn} = \frac{|\int \phi_{en}(z)\phi_{hm}(z)dz|^2}{\int |\phi_{en}(z)|^2 dz \times \int |\phi_{hm}(z)|^2 dz}$$

The area of the slab is denoted by A so that the volume is $A \cdot L$.

As the density of states is constant, the summation over in-plane wave vector $k_\parallel$ in Eq. (7.123) can easily be performed by using the property of δ-function. The absorption coefficient may be expressed now as

$$\alpha(E) = \frac{CA\rho_{2D}}{L} \sum_{m,n} H(E - E_g - E_{en} - E_{hm})I_{mn} \qquad (7.124)$$

where C is another constant. Miller et al (1986a) have shown that as the slab thickness is large the previous equation may lead to Franz–Keldysh oscillation characteristics of bulk semiconductor.

The nature of envelope functions and the positions of the subbands with respect to the positions in a QW without field are shown in Fig. 7.12. It is seen that the envelope functions become asymmetrical, pushed to the boundaries of the well and the shifts oppositely for electrons and holes. The variation of absorption coefficient as given by Eq. (7.124) should be step-like.

7.4.2.2   *Intersubband transitions*

Absorption from a lower conduction (or valence) subband to a higher subband in the presence of an applied electric field has led to infrared detectors in the mid IR range. The opposite transition occurs in a Quantum Cascade Laser (Faist et al 1994). In this section, a brief description of the characteristics of such intersubband transitions is given.

The absorption coefficient involves the following matrix element:

$$M_{if} = \frac{m_0(E_i - E_f)}{i\hbar} \int \phi_f^*(z)z\phi_i(z)dz \qquad (7.125)$$

where i and f denote quantities related to initial and final states. The matrix element is polarization dependent. The following conclusions may be drawn.

(1) The envelope functions are altered by the field which leads to a change in the overlap integral. The matrix element and the absorption coefficient are changed from the value without field.

(2) Although the ground subband energy decreases quadratically with field, energy of the next higher subbands increases slightly. The absorption peak occurs at a higher photon energy.

(3) As the envelope functions shift to the same side, the overlap function increases to increase the absorption.

(4) The broadening may also increase to negate the increase in absorption.

(5) Intersubband transitions, like $1 \to 3$ transition, forbidden for $F = 0$, are now allowed.

Experimental results by Harwit and Harris (1987) show a decrease in absorption due to increased line broadening caused by electric field. The peak however shifts to higher energy.

7.4.2.3 *Excitonic effects*

We first discuss qualitatively the effects of electric field applied parallel and perpendicular to the QW layer plane and then present the theory of electroabsorption for the more important case when the field is applied normal to the QW layer plane.

Field parallel to the QW layer plane   The effect of the electric field is to shift the exitonic absorption peak to lower energies with increasing electric field. This is due to hydrogenic Stark shift. In addition, the absorption spectra become broadened due to additional contribution from field ionization which shortens the lifetime of excitons. At a sufficiently large electric field excitonic signature is not visible in the absorption spectra. In this way, the behaviour of excitons is not much different from that exhibited by bulk excitons. However, in bulk semiconductors complete field ionization of excitons take place at a lower electric field. In QWs this occurs at a higher electric field due to larger binding energy of the 2D excitons. For further discussion and experimental spectra, the reader is referred to the paper by Miller et al (1985).

Field perpendicular to the QW plane   The effect of a perpendicular electric field on excitonic electroabsorption in a GaAs QW is studied by Miller et al (1986b) and the spectra are different for TE and TM polarizations. The spectra are shown in Fig. 17.5 of Basu (1997: Reading list) and the variation is qualitatively the same as shown in Fig. 7.13. In the former contributions come from both HH and LH bands, whereas in the latter only LH states contribute. In both the cases, increasing electric field causes shift of excitonic peaks to lower energies, and introduces more broadening. However, the broadening is not as large as in the parallel field situation, and excitons survive even at a field of the order of 2×10^5V/cm.

  The exciton peak should occur at an energy given by

$$E = E_g + E_{1e} + E_{1hh} - E_B \tag{7.126}$$

The field dependent subband energies E_{1e} and E_{1hh} change almost as F^2 and the exciton binding energy E_B is also field dependent.

Example 7.11 *Using Eq. (7.126), F = 10^5V/cm and L = 10 nm, the changes in subband energies for electrons and holes are respectively 10 meV and 3 meV. The binding energy is ∼ 10 meV and it decreases with increase in F. The shift in peak energy is primarily due to changes in subband energies.*

The effect is called the Quantum Confined Stark Effect (QCSE) as the Stark effect is prominent. Excitons survive at high field since the walls at the hetero interface prevent the electron and hole from being torn apart. Note that these walls do not exist in the parallel field configuration, thus making the field ionization of the exciton easier. The decreasing overlap between electrons and holes reduces the absorption oscillator strength.

Calculation of binding energy   Neglecting the in-plane kinetic energy due to centre-of-mass motion the Hamiltonian becomes

$$H = \frac{p_{ze}^2}{2m_e} + \frac{p_{zh}^2}{2m_h} + V_e(z_e) + V_h(z_h) + \frac{e^2}{4\pi\varepsilon|r^2 + z^2|^{1/2}} - \frac{1}{2m_\parallel}\nabla_\parallel^2 + eFz_e - eFz_h \quad (7.127)$$

Here z_e and z_h are, respectively, the electron and hole coordinates and $z = |z_e - z_h|$. The following variational solution with separable trial function is chosen:

$$\Phi(\mathbf{r}, z_e, z_h) = \phi_e(z_e)\phi_h(z_h)\phi_{eh}(\mathbf{r}) \quad (7.128)$$

The wavefunction for the electron under electric field satisfies the equation

$$\left[\frac{p_{ze}^2}{2m_e} + V_e(z_e) + eFz_e\right]\phi_e = E_e\phi_e \quad (7.129)$$

A similar equation may be written for holes. For an excitonic wavefunction we use a simple one-parameter variational wavefunction, using a variational parameter λ as used in Chapter 6, that is

$$\phi_{eh}(\mathbf{r}) = \left(\frac{2}{\pi}\right)^{1/2}\frac{1}{\lambda}\exp(-r/\lambda) \quad (7.130)$$

The expectation value is given by

$$\langle\Phi|H|\Phi\rangle = E_e + E_h + E_B \quad (7.131)$$

Using Eq. (7.130) the kinetic energy due to relative motion may be written as

$$H_{KE} = \langle\phi_{eh}|H_{KEr}|\phi_{eh}\rangle \quad (7.132)$$

Since it is difficult to calculate the potential energy term Miller et al (1985) used the wavefunctions as given by Eq. (7.112) valid for infinite well, but used effective well widths L_e and L_h, respectively for electrons and holes, in place of actual well width L. These effective well widths are so chosen that they give the same values of subband energies calculated by using finite barrier height and width L. Thus, $E_{e,h} = (\hbar^2/2m_{e,h})(\pi/L_{e,h})^2$. The potential energy term is now expressed as

$$E_{PE} = -\frac{e^2}{2\pi^2\varepsilon\lambda}N^2(\beta_e)N^2(\beta_h)\int_0^\infty\int_{\theta=0}^{2\pi}\int_{z_e=-L/2}^{L/2}\int_{z_h=-L/2}^{L/2}\cos^2\left(\frac{\pi z_e}{L_e}\right)\exp\left[-2\beta_e\left(\frac{z_e}{L_e}+\frac{1}{2}\right)\right]$$

$$\times\cos^2\left(\frac{\pi z_h}{L_h}\right)\exp\left[-2\beta_h\left(\frac{z_h}{L_h}+\frac{1}{2}\right)\right]\times\frac{rexp(-2r/\lambda)}{\left[(z_e-z_h)^2+r^2\right]^{1/2}}d\theta dr dz_e dz_h \qquad (7.133)$$

The normalization factors are defined in Eq. (7.113). It is easy to integrate over θ; integration over r yields

$$G(\gamma) = \frac{2}{\lambda}\int_0^\infty\frac{rexp(-2r/\lambda)dr}{[\gamma^2+r^2]^{1/2}} = \frac{2|\gamma|}{\lambda}\left\{\frac{\pi}{2}\left[H_1\left(\frac{2|\gamma|}{\lambda}\right)-N_1\left(\frac{2|\gamma|}{\lambda}\right)-1\right]\right\} \qquad (7.133\,a)$$

H_1 and N_1 are, respectively, the first-order Struve and Neumann functions (Miller et al 1985; Abramowitz and Stegun 1972). In order to evaluate E_{PE}, the authors used the values of H_1 and N_1 from tables and performed numerically the integration over z_e and z_h. The binding energy is obtained from the condition $\frac{\partial E_B}{\partial\lambda} = 0$ and the two other parameters β_e and β_h are obtained variationally by considering the two particles separately.

Electroabsorption and electrorefraction   The absorption spectra in MQWs under electric field have been calculated by different authors using various degrees of approximation. For the sake of illustration we briefly mention the work by Stevens et al (1988) on electroabsorption that successfully explains the experimental data.

It is assumed that each exciton is associated with a pair of electron and hole subbands (i and j). The overall absorption is due to contributions from different bound states associated with a pair of subbands and from continuum states. The expression, without considering broadening, is

$$\alpha(\hbar\omega,F) = \sum_{ij}\frac{\pi e^2\hbar}{2m_0\eta_r cV}f_{ex}(i,j)\delta[\hbar\omega - E_{ij}(F)]$$

$$- E_{Bij}(F)]\int_{E_{ij}(F)}^\infty\frac{\pi e^2\hbar}{2m_0\eta_r cV}f_{con}(i,j)\rho_{2D}K\{\hbar\omega, E_{Bij}(F)\}dE \qquad (7.134)$$

Here V is the volume, E_{ij} is the transition energy between ith electron and jth hole subbands, E_{Bij} is the exciton binding energy, f_{ex} and f_{con} are the oscillator strengths for 1s exciton state and continuum state transitions, respectively, ρ_{2D} is the joint DOS, and K is the 2D Sommerfeld factor. The expressions for the oscillator strengths are $f_{ex} = 2(p_{i,j})^2/(m_0 E_{ij})$ and $f_{con} = f_{ex}/12R_x$.

7.4.2.4   *Empirical relation for absorption coefficient*

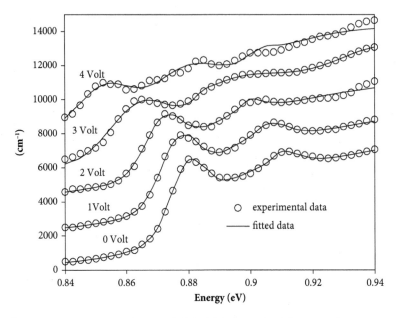

Figure 7.13 *Excitonic electro absorption in Si/SiGe MQWs. Each of the curves for 1 – 4 volts has been shifted by 2000 cm⁻¹ from the curve for immediate lower value of bias.*

It has been shown in an earlier work by Chemla et al (1984) that an empirical expression can satisfactorily model the experimental absorption data, see Fig. 7.13.

The expression involves 1s states of HH and LH excitons and continuum states taking broadening into account. In the following we illustrate this method by considering excitonic absorption in Ge QWs sandwiched between two strained $Si_{1-x}Ge_x$ barriers. The excitonic states are formed between HH and LH subbands and the first conduction subband due to direct Γ valley in Ge. It may be noted that in Ge direct Γ valley occurs only 140 meV above the indirect L valleys.

The following empirical expression that takes into account two discrete excitonic transitions (hh-e, and lh-e) as well as absorption by the 2D continuum states has been used (Basu et al 2009):

$$\alpha(\hbar\omega) = \alpha_h \, exp\left[-\frac{(\hbar\omega - \hbar\Omega_h)^2}{2(\hbar\Gamma_h)^2}\right] + \alpha_l \, exp\left[-\frac{(\hbar\omega - \hbar\Omega_l)^2}{2(\hbar\Gamma_l)^2}\right]$$

$$+ \frac{\alpha_c}{1 + exp\left(\frac{\hbar\Omega_c - \hbar\omega}{\hbar\Gamma_c}\right)} \times \frac{2}{1 + exp\left\{2\pi[|\hbar\Omega_c - \hbar\omega|/R_y]^{-1/2}\right\}} \qquad (7.135)$$

where $\hbar\omega$ is the photon energy, $\hbar\Omega$ denote the excitonic peak energy, Γ's represent the linewidths half-width at half-maximum (HWHM), α's are fitting parameters, R_y stands for exciton Rydberg, and subscripts h, l, and c correspond, respectively, to HH, LH, and continuum states. The two different continuum contributions from the HH and LH subbands have not been included separately.

The agreement between experimental curves (Kuo et al 2004) and the values calculated with suitable chosen parameters is excellent as shown in Fig. 7.9.

7.5 Excitonic characteristics in fractional dimensional space

It is natural to conclude that excitonic motion in a QW is not purely 2D in nature, since the thickness of the well is finite and the potential discontinuities at the heterobarriers are finite to make the envelope functions penetrate into the barrier layers. The relative motion of electrons and holes, which constitute the excitons, is at best anisotropic. The binding energy of excitons, as noted in earlier sections, equals neither the 3D (bulk) nor the 2D (QW with zero well thickness) values.

It has been stated in this chapter that for 3D and ideal 2D structures, analytical expressions for excitonic binding energy and absorption coefficients are available and the corresponding theory have already been presented. For actual QWs, the models, primarily based on variational approach, have been introduced. However, better models rely mostly on numerical calculations. Even then, these models can calculate the binding energies only, but are not suitable to obtain the correct absorption spectra. Chuang et al (1991) have established that Green's function approach can correctly reproduce the absorption spectra even in the continuum region. Again, though reliable the method is, it requires numerical work.

A very useful approach has been followed by a number of workers to obtain reliable expression for binding energy and absorption spectra in QWs and even in lower-dimensional structures like QWRs and QDs (He 1990, 1991; Christol et al 1993, 1995; Lefebvre et al 1993; Lohe and Thilagam 2004; Mathieu et al 1992; Pedersen et al 2016; Pedersen 2017). In these studies, the Coulomb interaction between electrons and holes in nanostructures, is characterized by a single-dimensionality parameter α. In type I structures α continuously vary from 2 to 3 and from 1 to 3 in QWRs. In other words, the dimension is a fraction.

In the following, we shall introduce the fractional dimensional space, the effective mass equation, and the method to calculate the binding energy and absorption spectra for QWs in terms of this fractional dimensional parameter α. Similar treatment for QWRs will be presented in Chapter 8.

7.5.1 Wannier equation in fractional dimension

The effective mass (Wannier) equation in α-dimensional (α-D) space can be written as

$$\left[-\frac{\hbar^2}{2\mu} \frac{1}{r^{\alpha-1}} \frac{\partial}{\partial r} \left(r^{\alpha-1} \frac{\partial}{\partial r} \right) - \frac{e^2}{4\pi \in r} - \frac{1}{2\mu r^2} \frac{\hbar^2}{\sin^{\alpha-2}\theta} \cdot \frac{\partial}{\partial\theta} \left(\sin^{\alpha-2}\theta \frac{\partial}{\partial\theta} \right) \right]$$

$$\psi(r,\theta) = E\psi(r,\theta) \quad (7.136)$$

Here, the relative motion is described in the generalized (r, θ) coordinate, and μ is the reduced mass. The envelope function is written as $\psi(r, \theta) = R(r) \cdot \Theta(\theta)$, where $\Theta(\theta)$ is an eigenfunction of L^2 with eigenvalues $l(l + \alpha - 2)\hbar^2$. Eq. (7.136) then leads to two separate equations. The angular equation has solutions in terms of Gegenbauer polynomials $C_l^{\alpha/2-1}(\cos\theta)$, which reduce to Legendre polynomials for $\alpha = 3$, and to Chebyshev polynomials for $\alpha = 2$ (Abramowitz and Stegun 1972). The normalization constants determine the absorption coefficients. The radial part of the envelope function gives rise to bound states defined by two integer quantum numbers n and l. The quantized energies are expressed as

$$E_n = -R_x/[n + (\alpha - 3)/2]^2 \tag{7.137}$$

For bulk $\alpha = 3$ and for ideal 2D $\alpha = 2$, and in both the cases the respective expressions already derived earlier are recovered.

The expression for the absorption coefficient for an α-D medium is given as

$$\alpha_{abs}(\hbar\omega) = \alpha_0 \left\{ \sum_{n=0}^{\infty} \frac{R_x \Gamma(n + \alpha - 2)}{(n - 1)!\left(n + \frac{\alpha-3}{2}\right)^{\alpha+1}} \delta(\hbar\omega - E_n) + \frac{\left|\Gamma\left(\frac{\alpha-1}{2} + i\gamma\right)\right|^2 e^{\pi\gamma} \gamma^{2-\alpha}}{2^{\alpha}\pi^{2-\alpha}\Gamma(\alpha/2)} Y(\hbar\omega) \right\} \tag{7.138}$$

In Eq. (7.138), Y(x) represents Heaviside step function, $\Gamma(x)$ is the Euler gamma function and $\gamma = \sqrt{R_x/\hbar\omega}$. The prefactor is expressed as

$$\alpha_0 = \frac{2^{2\alpha-1}\omega|d_{cv}|^2[\Gamma(\alpha/2)]^2\Gamma\{(\alpha - 1)/2\}}{\pi^{(\alpha-3)/2}\eta_r c R_x a_B^{\alpha} L^{2-\alpha}[\Gamma(\alpha - 1)]^3} \tag{7.139}$$

Here the symbols have the same significance as defined earlier, and $|d_{cv}|^2 = |\langle c|d|v\rangle|^2$, is the squared dipole matrix element for transition from VB to CB subbands, that includes the overlap between the envelope functions of electrons and holes and hence determine the selection rule.

The first term on the right-hand side of Eq. (7.138) represents absorption due to bound states and the second term accounts for the absorption by continuum states. It may be proved that the second term reproduces the joint DOS for α-D space at high energy $[E \to \infty, \gamma \to 0]$.

7.5.2   Binding energy in quantum wells

The exciton bound state energies and wavefunctions can be calculated as a function of spatial dimension α which describes the degree of anisotropy of the electron-hole interaction. To express α, a pertinent dimensionless parameter, β is chosen as

$$\beta = \left\langle \frac{|z_e - z_h|}{a_0} \right\rangle = \int\limits_{-\infty}^{+\infty} dz_e dz_h \frac{|z_e - z_h|}{a_0} |f_p{}^e(z_e)|^2 |f_q{}^h(z_h)|^2 \qquad (7.140)$$

β expresses the average electron-hole distance along the z-direction: the direction of quantum confinement, $f_p{}^e(z_e)$ and $f_q{}^h(z_h)$ are the envelope functions of electron and hole respectively corresponding to the pth (qth) electron (hole) quantum level and a_0 is the 3D effective Bohr radius.

The fractional dimension α may be related to β by a simple relation like

$$\alpha = 3 - e^{-\beta} \qquad (7.141)$$

Christol et al (1993) used $\beta = L/(2a_0)$, where L is the quantum well width. This is modified for wells of finite barrier. Considering a 1D motion of a particle of effective mass m^* in a quantum well of depth V and well width L, the bound state energy, E of the particle can be determined using the following transcendental equation:

$$k_w L_w = p\pi - 2\sin^{-1}\left[\frac{k_w/m_w}{\sqrt{(k_w/m_w)^2 + (k_b/m_b)^2}}\right] \qquad (7.142)$$

where $k_w = \sqrt{2m_w^* E}/\hbar$, $k_b = \sqrt{2m_w^*(V - E)}/\hbar$, and m_w^* and m_b^* are the effective masses of the particle in the well and in the barrier respectively.

For finite quantum wells, the spreading of envelope functions into the barrier is to be taken into consideration, and the dimensionless pertinent parameter is written as $\beta = L^*/2a_0^*$, where $L^* = (2/k_b) + L$ and $a_0^* = (\in/\in_0)(m_0/\mu^*) a_H$, and μ^* is the mean value of the 3D reduced mass of exciton and is given by

$$\frac{1}{\mu^*} = \frac{1}{m_e^*} + \gamma_1^* \qquad (7.143)$$

The mean values of electron effective mass and valence band parameters can be defined as

$$m_e^* = \beta_e m_{ew}^* + (1 - \beta_e)m_{eb}^* \qquad (7.144a)$$

and

$$\gamma_1^* = \beta_h \gamma_{1w} + (1 - \beta_h)\gamma_{1b} \qquad (7.144b)$$

Here two weighting parameters β_e and β_h are considered to account for the effective mass mismatch between the well and barrier material; and are expressed as $\beta_e = L/(2/k_{be} + L)$ and $\beta_h = L/(2/k_{bh} + L)$.

The binding energy of the confined exciton for finite quantum well is given by

$$E_b = \frac{E_0^*}{\left[1 - \frac{1}{2}e^{-d^*/2a_0^*}\right]^2}$$ (7.145)

where E_0^* is the mean value of the effective Rydberg energy for the 3D exciton.

In a 2D system, the heavy hole and light hole subbands split due to strong anisotropy and the in plane effective masses are taken as

$$m_{hh} = \frac{m_0}{\gamma_1 + \gamma_2} \ \text{and} \ m_{lh} = \frac{m_0}{\gamma_1 - \gamma_2}$$

The calculated values of dimensional-parameter α for GaAs QW with $Ga_{1-x}Al_xAs$ barriers for different values of composition x are shown in Fig. 7.14 as a function of well width. The values approach 3 for very small well width, corresponding to bulk $Ga_{1-x}Al_xAs$ barriers. As may be noted the parameter does not attain the ideal value of 2. The binding energy for the lowest 1s exciton involving the first electron and first hh subband is plotted in Fig. 7.15 against well width for two values of x. The binding energy increases first, reaches a maximum and then decreases. The calculated values, based on the simple expression, agree quite well with the values obtained from more refined calculation. The nature of variation of e1-lh1 exciton is the same.

Example 7.12 *The values of α and binding energy for GaAs QW with L = 7 nm are presented. The band gap of the alloy is calculated from $E_g(x)$ = 1.424 + 1.247xeV.*

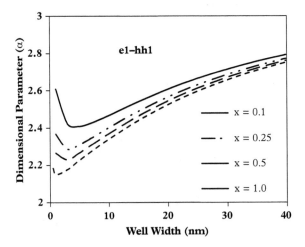

Figure 7.14 *Calculated dimensional parameter for different well widths for 1s HH excitons. Four values of alloy composition(x) are chosen.*

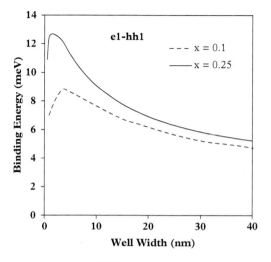

Figure 7.15 *Variation of binding energy of HH excitons in GaAs/AlGaAs QWs with well widths. Two different values of x (0.1 and 0.25) are chosen.*

For x = 0.25, ΔE_c = 0.230eV and ΔE_v = 0.109eV using a 65:35 ratio. The effective mass values are m_{ew} = 0.067 m_0, m_{hhw} = 0.5 m_0, m_{eb} = [(1–x_b).0.067 + x_b.0.15] m_0 and m_{hhb} = [(1–x_b).0.5 + x_b.0.79] m_0. The calculated values of subband energy are E_{c1} = 0.0432 eV, E_{vhh1} = 0.00886 eV. The parameter α = 2.346 and the binding energy is 10.2 meV. If we take the value of 4.8 meV for bulk GaAs as calculated in Example 7.1, the ideal value of binding energy in GaAs QW is 19.2 meV.

Problems

7.1 *Assume that the reduced mass in semiconductors is approximately equal to the electron mass. Hence prove that the excitonic binding energy increases almost linearly with increasing band gap.*

7.2 *Complete the following table by calculating the excitonic binding energy Ry and Bohr radius a_B of the materials, using parameter values given against each.*

Material	m_e	m_{hh}	ε	Ry (meV)	a_B(nm)
CdTe	0.1	0.4		10	7.5
CdSe	0.13	0.7		16	4.9
ZnSe	0.15	0.8		19	3.8
CuCl	0.4	2.4		190	0.7
GaN	0.17	1.4		28	2.1

7.3   *The electron effective mass, Luttinger parameters and refractive index for $Cd_xZn_{1-x}Se$ alloy are expressed as $m_e(x) = (0.16 - 0.03x)m_0$, $\gamma_1(x) = 4.30 + 0.65x$, $\gamma_2(x) = 1.14 + 0.22x$, $\gamma_3(x) = 1.84$, and $\eta(x) = 3.022 + 0.07x$. Calculate the binding energy of 1s exciton for x = 0, 0.2, 0.5, 0.8, and 1.0.*

7.4   *The absorption coefficients for the 1s and 2s states are equal at the 2s excitonic peak. Calculate the halfwidth corresponding to a Gaussian lineshape. The binding energy for 1s state is 4.7 meV and the halfwidths are the same for both the states.*

7.5   *Obtain the ideal 2D value of binding energy in $In_{0.53}Ga_{0.47}As$. The parameter values for InAs(GaAs) are: $\gamma_1 = 19.7(6.8)$, $\gamma_2 = 8.4(2.1)$, $\gamma_3 = 9.3(2.9)$.*

7.6   *The linewidth for infrared (IFR) scattering is expressed as $\Gamma_{IFR} = [(2\pi^2\hbar^2\Delta)/(L^3M)]I(\Lambda)$, where the fluctuation in well width is assumed to have Gaussian auto-correlation with mean height Δ and spatial length Λ and $I(\Lambda)$ is a function of Λ. Calculate the linewidth for GaAs using $\Delta = 0.283$ nm, $\Lambda = 10$ nm, and $I(\Lambda) = 0.6$.*

7.7   *Calculate the temperature variation of HWHM of lowest excitonic absorption peak for InGaAs QW using Eq. (7.49). use $\Gamma_0 = 2.3$ meV, $\Gamma_{LO} = 15.3$ meV and LO phonon energy of 35 meV. Compare your calculation with the data obtained by Livescu et al (1988) Free carriers and many body effects in absorption spectra of modulation-doped quantum wells. IEEE Journal of Qunatum Electronics 24: 1677–1689.*

7.8   *Assuming that the envelope functions are sin functions, calculate the matrix element in Eq. (7.109) and summing over all subbands arrive at the expression for change in subband energy given by Eq. (7.111).*

7.9   *Using the envelope function expressed by Eq. (7.112) show that the normalization factor is given by Eq. (7.113).*

7.10   *Derive the expressions for change in subband energy given by Eqs. (7.117) and (7.118) using the general expression for the envelope functions.*

7.11   *Show that the change in effective gap under electric field is effective gap $\Delta E_g = -3 \times 10^{-20}(m_e + m_h)F^2L^4 eV$, where F is in V/cm and L is in Å.*

7.12   *The wavefunctions in a parabolic QW under an electric field is displaced harmonic oscillator. Derive the expression for the interband matrix element.*

7.13   *Calculate binding energy for GaAs QW for dimensionality parameter of 2.2.*

7.14   *Prove that Eq. 7.4 reduces to the expression for continuum absorption when $\alpha = 3$ and 2.*

7.15   *Calculate the absorption coefficient in a GaAs QW for $\alpha = 2.3$. Assume half width = 5 meV.*

Reading List

Abramowitz M S and StegunI A (1972) *Handbook of Mathematical Function*, 9th edn. New York: Dover.

Bassani F and Parravicinni G P (1975) *Electronic States and Optical Transitions in Solids*. Oxford: Pergamon Press.

Basu P K (1997) *Theory of Optical Processes in Semiconductors: Bulk and Microstructures*. Oxford, UK: Oxford University Press.

Dutta S (1989) *Quantum Phenomena*. Reading, Mass: Addison Wesley.

Gradshteyn I S and Ryzhik I M (1980) *Tables of Integrals, Series and Products*, p. 316. London: Academic Press

Klingshirn C F (2012) *Semiconductor Optics*, 4th edn. Heidelberg: Springer.

Landau L D and Lifshitz E M (1975) *Quantum Mechanics*, 3rd edn. Oxford: Pergamon Press.

Schmitt-Rink S, Chemla D S, and Miller D A B (1989) Linear and nonlinear optical properties of semiconductor quantum wells. *Advances in Physics* 38: 89.

References

Ahn D and Chuang S L (1987). Calculation of linear and nonlinear intersubband optical absorptionin a quantum well with an applied electric field. *IEEE Journal of Quantum Electronics* 23: 2196–2204.

Bastard G, Mendez E E, Chang L L, and Esaki L (1982) Exciton binding energy in quantum wells. *Physical Review B* 26: 1974–1979.

Bastard G, Mendez E, Chang L L, and Esaki L (1983) Variational calculations on a quantum well in an electric field. *Physical Review B* 28: 32–41.

Basu P K (1990) Linewidth of free excitons in quantum wells: Contribution by alloy disorder scattering. *Applied Physics Letters* 56(12): 1110–1112.

Basu P K (1991) Effect of interface roughness on excitonic linewidth in a quantum well: Golden-rule and self-consistent-Born-approximation calculations. *Physical Review B* 44(16): 8798–8801.

Basu P K and Ray P (1991) Calculation of the mobility of two-dimensional excitons in a GaAs/Al x Ga 1– x As quantum well. *Physical Review B* 44(4): 1844–1851.

Basu P K, Das N R, Mukhopadhyay B, Sen G, and Das M K (2009) Ge/Si photodetectors and group IV alloy-based photodetector materials. *Optical and Quantum Electronics* 41(7): 567–581.

Chemla D S, Miller D A B, Smith P W, Gossard A C, and Wiegmann W (1984) Room temperature excitonic non-linear absorption and refraction in GaAs/AlGaAs multiple quantum well structures. *IEEE Journal of Quantum Electronics* 20: 265–275.

Christol P, Lefebvre P, and Mathieu H (1993) Fractional-dimensional calculation of exciton binding energies in semiconductor quantum wells and quantum-well wires. *Journal of Applied Physics* 74: 5626–5637.

Christol P, Lefebvre P, and Mathieu H (1994) A single equation describes excitonic absorption spectra in all quantum-sized semiconductors. *IEEE Journal of Quantum Electronics* 30(10): 2287–2292.

Chuang S -L, Schmitt-Rink S, Miller D A B, and Chemla D S (1991) Exciton Green's-function approach to absorption in a quantum well with an applied electric field. *Physical Review B* 43: 1500–1509.

Elliott R J (1957) Intensity of optical absorption by excitons. *Physics Review* 108:1384–1389.

Fang F F and Howard W E (1966) Negative field effect mobility on (100) silicon surfaces. *Physical Review Letters* 16: 797–799.

Faist J, Capasso F, Sivco D L, Sirtori C, Hutchinson A L, and Cho A Y (1994) Quantum cascade laser: An intersub-band semiconductor laser operating above liquid nitrogen temperature. *Electronics Letters* 30: 865–866.

Feldmann J, Peter G, Gobel E O, Dawson P, Moore K, Foxon C, and Elliot R J (1987) Linewidth dependence of radiative exciton lifetimes in quantum wells. *Physical Review Letters* 59: 2337–2340.

Franz W (1958) Einfluß eines elektrischen Feldes auf eine optische Absorptionskante. *Zeitschrift Naturforschung (a)* 13: 484.

Harwit A and Harris J S (1987). Observation of Stark shifts in quantum well intersubband transition. *Applied Physics Letters* 50: 685.

He X F (1990) Anisotropy and isotropy: A model of fraction-dimensional space. *Solid State Communications* 75: 111–114.

He X F (1991) Excitons in anisotropic solids: The model of fractional-dimensional space. *Physical Review E* 43: 2063–2069.

Keldysh L V (1958) The effect of a strong electric field on the optical properties of insulating crystals. *Soviet Physics JETP* 34(7): 788–790.

Kuhl J, Honold A, Schultheis L, and Yu C W (1989) In: Optical Switching in Low Dimensional Systems. H Haug and L Banyai (eds.) *NATO ASI Series, B, Physics* 194: 267. NY:Plenum Press.

Kuo Y H, Lee Y K, Ge Y, Ren S, Roth J E, Kamins T I, Miller D A B, and Harris J S Jr (2005) Strong quantum confined stark effect in Ge quantum well structures on Si. *Nature* 437: 1334–1336.

Lefebvre P, Christol P, and Mathieu H (1993) Unified formulation of excitonic absorption spectra in semiconductor quantum wells, superlattices and quantum wires. *Physical Review E* 48: 17308–17315.

Lohe M A and Thilagam A (2004) Weyl-ordered polynomials in fractional-dimensional quantum mechanics. *Journal of Physics A: Math General* 37: 61–81.

Mathieu H, Lefebvre P, and Christol P (1992) Excitons in semiconductor quantum wells: A straightforward analytical calculation. *Journal of Applied Physics* 72: 300–302.

Mathieu H, Lefebvre P, and Christol P (1992) Simple analytical method for calculating exciton binding energies in semiconductor quantum wells. *Physical Review E* 46: 4092–4101.

Miller D A B, Chemla D S, Damen T C et al (1985) Electric field dependence of optical absorption near the band gap of quantum well structures. *Physical Review B* 32: 1043–1060.

Miller D A B, Chemla D S, and Schmitt-Rink S (1986a) Relation between electro absorption in bulk semiconductors and in quantum wells: The quantum-confined Franz–Keldysh effect. *Physical Review B* 33: 69–76.

Miller D A B, Weinar J S, and Chemla D S (1986b) Electric-field dependence of linear optical properties in quantum well structures: Waveguide electro absorption and sum rules. *IEEE Journal of Quantum Electronics* 32: 1816–1830.

Pedersen T G (2017) Stark effect in finite-barrier quantum wells, wires, and dots. *New Journal of Physics* 19(043011): 1–10.

Pedersen T G, Mera H, and Nikolić B K (2016) Stark effect in low-dimensional hydrogen, *Physical Review A* 93: 013409(1-10).

Priester C, Allan G, and Lanoo M (1984) Wannier excitons in GaAs-Ga1−xAlxAs quantum-well structures: Influence of the effective-mass mismatch. *Physical Review B* 30: 7302.

Rudin S and Reinecke T L (1990) Exciton linewidths in semiconductor quantum wells. *Physical Review B* 41: 3017.

Rudin S, Reinecke T L, and Segall B (1990) Temperature-dependent exciton linewidths in semiconductors. *Physical Review B* 42: 112–118.

Saha, M N (1922) On the temperature ionization of elements of the higher groups in the periodic classification. *Philosophical Magazine* 44: 11–28.

Shinada M and Sugano S (1966) Interband optical transitions in extremely anisotropic semiconductors. I. Bound and unbound exciton absorption. *Journal of the Physical Society of Japan* 21: 1936–1946.

Singh J and Bajaj K (1985) Role of interface roughness and alloy disorder in photoluminescence in quantum-well structures. *Journal of Applied Physics* 57: 5433–5437.

Stevens J, Whitehead M, Parry G, and Woodbridge K (1988) Computer modeling of the electric field dependent absorption spectrum of multiple quantum well material. *IEEE Journal of Quantum Electronics* 24: 2007–2016.

Van Roosbroeck W and Shockley W (1954) Photon-Radiative recombination of electrons and holes in germanium. *Physical Review* 94: 1558–1560.

Weisbuch C, Dingle R, Gossard A C, and Weigmann W (1981) Optical characterization of interface disorder in GaAs-Ga1-xAlxAs multi-quantum well structures. *Solid State Communications* 38: 709–712.

8

Nanowires

8.1 Introduction

The simplest nanostructure, that is, the structure having nanometer size in a single direction, is a Quantum Well (QW). After the phenomenon of quantization of electron motion is firmly established in silicon (Si) metal-oxide semiconductor field-effect transistors (MOSFET)s in 1966, and in semiconductor heterostructures in early 1970s, a flood of activities started to discover new phenomena related to two-dimensional (2D) systems, to understand the related physics, to refine the growth technology, and to propose and realize novel devices and application areas.

The effect of reducing dimension of the structure in 2D and three dimensions (3D) was first investigated by Arakawa and Sakaki (1982) in their classic paper. As already mentioned in Chapter 1, structures having dimensions of the order of few tens of nanometers along two directions are called Quantum Wires (QWRs), which support 1D electron gas (1DEG). If the dimensions in all three directions are reduced, motion of electrons is inhibited in all three directions. Such a 0DEG is supported by a Quantum Dot (QD) or Quantum Box (QB). Arakawa and Sakaki studied the optical properties of QWRs and QDs and predicted lower threshold current density and better temperature sensitivity of lasers, at reduced dimensionalities.

Since their prediction, intense research activities started around the globe to grow and study the electronic and optoelectronic processes in wires and dots, to examine new devices, and to optimize their performances. The main emphasis was on semiconductor-based structures and the growth and fabrication are related to top-down processes, that is, epitaxy and etching. Field-induced quantization is also employed in some cases, for example, in nanowire field-effect transistors (FET)s and high electron mobility transistors. Since QDs support atom-like sharp energy levels, workers gave more attention to the studies related to QDs physics, as well as to the device applications. Optoelectronic phenomena and devices using QDs will be discussed in Chapter 9.

A few novel phenomena, like conductance quantization, Aharanov–Bohm effect, etc., have been observed by using semiconductor QWRs. However, the advantages predicted for QWR lasers were not convincingly demonstrated. The reason for poor performance has been attributed to the surface roughness scattering. A few types of QWRs, T shaped, V groove, etc., have been reported during the 1980s–1990s.

Semiconductor Nanophotonics. Prasanta Kumar Basu, Bratati Mukhopadhyay, and Rikmantra Basu, Oxford University Press.
© P.K. Basu, B. Mukhopadhyay, R. Basu (2022). DOI: 10.1093/oso/9780198784692.003.0008

An important class of semiconductor nanostructures, with cross-sections of 2–200 nm and high aspect ratio, that is, lengths starting from a fraction of a micrometer to several micrometers, emerged during the 1990s. These structures were initially called nanowhiskers, and later nanowires. In these structures also, electrons and holes are free to move along the longer direction, thus supporting a 1DEG or a one-dimensional hole gas (1DHG). These nanowires serve as the bridge between the nanoscopic and macroscopic worlds and facilitate the integration of nanoscale building blocks in electronic and photonic device applications. Conventional photonic circuits consist of components with high aspect ratio, like interconnects, waveguides, and so on (see, Yang 2005; Yang et al 2010) Semiconductor nanowires serve as an essential building block for Nanophotonics. The nanowires can be built by bottom-up approach including chemical synthesis. It can have different structures, like nanodiscs, nanorods, nanotubes resembling a hollow pipe, nanoribbon, nanobelt, and so on. It may also be fabricated with core-shell structure, in which the core material may be surrounded by another material of lower refractive index and higher bandgap, just like an optical fibre in which core glass is surrounded by another cladding glass. Materials synthesis has been possible in traditionally inaccessible compositional regions. Single-crystalline indium gallium nitride (InGaN) nanowires have been grown in which the bandgap can be tuned from the ultraviolet (UV) to the near infrared. Application areas of such nanowires with tunable electronic structures seem to lie in photovoltaics, solid-state lighting and solar-to-fuel energy conversion. In recent years, growth techniques for semiconductor nanowires are quite advanced, so that desired composition, heterojunctions and architectures can be readily synthesized. 1D nanostructures find use in various optoelectronic devices including photodetectors, chemical and gas sensors, waveguides, nonlinear optical converters, light emitting diodes (LED)s, and microcavity lasers.

We begin this chapter by introducing the basic optical processes in QWRs, like absorption, gain, and excitonic processes. We present the simplest treatments for each case to illustrate the ingredients of the theory. The methods of formation of QWRs by epitaxy and lithography are briefly outlined. This is followed by a brief description of growth, structures, size, and shapes of various nanowires including core-shell structures. Some novel phenomena related to QWRs, and some application areas of the structures are briefly mentioned to conclude this chapter.

8.2 Quantum wires: Preliminaries

A brief introduction to QWRs has been given in Chapter 2 (Section 2.10.3). In this section, we first elaborate the subband structure, density-of-states (DOS), interband absorption, and excitonic processes by using simple illustrative examples. More refined treatments will then follow.

8.2.1　Subbands in the ideal case

As already stated, in a QW structure the width of the well sandwiched between two barrier layers is of the order of de Broglie wavelength. In a rectangular QWR, both the width and height of the active region, along the y- and z-directions, are comparable to the de Broglie wavelength and the active region is covered by the barrier material on both sides. The length of the active region in the remaining x-direction is large. The electron motion is free along x but is restricted along y and z, thus giving rise to a 1DEG.

To illustrate the basic electronic properties, let us assume that the active region is made of a gallium arsenide (GaAs) parallel-piped of cross section $L_y \times L_z$ and length L_x, surrounded by $Ga_{1-x}Al_xAs$, as shown in Fig. 8.1. The potential discontinuities ΔE_c and ΔE_v in the respective bands appear at the heterointerfaces at $y = \pm L_y/2$ and $z = \pm L_z/2$.

Consider now the Schrodinger equation

$$\left[\frac{p_x^2 + p_y^2 + p_z^2}{2m_e} + V(y, z) \right] \phi(x, y, z) = E\phi(x, y, z) \tag{8.1}$$

The electron moves freely along the x-direction, as there is no confining potential, and therefore

$$\phi(x, y, z) = \frac{1}{\sqrt{L_x}} \chi(y, z) exp(ik_x x) \tag{8.2}$$

The envelope function $\chi(y, z)$ now satisfies the following 2D Schrodinger equation:

$$\left[\frac{p_y^2 + p_z^2}{2m_e} + V(y, z) \right] \chi_i(y, z) = E_i \chi_i(y, z) \tag{8.3}$$

In general, the envelope function cannot be written as a product of two functions, since the potential depends on both y and z. However, the problem is quite simplified by

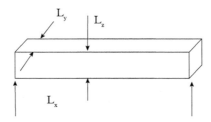

Figure 8.1 *Schematic diagram of a QWR with rectangular cross section. The width (along y) and height (along z) are respectively, L_y and L_z and the length is L_x.*

assuming infinite barrier heights at the four heterointerfaces and the envelope function becomes

$$\chi_i(y, z) = \frac{2}{\sqrt{L_y L_y}} \cos\left(\frac{n_y \pi y}{L_y}\right) \cos\left(\frac{n_z \pi z}{L_z}\right) \tag{8.4}$$

The expression for subband energies may easily be obtained following the treatment for the QW and is given by

$$E_i = E_{n_y, n_z} = \frac{\hbar^2 \pi^2}{2 m_e} \left(\frac{n_y^2}{L_y^2} + \frac{n_z^2}{L_z^2}\right) \tag{8.5}$$

where both the subband indices n_y and n_z are integers including 0, but both cannot be simultaneously zero $[n_y, n_z \neq 0, 0]$.

Eq. (8.4) indicates that electrons form standing wave patterns in both the y and z directions. The situation is similar to standing wave pattern of electromagnetic (EM) waves in a rectangular waveguide.

The dispersion relation for 1D electrons may be expressed as

$$E = E_{n_y, n_y} + \frac{\hbar^2 k_x^2}{2 m_e} \tag{8.6}$$

Similar expressions are obtained for subbands due to heavy-holes (HH) and light-holes (LH).

It is also easy to obtain expressions for the 1D DOS function which is given by

$$D_i(E) = D_i(k_x) 2 \left(\frac{dE}{dk_x}\right)^{-1} = \frac{g_s L_x}{\hbar \pi} \left\{\frac{m_e}{2(E - E_i)}\right\}^{1/2} \tag{8.7}$$

where g_s represents spin degeneracy and a factor 2 has been included to take into account two wavevectors $\pm k_x$. The plot of DOS, shown in Fig. 2.13 has singularities at the subband edges E_i, and is similar to the DOS function under a quantizing magnetic field.

We will now quote the expression for the number of electrons N_i per subband per unit length at 0 K, when the Fermi function is box-like. It takes the form

$$N_i = g_s \int_{E_i}^{E_F} D_i(E) f(E) dE = \frac{g_s}{\pi \hbar} \sqrt{2 m_e (E_F - E_i)} \tag{8.8}$$

8.2.2 Interband optical processes in quantum wires

We consider first transition from a hole subband to a conduction subband with a view to calculating gain or absorption coefficient. By assuming the **k**-conservation condition to be valid along the direction of free motion (*x*-direction), the absorption coefficient is expressed as

$$\alpha(\hbar\omega) = -\frac{B}{(c/\eta_r)} \sum_k (f_e - f_h)\delta\left(E_g + \frac{\hbar^2 k^2}{2m_r} - \hbar\omega\right) \tag{8.9}$$

where we have dropped the subscript *x* to the wavevector. The symbol E_g denotes the effective gap which is the bulk band gap plus the electron and hole subband energies. The factor *B* contains the polarization dependent momentum matrix elements, other fundamental constants and material parameters. The summation over *k* may be converted into an integral and, assuming $f_e - f_h = 1$, the integration may be carried out to yield

$$\alpha(\hbar\omega) = B_1(\hbar\omega - E_g)^{-1/2} \tag{8.10}$$

where

$$B_1 = \frac{e^2 A_{1D} C_{1D} \langle |p_{cv}|^2 \rangle (2m_r)^{1/2}}{2m_0^2 \varepsilon_0 \eta_r \hbar\omega c A} \tag{8.11}$$

The coefficients A_{1D} and C_{1D} have similar meanings to A_{mn} and C_{mn} in QWs, and *A* is the cross-sectional area of the wire. The polarization dependence of the matrix element is discussed by Corzine et al (1993).

As is evident from Eq. (8.10), the absorption coefficient is proportional to the joint DOS function, and hence the absorption shows first a singularity at photon energy equal to the effective band gap energy and then falls with increasing photon energy, and then shows again another singularity. This sharp feature is however smeared out due to the effect of broadening. The expression for the gain is written by replacing the δ-function in Eq. (8.9) by a Lorentzian with an intraband relaxation time τ_{in} as (Kapon 1993, see Reading List)

$$g(\hbar\omega) = B_1 \int (f_e - f_h) \frac{(\hbar/\tau_{in})dE}{(E - \hbar\omega)^2 + (\hbar/\tau_{in})^2} \tag{8.12}$$

Example 8.1 *We derive an approximate expression for the absorption coefficient of a QWR by replacing the δ-function in Eq. (8.9) by a Lorentzian with width γ. Taking $f_e - f_h = 1$, the integration over k may be performed analytically and we obtain (Ray and Basu 1993).*

$$g(\hbar\omega) = B_1 \left\{ \frac{\Delta + (\Delta^2 + \gamma^2)^{1/2}}{\Delta^2 + \gamma^2} \right\}^{1/2} \quad \Delta = \hbar\omega - E_g,$$

Let the cross-section of the GaAs wire be $L_y = L_z = 10$ nm, $m_e = 0.067\, m_0$, $m_h = 0.112\, m_0$, giving $m_r = 0.042\, m_0$. Also let $E_g = 1.6$ eV and $\gamma = 6.6$ meV ($\tau_{in} = 0.1$ ps). Then $\alpha = 6.4 \times 10^5 \mathrm{m}^{-1}$.

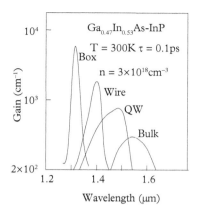

Figure 8.2 *Calculated gain spectra for InGaAs bulk, QW, QWR, and cubic QB. The thickness in the direction(s) of confinement is 10 nm. InP acts as the barrier material.*

The gain coefficient given by Eq. (8.12) may be calculated for a given carrier density. It is to be noted however, that the transparency carrier density is quite high in practical conditions, and therefore, occupancy of a number of subbands need be considered. The quasi-Fermi levels, probabilities f_e and f_h, and the gain spectra are then calculated by using numerical techniques.

As the DOS function is narrower in QWRs than in QWs, carriers injected into QWR are confined within a narrower spectral range. It leads one to conclude that the transparency carrier density, that is the carrier density needed for zero absorption, should be less in QWR lasers. The optical gain is also expected to be larger in QWR. A theoretical calculation of gain in various systems (Asada et al 1986) for $In_{0.53}Ga_{0.47}As/InP$ QB, QWR, QW, and bulk lasers with each side 10 nm wide is shown in Fig. 8.2. It is found that the peak gain in QWR laser is at least two times higher than that in a QW. The differential gain is also higher in QWRs. Based on these results, it is expected that QWR lasers would exhibit a lower threshold current, higher modulation bandwidth, and weaker dependence of threshold current on temperature. However, all these conclusions are based on the ideal situation. In practice, the wires show size fluctuations and interface roughness, which introduce a larger broadening of the spectra. This causes a detrimental effect on gain and threshold current.

8.3 Excitonic processes in quantum wires

In general, it is thought that both the excitonic binding energy and oscillator strength should increase when the dimension of the structure is further reduced. This means that both these values are expected to increase from their respective values in the QW.

Experimental results for excitonic processes in T and V shaped QWRs have been re-viewed by Akiyama (1998). In this section, we shall present first the theory of excitons in a quasi-1D system, followed by a representative theory for a practical wire structure.

8.3.1 Ideal theory

The calculation of the binding energy in QWRs is more difficult in comparison to its calculation in a bulk semiconductor or QW. This is due to the singularity of Coulomb potential between electrons and holes at the origin to make the binding energy for the lowest state divergent. Here we present a theory by Ogawa and Takagahara (1991) and quote their main conclusions.

The effective mass Schrodinger equation for the envelope function is written as

$$\left[-\frac{\hbar^2}{2m_e}\nabla_e^2 - \frac{\hbar^2}{2m_h}\nabla_h^2 + V_e(r_e) + V_h(r_h) + U(r_e, r_h)\right]\chi(r_e, r_h) = E'(r_e, r_h) \tag{8.13}$$

where V_e and V_h are confinement potential and U is the Coulomb potential. The bulk band gap is used as the reference level from which the eigen energies are measured. Considering free motion of the centre-of-mass (CM) along the x-direction, the envelope function of the exciton is written as

$$\chi(r_e, r_h) = exp(i\mathbf{K} \cdot \mathbf{X})f_e(y_e, z_e)f_h(y_h, z_h)\phi(r_e, r_h) \tag{8.14}$$

As before, $\mathbf{X}$ and $\mathbf{K}$ denote, respectively, the x-coordinate and the wavevector of the CM, f_e and f_h are, respectively, the envelope functions for the lowest electron and hole subbands, and ϕ is the envelope function for relative motion, which depends on $|x_e - x_h|$, the separation between the electron and hole coordinates along the x-direction. The envelope function f_e satisfies the following equation:

$$\left\{-\frac{\hbar^2}{2m_e}\left[\frac{\partial^2}{\partial y_e^2} + \frac{\partial^2}{\partial z_e^2}\right] + V_e(y_e, z_e)\right\}f_e(y_e, z_e) = E_e f_e(y_e, z_e) \tag{8.15}$$

The envelope function for hole, f_h satisfies a similar equation. Using Eq. (8.15), Eq. (8.13) may be rewritten as

$$\left[-\frac{\hbar^2}{2m_r}\frac{\partial^2}{\partial x^2} + U(r_e, r_h)\right]f_e f_h\phi(x) = \left[E' - E_e - E_h - \frac{\hbar^2 K^2}{2M}\right]f_e f_h\phi(x) \tag{8.16}$$

An effective Coulomb potential defined as follows:

$$V_{eff}(x) = \int dy_e dz_e dy_h dz_h U(r_e, r_h)|f_e|^2|f_h|^2 \tag{8.17}$$

is now introduced and Eq. (8.16) reduces to

$$\left[-\frac{\hbar^2}{2m_r}\frac{d^2}{dx^2} + V_{eff}(x) \right] \phi(x) = E\phi(x) \tag{8.18}$$

where

$$E = E' - E_e - E_h - \frac{\hbar^2 K^2}{2M} \tag{8.19}$$

Although the effective Coulomb potential does not have singularity, it is difficult to handle this type of potential. The following effective Coulomb potential with cusp-type cut-off is sometimes chosen:

$$V(x) = \frac{e^2}{4\pi\varepsilon_r\varepsilon_0(x)(|x| + x_0)} \tag{8.20}$$

When the dimensions of the y-and z-directions are of the order of the Bohr radius, Eq. (8.20) truly represents the effective potential. The bound and continuum solutions of Eq. (8.18), using Eq. (8.20), will now be presented.

8.3.1.1   Bound solutions

We express $E = -R_x/n^2$ and $u = 2(|x| + x_0)/(na_B)$. Here u is real, and n (>0) is a real parameter

$$\frac{d^2\phi(u)}{du^2} - \left[\frac{1}{4} - \frac{n}{u} \right]\phi = 0 \tag{8.21}$$

Eq. (8.21), known as Whittaker equation, has solutions in the form of confluent hyper-geometric functions. For $x_0/a_B \ll 1$, the parameter n related to the eigen energies are given by

$$n = n_v = \begin{cases} \nu + 2x_0/a_B \text{ odd case} \\ \nu - 1/\left[\ln\left(2x_0/\nu a_B\right)\right] \text{ even case} \end{cases} \quad \nu = 1, 2, 3\ldots.. \tag{8.22}$$

For the lowest eigenstate, one obtains

$$\ln\left[\frac{2x_0}{n_0 a_B} \right] + \frac{1}{2n_0} = 0 \tag{8.23}$$

as the cut-off decreases ($x_0 \to 0$), $n_0 \to 0$ and the binding energy becomes extremely large. In the limit $x_0 \to 0$, the bound states excluding the lowest one, are characterized by $n_y^{odd} = n_y^{even} = n = 1, 2, 3 \ldots$, and

$$E_n^{odd} = E_n^{even} = -R_x/n^2; n = 1, 2, 3 \ldots. \tag{8.24}$$

The oscillator strength for the *n*-th state is written in the following standard form:

$$f_n = \frac{2m_0\omega}{\hbar}|\langle c|e_\lambda \cdot \mathbf{p}|v\rangle|^2|\phi_n(0)|^2 \tag{8.25}$$

The values are different for different *n* as well as for different x_0. As $x_0 \to 0$, the oscillator strength for the lowest state (*n*=0), $f_0 \to \infty$, whereas it tends to 0 for $n = 1$. Even for finite x_0, the value is quite small for $n = 1$. The conclusion is that there is an enormously large oscillator strength for the lowest state.

8.3.1.2   Continuum states

We make the substitution $u = 2ik(|x|+x_0)$, where u is purely imaginary, and $\alpha = 1/(a_Bk)$. The 1D Schrodinger Eq. (8.18) may be written as

$$\frac{d^2\phi}{du^2} - \left[\frac{1}{4} + \frac{i\alpha}{u}\right] = 0 \tag{8.26}$$

The previous equation is again a Whittaker equation, and its solutions giving unbound states consist of proper confluent hypergeometric functions. The envelope functions are needed to express the absorption coefficient and Sommerfeld factors. The absorption coefficient without coulomb interactions is proportional to $(\hbar\omega - E_g)^{-1/2}$. The continuum states absorption coefficient is found to be less than the value predicted from the theory of interband absorption without excitonic effect.

8.3.2   Excitonic processes in a practical structure

The ideal theory of binding energy of the lowest excitonic state presented previously gives a very large binding energy. Calculation of excitonic binding energy, oscillator strength, and absorption coefficient need sophisticated methods. However, to obtain information of binding energy and the nature of its variation with changes in length and width in practical conditions, some approximate methods will be helpful. One such method is fractional dimensional analysis (Christol et al 1993) introduced in Section 7.5 for QWs. The method has also been applied to the QWR (Pedersen 2015, 2017).

In practical structures, the material forming the QWR is surrounded by cladding (barrier) layers. The envelope functions in the wire are not totally confined within the wire, but they leak into the barrier layers. The variation of binding energy with width is expected to follow the nature shown by Fig. 7.15 for a QW. For lower widths the envelope functions penetrate into the barriers and the binding energy tends to the 3D value for the barrier. The values increase and reach a maximum for maximum confinement at a critical length scale and then decrease as the envelope functions spread out, finally reaching the bulk 3D value. Therefore, not only the value of excitonic binding energy but also the critical confinement width are parameters of importance for QWRs.

In this section, a variational method developed by Koh et al (2001) will be presented to calculate the exciton binding energy, its nature of variation with length and width of the QWR, and to correlate the binding energy with the fractional dimensional parameter. The analysis applies for the general case of cylindrical quantum disc of radius of cross-section R and length L. Depending on the values of R and L, one may easily obtain the results for various limiting cases, (1) bulk (both R and $L>>a_B$, where a_B is the 3D exciton Bohr radius), (2) QW ($R>>a_B$, $L<a_B$), (3) QWR ($R<a_B$, $L>>a_B$), and (4) QD ($R<a_B$, $L<a_B$). The authors used effective mass approximation, finite values of confinement potential at the heterointerfaces, and a two-parameter variational wavefunction. The method will now be described.

8.3.2.1 Model calculation

The disc is in the form of a cylinder as shown in Fig. 8.3. The in-plane coordinates on the circular cross section with radius R are denoted by r_c ($c = e,h$), and the length/width of the cylinder/disc is denoted by L and is along the z-axis.

First of all, the ground state of a single particle is obtained by solving the effective mass equation. The mass of a particle is denoted by

$$m_c^{\parallel}(r_c) = m_c H(R - r_c) + m_c^b H(r_c - R) \tag{8.27a}$$

$$m_c^{\perp}(z_c) = m_c H(L/2 - |z_c|) + m_c^b H(|z_c| - R) \tag{8.27b}$$

Here, m_c and m_c^b are, respectively, the effective mases in the disc and the barrier and H denote the Heaviside step function. The in-plane confining potential and the same along the z-axis are expressed as

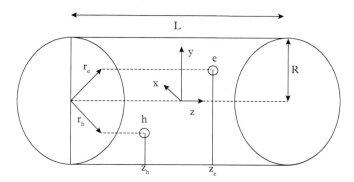

Figure 8.3 *Schematic diagram of the QWR/Quantum Disc. The coordinates are denoted by (r_c, z_c, $c = e$, h).*

$$V_c^{\parallel}(r_c) = 0 \; [V_c] \; \text{for} \; r_c < R \, [\text{otherwise}] \tag{8.28a}$$

and

$$V_c^{\perp}(z_c) = 0 \; [V_c], \text{if} \; |z_c| < L/2 \, [\text{otherwise}] \tag{8.28b}$$

We may write now the 2D and 1D effective mass Schrodinger equation (EMSE) as

$$\left\{ -\frac{\hbar^2}{2m_c^{\parallel}} \nabla_c^2 + V_c^{\parallel}(r_c) \right\} f_c(r_c) = E_c^{\parallel} f_c(r_c) \tag{8.29a}$$

$$\left\{ -\frac{\hbar^2}{2m_c^{\perp}} \nabla_c^2 + V_c^{\perp}(z_c) \right\} g_c(z_c) = E_c^{\perp} g_c(z_c) \tag{8.29b}$$

The ground state solutions of the envelope functions take the form

$$f_c(r_c) = \mathcal{J}_0(\theta_c r_c), \; r_c \leq R; \; f_c(r_c) = B_c K_0(\beta_c r_c), r_c > R \tag{8.30a}$$

$$g_c(z_c) = \cos(k_c z_c), \; |z_c| \leq L/2; \; g_c(z_c) = A_c \exp(-q_c z_c), |z_c| > L/2 \tag{8.30b}$$

Since the effective masses in the disc and the barrier are different, the quantities θ_c, β_c, k_c and q_c are determined from the boundary conditions requiring that the respective wavefunctions and their flux densities are continuous at the interfaces between the two materials.

For the full 3D motion, the product function $f_c(r_c)g_c(z_c)$ cannot be a solution of the 3D effective mass Schrodinger equation, since the 3D finite confinement potential, $V_c(\mathbf{r}_c) \neq V_c^{\parallel} + V_c^{\perp}$, consists of another potential that has been as a perturbation term. Thus,

$$V_c(\mathbf{r}_c) = V_c^{\parallel}(r_c) + V_c^{\perp}(z_c) + \delta V_c(r_c, z_c) \tag{8.31}$$

where

$$\delta V_c = 0, \text{if} \; r_c < R, |z_c| < L/2, \text{and} \; \delta V_c = -V_c, \text{otherwise} \tag{8.32}$$

The term δV_c is treated as a perturbation. The ground state energy of a particle in a quantum disc with finite confinement may be expressed as

$$E_c = E_c^{\parallel} + E_c^{\perp} - \frac{\langle f_c g_c | \delta V_c | f_c g_c \rangle}{\langle f_c g_c | f_c g_c \rangle} \tag{8.33}$$

8.3.2.2   *Exciton ground state*

The relative coordinate is written as $r = |\mathbf{r}_e - \mathbf{r}_h| = (r_e^2 + r_h^2 - 2r_e r_h \cos\theta)^{1/2}$. The Hamiltonian is then written as follows:

$$
\begin{aligned}
H = {}& -\frac{\hbar^2}{2m_e^\parallel}\left\{\frac{\partial^2}{\partial r_e^2} + \frac{1}{r_e}\frac{\partial}{\partial r_e} + \frac{r_e^2 - r_h^2 + r^2}{r_e r}\frac{\partial^2}{\partial r_e \partial r}\right\} \\
& -\frac{\hbar^2}{2m_h^\parallel}\left\{\frac{\partial^2}{\partial r_h^2} + \frac{1}{r_h}\frac{\partial}{\partial r_h} + \frac{r_h^2 - r_e^2 + r^2}{r_h r}\frac{\partial^2}{\partial r_h \partial r}\right\} \\
& -\frac{\hbar^2}{2m_r}\left\{\frac{\partial^2}{\partial r^2} + \frac{1}{r}\frac{\partial}{\partial r}\right\} - \frac{\hbar^2}{2m_e^\perp}\frac{\partial^2}{\partial z_e^2} - \frac{\hbar^2}{2m_h^\perp}\frac{\partial^2}{\partial z_h^2} \\
& + V_e(r_e, z_e) + V_h(r_h, z_h) - \frac{e^2}{4\pi\varepsilon\sqrt{r^2 + (z_e - z_h)^2}}
\end{aligned}
\tag{8.34}
$$

The in-plane reduced mass is $m_r(r_e, r_h)^{-1} = m_e^\parallel(r_e)^{-1} + m_h^\parallel(r_h)^{-1}$.

For the variational calculation, the following trial wavefunction is chosen:

$$
\psi_{ex} = F_e(r_e, z_e)F_h(r_h, z_h)\phi(r, z_e, z_h) = f_e(r_e)g_e(z_e)f_h(r_h)g_h(z_h)\phi(r, z_e, z_h)
\tag{8.35}
$$

where $\phi(r, z_e, z_h)$ is the wavefunction related to internal motion of electron-hole pairs. Workers used different forms of $\phi(r, z_e, z_h)$ involving one or two variational parameters. Koh et al (2001) used the following form, the justification for which is given in their work:

$$
\phi(r, z_e, z_h) = \exp\left[-\alpha\sqrt{r^2 + \gamma(z_e - z_h)^2}\right]
\tag{8.36}
$$

The ground state energy and wavefunction are determined as usual by minimizing the following functional with respect to both α and γ:

$$
\langle E_{ex}(\alpha, \gamma)\rangle = \frac{\langle\psi_{ex}|H|\psi_{ex}\rangle}{\langle\psi_{ex}|\psi_{ex}\rangle}
\tag{8.37}
$$

The binding energy is obtained from

$$
E_b = E_e + E_h - E_{ex}
\tag{8.38}
$$

Example 8.2 *The binding energy of the lowest exciton in a QWR is 6.25 times the bulk value. We may calculate the fractional dimensional parameter from the expression $E_b = R_x/[n + (\alpha_f - 3)/2]^2$, and obtain $\alpha_f = 1.8$.*

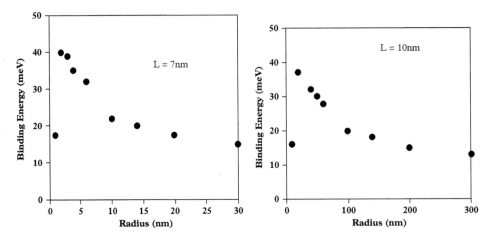

Figure 8.4 *Calculated binding energy for (a) L = 7 nm and (b) L = 10 nm.*
Source: reprinted figure from Koh T S, Feng Y P, Xu X, and Spector H N (2001) Excitons in semiconductor quantum discs. *Journal of Physics and Condensed Matter* 13: 1485–1498, with the permission of Institute of Physics, UK.

8.3.2.3 *Results*

Koh et al (2001) presented the calculated results for GaAs/gallium-aluminum-arsenide (GaAlAs) structures for different values of *R* and *L*. We include here some of the representative calculations relevant for QWRs.

Figures 8.4(a) and (b) show the plots of binding energy as a function of wire radius for two different values of the wire length: 7 and 10 nm, respectively. The curves exhibit the expected variation with radius; the binding energy starts from a low value, increases rapidly, reaches a peak and then decreases slowly and finally attains a smaller value. The values depend weakly on the length of the wire.

The nature of variation of the fractional dimensional parameter with wire radius is opposite to that shown in Fig. 8.4. The minimum value is nearly 1.5, an indication that the wire is not truly 1D. The values at the two ends of the radius exceed 2, meaning thereby that for very small and large values of radius, excitons approach the bulk character.

8.4   Classification of nanowires

QWRs or nanowire(NWR)s can be classified according to the methods of growing these structures. Fig. 8.5 shows different processes involved. There are two basic methods in growing the structures: by chemical synthesis; most of the NWRs are grown by using this. The study of NWRs gained momentum in the late 1990s or so. The NWRs may have two different configurations. In the first one, NWRs are surrounded by air. The second structure: the core-shell, consists of a NWR, termed as the core, surrounded by

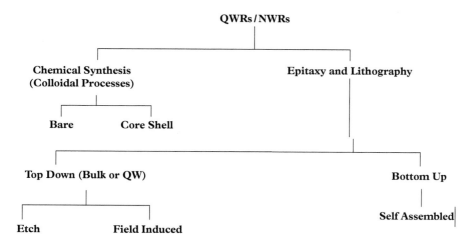

Figure 8.5 *Classification of QWRs and NWRs on the basis of growth processes.*

another shell material of higher bandgap. A cylindrical NWR looks like an optical fibre. Different core-shell structures will be introduced in the Section 8.5.

QWRs grown by epitaxy and lithography may be fabricated by using either the top-down or the bottom-up approach. The top-down method had been in use in the late 1980s after the advantages of 1D systems involving QWRs were pointed out by Arakawa and Sakaki (1982). Some of the top-down methods will be introduced in the next sub-section. The bottom-up approach uses growth by self-assembly, again to be illustrated in the Section 8.5.

8.5   Growth of quantum wires

Several workers took the challenging task of growing good quality QWRs after the appearance of the seminal paper by Arakawa and Sakaki (1982). Growth of NWRs was undertaken somewhat later. In the following section, the growth processes will be described briefly.

8.5.1   Growth of quantum wires

Initially, QWRs were grown by various processes (see Basu 1997 for some of these methods). However, the quality of the structures was not satisfactory to establish the predicted advantages of 1D systems, namely, sharp emission line, laser action with low threshold current density, better temperature sensitivity and so on. Detailed discussions of all the growth processes are beyond the scope of this book. In this section, only three methods which proved to be successful in giving good quality structures, will be introduced.

8.5.1.1 T-shaped quantum wires

Pfeiffer et al (1990) fabricated a GaAs/AlGaAs T-shaped QWR (T-QWR) by using a molecular beam epitaxy (MBE) machine and cleaved-edge-overgrowth (CEO) technique. The process consists of three steps. First a GaAs/gallium aluminum arsenide (AlGaAs) QW is grown on a [001] GaAs substrate. The sample is then cleaved to expose the surface. Finally, a GaAs well layer and then an AlGaAs barrier layer are grown on the cleaved surface. At the intersection of the T structure, as shown in Fig. 8.6, 1D electron and hole states are formed. The first QW forms the stem part of the T structure, whereas the second QW forms the arm part. Typical cross section of the QWR is 5.2 nm × 4.8 nm.

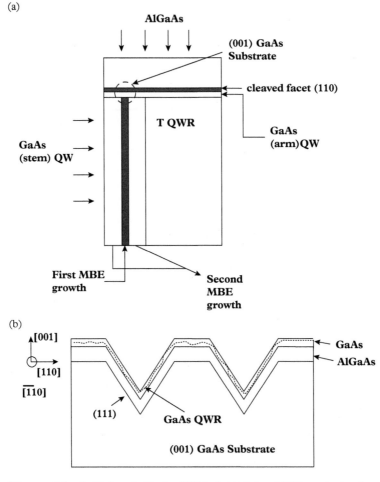

Figure 8.6 *A T-shaped GaAs QWR formed by MBE and cleaving. (b) A V-groove QWR formed by anisotropic etching and MBE.*

8.5.1.2　*V-groove quantum wires*

Kapon et al (1987) and Bhat et al (1988) fabricated GaAs/AlGaAs V-shaped QWR (V-QWR) by metal-organic vapor phase epitaxy (MOVPE) growth method. The process is illustrated in Fig. 8.6(b). First, using photolithography or electron-beam lithography a photoresist pattern is formed on a GaAs substrate. Then, V-shaped grooves were created by wet chemical etching, exposing two intersecting (111) A side-wall facets. In the final step, GaAs QWRs were fabricated at the V-groove bottom by growing GaAs/AlGaAs multilayers on this substrate.

8.5.1.3　*Self-assembled quantum wires*

Self-assembled QWRs or QDs can be grown by using Stranski–Krastanow growth mode, which requires a large lattice mismatch between the grown QWR/QD material and the substrate. InAs quantum nanostructures grown on InP material are of considerable interest due to their potential application as photonic devices operating at 1.3 and 1.55 μms. Self-assembled InAs/InP QWRs and QDs can be realized by MBE above a certain critical thickness, because of substantial (3.2%) lattice mismatch. Because of an anisotropy of built-in stress at the InAs/InP (001) interface, where the stress is higher in the [110] direction than along [$\bar{1}$10], QWRs are preferably grown rather than QDs. Self-assembly growth processes will be discussed in some detail in Chapter 9.

8.6　Nanowires

As pointed out in the Introduction, growth and study of the properties of nanowires starting from 1990s have heralded a new era in the field of nanoelectronics and Nanophotonics. Though the word wire indicates that the structures have cylindrical shape, other shapes are also possible, and the structures are called nanoribbons, nanobelts, nanowhiskers, nanodiscs, and so on. In this section, we shall briefly point out the growth processes and then discuss some of the special properties of the structures having high aspect ratio. Growth, properties, and applications of NWRs are discussed by various authors (Lu and Lieber 2006; Law et al 2004; Yan et al 2009; Yang 2005; Yang et al 2010; Picraux et al 2012; Sergey et al 2013; Ishikawa and Buyanova 2017; special issue of *Physica Status Solidi B* RRL 2013).

It may be pointed out at this stage, that an important member of the family of nanowires is the carbon nanotube (CNT), which has attracted attention from many workers due to its remarkable mechanical, electrical, and optical properties. Under suitable conditions, CNTs can act as semiconductors. However, to understand the structural details, band structures, semiconducting behavior of CNTs and their electronic and optical properties, an altogether different treatment from the principles presented so far, is needed. We do not include any further discussions about CNTs in this book.

8.6.1   Growth processes

NWRs are mainly synthesized by the bottom-up process. There exist several growth processes including vapour phase epitaxy, vapour-liquid-solid (VLS) growth process and solution-liquid-solid methods (Law et al 2004(a) and (b); Yang 2005; Yan et al 2010; Lieber 2011; Joyce et al 2011). In addition, a top-down approach, depending on lithography and etching, is also employed.

8.6.2   Types of nanowires

As shown in Fig. 8.4, NWR structures are classified into two categories: bare and core-shell. The bare NWR has a simple wire like structure made of only one type of semiconductor. The wire is surrounded by air. It is evident that the exposed surface of the NWR contains unsaturated bonds, surface states and unwanted adsorbed elements. Due to these, often the electronic and optical behaviour expected of the constituent semi-conductor nanostructure is masked by surface-related effects. To avoid these unwanted effects, heterostructure NWR structures are developed. The heterostructure NWR may be of three types. The core-shell structure consists of a core semiconductor surrounded by another semiconductor forming a radial heterostructure NWR. A schematic of this structure is shown in Fig. 8.7(a). In another type, an axial heterostructure NWR, as shown in Fig. 8.7(b), may be grown. The composition is thus modulated along the axis. Two different semiconductors are grown alternately along the growth axis. Finally, a heterostructure may be formed between the substrate and the NWRs grown on it vertically.

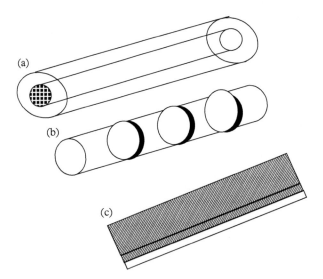

Figure 8.7 *Examples of different heterostructured NWRs;*
(a) radial (b) axial, and (c) planar in the form of nanoribbon.

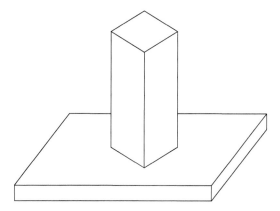

Figure 8.8 *A schematic diagram of an NWR-substrate heterostructure.*

In addition to the previously mentioned structures, in which the structures have mostly cylindrical shape, planar heterostructures are also grown. A typical example is a nanoribbon or a nanobelt, as shown in Fig. 8.7(c) (Lauhon et al 2002).

In the NWR-substrate heterostructures, NWRs can be epitaxially grown over a substrate having high degree of lattice mismatch. This has enabled growth of III–V NWRs on silicon (Si) substrate without having misfit dislocation. A schematic of such NWR-substrate heterostructure is shown in Fig. 8.8.

It has also been possible to grow branched or tree-like NWRs as well as a kinked nanowire with structurally coherent 'kinks' introduced in a controlled manner during axial elongation (Lieber 2011).

Fig. 8.9 shows a scanning electron microscope (SEM) image of zinc oxide (ZnO) NWRs grown by the VLS method.

8.7 Properties of nanowires

All the structures considered in this section under the category NWRs support 1D electron and hole gases moving freely along one direction only. The subband structures, DOS optical processes including excitonic effects have already been discussed in connection with QWRs. In this section, a few of the novel optical properties of NWRs will be pointed out and discussed.

8.7.1 Polarization anisotropy in absorption and emission

The large geometrical anisotropy possessed by semiconductor NWRs gives rise to anisotropic behaviour of light scattering, absorption, and emission. In this section, the polarization anisotropy in absorption and emission exhibited by individual NWRs will

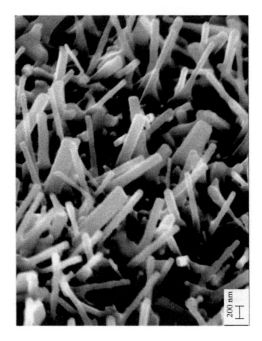

200 nm

Figure 8.9 *Zinc oxide NWRs grown by the VLS method.*
Source: SEM image courtesy of Anirban Bhattacharyya, Institute of Radio Physics and Electronics/Centre for Research in Nanoscience and Nanotechnology, University of Calcutta.

be discussed. It is assumed that the diameter of the NWRs is much smaller than the wavelength of incident and emitted radiation (Rivas et al 2008).

8.7.1.1 Absorption

The NWR is assumed to be an infinitely long cylinder. The EM radiation incident onto it may have its electric (E)-field parallel or perpendicular to the axis. Since the wavelength is large, the magnitude of the E-field is constant along the diameter. When the E-field is parallel to the axis, the continuity of the tangential component of the field at the interface between the wire and its surrounding requires that

$$\mathbf{E}_{in\parallel} = \mathbf{E}_{out\parallel} \tag{8.39}$$

where out refers to the incident field and in refers to the field inside the wire.

When the E-field vector is perpendicular to the axis, the normal component of displacement $\mathbf{D} = \epsilon_0\mathbf{E} + \mathbf{P}$ should be continuous. Therefore there is a depolarization or reduction of the E-field inside the wire expressed as $\mathbf{E}_{in\perp} = \mathbf{E}_{out\perp} - \mathbf{P}/\epsilon_0$. The field amplitude inside the wire is given by

$$E_{in\perp} = \frac{2}{1 + \epsilon(\omega)} E_{out\perp} \qquad (8.40)$$

where $\epsilon(\omega)$ is the permittivity contrast between the NWR material and its surrounding medium and $2/1 + \epsilon(\omega)$ is the depolarization factor. Since the absorbed energy is proportional to the square of the field inside, we may define the anisotropy of absorption in the following way:

$$\rho_{abs} = \frac{|E_{in\parallel}|^2}{|E_{in\perp}|^2} = \frac{|1 + \epsilon(\omega)|^2}{4} \qquad (8.41)$$

In general, the material forming the NWR has a larger permittivity than the surrounding material. It follows from Eq. (8.41) therefore that the absorption is quite larger when the incident field is polarized parallel to the NWR axis.

Example 8.3 *Taking a value of $\epsilon(\omega) = 12$ at the operating wavelength, the anisotropy becomes 42.*

Experimentally, the absorption is measured by integrating the photoluminescence spectra. For an InP NWR of 30 nm diameter and of a few μm in length, the absorption anisotropy was found to be as high as 20, when the NWR was illuminated with a laser beam of $\lambda = 458$ nm.

8.7.1.2 Emission

Like absorption, light emission from a NWR shows anisotropy when the polarization is parallel and perpendicular to the axis. (Ruda and Shik 2005) considered an array of dipoles randomly oriented in a dielectric cylinder and calculated the Poynting vector outside the cylinder. In the recombination process, information of the polarization of excitation is lost due to inelastic relaxation of photoexcited electrons. The emission from randomly oriented dipoles is thus incoherent. The anisotropy contrast of emission has been expressed by Ruda and Shik by

$$\rho_{em} = \frac{I_{em\parallel}}{I_{em\perp}} = \frac{2 + |\epsilon(\omega) + 1|^2}{6} \qquad (8.42)$$

where I stands for the intensity of emitted light.

The measured anisotropy for an InP NWR at 458 nm was 4.6, lower than the measured anisotropy in absorption.

8.7.1.3 Anisotropy in radiative lifetime

The emission from the NWR is due to excitonic recombination and it obeys the rate equation $dn/dt = -n/\tau_r$, where τ_r is the recombination lifetime. Since the emission depends on the polarization, the radiative lifetime should be different when the exciton dipoles are aligned parallel and perpendicular to the axis of the NWR. Following Eq.

(8.41), the lifetime is $(1 + \epsilon)^2/4$ times longer when the dipoles are aligned perpendicular to the axis in comparison to lifetime when the dipoles are parallel to the axis.

Example 8.4 *Take the example of GaAs NWR in which $\epsilon = 13$. The lifetime is nearly 40 times higher with dipoles perpendicular to the axis.*

The effect is similar to a reverse Purcell effect in which lifetime is suppressed as the field is reduced. The Purcell effect will be introduced in Chapter 11.

The polarization anisotropy discussed previously gives rise to an interesting effect related to crystal structure. InP usually has a zinc blende (ZB) structure, but under certain growth conditions, InP NWRs may have wurtzite (WZ) structure. ZB InP nanowires show emission with polarization parallel to the axis. However, for WZ InP wires, emission is polarized perpendicular to the axis. This occurs because in WZ InP NWRs, the lowest exciton is A type and recombination is only allowed when the E-field is polarized perpendicular to the c-axis, that is, the growth axis.

Experimental results related to lifetime and dependence of emission on crystal structure are described in the review paper by Joyce et al (2011).

8.7.2   Giant birefringence

We now consider coherent propagation of light in a medium consisting of an ensemble of NWRs. Coherent propagation means elastically scattered light in the forward direction, and its propagation through ensembles of scatterers is described by the propagation in an effective medium that is considered homogeneous. The medium contains anisotropic scatterers which are aligned along certain directions, the homogenization performed by any effective medium model leads to the following permittivity tensor:

$$\bar{\epsilon} = \begin{bmatrix} \epsilon_1 & 0 & 0 \\ 0 & \epsilon_2 & 0 \\ 0 & 0 & \epsilon_3 \end{bmatrix} \tag{8.43}$$

where ϵ's are permittivities along the principal axes. A medium with this characteristic is called birefringent, in which light propagating through it has different refractive index (RI)s depending on polarization. A medium is called biaxial when $\epsilon_1 \neq \epsilon_2 \neq \epsilon_3$, and uniaxial when $\epsilon_1 = \epsilon_2 \neq \epsilon_3$. In a uniaxial medium, ϵ_1 and ϵ_2 are called ordinary permittivity, while ϵ_3 is called the extraordinary permittivity.

Since the absorption and emission of light by nanowires are polarization dependent, it is to be expected that the scattering cross-section for light polarized parallel to the long axis of the semiconductor NWRs is larger than the same for light polarized perpendicular to this axis. Furthermore, NWRs can be grown epitaxially on crystalline substrates along well-defined crystallographic directions. If this direction of growth is unique, layers of aligned NWRs form a uniaxial birefringent material having the following effective permittivity tensor:

$$\bar{\epsilon} = \begin{bmatrix} \epsilon_\perp & 0 & 0 \\ 0 & \epsilon_\perp & 0 \\ 0 & 0 & \epsilon_\| \end{bmatrix} \qquad (8.44)$$

The symbols $\perp$ and $\|$ are used to signify that light is polarized perpendicular and parallel to the long axis of the wire, respectively. The former is the ordinary and the latter is the extraordinary effective permittivity.

Let the E-field of the wave propagating in the medium be parallel to all the interfaces of the nanowires. In this case, the polarization is parallel to their long axis. The effective extraordinary permittivity is given by

$$\epsilon_\| = \epsilon_{\|nw} f + \epsilon_m (1 - f) \qquad (8.45)$$

where the subscript *nw* refers to extraordinary permittivity of the NWR material, and f is the volume fraction occupied by NWRs embedded in the dielectric matrix of permittivity ϵ_m.

In the long-wavelength limit, that is, when the wavelength of the light propagating in the medium is much larger than the diameter of the nanowires and the mean distance between nanowires, and for nanowires filling fractions smaller than 50%, the ordinary permittivity $\epsilon_\perp$ approaches the Maxwell–Garnett effective permittivity

$$\epsilon_\perp \approx \epsilon_m \left(1 + \frac{2f\alpha}{1 - f\alpha} \right) \qquad (8.46)$$

where $\alpha = (\epsilon_\| - \epsilon_m)/\epsilon_\| + \epsilon_m$ is the depolarization factor. A birefringence parameter is introduced as follows:

$$\Delta\eta = \eta_\| - \eta_\perp = \sqrt{\epsilon_\|} - \sqrt{\epsilon_\perp} \qquad (8.47)$$

The birefringent parameter for gallium phosphide (GaP) NWRs has been obtained experimentally (see Rivas et al 2013 and references therein). It has been found that $\Delta\eta$ is large for short NWRs and its value decreases with increase in the length of the wire.

8.7.3   Engineering light absorption

Light absorption in a NWR can be tailored by changing the wavelength, diameter of the wire, and surrounding medium as in the core-shell NWRs. This tunability with wavelength was first demonstrated by Cao et al (2009) using a Ge NWR. Recent study of the interaction between nanowires and visible light has shown resonances that promise light absorption/scattering engineering for photonic applications (Pura et al 2018). It has been shown that light couples to NWRs in different ways depending on the NWR diameter, NWR composition, light wavelength, and the dielectric mismatch between the

NWR and the surrounding media. All these variables allow tuning the optical properties of the NWRs. One of the most relevant properties concerning the light/NWR interaction is the ability of NWRs to enhance the optical absorption/scattering for certain NWR diameters. Pura et al have grown heterostructured nanowires, both axial and radial, and also modulated the doping level and the surface conditions, among other factors than can affect the light/NWR interaction. It is shown by experiment and modelling that E-field is greatly enhanced in the axial heterojunction. For radial heterojunction NWRs, absorption shows resonances for particular values of wire diameter. These findings are useful for enhancing the performance of photonic devices, like solar cells and photodetectors.

8.7.4 Novel excitonic effects: Laser cooling

The idea of cooling matter by strong light was introduced long ago. The cooling of a semiconductor by shining it with strong laser light has been the subject of investigation over a few decades. The basic principle of laser cooling and progress in the area is discussed by several authors (Nemova and Kashyap 2010; Seletskiy et al 2016). When semiconductors are illuminated by strong light, hot excitons are created. Thermalization of excitons by emission of longitudinal optical(LO) phonons drains out heat from the sample and cools it. The strong exciton-phonon scattering in cadmium sulfide (CdS) nanobelt has effected a net cooling by 40 K from 290 K when pumped by a 514 nm laser and has opened a new way to optical refrigeration (Xu et al 2014).

8.7.5 Excitons under an electric field

An external E-field in QWs gives rise to several phenomena such as modulation of absorption coefficient, broadening of exciton absorption line, shift of the absorption peak, dissociation of excitons, etc., as discussed in Chapter 7 (see Schmitt Rink et al 1987). Investigation of excitons in the presence of an extrinsic E-field can provide useful excitonic information for the nanowires and nanobelts. The applied E-field changes the absorption properties both via Franz–Keldysh Effect and Quantum Confined Stark Effect (QCSE) (Pederseon 2015, 2017).

8.8 Applications of nanowires

The development of semiconductor NWRs has opened up the possibility of realizing ultra-small photonic devices, which include highly efficient coherent light sources, optical modulators, photodetectors, and solar cells. In this section, some progress in such developments will be briefly mentioned, giving some representative references.

8.8.1 Light sources

Semiconductor NWR lasers generate highly localized, intense optical field of narrow linewidth, which can be efficiently coupled to highly integrated nanophotonic elements and optical circuits. The high refractive index (RI) of NWR material allow downsizing of the device dimensions to only a few hundreds of nm. The geometry favours low-loss optical waveguiding and optical recirculation in the active region. The materials used so far are zinc oxide (ZnO), Gr. -III nitrides, II/VI materials and III–V NWRs. By growing epitaxially InGaAs nanoneedles directly on Si, lasing at IR has been observed at low temperature. This seems useful to have electronic-photonic integration on Si platform. Lasing has been observed by using optically pumped GaSb and core-shell NWRs using GaAs-GaAsP. A GaAs/AlGaAs core-shell NWR laser has been reported that emit IR radiation even at room temperature (Mayer et al 2013; see references quoted therein for representative work in other material systems).

8.8.2 Photodetectors

An NWR photodetector/switch has been realized by using NWRs in a horizontal transistor configuration, in which the incident optical flux changes the conductivity of the NWR from insulating to a conducting one. The materials used are ZnO, InP, and CdS (see references cited by Yan et al 2009).The responsivity of typical InP nanowire photodetectors is of the order of 3000 A/W. To obtain higher responsivity that is needed to detect low-level light as in single-photon detection, avalanche multiplication is preferred. High responsivity with detection limits less than 100 photons has been found in crossed silicon-cadmium sulfide (CdS) NWRs. The sensitivity for NWR photodetectors is comparable to that reported for planar avalanche photodetector (APD)s. However, the response time is higher. Core-shell NWRs are being studied for better performance. NanoAPDs find application in many diverse areas, such as, nano positioning, integrated Photonics, and near-field detection, real-time observation of single-protein dynamics with integrated nanoAPDs in microfluidics.

8.8.3 Solar cells

Miniaturized solar cells can be readily incorporated into nanophotonic systems to act as an integrated power source. The NWR based photovoltaic sources have a number of advantages, listed as follows. (1) Absorption takes place over the whole length of the NWR, while the distance covered by carriers for being collected is quite low. (2) Presence of interpenetrating heterojunction allows efficient carrier extraction after photogeneration. (3) Using high density NWR arrays, strong light trapping can be achieved. (4) Using different size and compositions, the material properties can be tailored, and cell efficiencies can be optimized.

Different materials systems and results obtained for NWR based solar cells are reviewed by Yan et al (2009) and Joyce et al (2013) (see also Zhang et al 2016: Reading list).

8.8.4 Waveguides

NWRs can be used as light sources. In order to perform logic operations in computing, communication and sensing, as envisaged in future photonic integrated circuits (PICs), the photons generated must be efficiently captured and delivered to other photonic devices. Photodetectors, frequency converters, filters and switches form a few examples of such devices. With the development of NWR subwavelength waveguides, on-chip routing of optical signals to perform complex tasks has been feasible. Chemically synthesized binary oxide NWRs have several advantages over waveguides formed by lithographic processes. NWR waveguides are 1D, are single crystals, large refractive index, low surface roughness, and high flexibility. Different material systems can be used to grow such waveguides, which may operate both above and below the diffraction limit. Tin dioxide (SnO_2) nanoribbons have proved to be excellent subwavelength waveguides (see Yan et al 2009). NWR waveguides have the same structure of optical fibres, have a high refractive index (n = 2.1), with typical dimensions of 100–400 nm. They can efficiently guide their own visible photoluminescence and visible/UV emissions from other nanowires and fluorophores. Typical loss values cover the range from 1–8 dB mm^{-1} for wavelengths of 450–550 nm, depending on the ribbon's cross-sectional area. The values are higher than those of conventional optical fibres. However, for integrated planar photonic applications such as short-distance on-chip signal routing and distribution, the high values do not matter much. Nanoribbons of different cut-off frequencies can be integrated with a common multimode core waveguide ribbon to produce an optical router based on input colour. Nanowire architectures could be potentially used in integrated optical logic and all-optical switching. Semiconductor NWRs, due to their high RI, can efficiently guide light through water and other liquid media, and thus have an edge over silica fibres. This proves to be helpful when semiconductor NWRs are interfaced with liquid media for on-chip chemical or biological spectroscopic analysis.

8.8.5 Other applications

Laser cooling or optical refrigeration has been discussed briefly in Section 8.7.3. Sergey et al (2013) discussed quantum computing based on semiconductor nanowires.

Problems

8.1 *Obtain the expression for the DOS function for 1DEG, as given by Eq. (8.7).*

8.2 *The density-of-states in 0D system is given by $\rho_{oD}(E) = \sum_n \delta(E - E_n)$. Using n as a continuous variable derive the expression for DOS in 1D system. Repeat the process to arrive at the expressions for 2D and bulk semiconductors.*

8.3 *Obtain an expression relating the number of electrons in a QWR with the Fermi energy, at 0K.*

8.4 *Show that at low temperature when the Fermi function is box-like the 1D carrier density is proportional to $(E_F - E_i)^{1/2}$. Obtain the complete expression for n_{1D}.*

8.5 *Show that the density of electrons in a 1D wire under non degenerate condition may be expressed as $n_{1D} = (2m_e k_B T/\pi \hbar^2)^{1/2} exp[(E_F - E_i)/k_B T]$.*

8.6 *Assume that the 1D potential in a QWR is parabolic and the subband separation is 10 meV. The Fermi energy is 35 meV above the bottom of the lowest subband. Find the number of electrons per unit length at 0K if the electron effective mass is 0.025 m_0.*

8.7 *The potential in a QWR is expressed as $V(x) = 1/2 m_e \omega_0^2 x^2$. Calculate the ground subband energy taking $m_e = 0.014\, m_0$.*

8.8 *The potential in a GaAs QWR is $V(x) = \frac{1}{2} K x^2$, $K = 7.44 \times 10^{-7}\ \text{Jm}^{-2}$. The fermi level is $E_F = V\left(\frac{W}{2}\right)$, where the width of the channel $W = 160\ nm$. Calculate the electron density in the channel and the number of subbands occupied. Assume $T = 0\ K$.*

8.9 *Calculate the threshold current in a QWR laser using the following parameters: $n_{tr} = 10^{18}\ cm^{-3}$, $\tau = 3\ ns$, $a = 10^{15}\ cm^{-2}$, $\alpha_{sc} = 1\ cm^{-1}$, $\alpha_{fc} = 0.1\ cm^{-1}$, $L = 200\ \mu m$, $R = 0.99$, area $= 50\ nm^2$. Assume unity mode confinement factor $= 0.3$, quantum efficiency $= 0.8$, and $g = a(n - n_{tr})$.*

8.10 *The intersubband transition energy is expressed as $\omega_r^2 = \omega_{10}^2 + \omega_{10} e^2 N_{1D}/(\pi \varepsilon_s \varepsilon_0)$, where ω_{10} is the shift due to depolarization in presence of carrier density N_{1D}. take $N_{1D} = 10^6\ cm^{-1}$, $\varepsilon_s = 16$. Calculate by using potential given in Problem 8.5.*

8.11 *Obtain a plot of dimensional parameter versus wire radius from the calculated values of the binding energy of a NWR shown in Fig. 8.4(a).*

8.12 *Obtain a plot of $\Delta \eta$ versus f for a semiconductor with permittivity 16 surrounded by air, according to Eq. (8.47).*

Reading List

Basu P K (1997) *Theory of Optical Processes in Semiconductors: Bulk and Microstructures.* Oxford, UK: Oxford University Press.

Basu P K, Mukhopadhyay B, and Basu Rikmantra (2015) *Semiconducor Laser Theory.* Boca Raton, FL, USA: CRC Press.

Harrison P (2000) *Quantum Wells, Wires and Dots: Theoretical and Computational Physics.* Chichester: John Wiley & Sons, Ltd.

Haug H and Koch S W (1993) *Quantum Theory of the Optical and Electronic Properties of Semiconductors*, 2nd edn. Singapore: World Scientific.

Kapon E (1993) Quantum wire semiconductor lasers. In: *Quantum Well Lasers*, P S Zory (ed.) Chaper 10, pp. 461–500. New York: Academic Press.

Mitin V, Kochelap V, and Strocio M A (1999) *Quantum Heterostructures.* NY: John Wiley.

Moral A F, Dayeh S A, and Jagadish C (eds.) (2015) *Semiconductor Nanowires I: Growth and Theory, vol 93 of Semiconductor and Semimetals.* Waltham, MA, USA: Academic (Elsevier).

Singh J (2003) *Electronic and Optoelectronic Properties of Semiconductor Structures.* Cambridge, New York: Cambridge University Press.

Weisbuch C and Vinter B (1991) *Quantum Semiconductor Structures*. San Diego: Academic Press.

Zheng A, Zheng G, and Lieber C (2016) *Nanowires: Building Blocks for Nanoscience and Nanotechnology*. Switzerland: Springer.

References

Akiyama H (1998) One-dimensional excitons in GaAs quantum wires. *Journal of Physics: Condensed Matter* 10(1998): 3095–3139.

Arakawa Y and Sakaki H (1982) Multidimensional quantum well laser and temperature dependence of its threshold current. *Applied Physics Letters* 40: 939–941.

Asada M, Miyamoto Y, and Suematsu Y (1986) Gain and the threshold of three-dimensional quantum-box lasers. *IEEE Journal of Quantum Electronics* 22: 1915–1921.

Bhat R, Kapon E, Hwang D M, Koza M A, and Yun C P (1988) Patterned quantum well heterostructures grown by OMCVD growth on nonplanar substrates: Applications to extremely narrow SQW lasers. *Journal of Crystal Growth* 93: 850–856.

Cao L, White J S, Park J -S, Schuller J A, Clemens B M, and Brongersma M L (2009) Engineering light absorption in semiconductor nanowire devices. *Nature Materials* 8: 643–647.

Christol P, Lefebvre P, and Mathieu H (1993) Fractional-dimensional calculation of exciton binding energies in semiconductor quantum wells and quantum-well wires. *Journal of Applied Physics* 74(9): 5626–5637.

Corzine S W, Yan R H, and Coldren L A (1993) *Quantum Well Lasers*. P S Zory (ed.) Chapter 1, pp. 17–98. NY: Academic Press.

Focus on semiconductor nanowires. *Rapid Research Letters Physica Status Solidi* 7(10): 683–925 (Special Issue).

Ishikawa F and Buyanova I (2017) *Novel Compound Semiconductor Nanowires: Materials, Devices, and Applications*. Pan Stanford; CRC Press.

Joyce, H J, Gao Q, Hoe Tan H, Jagdish C et al (2011) III-V semiconductor nanowires for optoelectronic device applications. *Progress in Quantum Electronics* 35: 23–75.

Kapon E, Tamargo M C, and Hwang D M (1987) Molecular beam epitaxy of GaAs/AlGaAs superlattice heterostructures on nonplanar substrates. *Applied Physics Letters* 50: 347–350.

Koh T S, Feng Y P, Xu X, and Spector H N (2001) Excitons in semiconductor quantum discs. *Journal of Physics and Condensed Matter* 13: 1485–1498.

Lauhon L J, Gudiksen M S, Wang D, and Lieber CM (2002) Epitaxial core–shell and core-multishell nanowire heterostructures. *Nature* 420: 57–61

Law M, Goldberger J, and Yang P (2004a) Semiconductor nanowires and nanotubes (2004) *Annual Review of Material Research* 34: 83–122

Law M, Sirbuly D J, Johnson J C, Goldberger J, Saykally R J, and Yang P (2004b) Nanoribbon waveguides for subwavelength photonics integration. *Science* 305(27): 1269–1273.

Lieber C M (2011) Semiconductor nanowires: A platform for nanoscience and nanotechnology. *Materials Research Society Bulletin* 36: 1052–1063.

Lu W and Lieber C M (2006) Semiconductor nanowires. *Journal of Physics D: Applied Physics* 39: R387–R406.

Mayer B, Rudolph D, Schnell J, Morkötter S, Winnerl J, Treu J, Müller K, Bracher G, Abstreiter G, Koblmüller G, and Finley J (2013) Lasing from individual GaAs-AlGaAs core-shell nanowires up to room temperature. *Nature Communications* 1–7.

Nemova G and Kashyap R (2010) Laser cooling of solids. *Reported Progress in Physics* 73(086501): 1–20.

Ogawa T and Takagahara T (1991) Optical absorption and Sommerfeld factors of one-dimensional semiconductors: An exact treatment of excitonic effects. *Physical Review B* 44: 8138–8145.

Pedersen T G (2015) Analytical models of optical response in one dimensional semiconductors. *Physics Letters A* 379: 1785–1790.

Pedersen T G (2017) Stark effect infinite-barrier quantum wells, wires, and dots. *New Journal of Physics* 19(043011): 1–10.

Pfeiffer L, West K W, Strömer H L, Eisenstein J P, Baldwin K W, Gershoni D, and Spector J (1990) Formation of a high-quality two-dimensional electron gas on cleaved GaAs. *Applied Physics Letters* 56: 1697–1699.

Picraux S T, Yoo J, Campbell I H, Dayeh S A, and Perea D E (2012) Semiconductor nanowires for solar cells. In: *Semiconductor Nanostructures for Optoelectronic Devices, NanoScience and Technology*, G -C Yi (ed.) Chapter 11, pp. 297–328. Berlin Heidelberg: Springer-Verlag.

Pura J L, Anaya J, Souto J, Prieto A C, Rodrıguez A, Rodrıguez T, Periwal P, Baron T, and Jimenez J (2018) Electromagnetic field enhancement effects in group IV semiconductor nanowires. A Raman spectroscopy approach. *Journal of Applied Physics* 123(114302): 1–8.

Ray P and Basu P K (1993) Peaked nature of excitonic absorption in quantum well wires of indirect-gap semiconductors. *Physical Review B* 48: 11420–11423.

Rivas J G, Muskens O L, Borgström M T, Diedenhofen S L, and Bakkers E P A M (2008) Optical anisotropy of semiconductor nanowires. In: *One-Dimensional Nanostructures*, Zhiming M Wang (ed.) Chapter 6, pp. 127–145. New York: Springer.

Ruda H E and Shik S (2005) Polarization-sensitive optical phenomena in semiconducting and metallic nanowires. *Physical Review B* 72(115308): 1–11.

Schmitt Rink S, Miller D A B, and Chemla D S (1987) Theory of the linear and nonlinear optical properties of semiconductor micro crystallites. *Physical Review B* 35: 8113.

Seletskiy D V, Epstein R, and Sheik-Bahae M (2016) Laser cooling in solids: Advances and prospects. *Reports on Progress in Physics* 79(096401): 23.

Sergey M, Frolov, Plissard S R, Nadj-Perge S, Kouwenhoven L P, and Bakkers E P A M (2013) Quantum computing based on semiconductor nanowires. *Materials Research Society Bulletin* 38: 809–815.

Xu X, Zhang Q, Zhang J, Zhou Y, and Xiong Q (2014) Taming excitons in II–VI semiconductor nanowires and nanobelts. *Journal of Physics D: Applied Physics* 47(394009): 1–14.

Yan R, Gargas D, and Yang P (2009) Nanowire photonics. *Nature Photonics* 3(October): 569–576.

Yang P (2005) The chemistry and physics of semiconductor nanowires. *Materials Research Society Bulletin* 30: 85–91.

Yang P, Yan R, and Fardin M (2010) Semiconductor nanowire: What's next? *Nano Letters* 10: 1529–1536.

Zheng A, Zheng G, and Lieber C (2016) *Nanowires: Building Blocks for Nanoscience and Nanotechnology*. Switzerland: Springer.

9

Nanoparticles

9.1 Introduction

The structure, electronic, and optical properties of the simplest nanostructure, a Quantum Well (QW), in which carrier motion is inhibited in one dimension, have been discussed in Chapter 6. Next comes the Quantum Wire (QWR) or nanowires, in which carrier motion is one-dimensional (1D). Their properties are discussed in Chapter 7.

We now present the basic properties and optical processes in semiconductor Quantum Dots (QD)s in which the carrier motion is restricted in all three dimensions giving rise to a zero-dimensional (0D) electron gas (0DEG) (or a 0D hole gas). The preliminary concept for structures and quantization phenomena in these nanostructures have been given in Introduction. In this chapter, we make an elaborate discussion on QDs made of semiconductor materials. These are also termed as semiconductor nanoparticles. Solid state physicists view QDs or nanocrystals as a member of the low-dimensional structures. As already stated, confinement in multi dimensions was first proposed by Arakawa and Sakaki (1982).

A different way of looking into nanocrystals is followed by molecular physicists, who consider them as large molecules. Nanocrystals or nanoparticles have been in use in human civilization from a very ancient time. The Lycurgus cup, a fourth-century Roman glass kept in the British Museum, used Au and Ag nanoparticles dispersed in glass. It appears red when lit from behind and green when lit from the front. Artisans in churches in Europe and in other countries used tiny particles of Au, Ag, Cu, etc., in glass panes to form pictures of different people, flowers, and so on. It is now known that the colour emitted by nanoparticles depends on their sizes. Nanoparticles dispersed in glass matrices by controlled diffusion have been used by several commercial firms to produce colour cut-off filters and photochromic glasses. QDs in glass crystals were first studied by Ekimov et al (1980) in Russia and in colloidal solutions by Brus (1982) in the USA. Brus observed that the wavelength of emission or absorption changed over a period of days as the crystals grew. He ascribed this to the quantum confinement effect. Efros and Efros (1982) also did pioneering work on nanocrystals.

Unlike in QWs or QWRs, there is no translational motion in QDs. The ideal energy levels are as sharp as atomic energy levels, and the density-of-states (DOS) function is a series of δ-functions. Therefore, only a finite number of electrons and holes may be

Semiconductor Nanophotonics. Prasanta Kumar Basu, Bratati Mukhopadhyay, and Rikmantra Basu, Oxford University Press.
© P.K. Basu, B. Mukhopadhyay, R. Basu (2022). DOI: 10.1093/oso/9780198784692.003.0009

created in a QD. In practice, it is not a single QD or a single nanocrystal, rather an assembly of such particles, are grown by using epitaxy, or self-assembly, or by using other chemical methods, in which the particles are dispersed in a host matrix. The ensemble has fluctuations in size and shape due to distribution in strain, environment, defect distribution and many others. The observed emission and absorption spectra show inhomogeneous broadening due to such fluctuations.

In the present chapter, we first introduce the basic electronic properties of QDs, the quantized subbands, DOS function, absorption, gain and emission properties by following the simple particle-in-a box picture. The effect of size distribution is then mentioned. The difference between strong and weak confinement regimes is then pointed out. The elementary theory of excitonic processes in 0D is then developed, mentioning briefly more refined theories and then empirical methods. A classification of nanocrystals based on growth processes like top-down and bottom-up methods is then given, mentioning briefly some of the bottom-up processes. The core-shell nanostructures are next illustrated.

QDs play a pivotal role in the area of Nanophotonics. As the energy levels in QDs or nanoparticles are sharp, like atomic energy levels, many interesting physical phenomena related to quantum optics in atomic systems, are reproduced or are being reproduced by using QDs. One such example is cavity quantum electro dynamics (CQED), which was first observed using atomic systems. The phenomena are now observed in QD systems. Slow light phenomena, Bose–Einstein condensation (BEC), and many such processes are now studied using QDs or nanoparticles. In addition, emission and detection of single photon, quantum computing, encryption and several such applications involving QDs, currently embody a very active and fruitful area of research and development from both fundamental and application points of view. A few of these phenomena and applications using QDs will be studied in the next few chapters. The present chapter therefore aims at providing the background to understand the previously mentioned phenomena.

9.2 Quantum dots

It is useful to compare the size and other structural properties of semiconductor QDs with the similar parameters for atoms, molecular clusters and bulk materials. The length scales of interest are lattice constant a_L, exciton Bohr radius a_B and wavelength of photons λ. Table 9.1 summarizes the typical size, number of atoms contained, and usual model to describe electronic and optical properties of the substance (see Gaponenko 2010 for more detailed description).

Example 9.1 *The lattice constant in most semiconductors is typically 5 Å (0.5 nm). The number of atoms in a cubic quantum box (CQB) having 3 nm length on each side is $6^3 \sim 216$.*

Table 9.1 *Size, number of atoms, and models used.*

Type of structure	Typical length a	No of atoms	Model
Atom	Atomic radius	A few	Quantum theory
Molecular cluster	$a < a_L$	3–10^2	Quantum Chemical Theory
QD Nanostructures	$a_L \ll a \ll \lambda$	10^2–10^6	Effective mass approximation
Bulk	$a \gg a_L$	$> 10^6$	Quantum theory of solids

As stated in the introduction, nanostructures including QDs can be grown either by top-down or bottom-up approach. In the former, size of bulk material can be aggressively scaled down to achieve QDs. On the other hand, molecular clusters can be made oversized by adding more molecules to make a QD. The properties of QDs can be understood by using either quantum chemical theory or effective mass approximation. The results of the two theories agree with each other.

9.3   Quantum dot growth mechanisms and structures

We first discuss briefly how the QD arrays are grown by using epitaxy. The typical shape and size of the dots grown by the process are mentioned, and their influence on electronic structures and optical spectra is pointed out.

9.3.1   Growth of quantum dots

It has been mentioned already that both epitaxy and patterning are used for the fabrication of QD structures. The epitaxially grown layer may have the same lattice constant as of the substrate. It may also have slight lattice mismatch so as to develop strain in the layer. The reader is referred to the paper by Oura et al (2003) for a detailed understanding of the various growth processes.

Fig. 9.1 illustrates the different growth processes of a layer on a substrate for almost lattice matching condition, a highly lattice mismatched pair and for an intermediate condition. The lattice matched or nearly lattice matched 2D layer by layer growth on a substrate is known as Frank-van der Merwe growth and is illustrated in Fig. 9.1(a). In this case, an entire atomic layer is grown first followed by the next atomic layer. In this way the desired material grows uniformly. The growth of a highly lattice-mismatched (highly strained) layer on a substrate is known as Volmer–Weber growth. An example is the growth of indium arsenide (InAs) on silicon (Si). In this case, the strain energy is so high that the growth of a uniform layer is interrupted. The grown material is in the form

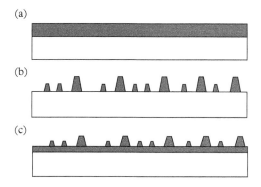

Figure 9.1 *Illustration of (a) Frank–van der Merwe growth, (b) Volmer–Weber growth, and (c) Stranski–Krastanov growth.*

of islands randomly positioned on the substrate and the clusters or islands have random sizes, as illustrated in Fig. 9.1(b).

In between the two extremes, an intermediate growth process, known as the Stranski–Krastanov (SK) process exists, and the process is shown in Fig. 9.1(c). Initially, the strain energy is not too large, permitting 2D layer growth. After a few monolayers are grown, the strain energy reaches a higher level favouring the growth of 3D islands. The grown islands are called *self-assembled* QDs and the thin 2D layer grown initially is called the *wetting layer.*

In the SK process, growth of the dots starts only when the wetting layer reaches a critical thickness. For example, for growth of InAs QDs on gallium arsenide (GaAs), the critical thickness is about 1.3 monolayer (ML). With the addition of more materials, InAs islands begin to form, typical areal density of the islands being $> 10^{12}$ cm^{-2}. Additional dots are not, however, formed by adding more material, but the size of the islands expand containing dislocations.

The QDs formed by self-assembly, as shown in Fig. 9.1(c), are in the form of pyramidal or conical bumps. Typical heights of the bumps are about 3–4 nm and their extension in the base is about 25 nm; this is also the typical base diameter for conical shaped dots. Each dot is, however, isolated. As the wetting layer provides a high barrier for the electrons in a dot, each dot is decoupled from its neighbours. The distributions of dot size, and geometry, and spacing between adjacent dots are random. The irregular spacing between adjacent dots leads to an irregular probability of tunneling. Moreover, due to random size distribution of the dots, the discrete energy levels of the electrons and holes show appreciable distribution. As a result, there is a large inhomogeneous broadening of the emission spectra.

9.4   Zero-dimensional systems

The electrons and holes in a QD or nanocrystals (NC) cannot move freely along any directions. They therefore behave as a 0D electron (or hole) gas. In the following, some of the basic properties of 0D systems will be described.

9.4.1   Wave functions and density-of-states

9.4.1.1   Rectangular box

We consider first the ideal example of a rectangular semiconductor QD with dimensions L_x, L_y, and L_z. The barriers at the edges of the box are assumed to have infinite heights. The electron and hole wavefunctions are, as usual, written as

$$\phi_{lmn} = \phi_l(x)\phi_m(y)\phi_n(z) \tag{9.1}$$

where $\phi_l(x) = \sqrt{2/L_x}\sin(k_x x)$, and $\phi_m(y)$ and $\phi_n(z)$ can be expressed in a similar way. The wavenumbers are discrete, such that. The discrete eigen energies are expressed as

$$E_{lmn} = E_g + E_{xl} + E_{ym} + E_{zn} \tag{9.2}$$

where $E_{xl} = (\hbar^2/2m_e)(\pi l/L_x)^2$, and E_{ym} and E_{zn} are similarly expressed. The 0D DOS is written as

$$\rho^{0D}(E) = 2\sum_{l,m,n}\delta(E - E_{lmn}) \tag{9.3}$$

The factor 2 is for spin degeneracy and the DOS function is thus a series of Dirac δ-function.

Example 9.2   *The energy of n = 1,1,1 state of a GaAs cubic QB with each side L = 10 nm is obtained from $E_{111} = (3\hbar^2\pi^2/2m_eL^2) =$ **0.168** eV with $m_e = 0.067\ m_0$.*

Example 9.3   *The energy difference between n = 211 and n =111 states using the same parameters as in Example 9.2 is 0.168 eV. The absorption and emission between these two states occur at mid infrared region with a wavelength of 7.38 μm.*

Example 9.4   *The energy of the second heavy-hole (HH) subband may easily be obtained from 0.438 (m_e/m_{hh}) and is 0.0652 eV. Therefore, the transition energy is 1.93 eV, for Eg = 1,426 eV. The corresponding emission wavelength is 642 nm.*

9.4.1.2   Spherical dot

A spherically shaped dot of radius R is now assumed to be embedded in a higher band gap material and the conduction band offset is denoted by ΔE_c. The potential is zero within the dot ($r < R$). Considering the simplest case of zero angular momentum (Schiff 1968), the wavefunction may be written as a product of radial component and an angular component. The radial component R(r) satisfies the equation

$$\frac{1}{r^2}\frac{d}{dr}\left(r^2\frac{dR}{dr}\right) + \left\{\frac{2m_e}{\hbar^2}[E - V(r)] - \frac{\lambda}{r^2}\right\}R = 0 \tag{9.4}$$

Putting $R(r) = \phi(r)/r$, one may arrive at

$$-\frac{\hbar^2}{2m_e}\frac{d^2\phi}{dr^2} + \left[V(r) + \frac{l(l+1)\hbar^2}{2m_er^2}\right]\phi = E\phi \tag{9.5a}$$

When $l = 0$, Eq. (9.5a) reduces to

$$-\frac{\hbar^2}{2m_e}\frac{d^2\phi}{dr^2} + \Delta E_c\phi = E\phi, \quad r > R$$
$$-\frac{\hbar^2}{2m_e}\frac{d^2\phi}{dr^2} = E\phi, \quad r < R \tag{9.5b}$$

It follows easily that the solution of Eq. (9.5b) is similar to that obtained in a 1D particle-in-a box problem. Therefore, the eigenvalues are obtained by solving the equation

$$\alpha\cot\alpha R = -\beta \tag{9.5c}$$

where $\alpha = \sqrt{2m_e(\Delta E_c - E)/\hbar^2}$ and $\beta = \sqrt{2m_eE/\hbar^2}$.

There is no solution unless $\Delta E_cR^2 > \pi^2\hbar^2/8m_e$. This defines a critical radius of the dot as follows:

$$R_c^2 > \frac{\pi^2\hbar^2}{8m_e\Delta E_c} \tag{9.6}$$

The critical radii for electrons and holes are therefore different. This gives rise to three different situations: (1) a range of R where no particle has bound states, (2) a range when only electrons have quantized motion and (3) a range where both electrons and holes have quantized states (Vahala 1988).

Example 9.5 *Assuming band offset value $\Delta E_c = 230$ meV and $m_e = 0.067 m_0$ for a GaAs QD, the critical radius is 2.5 nm.*

The following form of envelope function is obtained for a QD of radius R:

$$\psi_{nlm}(r) = Y_{lm}(\theta,\phi)\frac{1}{R}\left(\frac{2}{r}\right)^{\frac{1}{2}}\frac{\mathcal{J}_{l+\frac{1}{2}}(k_{nl}r)}{\mathcal{J}_{l+\frac{3}{2}}(k_{nl}r)}U_c(r) \tag{9.7}$$

The energy of the electrons is given by

$$E_{nl} = E_g + \hbar^2 k_{nl}^2/2m_e \tag{9.8}$$

where $-1 \leq m \leq 1, l = 0, 1, 2, \ldots; n = 1, 2, 3, \ldots$ and $\mathcal{J}_{l+1/2}(k_{nl}R) = 0$.

Here, $\mathcal{J}_{\mu}$ is the Bessel function, Y_{lm} s are normalized spherical functions (Abramowitz and Stegun 1972) and U_c is the cell-periodic part of the Bloch function. Similar equation may be used for holes. It should be noted that the envelope function is forced to zero at the surface of the QD.

Example 9.6 *For a spherical dot, the subband energy is expressed as $E_n = (n^2 \hbar^2 \pi^2 / 2m_e R^2)$. The energy of the first subband $n = 1$ for $R = 6.2$ nm is 0.146 eV. The energy difference between $n = 2$ and $n = 1$ subbands is therefore 0.438 eV.*

9.4.2   Optical processes

We consider a quantum box (QB) of cubic shape with each side length a. The absorption coefficient for this box is expressed as (Wu et al 1987)

$$\alpha = \frac{A}{a^3} \sum_{n^2} g(n^2) \delta[\hbar\omega - E_g - (\pi^2 \hbar^2 n^2 / 2m_r a^2)], \quad A = \frac{2\pi e^2 |p_{cv}|^2}{m_0^2 n_r \varepsilon_0 c\omega} \tag{9.9}$$

$g(n^2)$ is the degeneracy of the energy level determined by n^2. Only transitions satisfying $\Delta n = 0$ are allowed. Eq. (9.9) conforms to the expected result that the interband absorption spectra consist of a series of discrete lines representing the reduced DOS function of a 0D system. Ideally, the discrete lines will occur at photon energies

$$\hbar\omega = E_g + \left(\frac{\pi^2 \hbar^2 n^2}{2m_r a^2}\right) \tag{9.10}$$

There are several factors that broaden the discrete absorption lines. Consider for example, a situation in which the family of dots have varying sizes whose distribution follows a Gaussian distribution described by

$$P(a) = \frac{1}{D\sqrt{2\pi}} exp\{-(a - a_0)^2 / 2D^2\} \tag{9.11}$$

In Eq. (9.11)a_0 is the average side length and the standard deviation D is given by $D^2 = \left\langle (a - a_0)^2 \right\rangle$. Using Eqs. (9.9) and (9.11) the absorption coefficient for a non-uniform dot size distribution is

$$\alpha = \frac{A}{D} \frac{1}{\sqrt{2\pi}} \sum g(n^2) \int_0^\infty a^{-3} exp\left\{-(a - a_0)^2 / 2D^2\right\} \delta[\hbar\omega - E_g - (\pi^2 \hbar^2 n^2 / 2m_r a^2)] da \tag{9.12}$$

Using the following reduced variables:

$$x^2 = \frac{\hbar\omega - E_g}{\pi^2 \hbar^2 / 2m_r a_0{}^2}, \text{ and } \sigma = \frac{D}{a_0} \tag{9.13}$$

the absorption coefficient may be expressed as

$$\alpha = \frac{\beta}{a_0} \sum_{n^2} \frac{g(n^2)}{\sigma n^2} exp[-\{(n/x) - 1\}^2 / 2\sigma^2], \quad \beta = \left(\frac{1}{\sqrt{2\pi}}\right)\left(\frac{Am_r}{\pi^2 \hbar^2}\right) \tag{9.14}$$

The absorption spectra, according to Eq. (9.14), consist of a set of absorption peaks, with peak positions occurring at $n/x = 1$. The peak positions are not, however, affected by dot size distribution. The linewidth is given by

$$\Gamma = 4\sigma n^2 \left(\frac{\pi^2 \hbar^2}{2m_r a_0{}^2}\right) = 4\sigma(\hbar\omega - E_g). \tag{9.15}$$

The line broadening is also governed by other processes including phonon scattering (Schmitt-Rink et al 1987; Flytzanis et al 1991).

9.4.3 Polarization dependent momentum matrix element

The momentum matrix element given by Eq. (9.9) is, like QW and QWR, polarization dependent. In the following, we discuss the analysis by Asada et al (1986) who considered the polarization dependence of the dipole moment. Consider the following form of the envelope function:

$$\chi_{nz} = A[exp(ik_z z) \pm exp(-ik_z z)] \quad k_z = \frac{(2m_{el}E_{nz})^{\frac{1}{2}}}{\hbar} \tag{9.16}$$

We may use similar expressions for quantized k_x and k_y. Let the orientation of k vector be denoted by (θ, ϕ) with reference to rectangular coordinate system, so that $\theta = 0$ for the z-axis, and $\phi = 0$ for the x-axis. The eight directions of k are $(\theta, \pm\phi)$, $(\theta, \pi \pm \phi)$, $(\pi - \theta, \pm\phi)$, $(\pi - \theta, \pi \pm \phi)$. It may be noted that $\cos^2\theta = k_z^2/k^2 = E_{nz}/E_{lmn}$ and $\cos^2\phi = k_x^2/(k^2 - k_z^2) = E_{lx}/(E_{lmn} - E_{nz})$.

The dipole moment is given by

$$R_{ch} = \langle \psi_{clmn} | e\mathbf{r} | \psi_{hl'm'n'} \rangle = \delta_{ll'}\delta_{mm'}\delta_{nn'} \sum_{\mathbf{k}} \langle U_c | e\mathbf{r} | U_h \rangle \tag{9.17}$$

where the summation is over the previous eight wavevectors. For one k, the components of the dipole moments are written as

$R(\cos\theta\sin\phi - j\sin\phi)$: x-direction; $R(\cos\theta\sin\phi + j\cos\phi)$: y-direction; and $-R\cos\theta$: z-direction.

The square of the effective dipole moment is

$$\langle R_{ch}^2 \rangle = R^2(\cos^2\theta\sin^2\phi + \cos^2\theta)\delta_{ll'}\delta_{mm'}\delta_{nn'} = R^2[(k_y^2 + k_z^2)/k^2]\delta_{ll'}\delta_{mm'}\delta_{nn'} \qquad (9.18)$$

The quantized energy levels depend on the shape of the QB, which in turn determines the values of θ and ϕ and hence $\langle R_{ch}^2 \rangle$. In any case $\langle R_{ch}^2 \rangle < R^2$, and in the special case of cubic QB $\langle R_{ch}^2 \rangle = (2/3)R^2$, the value for bulk material. The interaction energy $\mathbf{R}_{ch} \cdot \mathbf{E}$ becomes a maximum when E is parallel to the longest side of the box.

9.4.4   Linear gain

The expression for linear gain is obtained by following the methods for bulk and QW and it may be written as

$$g = \frac{\omega}{\eta_r \varepsilon_0 c} \sum_{lmn} \int_0^\infty \langle R^2 \rangle_{ch} \frac{\rho^{0D} (f_c - f_v) (\hbar/\tau_{in})}{(E_{ch} - \hbar\omega)^2 + (\hbar/\tau_{in})} dE_{ch} \qquad (9.19)$$

$$\rho^{0D} = \frac{1}{L_x L_y L_z} 2\delta(E_{ch} - E_{clmn} - E_{hlmn} - E_g) \qquad (9.20)$$

The injected electron density is

$$N = \sum_{lmn} 2[1 + \exp(E_{clmn} - F_e)/k_B T]^{-1}(L_x L_y L_z)^{-1} \qquad (9.21)$$

The quasi-Fermi level for electrons is denoted by F_e, and the hole concentration may be similarly expressed in terms of quasi-Fermi level F_h.

The calculated values of gain spectra for a QB of 10 nm size have already been shown in Fig. 8.2. The spectra have the sharpest feature for QB. The emission processes are governed by k-selection rule. However the DOS in bulk and QW are such that the injected carriers occupy a wider range in k-space. This range is reduced in QWRs, and in QDs the DOS function is sharpest and has the ideal δ-function characteristics. The emission spectra are proportional to the joint DOS function ideally. In practical situation, several factors, like scattering by phonons, size fluctuations, etc., contribute to the broadening of the emission like.

As already indicated, the momentum matrix element depends on the polarization of the electric field and also on the shape of the box. As a special case, when the box is a cube, it equals the value for bulk material. For rectangular shaped box, the matrix element becomes maximum when the electric field vector is parallel to the longest side.

The peaked absorption and gain in a QB with highest value in comparison to the values in QWs and QWRs is due to the concentration of carriers within a very small energy interval in 0D. The energy levels look like very sharp atomic levels and the resonant transitions occur between two levels.

9.5   Deviation from simple theory: Effect of broadening

The Lorentzian function appearing at the expression for gain given by Eq. (9.19) is due to homogeneous broadening. Its origin lies in the finite lifetime of the spontaneous emission process and lifetime due to electron-phonon scattering in an individual QD. Both finite lifetime and scattering result in a broadening of the energy levels of QDs. The scattering process is mainly due to electron-LO phonon interaction. The linewidth for homogeneous broadening is low, typically a meV or even a fraction of it.

The major contribution to linewidth is from inhomogeneous broadening. Even using careful growth mechanism, it is difficult to grow dots having the same base dimension and height. The size fluctuation from dot-to-dot results in a distribution of energy levels. The following example illustrates this.

Example 9.7 *Let a rectangular QB have dimensions $d_x = 5$ nm, $d_x = 10$ nm and $d_z = 20$ nm. Let there be a change δd_x, such that $\delta d_x / d_x = 0.1$. This produces a change $\delta E_{lmn} = -2(\hbar^2/2m_e)(\pi^2/d_x^3)\delta d_x$ in the subband energy, if infinite barrier approximation is used. The change in subband energy is then 20%.*

The distributions of electronic subband energies for the ground and excited states due to size fluctuation, which are approximated by Gaussian distributions, are shown in Fig. 9.6. There will be a similar distribution of hole energies. Consequently, different dots will emit different wavelengths. Each individual dot has its own homogeneous broadening. The resulting emission spectra are now displayed in Fig. 9.2.

The emission spectra from an individual QD are quite narrow due to homogeneous broadening as shown in Fig. 9.3. However, the overall emission spectra from the ensemble are the independent sum of the individual spectra. Due to distribution of emission wavelengths caused by size fluctuation, the overall spectra are inhomogeneously broadened, and the linewidth may be as large as 20–30 meV. The homogeneous and inhomogeneous broadenings of lines are illustrated in Fig. 9.3.

The calculation of gain and threshold current of an assembly of QD lasers must therefore take into account the large inhomogeneous broadening, in addition to the homogeneous broadening.

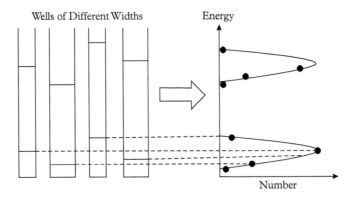

Figure 9.2 *Illustration of size distribution of QDs and consequent distribution of subband energies for ground and excited states in the conduction band. The distribution of the number of dots in a small energy interval is shown in the right part of the figure.*

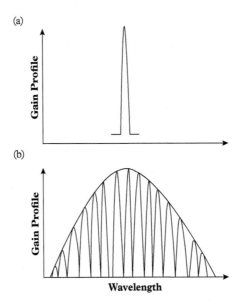

Figure 9.3 *Gain profiles of (a) single QD showing homogeneous broadening, and (b) ensemble of self-assembled QDs emitting at different wavelengths due to size distribution.*

9.6   Quantum dot lasers: Structure and gain calculation

The expression for gain in 0D systems, Eq. (9.19), has been derived for an ideal system in which transitions occur between two discrete states in the CB and VB in a single QD.

However, the expression becomes more involved in practical systems. First, instead of a single QD, an assembly of QDs formed in multiple layers are grown to enhance the volume of active region. In addition, the exact process leading to gain in the structures is not the interband process as considered before but includes excitonic processes also. In this section, we present a schematic diagram of a practical QD laser and describe its features. Next, we discuss the gain mechanism and quote the relevant expressions and examine the theoretical results in light of experimental data.

9.6.1 Structures for enhanced carrier collection

The QD assembly is grown within a thin QW layer. It is evident then that the volume of the active region formed by the assembly is drastically reduced, as illustrated by the following example.

Example 9.8 *Suppose a single layer of QD array, each having cylindrical shape with a base diameter of 25 nm and height 4 nm and areal density of the array as 5×10^{10} cm^{-2}, is grown in a QW layer of thickness 50 nm. The active region volume is thus $\pi \cdot (25/2)^2 \cdot 4 \cdot 10^{-27} \cdot 5 \cdot 10^{14} \cdot W \cdot L\, m^3 \approx 10^{-9}\, W \cdot L$, where W and L are, respectively, the width and length of the QW. With a QW of thickness 50 nm and same W and L, the ratio is 1/50.*

The active volume is usually increased by growing a few number of layers. The formation of dots in the subsequent layers is aided by the strain field from the underlying layers. However, addition of a greater number of layers increases the size variation of the dots, causes build-up of strain and introduction of more defects and recombination centres. To suppress the strain, strain compensation techniques need to be employed. A schematic diagram of practical QD lasers embedded in different QW layers is shown in Fig. 9.4. The active layers are sandwiched between two optical confinement layers (OCLs).

The capture of the carriers injected into the active layer by the dots is not much effective due to their small volume and large separation between the QDs. The carrier capture is effective in the vicinity of the p and n layers and increased carrier collection is accomplished by using the dot-in a-well (DWELL) structure, the band diagram of which is shown in Fig. 9.5(a). The QW itself captures a high density of carriers and then supplies the carriers to the dots. In another method, a tunnel injection structure, a thin tunnel barrier separates a reservoir of carriers in a QW from the QDs. Carriers from the well tunnel into the dot energy states. The band diagram is shown in Fig. 9.5(b).

9.6.2 Gain calculation

Gain in QD lasers has been calculated by several authors by using different models. A very simple model of gain calculation has been presented in Section 9.3.4. In this sub-section, a representative model is presented, which considers three different situations: (1) recombination by free carriers, (2) excitonic recombination, and (3) recombination of both free carriers and excitons (Dikshit and Pikal 2004)

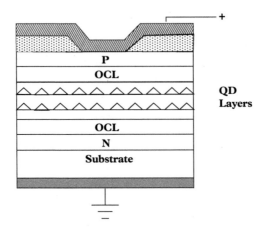

Figure 9.4 *Schematic view of QD laser. Multiple layers containing QDs are sandwiched between two optical confining layers (OCLs).*

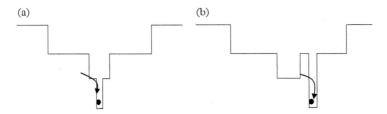

Figure 9.5 *Band diagrams illustrating enhanced carrier collection by dots: (a) DWELL Structure, and (b): tunnel injection structure.*

In the model, the allowed energy levels are calculated by using a simple harmonic approximation. The threshold gain is determined from the internal loss and mirror loss. By choosing a particular carrier distribution, the quasi-Fermi level to achieve the threshold gain can be determined and can be used to calculate the threshold carrier density and the threshold current.

9.6.2.1 *Expressions for energy levels and gain*

The band diagram of the structure considered is shown in Fig. 9.6. The energy levels for the pyramidal QD are calculated by using the harmonic oscillator Hamiltonian and the corresponding energy levels are given by Eq. (9.17) in Basu et al (2015).

Example 9.9 *Consider growth of InAs dots (band gap = 0.847 eV) with InGaAs wetting layer (E_g = 1.25 eV). Using $\Delta E_c = 0.8\Delta E_g$ and $\Delta E_v = 0.2\Delta E_g$, the level separations for electrons and holes, respectively, are 76 meV and 8 meV, the number of confined electron and hole levels are, respectively, 3 and 8.*

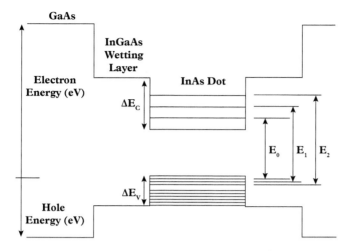

Figure 9.6 *Band diagram of QD structure.*

The excited states have a higher degeneracy leading to a higher DOS. The QDs are assumed to have pyramidal shape in which height is less than the base; the degeneracy of the first excited state and the second excited states are 2 and 3, respectively.

The carrier density in the QDs is given by

$$n = \int_{E_g}^{\infty} \sum_i 2N_{QD}G(E)f_c(E)dE \tag{9.22}$$

where N_{QD} is the areal dot density, i is the energy level, f_c (E) is the Fermi distribution function for electrons for which the quasi-Fermi level is F_n and the factor 2 is for spin degeneracy of each energy level. The following Gaussian form is assumed to describe the inhomogeneous broadening due to size fluctuations,

$$G(E) = \frac{1}{\sqrt{2\pi}\sigma}exp\left[\frac{-1}{2\sigma^2}(E - E_i)^2\right] \tag{9.23}$$

where σ is related to the size fluctuations. Similar formulas can be written for the holes.

The gain at a particular injection (quasi-Fermi level) is a product of the maximum gain and the inversion factor, $[f_c + f_v - 1]$, where f's are the Fermi functions. The expression for the maximum gain, considering contributions from all possible transitions, is

$$g_{max} = \sum_{i,j} \frac{\Gamma\pi e^2 \hbar |M_b|^2 N_{QD}N_L}{cm_0\varepsilon_0\eta_r E_{i,j}t_{QD}} \langle\psi_i|\psi_j\rangle G(E_{i,j})S(E_{i,j}) \tag{9.24}$$

where Γ is the optical confinement factor, η_r is the RI, $\langle \psi_i | \psi_j \rangle$ is the wavefunction overlap for transition $i \rightarrow j$, $|M_b|^2$ is the bulk momentum matrix element, N_L is the number of QD layers, and $S(E_{i,j})$ is the homogeneous broadening for a given transition.

It is assumed that only transitions having the same quantum number are allowed, i.e. $n_{x,e} = n_{x,h}, n_{y,e} = n_{y,h}$ and $n_{z,e} = n_{z,h}$. The current density is calculated by considering the contributions from the dot layers as well as the wetting layer and is given by

$$J = \frac{eN_L}{\eta_i}\left(\frac{t_{QD}n_{QD}}{\tau_{QD}} + \frac{t_W n_W}{\tau_W}\right) \tag{9.25}$$

Here η_i is the injection efficiency, t is the thickness of the respective layers, n is the volume density of carriers and τ is the recombination lifetime, the subscripts QD and W refer, respectively, to the dot layer and wetting layer. The recombination lifetime in the wetting layer is given by the usual expression $\tau^{-1} = A + Bn + Cn^3$.

Example 9.10 *Using Eq. (9.24) and the values of parameters for InAs-InGaAs QDs: $\Gamma = 0.02$, $|M_b|^2 = 20eV$, $\langle \psi_i | \psi_j \rangle = 0.2$, $N_{QD} = 3.10^{14} \ m^{-2}$, $N_L = 2$, $n_r = 3.4$, $E_{i,j} = 0.937$ eV, $t_{QD} = 3 \ nm$, and $\sigma = 40 \ meV$, and neglecting homogeneous broadening, the value of g_{max} is 11 cm^{-1}.*

9.6.2.2 Carrier distribution, gain, and threshold

The values of gain and threshold current densities are now estimated by considering (1) free carriers only, (2) excitonic distribution, and (3) presence of both free carriers and excitons.

A. Free carrier model   In this case, first a separation of Fermi levels is chosen, and then the concentrations of free electrons and holes are calculated assuming charge neutrality in the dots and using Eq. (9.22). Larger hole population in the higher states results, as the hole levels are closely spaced, and the degeneracy of the levels increases with the level index. On the other hand, the excited electron states are not occupied significantly in spite of larger degeneracy, since the level separation is larger.

The gain spectrum is calculated from the expression

$$g_{free}(E) = g_{max}[f_c(E, F_n) + f_v(E, F_p) - 1] \tag{9.26}$$

The authors calculated the threshold gain by assuming $\alpha_i = 2$ cm^{-1}, mirror reflectivity of 0.32, cavity length = 3.8 mm, and optical confinement factor of 0.01/per layer, and the value is 5.15 cm^{-1}. As a greater number of hole excited states are occupied, it is difficult to increase the hole Fermi function and to make the inversion large. Dixit and Pikal found that the gain in the ground state is below the threshold gain. It is not possible therefore to have laser action for transition between the pair of states labeled by $m = 0$, as well as by $m = 1$. The calculated gain exceeds the threshold only for $m = 2$ state. The excited holes do not directly increase the threshold current, as there are few excited electrons to recombine with. The calculated threshold current density of ~ 270 A/cm^2 is too high compared with the experimental values for lasers having dimensions and structures as assumed in the model.

B. Exciton model   Next it is assumed that all the carriers injected into the dots form excitons, and an excitonic Fermi level, instead of separate Fermi levels for electrons and holes, should be used to calculate the excitonic distribution based on Fermi–Dirac statistics. In this case, the holes follow the electron distribution. For the same injection level as in Case A, the higher occupation of excited hole states is reduced. Since the holes follow the electrons, $f_c = f_v$, and the expression for the excitonic gain becomes

$$g_{exc}(E) = g_{max} \times [2f_{ex} - 1] \tag{9.27}$$

The dimensions of the InAs dot ~ 25 nm are smaller than the bulk exciton radius~ 50 nm and the wavefunction overlap is ~ 1. The calculated ground state gain exceeds the threshold gain and lasing occurs from this state. The calculated threshold current density is 20 A/cm^2. By repeating the calculation for different temperatures, a characteristic temperature $T_0 = 550$ K is obtained, which is too high compared to experimental value.

C. Excitons and free carriers   In this case, both free carriers and excitons are assumed to coexist, and the gain is expressed as

$$g(E) = g_{exc}(E) \times \left(\frac{n_{ex}}{n}\right) + g_{free}(E) \tag{9.28}$$

where n_{ex} and n denote the exciton and total carrier density, respectively and g_{free} is the gain from free carriers. The ratio n_{ex}/n depends on the exciton binding energy. The ratio is calculated by using the following modified form of the Saha equation (Snoke and Crawford 1995).

$$\frac{n_{ex}}{n} = 1 - \sqrt{\frac{n_{ex}}{n} \frac{1}{exp(E_B/k_B T) - (1 + E_B/k_B T)}} \tag{9.29}$$

where E_B is the binding energy for a particular transition. For $E_B \gg k_B T$,the previous equation reduces to the standard form of the Saha equation, noting that $n = n_{ex} + n_e$, where n_e is the electron density.

Example 9.11 *Assuming $E_B = 20$ meV, solving the quadratic equation and putting $k_B T = 25.8$ meV at room temperature, the ratio $n_{ex}/n = 0.233$.*

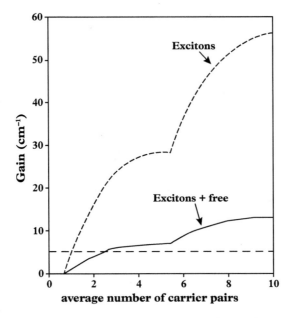

Figure 9.7 *Peak gain using excitons only (short, dashed line) and both excitons and free carriers (solid line). The horizontal dashed line represents the total loss of 5.15 cm^{-1} to be overcome at threshold.*

Figure 9.7 shows a plot of the peak gain considering free carriers and excitons, and excitons only, versus average number of carrier pairs per dot. The peak gain is reduced due to sizable fraction of carriers in the ground state. The threshold current density is now 44 A/cm^2, which fairly agrees with the experimental value. The calculated value of $T_0 = 85$ K is in excellent agreement with the observed value of 83 K for undoped samples (Shchekin and Deppe 2002).

The model has been extended for p-doped QDs, in which acceptors increases the free hole density. The quasi-Fermi level for holes is thus pushed down, increasing the inversion factor and thereby making the ground-state gain larger. The characteristic temperature also rises significantly in agreement with the experimental observation.

9.7   Intersubband transitions

Intersubband transitions involving two discrete levels within the same band, either conduction band (CB) or valence band (VB) have been introduced for QWs in Section 6.4. Similar transitions between two levels in a QD have been investigated by several workers (see Bhattacharya et al 2006; Stiff-Roberts 2009, for example).

We consider transitions between two discrete levels in the CB. The expression for the intersubband absorption may be obtained by using the Fermi Golden Rule and is given by

$$\alpha\left(\hbar\omega\right) = \frac{\pi e^2 \hbar}{\varepsilon_0 \eta_r c m_0^2 V_{av}} \sum_{fi} \frac{1}{\hbar\omega} \left|\mathbf{a}.\mathbf{p}_{fi}\right|^2 \quad N\left(\hbar\omega\right) \tag{9.30}$$

The symbols have the usual meanings and $N(\hbar\omega)$ is the intersubband joint DOS for QDs, given by

$$N(\hbar\omega) = \frac{1}{\sqrt{2\pi}\sigma} exp \frac{\left(E_{fi} - \hbar\omega\right)^2}{2\sigma^2} \tag{9.31}$$

The initial and final states are denoted by i and f, respectively, E_{fi} is the energy separation between them and σ is the gaussian linewidth for transition.

The absorption spectra are very sharp due to the 0D nature of the electron gas. A typical value of absorption peak is indicated in Example 9.3. The electrons are concentrated near the CB subband and the coefficient is almost independent of temperature. However, at higher temperatures, electrons occupy more than a single discrete state including the continuum states. The advantages due to quantum confinement is therefore lost.

9.8   Excitonic processes in quantum dots

Many interesting physical phenomena are exhibited by excitons in QDs. However, a detailed theoretical understanding of exciton binding energy, absorption, and luminescence processes require detailed knowledge of the band structure, particularly of the valence band, and we need to consider the finite barrier effect. Furthermore, the values depend on the shape and size of the dots and on surface polarization. The exchange interaction introduces fine structures in the excitonic levels and influences the radiative lifetimes.

In the present section, we give a simplified description of the quantum confinement effects on excitons in QDs using effective mass approximation, particle-in-a-box model with infinite barriers, and isotropic effective masses for electrons and holes. A spherical QD is considered. The model is due to Efros and Efros (1982) and it gives analytical expressions for two limiting cases: the *weak confinement* and the *strong confinement* limits (Gaponenko 2010).

9.8.1   Weak confinement regime

The treatment applies to larger QDs. The dot radius, a, is small but still a few times larger than the exciton Bohr radius, a_B. In this case, the motion of the centre-of-mass (CM) of the exciton is quantized. The energy of an exciton is expressed as

$$E_{nml} = E_g - \frac{Ry}{n^2} + \frac{\hbar^2 \chi_{ml}^2}{2Ma^2}, \qquad \text{n, m, l} = 1, 2, 3, ... \tag{9.32}$$

where χ_{ml} 's are the roots of the Bessel function, and $M = m_e + m_h$ is the CM of excitons. It is to be mentioned here that, an exciton in a spherical QD is characterized by the quantum number n describing its internal states arising from the Coulomb electron-hole interaction and by two additional numbers, m and l, describing the states connected with the CM motion in presence of the external potential barrier featuring spherical symmetry. To distinguish the 'internal' and the 'external' states, we shall use capital letters and small letters respectively.

For the lowest $1S1s$ state ($n = 1, m = 1, l = 0$) the energy is expressed as

$$E_{1S1s} = E_g - Ry + \frac{\hbar^2 \pi^2}{2Ma^2} = E_g - Ry\left[1 - \frac{m_r}{M}\left(\frac{\pi a_B}{a}\right)^2\right] \tag{9.33}$$

where the electron-hole reduced mass is given by $m_r = m_e m_h/(m_e + m_h)$.

In Eq. (9.33), the value $\chi_{10} = \pi$ has been used. Therefore the first exciton resonance in a spherical QD experiences a high energy shift by the value

$$\Delta E_{1S1s} = \frac{m_r}{M}\left(\frac{\pi a_B}{a}\right)^2 Ry \tag{9.34}$$

As we consider $a \gg a_B$, the previous expression gives a small value compared to Ry which justices the term 'weak confinement'.

Example 9.12 *For copper chloride (CuCl), a member of I–VII compound semiconductor, the exciton Bohr radius is 0.7 nm and exciton Rydberg is 200 meV. The mass values are $m_e = 0.4m_0$ and $m_{hh} = 2.4m_0$ The energy shift for a = 7 nm is 2.82 meV, which is quite small [$E_g = 3.45$ eV].*

Considering that photon absorption can create an exciton with zero angular momentum only, the absorption spectrum will consist of a number of lines corresponding to states with $l = 0$, Then the absorption spectrum can be derived from Eq. (9.32) with $\chi_{m0} = \pi m$ and is given by

$$E_{nm} = E_g - \frac{Ry}{n^2} + \frac{\hbar^2 \pi^2 m^2}{2Ma^2}, \qquad n, m = 1, 2, 3, ... \tag{9.35}$$

9.8.2   Strong confinement regime

The strong confinement limit corresponds to the condition that the dot radius is several times smaller the Bohr radius, i.e. $a \ll a_B$. The confined electrons and holes do not have any hydrogen like excitonic bound state. Here we consider the quantization of an electron and a hole motion separately and for the free electron and hole in a spherical potential box will be as follows:

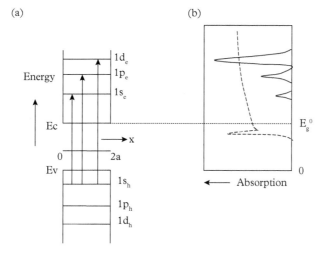

Figure 9.8 *(a) Energy levels in a spherical QD under strong confinement; (b) Absorption spectra consisting of spikes. The spectra for bulk are shown by the dashed curve and contains discrete excitonic peaks.*

$$E^e_{ml} = E_g + \frac{\hbar^2 \chi^2_{ml}}{2m_e a^2}, E^h_{ml} = -\frac{\hbar^2 \chi^2_{ml}}{2m_h a^2} \qquad (9.36)$$

Here the top of the valence band has been considered as the origin of the energy scale. The energy levels are shown in Fig. 9.8. The lowest states of electrons and holes are larger than R_y.

The selection rules allow optical transitions to take place between electron and hole states with the same principal n and orbital l quantum numbers. Therefore, the absorption spectrum reduces to a set of discrete bands peaking at the energies

$$E_{nl} = E_g + \frac{\hbar^2 \chi^2_{nl}}{2\mu a^2} \qquad (9.37)$$

For this reason, the QDs in a strong confinement limit are called 'artificial atoms', as the QDs exhibit discrete optical spectrum controlled by the number of atoms, i.e. by the size, whereas an atom has a discrete spectrum controlled by the number of nucleons.

However, an independent treatment of the electron and hole motion is by no means justified here and the problem including two particle Hamiltonian with the two kinetic energy terms, Coulomb potential and the confinement potential should be considered.

Two-particle Schrodinger equation will be

$$H = -\frac{\hbar^2}{2m_e}\nabla_e^2 - \frac{\hbar^2}{2m_h}\nabla_h^2 - \frac{e^2}{4\pi\varepsilon\,|r_e - r_h|} + U(r) \tag{9.38}$$

After treating this problem by a variational approach, the energy of the ground electron-hole pair state (1s1s) can be expressed as

$$E_{1s1s} = E_g + \frac{\pi^2\hbar^2}{2\mu a^2} - 1.786\frac{e^2}{4\pi\varepsilon_0\varepsilon_r a} \tag{9.39}$$

Here $e^2/4\pi\varepsilon_0\varepsilon_r a$ describes the effective Coulomb electron-hole interaction in a medium with dielectric permittivity ε_r. Comparing this term with the exciton Rydberg energy $Ry = e^2/8\pi\varepsilon a_B$ and remembering $a \ll a_B$, it can be concluded that Coulomb interaction vanishes in small QDs. Moreover, the Coulomb contribution to the ground state energy is greater than in bulk crystal. This is the main difference of QDs compared to crystals, QWs and QWRs where Coulomb energy of a free electron hole pair is zero. Thus, an elementary excitation in a QD can be classified as an exciton.

The exciton lowest state energy measured as a deviation from the bulk bandgap energy in the strong confinement limit can be expressed as

$$E_{1s1s} - E_g = \left(\frac{a_B}{a}\right)^2 Ry\left[A_1 + \frac{a}{a_B}A_2 + \left(\frac{a}{a_B}\right)^2 A_3 + ...\right] \tag{9.40}$$

A_1 is described by the roots of Bessel function, A_2 corresponds to the Coulomb term. Summarizing all findings relevant to the ground state, the energy of the first absorption peak will be as follows:

$$E_{1s1s} = E_g + \pi^2\left(\frac{a_B}{a}\right)^2 Ry - 1.786\frac{a_B}{a}Ry - 0.248Ry \tag{9.41}$$

Generally, the terms in the right part of Eq. (9.41) successively reduce in absolute value, though for small nanocrystals of narrow band semiconductors with small electron effective mass the second term describing the sum of the kinetic energies of the electron and a hole can be comparable to, or even greater than, the original band gap energy E_g of the parent bulk crystal. The optical absorption shift energy E_{1s1s} in terms of a dimensionless dot radius a/a_B manifests as the universal material dependent law for size dependent optical absorption if the photon energy is measured in dimensionless units E_{1s1s}/Ry. This law is plotted in Fig. 9.9.

9.8.3   Exciton binding energy

The binding energy of excitons in QDs is calculated by using simple methods like variational method as well as more sophisticated methods involving numerical work. Even

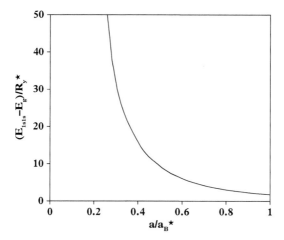

Figure 9.9 *Universal plot of normalized* E_{1s1s} *versus normalized dot radius following Eq. (9.41)*

the variational method is not completely analytical, that is, free from numerical exercise. Furthermore, the shape of the QD dictates the number of variational parameters to be considered in solving the problem. We first present a very simple variational method applicable for spherical QD for the sake of illustration.

9.8.3.1 Spherical quantum dots

The method has been presented by Sánchez-Cano et al (2008) to explain their experimental results for $Ga_{1-x}In_xAs_ySb_{1-y}$ spherical QDs grown in an GaSb matrix. The resulting heterostructure is grown in type I configuration, that is, the quaternary QD having lower band gap has band offsets ΔE_c and ΔE_v, respectively in the conduction and valence bands at the heterointerfaces. The effective mass Hamiltonian for the exciton is written as

$$H = -\frac{\hbar^2}{2m_e}\nabla_e^2 - \frac{\hbar^2}{2m_h}\nabla_h^2 - \frac{e^2}{4\pi\varepsilon_0\varepsilon_r\,|r_e - r_h|} + V_e\,(r_e) + V_h\,(r_h) \qquad (9.42)$$

where ε_r is the relative permittivity of the QD material, and the other symbols are already defined. The confining potentials are given as

$V_e(r_e) = 0 \ (r_e \langle R), = \Delta E_c\,(r_e)\, R)$ and $\ V_h(r_h) = 0 \ (r_h \langle R), = \Delta E_v(r_h)\, R)$

where R is the radius of the QD.

As an exact solution of Eq. (9.42) is difficult to obtain; a variational approach has been employed by the authors. For this purpose, the following trial function is assumed for the ground state exciton wavefunction inside and outside the spherical QD:

$$\psi_{in} = A(R - \alpha r_e)(R - \alpha r_h)\exp\{-\alpha(r_e - r_h)\}\text{for } r_e, r_h \leq R \qquad (9.43a)$$

and

$$\psi_{out} = \frac{A(1-\alpha)^2 R^4}{r_e r_h} \exp\{-2(\alpha-\beta)R - \beta(r_e + r_h)\} \text{for } r_e, r_h \geq R \tag{9.43b}$$

In Eq. (9.43) A is the normalization constant and α and β are the variational parameters, which are to be determined by minimizing the total energy. Applying the boundary condition

$$\frac{1}{m_{rin}} \frac{1}{\psi_{in}} \frac{\partial \psi_{in}}{\partial r_j}\Big|_R = \frac{1}{m_{rout}} \frac{1}{\psi_{out}} \frac{\partial \psi_{out}}{\partial r_j}\Big|_{Rj} = e, h \tag{9.44}$$

where m_{rin} and m_{rout} are the reduced masses inside and outside the sphere, respectively, the two variational parameters may be related as

$$\beta = \frac{q[\alpha + \alpha(1-\alpha)R] + \alpha - 1}{(1-\alpha)R} \quad q = m_{rout}/m_{rin} \tag{9.45}$$

The lowest electron and hole subband energies are obtained from the solution of the following transcendental equation:

$$\left[\frac{2m_j(\Delta E_A - E_{1j})}{\hbar^2}\right]^{1/2} \cot\left\{R\left[\frac{2m_j(\Delta E_A - E_{1j})}{\hbar^2}\right]^{1/2}\right\} = -\left[\frac{2m_j E_{1j}}{\hbar^2}\right]^{1/2} \tag{9.46}$$

In the previous equation, when $A=C$, $j = e$ and when $A = V$, $j = h$.

It follows easily that there exists a critical radius for the existence of a bound state for each particle, given by

$$R_{cj} = \left[\frac{\pi^2 \hbar^2}{8m_j \Delta E_A}\right]^{1/2} \tag{9.47}$$

The binding energy E_{1b} of 1s exciton state is obtained from the following:

$$E_{1b} = E_{ex} - E_{1e} - E_{1h} \tag{9.48}$$

As already stated, Sánchez-Cano et al applied the theory to calculate the subband energies and HH and light-hole (LH) exciton binding energies for gallium indium arsenide antimonide (GaInAsSb) QDs as a function of dot radius R. However, this simple theory can be applied to QDs made of other semiconductors. Therefore, instead of presenting the specific results, we discuss the broad features.

The subband energies increase with decreasing dot radius and attain a limiting value at critical radius. The larger the value of the barrier potential is, the larger the subband energies become. The exciton binding energy first increases with increasing dot radius, attains a peak, and then decreases. The qualitative variation is shown in Fig. 9.10. The

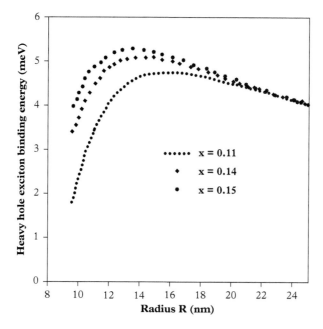

Figure 9.10 *Qualitative variation of exciton binding energy in spherical QDs for different values of barrier potentials.*

Source: reprinted figure from Sánchez-Cano R, Tirado-Mejía L, Fonthal G, Ariza-Calderón H, and Porras-Montenegro N. (2008). Exciton recombination energy in spherical quantum dots grown by liquid-phase epitaxy. *Journal of Applied Physics* 104(113706): 1–4, with the permission of AIP Publishing.

binding energy is higher for higher In concentration which gives a higher value of band offset. For a large dot radius all the curves merge with each other.

The binding energies for excitons in spherical QDs made of cadmium sulfide (CdS), cadmium selenide (CdSe), lead sulfide (PbS), and cadmium telluride (CdTe), embedded in organic materials and silicate glass have been calculated by similar variational function as Eq. (9.43) (Marin et al 1998). The trend follows a similar pattern as in Fig. 9.10. In all cases, satisfactory agreement has been obtained.

A variational calculation of the excitonic binding energy in a QD having the shape of truncated cone has been presented by Ferreira and Bastard (2015). The authors have also presented other methods of calculation and other aspects of excitons in a self-assembled QD.

The methods of calculation of excitonic parameters in fractional dimensional space have been mentioned in earlier chapters. Similar works for nanoparticles with and without field has been described by Pedersen (2010: 2017). Quantum Confined Stark Effect (QCSE) in semiconductor QDs has been treated in the work by Men et al (1995).

9.8.4   Absorption and luminescence

Apart from variational methods, more refined theories are available in the literature to calculate the binding energy, absorption oscillator strength, and luminescence lifetime. These methods include, among others, detailed band structure obtained under effective mass approximation using multiband **k.p** perturbation theory. Almost all these theories rely on numerical methods. It is beyond the scope of this text to include such theories. As discussed in earlier chapters, simple fractional dimensional analysis has been employed by several workers to understand various excitonic processes in QDs.

In this section, we shall mention a few of the important changes in the excitonic properties as the dimension is progressively reduced to reach 0D state. The treatment is due to Wu et al (2017) that relies on variational method. Though the results are obtained for GaAs QWs, QWRs, and QDs with AlGaAs as a barrier, the conclusions are almost general.

9.8.4.1   A. Binding energy

The variation of binding energy with size of the nanostructure follows the same pattern. It first increases, reaches a maximum and then decreases. The maximum occurs due to strongest Coulomb interaction and the decrease at the low values of the length parameter is due to the penetration of wavefunctions into the barrier and reduced wavefunction confinement. As the confinement increases with reduced dimensionality, the maximum value of binding energy occurs in QDs. The maximum values of binding energy are 2, 4.8, and 8.8 times the value in bulk material.

9.8.4.2   B. Size scaling of the optical gap

The optical gap E_{ex} is almost constant at large size but increases with decreasing size (L) in all nanostructures, the rate of increase at low values of L being largest for QDs. The dependence of optical gap in nanostructures is usually expressed by $E_{ex} = E_g + (A/L^B)$, where A and B are fitting parameters. It has been found by Wu et al (2017) that a power law in the form

$E_{ex} = E_g + \{A/exp(L^B)\}$ could fit the observed results better for all sizes of the three nanostructures. The exponent B lies in the range $0 < B < 1$.

9.8.4.3   C. Bohr radius

The quantum confined exciton Bohr radius is defined as $a_{ex} = \langle \psi_{ex} | 1/r_{eh} | \psi_{ex} \rangle^{-1}$, where ψ_{ex} is the wavefunction of the ground state exciton and $r_{eh} = |r_e - r_h|$ is the electron hole separation. Plots of a_{ex} versus L show that the radius decreases rapidly with decreasing L, reaches a minimum and then increases. The curve of Bohr radius in QD always occurs below the curve for QWR which in turn is lower than the curve for QW. The reason lies in the stronger localization of excitons and a reduced e-h separation. An interesting observation is that the binding energy increases almost linearly with $1/a_{ex}$.

9.8.4.4   *D. Effective dimensionality*

In Chapter 08 the calculation of exciton binding energy in quantum disc has been presented (Koh et al 2001). From the results, a fractional dimension may be defined as $d_{eff} = 1 + 2\sqrt{a_{ex}/a_B}$, where a_B is the bulk Bohr radius. It has been found that the effective dimensionality depends on the size of the crystal and a dimensional crossover is a possibility. However, Wu et al proposed a new expression $d_{eff} = 3 - n \cdot \exp(-a_{ex}/a_B)$, where n = 0, 1, 2, and 3 are for bulk materials, QWs, QWRs and QDs, respectively.

9.8.4.5   *E. Oscillator strength*

It is useful to define a normalized oscillator strength $\overline{f_{ex}}$, per unit volume of nanostructures, in the following way:

$$\overline{f_{ex}} = \frac{E_p}{E_{ex}L^n}\left|\int \psi_{ex}(\mathbf{r},\mathbf{r})d\mathbf{r}\right|^2 \tag{9.49}$$

where the exponent n = 0 for bulk, and n= 1,2,3 in L^n denote respectively the width of QW, the cross-sectional area of QWR, and the volume of the QD. E_p = 25.7 eV is the Kane energy in GaAs, and $\left|\int \psi_{ex}(\mathbf{r},\mathbf{r})d\mathbf{r}\right|^2$ is the overlap integral that represents the probability of finding an electron and a hole at the same position.

The normalized exciton oscillator strength has a high value for low L, but decreases rapidly with increasing L, and at large L, the values for all dimensions merge with one another. At L = 4 nm the values are 87.7 (0D), 30.8 (1D), and 6.1(2D) times larger than the bulk value. The values indicate that there is an increase in e-h spatial overlap and a decrease in Bohr radius. For L = 20 nm, however, the values are 1.8 (QD), 1.5 (QWR) and 1 (QW) times the bulk value.

The normalized oscillator strength with Coulomb interaction is larger than the oscillator strength without the interaction in all the three nanostructures. This Coulomb enhancement effect has already been established for the case of QWs.

9.8.4.6   *F. Radiative lifetime*

The exciton radiative lifetime depends directly on exciton effective volume and inversely on the oscillator strength, and thus $\tau_{ex} \propto V_{ex}/f_{ex}$. The exciton effective volume, or the 'coherence volume' V_{ex} = 1, $\langle z \rangle$, a_{1D}^2, and a_{0D}^3, respectively, for bulk, QWs, QWRs and QDs. Here $\langle z \rangle$ is the extension of the exciton along the z-direction of QWs, a_{1D}^2 is the same in the x-y plane and a_{0D}^3 is the exciton extension in QDs.

At very low temperatures, the radiative lifetime of bulk excitons has been found to be around 3.3 ns. As the excitons are squeezed more and more when the dimensionality is reduced, there is a rise in oscillator strength. Therefore, the radiative lifetime decreases progressively from QW to QWR and becomes minimum for QDs. This has been found in the results calculated by Wu et al. For very low values of L, the lifetime increases substantially, as the exciton wavefunction spreads into the barrier. As a result, the exciton oscillator strength decreases and the exciton effective volume increases.

9.8.5   Quantum confined Stark effect

The effect of an applied electric field in a semiconductor has been discussed in Section 7.4, and the effect in QWs has been mentioned in Section 7.4.2. Of particular interest is when the field is applied perpendicular to the QW layer plane. The primary effect is to shift the envelope functions of electrons and holes in opposite directions, with the result that effective gap decreases and excitonic oscillator strength diminishes. The excitonic peak is redshifted more with increase in the field. This effect is termed as quantum confined Stark effect (QCSE).

QCSE in QDs and colloidal NCs has been investigated by several authors both theoretically and experimentally. It is not possible to present the theoretical treatment here because of limited space. In essence, the subband energies decrease with field and change with dot radius (Pokutnyi et al 2004). A detailed theoretical treatment for excitons in spherical QDs made of CdSSe is given by Uhrig et al (1991). See also Scholes G D (2008) for further information on optical properties.

9.9   Classification of nanocrystals

We now make a classification of nanocrystals based on growth processes. The scheme is shown in Fig. 9.11. Basically, three methods are employed to grow nanocrystals. First, the oldest one, is the growth by controlled diffusion of particles into a glass matrix. The second one, which has assumed importance in recent years, is the growth of nanocrystals in colloidal form in solutions and polymers. This is chemical synthesis process. The third and the last one is growth by epitaxy and lithography.

The chemical synthesis produces QDs or nanocrystals in two different forms. First is the bare/organic class, in which NCs are exposed to air or at best surrounded by an organic coat. This forms a branch of inorganics in organics nanotechnology. The second category is core-shell structure, in which a semiconductor NC which forms the core is surrounded by another semiconductor acting as a shell.

The process based on epitaxy and lithography depends on molecular beam epitaxy (MBE) or metal organic chemical vapour, deposition (MOCVD) growth mechanism. It can be either top-down or bottom-up. The top-down process was followed in earlier days. Starting from bulk material and QWs, wet or dry etching techniques are employed to reduce the dimensions in other directions. In a different way, the dimensions are reduced along two directions by etching and application of a strong electric field along the third dimension which quantizes the motion in that direction. The bottom-up approach yields two types of QDs. The interfacial fluctuation QDs (IFQDs) form 'naturally' when the width of a narrow QW is formed by a monolayer. The other type, the self-assembled QDs (SAQDs) form when a semiconductor material A is deposited on a highly lattice mismatched layer of another semiconductor B. In order to relieve the strain A forms isolated islands of pyramidal, hemispherical or truncated pyramidal shape on B. The growth of SAQDs by Stransky–Kastranow growth process has already been introduced in Section 9.3.

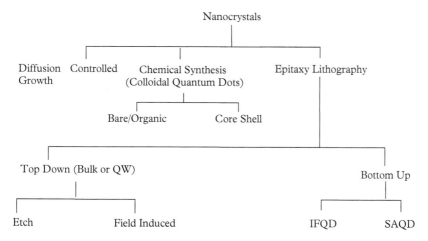

Figure 9.11 *Classification of NCs according to growth processes.*

9.10   Synthesis of nanocrystals

We now discuss briefly the methods to grow nanocrystals (NC)s. Three main processes will be discussed: growth by controlled diffusion into glass matrix, synthesis of colloidal NCs, and growth by self-assembly.

9.10.1   Growth by controlled diffusion

The technique of growth of semiconductor NCs in glass matrix has been known over many decades. Photochromic glasses have been developed for commercial use at a later period. Photochromic lenses darken when exposed to a certain type of light, most commonly ultraviolet (UV) light. By embedding NCs of silver halides into glass, the photochromic properties are obtained. Colour cut-off filters and photochromic glasses are now manufactured by different companies for commercial use. These glasses contain nm-sized crystallites of solid solutions of II–VI semiconductor alloys like Cd (x) Se (1–x), as well as NCs of I–VII compounds like CuCl, copper bromide (CuBr), and silver bromide (AgBr).

   The growth of NCs in glass matrices, typically borosilicate or silicate glass, has been investigated by Ekimov (1996). The process relies on the diffusion-controlled phase decomposition of oversaturated solid solutions. Three different methods have been employed. The first one involves oversaturated solution of semiconductor material in glass matrix by co-melting. Recent techniques employed involve high-frequency co-sputtering of glass and semiconductor material and ion implantation of semiconductor material into glass matrix.

   A process of diffusion-controlled phase decomposition is responsible for the growth of NCs. The sample containing supersaturated semiconductor in glass is annealed at a temperature high enough to cause diffusion into the matrix, but low enough to ensure that the solution remains oversaturated. The process goes on in three stages. In the first, small nuclei are formed. In the second stage, a monotonic growth of NCs occurs as the atoms cross the nucleus-matrix interface. During this process, the volume of semiconductor phase increases, but the degree of super saturation decreases. When the NCs are large and super saturation does not exist, there is a mass transfer from smaller particles to larger particles by diffusion.

9.10.2 Growth of colloidal nanocrystals

Typically the synthesis of colloidal NCs requires three components, precursors, organic surfactants and solvents, which may be surfactants as well. As the reaction medium is heated to a sufficiently high temperature, a chemical process transforms the precursors to active atomic or molecular species, called monomers. The formation of NCs then follows in two steps, and their growth is influenced by the presence of surfactants. In the first step, nucleation of an initial *seed* occurs; the precursors decompose or react at a high temperature and form a super saturation of monomers. The nucleation of NCs then follows, and the nuclei grow by using additional monomers present in the reaction medium. The grown NCs may be crystalline solids if the constituent atoms form an ordered state and anneal during growth. The temperature must be hot enough to allow rearrangement of atoms and anneal during growth. To grow small crystals, a lower melting temperature is sufficient. Further details of the growth kinetics may be found in Yin and Alivisatos (2005).

9.10.3 Self-assembly

Growth of NCs by self-assembly has already been discussed in Section 9.3. The wetting layers and assembly of QDs may be grown by epitaxy, particularly by MBE. The QD materials and the substrate and wetting layers have many different combinations. A SEM image of nanodots of III–V nitrides, grown by plasma enhanced MBE, is shown in Fig. 9.12.

9.11 Core-shell structures

Synthesis of colloidal semiconductor NCs described in Section 9.10.2 indicates that such NCs have an inorganic (semiconductor) core, surrounded by an organic outer layer or shell formed by surfactant molecules (ligands). This gives rise to a very high population of surface atoms and ligands play a crucial role in passivation of the core by the organic layer. As a consequence, there appear a significant amount of surface related trap states, which serve as fast non-radiative deactivation of photogenerated charge carriers. The quantum efficiency of luminescence from the NCs is therefore drastically reduced.

The surface related passivation by organic surfactants and related disadvantages are mitigated by surrounding the core material by another inorganic semiconductor. The resulting structure is called the *core-shell* NC. In essence, a heterojunction is formed

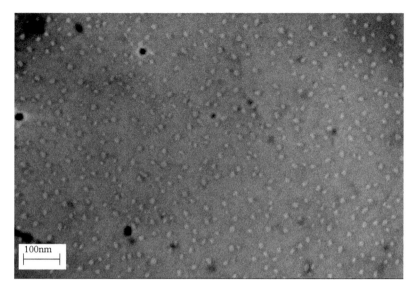

Figure 9.12 *III-Nitride Nano-dots grown by PA-MBE.*

Source: FESEM Image courtesy of Anirban Bhattacharyya, INRAPHEL/ CRNN, University of Calcutta.

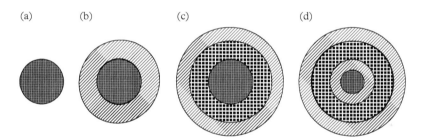

Figure 9.13 *Core shell structures with single shell and multiple shells. (a) bare (organically passivated) QD, (b) a single core shell, (c) multiple core shell and (d) onion like core shell*

between the interface of the core and shell semiconductors. The structures can be grown by chemical synthesis process (see, for example, Reiss et al 2009 and references therein).

Different types of NCs having bare (organically passivated), single shell, multiple shells, and onion-like core shell structures are illustrated in Fig. 9.13. Fig. 9.13(a) shows the bare (organically passivated) QD. Next comes the simple core-shell, a single shell, structure shown in Fig. 9.13(b). The next structure is multiple shell as in Fig. 9.13(c). The core is surrounded by a shell which is somewhat wider to form a QW. This structure is then surrounded by another shell layer. Last in this list comes the onion like structure, as shown in Fig. 9.13(d), in which, the core and shell layers alternate.

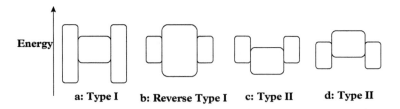

Energy

a: Type I   b: Reverse Type I   c: Type II   d: Type II

Figure 9.14 *Band alignment of different types of core shell structure. The top and bottom horizontal lines in each rectangle indicate the CB and VB edges, respectively.*

The band alignments in core-shell structure may take different forms as shown in Fig. 9.14. The arrangements are more or less the same as in a single or double heterostructure as illustrated in Fig. 2.6. Three cases are shown in Fig. 9.14: type I, reverse type I and type II. In type I, the core has smaller band gap than the shell and both electrons and holes are partially or totally confined in the core (Fig. 9.14(a)). In reverse type I, (9.14(b)), the shell material has a lower gap and depending on the thickness of the shell, the holes and electrons are confined totally or partially within the shell. In type II configurations (9.14(c) and (d))), either the VB edge or the CB edge of the shell material lies in the bandgap of the core.

Examples of type I structure are too many: II–VI (CdS/ZnS, CdSe/zinc selenide (ZnSe), III–V (GaAs, InP and their alloys), IV–VI (PbSe/PbS, PbSe/PbSeS alloy). (Reiss et al 2009). Type II core-shell structures have been realized using CdTe/CdSe and CdSe/ZnTe. Reverse type I is fabricated by using ZnSe core and CdSe shell.

For further information, the reader is referred to articles by Chen et al (2015) and Ivanov et al (2007) for example.

9.12   Bright and dark excitons

The formation of excitons in this and earlier chapters has been explained by considering direct electron-hole Coulomb interaction. However, some experimental results on luminescence in high-quality CdSe QDs indicated an extremely large recombination lifetime (Efros 1992). It prompted workers to seek a correct explanation and investigations revealed the role of short-range electron-hole exchange interaction. The reader is referred to the paper by Efros et al (1996) and Brus (1984) and references therein for detailed account and relevant theoretical and experimental work.

We describe here the essential outcome of these investigations, as detailed treatment is beyond the scope of this book. As shown in Fig. 9.15, $|G\rangle$ denotes the ground state of the crystal with no exciton. The exchange interaction splits the exciton ground state (1s) into two states: (1) the bright exciton state $|B\rangle$ and (2) the dark exciton state $|D\rangle$. The bright state is formed when the electron and hole spins are antiparallel. Excitons can recombine radiatively from this state and emit a photon. On the other hand, when the

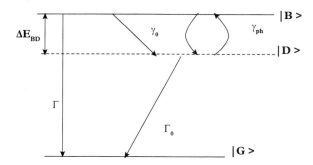

Figure 9.15 *Schematic diagram showing energy splitting of ground exciton level and radiative and non-radiative transitions between B and D excitons and the ground state.*

spins for both electron and hole are parallel, a dark exciton state is formed, and radiative recombination is no longer possible (Kodriano et al 2014).

The energy splitting, ΔE_{BD}, between the bright and dark states, as shown in Fig. 9.15 is obtained to a very good approximation within the first-order perturbation theory (Wu et al 2017) as follows:

$$\Delta E_{BD} = \pi a_B^3 \mathcal{J} \int dr |\psi_{ex}(\mathbf{r}, \mathbf{r})|^2 \tag{9.50}$$

where, $\mathcal{J}$ is the exchange energy of the $1s$ bulk exciton and its value is 0.03 meV in GaAs.

Since the splitting energy depends on the overlap of the envelope functions, the variation of ΔE_{BD} with the length parameter follows the same pattern for nanostructures, for example, the plot shown in Fig. 9.10 for QDs. As the length is reduced, the overlap increases increasing the value of splitting, reaches a maximum and then decreases as the envelope functions penetrate into the barrier. The value of maximum ΔE_{BD} is highest for QDs, while for bulk it is quite low ~ 0.03 meV. The maximum value may exceed 1 meV in typical QDs made of GaAs for example.

Different transitions between bright or dark states and ground state and the corresponding lifetimes are also indicated in Fig. 9.15. The lifetime for B→G radiative transition has a lifetime Γ_0~ 1 ns, while the lifetime for D→G transition is ~ 1 μs, the large value is characteristic of a typical non-radiative process. The inverse lifetime for B→D transition at 0K is γ_0 (10^8 s^{-1}) and transitions between these states occur via phonon absorption and emission with rate denoted by γ_{ph}.

9.13   Biexcitons and trions

In simple terms, excitons are considered as an analog of hydrogen atom, or more precisely a positronium atom in which the electron revolves around a positron. This analogy can be pushed further. A hydrogen molecule forms by combining two H-atoms with opposite electron spin. Similarly two positronium atoms can bind together to form a positronium molecule. It has been found that two excitons can form an excitonic molecule called a biexciton (Klingshirn 2012).

The binding energy of biexciton, expressed in terms of excitonic Rydberg energy depends on the mass ratio of electrons and holes. For hydrogen atom, in which the ratio $\sigma = m_e/m_h \approx 0$, the value $E^b_{biex}/R_x \approx 0.3$. The ratio decreases monotonically with increasing σ, reaches a minimum and then increases to attain a value of 0.3 when $\sigma \to \infty$.

The dispersion relation is expressed as

$$E_{biex}(\mathbf{k}) = 2(E_g - E^b_{ex}) - E^b_{biex} + \frac{\hbar^2 \mathbf{K}^2}{4M_{ex}} \qquad (9.51)$$

It is assumed that the effective mass of the biexciton is twice the exciton mass.

Biexcitons are found in QWs, QWRs, and QDs. As the dimensionality gets reduced, the binding energy of biexcitons increases. Theoretical results on biexciton binding energy for cadmium sulfide (CdS) QDs in glass matrix are reported and existence of stable biexcitons is demonstrated by Hu et al (1990). Results for indium arsenide (InAs)/gallium arsenide (GaAs) self-assembled quantum dots (SAQD)s are presented by Sarkar et al (2006).

Theoretical calculations have predicted another type of quasi particles called trions in quantum nanostructures. Trions are charged excitons or biexcitons, that contain two electrons and a hole or vice versa. Their existence is demonstrated in moderately (modulation) p- or n-doped QWs.

9.14 Applications

QDs or NCs find a number of applications in the areas of basic physics as well as of device and systems applications. Some of the use of QDs to observe and demonstrate basic physical phenomena like cavity QED, Bose Einstein Condensation (BEC), and single photon generation, etc., will be covered in later chapters. In this section, we point out briefly some device applications of QDs/NCs.

9.14.1 Semiconductor quantum dot lasers and amplifiers

Some representative structures, gain mechanisms and performance parameters of QD lasers have already been discussed in Section 9.6. However, theoretical predictions that the threshold current density will be lower than in QW lasers and the characteristic temperature T_0 will be infinite, have not been achieved so far. The reader is referred to the Nov–Dec 2017 issue of *IEEE Journal of Selected Top Quantum Electronics* for recent developments.

9.14.2 Quantum dot infrared photodetector

Intersubband transitions in QDs have been discussed in Section 9.7. It has been pointed out earlier in connection with QWs that intersubband transitions are exploited to realize Quantum Well Infrared Photodetectors (QWIP)s operating at mid IR region. Similarly, QDs are used to make QDIPs. The transitions may occur between bound and continuum states or between bound-to-bound states or between two different bound states belonging to two different bands like HH and LH states.

An advantage of QDIPs over QWIPs is that the former is insensitive to polarization of the incident light. In a QW only the component of radiation with electric field vector parallel to the growth direction causes absorption. This necessitates an additional component for unpolarized light. Such a complication is avoided in QDIPs.

QDIPs in the field of infrared (IR) detection prove to be an important device for high-temperature, low-cost, high-yield detector arrays required for military applications. As they operate at high-operating temperature ($\geq$ 150 K), photodetectors and focal plane arrays can be cooled by thermo electric coolers, less costly than massive cryogenic dewars and Stirling cooling systems. QDIPs are therefore well-suited for detecting mid-IR light at elevated temperatures (Bhattacharya et al 2006; Stiff- Roberts 2009; Barve et al 2010).

9.14.3 Quantum dot solar cells

The operation of a solar cell may be understood by considering a simple p-n junction structure. When light from the sun falls on the material, electron-hole pairs are created by photo absorption. The build in field separates the electrons and holes, after which the carriers flow in opposite directions. A potential difference, called the open circuit voltage, is created between the two electrodes. Connecting the electrodes causes a short circuit current to flow. A load connected between the terminals derives current across it.

The first-generation solar cells used bulk materials and second-generation cells used other material systems connected at tandem. The third-generation cells are made up of QDs embedded in supporting matrix using bottom-up approach, or configured as thin films having inorganic layers, organic dyes and organic polymers deposited on suitable substrates. The structure consists of a stack of p-n junctions of low-D semiconductor structures. Sometimes QD layers of lower band gap material are inserted to form an *intermediate band* to absorb solar power of lower energies than the band gap of the main QD layers (Luque et al 2012; Etgar 2013; Aeberhand 2018).

Thus far, several semiconductor QDs have been used as light harvesters in photovoltaic devices. The list includes CdS, CdSe, CdTe, $CuInS_2$, Cu_2S, PbS, PbSe, InP, InAs, silver sulphide (Ag_2S), bismuth sulphide ($Bi_2S_{3)}$, Sb_2S_3, and organo lead halide perovskites.

9.14.4 Colour filters, photochromic glasses

The growth of NC materials and their use in colour filters and as photochromic glasses have been discussed briefly in Section 9.10.1.

Problems

9.1 *Calculate the value of R_c using $\Delta E_v = 150\ meV$ and $m_h = 0.5\ m_0$. Using 2.5 nm for electrons, find the ranges for conditions (1)–(3).*

9.2 *Calculate the value of absorption coefficient in an array of GaAs QDs by using Eq. (9.9). Use $a_0 = 3\ nm$ and $D = 0.05$. Also calculate the absorption linewidth.*

9.3 *Calculate the peak value of gain coefficient in a cubic GaAs QD, 10 nm wide in each side for $n = p = 10^{10}\ cm^{-1}$ and $\tau = 0.1\ ps$.*

9.4 *Calculate and plot the ratio n_{ex}/n for different values of exciton binding energies ranging from 4 to 60 meV using Eq. 9.29.*

9.5 *Calculate the transition energy for heavy holes in GaAs QD of radius 10 nm, using Eq. (9.41).*

9.6 *Using Eq. (9.41), calculate and plot the effective band gap or the photon energy of emission for QDs of different radius. Use GaAs, CdS, CdSe, and ZnSe as QD materials.*

9.7 *Draw a band diagram of a core-shell structure using CdSe as core (Eg = 1.74 eV) and CdS as shell (Eg = 2.42 eV). The alignment is type I and the band offsets are in 60/40 ratio.*

9.8 *Give a sketch of envelope functions for electrons and holes in a type II core-shell structure for small and large core diameter. Explain the nature of the functions for two different radii.*

9.9 *Calculate the splitting energy between bright and dark excitons in a GaAs QD given $J = 0.3\ eV$. Assume that the overall integral in Eq. (9.50) is 0.6.*

9.10 *Prove that the rates of downward and upward transitions between bright and dark exciton states follow the ratio $(n_q+1)/n_q$.*

9.11 *Assume that the potential in a spherical QD varies parabolically and reaches the band offset value at the radius of the dot. Using this information, but still considering that the potential increases outside the barrier to assume infinite value, obtain an expression for the effective band gap of the QD in terms of the radius R_0.*

Reading List

Abramowitz M and Stegun I A (1975) *Handbook of Mathematical Functions.* New York: Dover Publications.

Bhattacharya P, Ghosh S, and Stiff-Roberts A D (2004) Quantum dot optoelectronic devices. *Annual Review of Matererials Research* 34: 1–40.

Bhattacharya P and Mi Z (2007) Quantum dot optoelectronic devices. *Proceedings of the IEEE* 95(9): 1723–1740.

Bimberg D, Grundmann M, and Ledentsov N (1999) *Quantum Dot Heterostructures.* Chichester, UK: Wiley.

Bimberg D and Pohl U W (2011) Quantum dots: Promises and accomplishments. *Materials Today*14(9): 388–397.

Blood P (2009) Gain and recombination in quantum dot lasers. *IEEE Journal of Selected Topics in Quantum Electronics* 15(3): 808–818.

Coleman J J, Young J D, and Garg A (2011) Semiconductor quantum dot lasers: A tutorial. *IEEE Journal of Lightwave Technology* 29(4): 499–510.

Gaponenko S (2010) *Introduction to Nanophotonics*. Cambridge: Cambridge University Press.

Kapon E (1997) Quantum wire and quantum dot lasers. In: *Semiconductor Lasers*, Eli Kapon, (ed.) Chapter 4, pp. 291–360. San Diego: Academic Press.

Klingshirn C F (2012) *Semiconductor Optics*, 4th edn. Heidelberg: Springer.

Ledentsov N (2011) Quantum dot lasers. *Semiconductor Science Technology* 26: (014001): 1–8.

Schiff L (1968) *Quantum Mechanics*. New York: McGraw-Hill.

Smyder J A and Krauss T D (2011) Coming attractions for semiconductor quantum dots. *Materials Today*14(9): 382–387.

Sugawara M (ed.) (1999) *Self-Assembled InGaAs/GaAs Quantum Dots, Semiconductors and Semimetals*, Vol. 60. San Diego, CA: Academic Press.

Tartakovskii A (ed.) (2012) *Quantum Dots: Optics, Electron Transport and Future Applications*. Cambridge: Cambridge University Press.

Tredicucci A (2009) Quantum dots: Long life in zero dimensions. *Nature Materials* 8: 775–776.

Warburton R J (2002) Self-assembled semiconductor quantum dots. *Contemporary Physics* 43: 351–364.

References

Aeberhand U (2018) Nanostructure solar cells. In: *Handbook of Optoelectronic Device Modeling and Simulation*, J Piprek (ed.) Chapter 41, pp. 441–474. Boca Raton, FL: CRC Press.

Arakawa Y and Sakaki H (1982) Multidimensional quantum well laser and temperature dependence of its threshold current. *Applied Physics Letters* 40: 939.

Asada M, Miyamoto Y, and Suematsu Y (1986) Gain and threshold of three-dimensional Quantum Box lasers. *IEEE Journal of Quantum Electronics* 22: 1915–1921.

Barve A V, Lee S J, Noh S K, and Krishna S (2010) Review of current progress in quantum dot infrared photodetectors. *Laser Photonics Review* 4(6): 738–750.

Basu P K, Mukhopadhyay Bratati, and Basu Rikmantra (2015) *Semiconductor Laser Theory*, Ch 9. Boca Raton, FL, USA: CRC Press (Taylor and Francis).

Bhattacharya P, Su X, Chakrabarti S, Stiff-Roberts A D, and Fischer C H (2006) Intersuband transitions in quantum dots. In: *Intersubband Transitions in Quantum Structures*, R. Paiella (ed.) Chapter 8, pp. 315–345. NY: McGraw-Hill.

Brus L E (1984) Electron–electron and electron-hole interactions in small semiconductor crystallites: The size dependence of the lowest excited electronic state. *Journal of Chemistry and Physics* 80: 4403–4409.

Chen G, Ågren H, Tymish, Ohulchanskyya Y, and Prasad P N (2015) Light upconverting core–shell nanostructures: Nanophotonic control for emerging applications. *Chemical Society Reviews* 44: 1680–1713.

DikshitA A and Pikal J M (2004) Carrier distribution, gain and lasing in 1.3 um InAs-InGaAs quantum-dot lasers. *IEEE Journal of Quantum Electronics* 40(2): 105–112.

Etgar L (2013) Semiconductor nanocrystals as light harvesters in solar cells. *Materials* 6(6): 445–459.

Efros Al A and Efros A L (1982) Interband absorption of light in a semiconductor sphere. *Fiz. Tekh. Poluprovodn.* 16(722): 1209.

Efros Al L (1992) Luminescence polarization of CdSe microcrystals. *Physical Review B*46: 7448.

Efros Al L, Rosen M, Kuno M, Nirmal M, Norris D J, and Bawendi (1996) Band-edge excitons in quantum dots of semiconductors with a degenerate valence band: Dark and bright exciton states. *Physical Review B*, 54(7): 4843–4856.

Ekimov A (1996) Growth and optical properties of semiconductor nanocrystals in a glass matrix. *Journal of Luminescence* 70: 1–20.

Ekimov A I, Onushchenko A A, and Tsekhomskii (1980) Exciton light absorption by CuCl microcrystals in glass matrix. Soviet Glass Physics and Chemistry 6: 511–512.

Ferreira R and Bastard G (2015) The dot and its environment. In: *Capture and Relaxation in Self-Assembled Semiconductor Quantum Dots*, Chapter 1, pp. 1–45. UK: Morgan & Claypool Publishing.

Flytzanis C, Hache F, Klein M C, Ricard D, and Roussignol P H (1991) *Nonlinear Optics in Composite Materials: 1. Semiconductor and Metal Crystallites in Dielectrics: 1. Semiconductor and Metal Crystallites in Dielectrics, Progress in Optics*, Vol. 29, pp. 321–411. New York: Elsevier.

Hu Y Z, Koch S W, Lindberg M, Peyghambarian N, Pollock E L, and Abraham F F (1990) Biexcitons in semiconductor quantum dots. *Physical Review Letters* 64(15): 1805–1807.

IEEE Journal of Selected Topics in Quantum Electronics 23(6)Nov–Dec. Special issue on semiconductor laser: 0200603–8200810.

Ivanov S A, Piryatinski A, Nanda J, Tretiak S, Zavadil K R, Wallace W O, Werder D, and Klimov V I (2007) Type-II core/shell CdS/ZnSenanocrystals: Synthesis, electronic structures, and spectroscopic properties. *Journal of the American Chemical Society* 129: 11708–11719.

Kodriano Y, Schmidgall E R, Benny Y, and Gershoni D (2014) Optical control of single excitons in semiconductor quantum dots. *Semiconductors in Science and Technology*29(053001): 23.

Koh T S, Feng Y P, Xu X, and Spector H N (2001) Excitons in semiconductor quantum discs. *Journal of Physics: Condensed Matter* 13: 1485–1498.

Luque A, Marti A, and Stanley C (2012) Quantum dot intermediate band solar cells. *Nature Photonics* 6: 146–152.

Marin J L, Riera R, and Cruz S A (1998) Confinement of excitons in spherical quantum dots. *Journal of Physics and Condensed Matter* 10: 1349–1361.

Men G W, Lin J Y, and Jiang H X, and Chen Z (1995) Quantum-confined Stark effects in semiconductor quantum dots. *Physical Review B* 52(8): 5913–5922.

Oura K, Katayama M, Zotov A V, Lifshits V G, Saranin A (2003) Growth of thin films. In: *Surface Science. Advanced Texts in Physics*. Berlin, Heidelberg: Springer. Available at: https://doi.org/10.1007/978-3-662-05179-5_14.

Pedersen T G (2010) Excitons on the surface of a sphere. *Physical Review B* 81(233406): 1–4.

Pedersen T G (2017) Stark effect in finite-barrier quantum wells, wires, and dots. *New Journal of Physics* 19(043011): 1–10.

Pokutnyi S I, Jacak L, Misiewicz J, Salejda W, and Zegrya G (2004) Stark effect in semiconductor quantum dots. *Journal of Applied Physics* 96: 1115–1119

Reiss P, Protie're M, and Li L (2009) Core/shell semiconductor nanocrystals. *Small* 5(2): 154–168.

Sánchez-Cano R, Tirado-Mejía L, Fonthal G, Ariza-Calderón H, and Porras-Montenegro N (2008) Exciton recombination energy in spherical quantum dots grown by liquid-phase epitaxy. *Journal of Applied Physics* 104(113706): 1–4.

Sarkar D, van der Meulen H P, Calleja J M, Becker J M, Haug R J, and Pierz K (2006) Exciton fine structure and biexciton binding energy in single self-assembled InAs/AlAsInAs/AlAs quantum dots. *Journal of Applied Physics* 100: 023109.

Schmitt-Rink S, Chemla D S, and Miller D A B (1989) Linear and nonlinear optical properties of semiconductor quantum wells. *Advances in Physics* 38(2): 88–188.

Scholes G D (2008) Controlling the optical properties of inorganic nanoparticles. *Advanced Functional Materials* (18): 1157–1172.

Shchekin O B and Deppe D G (2002) 1.3 µm InAs quantum dot laser with T = 161 K from 0 to 80 C. *Applied Physics Letters* 80(18): 3277–3279.

Snoke D W and Crawford J D (1995) Hysteresis and Mott transition between plasma and insulating gas. *Physical Review E* 52(6): 5796–5799.

Stiff-Roberts A D (2009) Quantum-dot infrared photodetectors: A review. *Journal of Nanophotonics* 3(031607): 1–17.

Uhrig A, Banyai L, Gaponenko S, Worner A, Neuroth N, and Klingshirn C (1991) Linear and nonlinear optical studies of CdSSe quantum dots. *Zeitschrift fur Physik* D 20: 345–348.

Vahala K J (1988) Quantum box fabrication tolerance and size limits in semiconductors and their effect on optical gain IEEE. *Journal of Quantum Electronics* 24(3): 523–530.

Wu W Y, Schulman J N, Hsu T Y, and Efron U (1987) Effect of size nonuniformity on the absorption spectrum of a semiconductor quantum dot system. *Applied Physical Letters* 51: 710–712.

Wu S, Cheng L, and Wang Q (2017) Excitonic effects and related properties in semiconductor nanostructures: Roles of size and dimensionality. *Materials Research Express* 4(085017): 1–13.

Yin Y and Alivisatos A P (2005) Colloidal nanocrystal synthesis and the organic-inorganic interface. *Nature* 437: 664–670.

10

Optical microcavities

10.1 Introduction

The concept of the optical cavity was introduced a long time ago. A cavity with finite dimensions was assumed to contain the emitter in developing the theory of blackbody radiation. In this work, Rayleigh (1900), and Jeans (1905) determined the number of modes in a cavity having regular shapes. The density-of-modes (DOM) was considered in a number of subsequent investigations, including the work by Bose (1924), who developed the first Quantum Theory of Radiation.

The use of Fabry–Perot (FP) resonators (Fabry and Perot 1899; Perot and Fabry 1899) in spectroscopy has been known to the scientific community for a long time. However, the most important use of the FP resonator or cavity is in lasers in which the mirrors provide the positive feedback to the enclosed gain medium in order to achieve self-sustained oscillation. In place of mirrors localized at the two ends of a FP resonator, multiple layers of dielectrics or semiconductors are commonly employed to increase the reflectivity of the mirrors. The multiple layers act as Bragg mirrors and two such Bragg mirrors form the cavity in the vertical cavity surface emitting lasers (VCSELs) (see Basu et al 2015 for structures and operation). In a sense Bragg mirrors resemble the conventional FP cavity.

Many other applications of optical resonators may be found in many branches in physics. Apart from providing feedback on lasers, resonators provide the filtering required for high-resolution optical spectroscopy.

The most important application of optical cavity in the modern era of Photonics and quantum optics is to provide strong interaction between light and matter. For that purpose, the cavity dimensions must be comparable to the wavelength. This means that the volume occupied by a cavity should be of the order of $(\lambda/\eta)^3$, where λ is the wavelength and η is the refractive index of the material enclosed. At optical frequencies, the wavelength is in the micrometre range and the cavity are rightly called a microcavity.

We have noticed already that the development of semiconductor nanostructures has enabled the control of electronic properties of matter. Microcavities in a similar way facilitate control of the optical modes, their dispersion relation, and the DOM in a very fundamental way. Semiconductor nanostructures, embedded in a microcavity or even in a nanocavity (of subwavelength dimensions), form a good playground to observe

Semiconductor Nanophotonics. Prasanta Kumar Basu, Bratati Mukhopadhyay, and Rikmantra Basu, Oxford University Press.
© P.K. Basu, B. Mukhopadhyay, R. Basu (2022). DOI: 10.1093/oso/9780198784692.003.0010

interesting quantum optic phenomena. The interaction of active or reactive materials, including semiconductor nanostructures, with the modal fields of optical microresonators leads to novel physical phenomena in the domain of basic research, such as, Cavity Quantum Electrodynamics (CQED) experiments, inhibition, and enhancement of spontaneous emission, nonlinear optics, bio-chemical sensing, and quantum information processing (Yokoyama 1992, Yokoyama and Ujihara 1995; Yamamoto 1993; Chang and Campillo 1996; Vahala 2003).

In addition to basic science related activities, development of photonic devices based on optical microresonators, that strongly confine photons and electrons, is both challenging and rewarding for next-generation, compact-size, low-power, and high-speed photonic circuits. The microresonator shape, size, or material composition can be tailored to have it tuned to support a spectrum of electromagnetic modes with the required polarization, frequency, and emission patterns. This opens up the possibility for developing new types of photonic devices, like light emitting diodes, low threshold microlasers, ultra-small optical filters, and switches for wavelength division multiplexed (WDM) networks, colour displays, etc., to name a few.

The present chapter begins with a few fundamental properties of cavity resonators. The most elementary optical resonator, the FP resonator is then introduced. The microresonators of current interest, namely, ring resonators, resonators supporting whispering gallery modes, and photonic crystal microresonators are then introduced in the same order. The simplest theory of operation of these types is presented first, followed by the outline of electromagnetic (EM) theory. Different types in each category with representative structures are briefly mentioned and materials used, and essentials of device fabrication technologies are discussed. At the end, some application areas are listed.

Example 10.1 *We first get an idea of the volume for a microcavity given by* $(\lambda/\eta)^3$. *Let the material be gallium arsenide(GaAs) having refractive index(RI) = 3.6. Let the wavelength of operation be 0.85 μm. The volume is 0.013 μm³.*

10.2 Cavity fundamentals

We first discuss some elementary properties of cavity resonators in general and a few parameters of interest (Chang and Campillo 1996; Yariv and Yeh 2007; Gaponenko 2010; Deen and Basu 2012; Basu et al 2015).

10.2.1 Density-of-modes

In Chapter 4, we developed the wave equation in a rectangular cavity having dimensions quite large compared to the wavelength of interest. The EM wave in the large cavity is found to have different modes characterized by three integers (m, n, q), where the integers denote the number of half wavelengths contained within the length along the respective directions. The number of modes per unit volume between angular frequencies ω and $\omega + d\omega$ has been calculated and will now be reproduced:

$$p(\omega)d\omega = \frac{\omega^2 d\omega}{\pi^2 c^3} \tag{10.1}$$

It is easy now to express the number of modes in a volume V of a medium with refractive index(RI)η in the frequency range between ν and $\nu + d\nu$ as

$$P(\nu)d\nu = \frac{8\pi \nu^2 \eta^3 d\nu V}{c^3} \tag{10.2}$$

Example 10.2 *We consider a cavity of volume 1 cm³, frequency 10^{14} Hz, linewidth $\Delta\nu = 10^{10}$ Hz, and $\eta = 1$. The number of modes is 9.3×10^7.*

Example 10.3 *The frequency spacing between adjacent modes in a cavity of length L is $\Delta\nu = c/2\eta L$. Let $L = 300$ μm and $\eta = 3.6$. Then $\Delta\nu = 0.14\times10^{12}$ Hz.*

10.2.2   Quality factor and loss

The optical beam becomes trapped in an optical cavity, and it forms a mode of the resonator. In an ideal case, the beam should remain trapped within the resonator for an indefinite period. There are several loss mechanisms, however, by which the beam leaks out of the structure. The measure of the fraction of this loss is called the quality factor or Q, which is defined as

$$Q = \omega \frac{energy\ of\ the\ field\ stored\ by\ resonator}{power\ dissipated\ by\ the\ resonator} \tag{10.3}$$

where ω is the angular frequency of the wave. Let us consider a simple resonator formed by two plane parallel mirrors separated by a distance L along the z-direction. The planes are assumed to be perfectly reflecting. Then following the arguments presented in Section 4.2.1, the electric field within the medium may be expressed as:

$$E(z, t) = E_0 \sin\omega\ t\ \sin k\ z \tag{10.4}$$

where E_0 is the amplitude of the electric field and k is the wavenumber. The average electrical energy stored within the resonator is given by

$$U_{elec} = \frac{A\varepsilon}{2T} \int_0^L \int_0^T E^2(z, t)\ dzdt \tag{10.5}$$

where A is the cross-sectional area of the resonator, ε is the permittivity of the medium enclosed by the mirrors and $T = 2\pi/\omega$ is the period of the wave. Using Eq. (10.4); Eq. (10.5) may be written as

$$U_{elec} = \frac{1}{8}\varepsilon E_0^2 V \tag{10.6}$$

Now the magnetic energy stored in the medium equals the electrical energy stored and therefore the total stored energy is

$$U = \frac{1}{4}\varepsilon E_0^2 V \tag{10.7}$$

Under steady state, the input power equals the dissipated power. If P denotes the dissipated power, then the quality factor defined by Eq. (10.3) is expressed as

$$Q = \frac{\omega \varepsilon E_0^2 V}{4P} \tag{10.8}$$

The peak electric field within the resonator is given by

$$E_0 = \sqrt{\frac{4QP}{\omega \varepsilon V}} \tag{10.9}$$

The peak electric field can therefore be increased by increasing the Q of the cavity, by increasing the power input and by decreasing the cavity volume. For strong light-matter interaction a large electric field is needed. This is made possible by using a very high Q microcavity having volume $\sim \lambda^3$ as we shall see later in this and the following chapters.

Example 10.4 *The peak electric field in a cavity of volume 0.1 mm³ having Q = 500 for input power of 10 mW, at an angular frequency of 10^{15} Hz is 1.37 × 10^3 V/m. If the Q is increased to 10^5 the value becomes 2.38 × 10^4 V/m. The relative permittivity of the material in the cavity is assumed to be 12.*

10.2.3 Quality factor and lifetime of photons

The quality factor Q defined in Section 10.2.2 is related to the losses in the optical resonator. It is related to the lifetime of EM radiation within the cavity, as well as to the linewidth of the resonance frequency or wavelength. To obtain the relationships, let us denote the energy stored in a mode of the resonator by $\mathcal{E}$. The decay of this energy is described in terms of photon lifetime τ_{ph} by

$$\frac{d\mathcal{E}}{dt} = -\frac{\mathcal{E}}{\tau_{ph}} \tag{10.10}$$

The time variation of the stored energy in the resonator may therefore be expressed as $E(t) = \mathcal{E}(0)exp(-t/\tau_{ph}) = \mathcal{E}(0)exp(-\omega t/Q)$. If the intensity decreases by a fraction γ per pass in the resonator of length L, then the fractional loss per unit time is $c\gamma/\eta L$, and we may write

$$\frac{d\mathcal{E}}{dt} = -\frac{c\gamma}{\eta L}\mathcal{E} \tag{10.11}$$

Comparing Eqs. (10.10) and (10.11) the photon lifetime may be expressed as

$$\tau_{ph} = \frac{\eta L}{c\gamma} \tag{10.12}$$

Let the resonator be made of two plane parallel mirrors of reflectivities R_1 and R_2 and the average loss in the medium enclosed by the mirrors be α. The average loss per pass is then $\gamma = \alpha L - ln\sqrt{R_1 R_2}$, so that

$$\tau_{ph} = \frac{\eta L}{c\left[\alpha - (1/L)\,ln\sqrt{R_1 R_2}\right]} \tag{10.13}$$

The quality factor of the resonator is defined as

$$Q = \frac{\omega\mathcal{E}}{P} = -\frac{\omega\mathcal{E}}{d\mathcal{E}/dt} \tag{10.14}$$

where $P = -\frac{d\mathcal{E}}{dt}$ is the power dissipated. Comparing Eqs. (10.10) and (10.14), we obtain

$$Q = \omega\tau_{ph} \tag{10.15}$$

The Q factor is related to the full width at half maximum (FWHM) of the resonators response curve, assumed to be Lorentzian, as

$$\Delta\nu_{1/2} = \frac{\nu}{Q} = \frac{1}{2\pi\tau_{ph}} = \frac{c\left[\alpha - (1/L)\,ln\sqrt{R_1 R_2}\right]}{2\pi\eta} \tag{10.16}$$

Eq. (10.13) has been used to arrive at the last equality in Eq. (10.16).

Example 10.5 *The photon lifetime of an FP resonator is calculated with the following values: $L = 100$ nm, $\alpha = 5$ cm^{-1}, R1= 0.999, R2 = 0.995, and $\eta = 3.5$. The calculated value is 0.38 ps. The corresponding value of Q at a wavelength of 1550 nm is 459.*

10.2.4 Losses in resonators

In general, there exist several loss processes in a resonator and the intrinsic quality factor Q_{int} is related to the quality factors due to individual loss processes in the following way:

$$Q_{int}^{-1} = Q_{mat}^{-1} + Q_{surf}^{-1} + Q_{scatt}^{-1} + Q_{bend}^{-1} \qquad (10.17)$$

Here, Q_{mat}^{-1} arises due to intrinsic material absorption, Q_{surf}^{-1} occurs due to surface absorption losses, which may be due to surface coatings or adsorbed material, or due to other contaminants. An example of adsorbed material is water in silica resonators. Q_{scatt}^{-1} denotes scattering losses, both intrinsic and related to the surface of the cavity. A surface related scattering occurs due to imperfections in the form of surface roughness, as the surface can never be realized in an atomically smooth form. The term Q_{bend}^{-1} is related to bending loss (or whispering gallery, or tunnel, or radiation loss).

The Q factor denoted by Q_{mat} is associated with material absorption and bulk Rayleigh scattering in the material used in the microresonator. An approximate expression

$$Q_{mat} = \frac{4.3 \times 10^3}{\beta} \frac{2\pi\eta}{\lambda} \qquad (10.18)$$

may be used to estimate the value. Here β is in dB/km.

Example 10.6 *For silica glass used in optical fibre $\beta = 0.2$ dB/km and taking $\eta = 1.5$ and $\lambda = 1.55\mu m$, $Q_{mat} = 1.3 \times 10^{11}$.*

The experimental Q factor is much lower $\sim 10^8$–10^9, as the contribution from surface contaminants is to be taken into account.

Different authors have given different expressions for the contribution for surface scattering. Here we reproduce the expression by Vernooy et al (1998):

$$Q_{ss} \approx \frac{3\eta^2(\eta^2 + 2)^2 \lambda^{7/2} d^{1/2}}{(4\pi)^3(\eta^2 - 1)^{5/2}\sigma^2 B^2} \qquad (10.19)$$

Here d is the diameter of the resonator, σ and B are, respectively, the rms value and the correlation length of the of inhomogeneity.

Example 10.7 *considering a micropillar made of GaAs, let $\sigma = 2$ nm, $B = 5$ nm, $d = 100$ nm, $\lambda = 1.55 \mu m$, $\eta = 3.6$. The value of $Q_{ss} = 1.3 \times 10^8$.*

Bending losses at the curved interfaces may be explained by ray optic theory. The ray suffers total internal reflection at planar surface, but at the curved surface the angle of incidence is lower than the critical angle. As a result, there is incomplete total internal reflection and some part of the energy leaks into the surrounding lower refractive index(RI) material. The quality factor Q_{bend} increases exponentially with the radius of the resonator. For a spherical resonator, $Q_{bend} \propto exp(4\pi\eta a/\lambda_0)$ where a is the radius of the sphere.

In actual practice, a cavity is coupled to other optical structures like a prism or a waveguide. The losses occurring in these external components influence the total (or loaded) quality factor. If Q_{ext} denotes the quality factor due to external modes, the total quality factor is expressed as

$$Q_{tot}^{-1} = Q_{int}^{-1} + Q_{ext}^{-1}$$
(10.20)

10.2.5   Finesse

Finesse, another important performance parameter for resonators, is defined as the ratio of FSR and the linewidth FWHM of resonances. Thus,

$$\mathcal{F} = \frac{FSR}{\Delta\nu_{\frac{1}{2}}}$$
(10.21)

For good resonators a large value of finesse is desirable. When used as a tunable interferometer, finesse denotes how many wavelengths can be distinguished by it. Finesse therefore is an indication of the effective resolution of the resonator as an interferometer. Finesse is also related to the quality factor.

10.2.6   Mode volume

A prime requirement for a good microcavity is its low mode volume. For a field eigenmode it is defined as the ratio of total energy of that mode stored within the resonator and the maximum energy density of it. If $U(r)$ denotes the energy density at a position r, then the mode volume is expressed as

$$V_{eff} = \frac{\text{Stored energy}}{\text{Maximum energy density}}$$

$$= \frac{\int_V U(r)d^3r}{\max(U(r))}$$
(10.22)

It is well known that for EM field

$$U(r) = \left(\frac{1}{2}\right)\left[\varepsilon_0\varepsilon(r)|E(r)|^2 + \frac{|B(r)|^2}{\mu_0\mu(r)}\right]$$
(10.23)

where $\varepsilon(r)$ is the relative permittivity and $\mu(r)$ is the relative permeability, and ε_0 and μ_0 are, respectively, the permittivity and permeability of vacuum. It may be noted that the spontaneous and stimulated emission rates are inversely proportional to the mode volume and the threshold current density in a laser is directly proportional to it. From all these considerations, smaller mode volume is most desirable. Smaller mode volume means higher free spectral range (FSR).

10.3 Fabry–Perot resonators

The simplest form of optical cavity is a FP resonator or etalon, which has been in use in spectroscopy over a long time. It is also used in lasers, as fast tunable filters, and in several other applications. The cavity is formed by two plane parallel mirrors. Light entering into the cavity suffers multiple reflections at the two mirrors. A particular wavelength, determined by the separation between the mirrors, is transmitted through a mirror, while all other wavelengths interfere destructively. The mirror separation may be altered mechanically or by changing the RI of the material within the cavity.

A schematic diagram of a FP resonator, formed by two planes parallel and highly reflecting mirrors M_1 and M_2, is shown in Fig. 10.1. The two mirrors have (field) reflection and transmission coefficients, denoted by (r_1, t_1) and (r_2, t_2), respectively. To understand the operation of the resonator, we consider that an electric field of amplitude E_i is incident at the left-hand mirror M_1. The field enters into the cavity and propagates to the second mirror, where it is partially transmitted and partially reflected. The reflected field enters into the cavity and propagates backwards and the processes repeat. The fields of the various reflected and transmitted beams at the two mirrors are shown by parallel lines. Summing over the electric fields of the transmitted rays coming out of mirror M_2, the total transmitted field, E_t, may be expressed as (Verdeyen 1995)

$$E_t = E_i t_1 t_2 e^{-i\theta}[1 + r_1 r_2 e^{-i2\theta} + (r_1 r_2)^2 e^{-i4\theta} + \ldots \ldots] \tag{10.24}$$

where $\theta = kL$, $k = \omega \eta_g / c$, η_g is the group index and L is the separation between the two mirrors. The infinite series may be summed to give the expression for E_t.

The ratio of output to input optical powers of the resonator, known as the transmitivity of the FP resonator, may be expressed as

$$T = \frac{|E_t|^2}{|E_i|^2} = \frac{I_{out}}{I_{in}} = \frac{(1 - R_1)(1 - R_2)}{\left(\sqrt{R_1 R_2}\right)^2 + 4\sqrt{R_1 R_2}\sin^2\theta},$$

$$\theta = \left(\frac{2\pi f \eta_g}{c}\right)L, R_1 = r_1^2 = 1 - t_1^2, R_2 = r_2^2 = 1 - t_2^2 \tag{10.25}$$

where the R's are power reflection coefficients. Fig. 10.2 gives the plots of transmittance of the etalon as a function of wavelength for different values of reflectivities. Larger

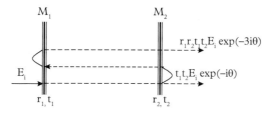

Figure 10.1 *Schematic diagram of a FP resonator.*

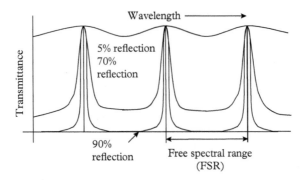

Figure 10.2 *Transmittance of a FP resonator.*

the reflectivity is sharper and becomes the peaks. The peaks occur at the successive longitudinal mode frequencies given by

$$f = f_m = \frac{mc}{2\eta_g L} \qquad (10.26)$$

where m is an integer. The tuning range of the resonator specifies the range of wavelengths over which it may act. In etalons, the transmittivity repeats itself after a certain period, referred to as the FSR. The frequency spacing between two successive transmission peaks, or FSR, is easily obtained from Eq. (10.26) and given by

$$\Delta f = \frac{c}{2\eta_g L} \qquad (10.27)$$

The finesse of the resonator is related to the reflectivities of the mirrors (assumed equal for both the mirrors) by

$$\mathcal{F} = \frac{\pi\sqrt{R}}{1 - R} \qquad (10.28)$$

It defines the sharpness of the transmission spectra.

The maximum value of transmittivity is 1. This occurs when $\theta = (2\pi f\eta_g/c)L = m\pi$, so that an integral number of half wavelengths fits into the resonator. The minimum transmittivity has the value

$$T = \frac{(1 - R_1)(1 - R_2)}{(1 - \sqrt{R_1 R_2})^2 + 4\sqrt{R_1 R_2}} = \left(\frac{1 - R}{1 + R}\right)^2 \qquad (10.29)$$

where the last equality is valid when $R_1 = R_2 = R$. Eq.(10.29) is valid for a resonator without any loss. When the gain (or loss) per pass is G, it can be modified as

$$T = \frac{(1-R)^2 G}{(1-GR)^2 + 4GR sin^2\theta} \tag{10.30}$$

If the loss is due to material absorption, then $G = exp(-\alpha L)$, and the maximum value of T is less than unity.

Example 10.8 *let $L = 300\ \mu m$ and $\alpha = 10\ cm^1$. Then $G = 0.741$. Taking $R = 0.99$, $T_{max} = 1.056 \times 10^{-3}$.*

Example 10.9 *The free spectral range is calculated for a FP resonator having separation between the mirrors as 0.1 mm and filled with a material with group index $\eta_g = 1.5$. Using Eq. (10.27) and the relation $\Delta\lambda/\lambda = \Delta f/f$, $\Delta\lambda_{FSR} = \lambda^2/2n_g L$. Thus for $\lambda = 1.55$ μm, $\Delta\lambda_{FSR} = 8\ nm$.*

10.4   Bragg gratings and Bragg mirrors

FP resonators discussed in Section 10.3 are used in edge emitting lasers to provide proper feedback to the amplified EM wave in presence of dominant stimulated emission. In vertical cavity surface emitting lasers(VCSEL)s, the active region is enclosed between two Bragg reflectors or mirrors. The Bragg mirror or grating consists of alternate layers of two different dielectrics or semiconductors. If the growth direction is the z-direction, a periodic modulation of RI thus occurs along this direction. The grating planes are of a constant period, and the phase fronts of the EM wave are perpendicular to z. The propagating light is scattered by each grating plane. The nature of the reflected light is determined by the Bragg condition, which is expressed as follows:

$$\lambda_B = 2\eta\Lambda. \tag{10.31}$$

where λ_B is the Bragg wavelength, Λ is the grating period and η is the RI of the medium.

For wavelengths different from the Bragg wavelength, the reflected light from each of the planes becomes progressively out of phase and there is no back reflection due to destructive interference. When the Bragg condition is satisfied, that is, $\lambda = \lambda_B$, the reflected lights from each grating plane add constructively in the backward direction and the reflectivity shows a peak.

The Bragg condition may be derived from coupled mode theory. Here we provide a simple theory based on the principles of energy and momentum conservation. The energy conservation dictates that the energies of forward (f) and back-reflected (b) photons must be equal, leading to the condition $\omega_f = \omega_b$. Momentum conservation requires that the incident wave vector β_f and the grating wave vector β_B should be equal to the back-reflected wave vector β_b. Thus,

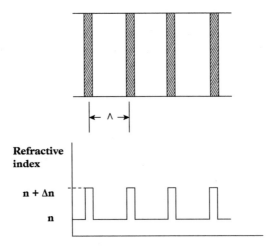

Figure 10.3 *A periodic structure forming a Bragg grating and the RI variation.*

$$\beta_f + \beta_B = \beta_b \tag{10.32}$$

The grating wave vector is $(2\pi/\Lambda)$, and is along the direction normal to the grating planes. The back-reflected wave vector is equal in magnitude but opposite in direction to the incident wave vector. Therefore Eq. (10.32) may be written as

$$2\left(\frac{2\pi\eta}{\lambda_B}\right) = \frac{2\pi}{\Lambda} \tag{10.33}$$

from which the following first-order Bragg condition, Eq. (10.31) is obtained.

The *Bragg wavelength* refers to free space and η is the RI of the guide.

Example 10.10 *A Bragg grating is to be designed for a single mode laser operating at 1550 nm. The material is InGaAsP. The RI is 3.39. The required grating period obtained from Eq. (10.31) is $\Lambda = 228.6$ nm.*

The reflectivity increases with the increase of index perturbation $\Delta\eta$. The nature of the reflectivity spectrum is displayed in Fig. 10.4 as a function of wavelength detuning. The bandwidth of the reflectivity spectrum is approximately given by

$$\Delta\lambda = \lambda_B\alpha\sqrt{\left(\frac{\Delta\eta}{\eta}\right)^2 + \left(\frac{1}{N}\right)^2} \tag{10.34}$$

where N is the number of grating planes. The parameter α is ~ 1 for strong gratings and is ~ 0.5 for weak gratings. The side lobes of the resonance shown in Fig. 10.4 are due to multiple reflections to and from opposite ends of the grating region.

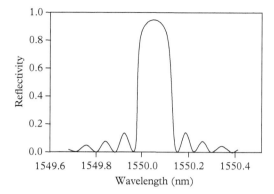

Figure 10.4 *Typical reflectivity variation of an FBG.*

Example 10.11 *We will now make an estimate of the bandwidth of the reflectivity spectrum. Let $L = 0.5$ cm, $\lambda_B = 1550$ nm, $\eta = 1.5$ for silica and change in RI $\Delta\eta = 5\times10^{-4}$. From Eq. (10.31) the grating period $\Lambda = 516.67$ nm. The number of grating planes is $N = L/\Lambda = 8900$. Assuming $\alpha = 1$ the bandwidth $\Delta\lambda = 0.31$nm.*

The Bragg resonance wavelength is affected by strain and temperature. Strain changes the grating spacing and, through the strain-optic effect, changes the mode index. Similar effects are also produced by temperature, the more dominant being the change of RI by thermo-optic effect. The expected change of λ_B of a germanosilica fibre with temperature is 13.7 pm/C at around 1,550 nm.

Example 10.12 *An estimate of the change of wavelength due to the TO effect is made by using Eq. (10.31). Taking differentials, one may write:*

$$\Delta\lambda_B = \lambda_B \left[\frac{1}{\Lambda} \frac{\partial\Lambda}{\partial T} + \frac{1}{\eta} \frac{\partial\eta}{\partial T} \right] = \lambda_B [\alpha + \zeta]\Delta T.$$

The first term is the thermal expansion coefficient, and the second term is the TO coefficient. The values are 0.55×10^{-6} for silica and 8.6×10^{-6} for Germania-doped silica fibre. The change in wavelength at 1550 nm is 14.2×10^{-12} m /°C. The slightly larger value than 13.7 pm/°C arises because parameters for two different materials are used.

The grating structures may be classified into three distinct types: (1) the common Bragg reflector, (2) the blazed Bragg grating, and (3) the chirped Bragg grating.

The common Bragg reflector has a constant pitch, and its reflectivity characteristics depend on the period of the grating and the amplitude of RI modulation. Its reflective character can be converted into a transmissive character when the gratings are inserted in the two arms of a Mach–Zehnder interferometer (MZI). A proper cascading of two Bragg gratings can result in a bandpass filter.

10.5   Resonators

Bulk optical resonators, including FP cavity, have large size and weight and offer align-ment and stability problems. Their use in photonic circuits is thus much limited. In integrated optics, ring resonators form an important component. The schematic dia-gram of a ring resonator is shown in Fig. 10.5. The resonator has the form of a circular ring, is excited by the straight guide, and a fraction of the input power is coupled to the ring. The light circulates through the ring and suffers a phase shift $\Delta\varphi = \beta L$ for one round trip, where β is the propagation constant in the ring, and L is the optical path length. When this phase shift equals an integral multiple of 2π, the device acts as a resonator and the resonance condition is given by

$$\Delta\varphi = \beta L = 2m\pi \tag{10.35}$$

where m is an integer. Using $L = 2\pi R$, where R is the radius of the ring, and expressing β in terms of effective index N, the resonance condition becomes

$$\lambda = \frac{2\pi NR}{m} \tag{10.36}$$

We will now use coupled mode theory to analyze the ring resonator. Let the input and output electric fields of the straight guide are denoted by E_i and E_0, respectively. The electric field coupled into the ring is B_i and that after covering length L is B_0. The steady-state input and output fields are written as

$$E_0 = (1-\gamma)^{1/2}[E_i cos(\kappa L_c) - jB_0 sin(\kappa L_c)] \tag{10.37a}$$

$$B_i = (1-\gamma)^{1/2}[-jE_i sin(\kappa L_c) + B_0 cos(\kappa L_c)]. \tag{10.37b}$$

In Eq. (10.37), κ is the mode coupling coefficient, γ is the intensity insertion loss coeffi-cient, and L_c is the coupling length. It is assumed that the propagation constants are the same, β, in straight and ring guides. The electric field after a round trip B_0 is expressed as

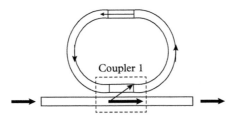

Figure 10.5 *Structure of a ring resonator.*

$$B_0 = B_i exp[(-\alpha L/2) - j\beta L] \tag{10.38}$$

where α is the intensity attenuation coefficient of the ring.

The field transmittance of the ring resonator is then expressed by using Eqs. (10.37) and (10.38) as

$$\frac{E_0}{E_i} = (1-\gamma)^{\frac{1}{2}} \left[\frac{cos(\kappa L_c) - (1-\gamma)^{\frac{1}{2}} exp[(-\alpha L/2) - j\beta L]}{1 - (1-\gamma)^{\frac{1}{2}} cos(\kappa L_c) exp[(-\alpha L/2) - j\beta L]} \right] \tag{10.39}$$

We introduce two new parameters defined as

$$x = (1-\gamma)^{\frac{1}{2}} exp(-\alpha L/2), y = cos(\kappa L_c), \text{and } \phi = \beta L, \tag{10.39}$$

The intensity transmittance of the optical ring resonator may be expressed as

$$T(\phi) = \left|\frac{E_0}{E_i}\right|^2 = (1-\gamma)\left[1 - \frac{(1-x^2)(1-y^2)}{(1-xy)^2 + 4xysin^2\left(\frac{\phi}{2}\right)}\right] \tag{10.40}$$

The transmittance characteristics of the ring resonator show similar maxima and minima as in a FP resonator (see Fig. 10.2). The minimum transmission occurs when

$$\phi = \beta L = 2m\pi \tag{10.41}$$

as noted before. The maximum and minimum transmittances are given by

$$T_{max} = (1-\gamma)\frac{(x+y)^2}{(1+xy)^2} \tag{10.42}$$

$$T_{min} = (1-\gamma)\frac{(x-y)^2}{(1-xy)^2} \tag{10.43}$$

These equations indicate that, in order to maximize T_{max} and to minimize T_{min}, $x \cong y \cong 1$ should be satisfied. The FWHM $\delta\phi$ and finesse F are given by

$$\delta\phi = \frac{2(1-xy)}{\sqrt{xy}} \tag{10.44a}$$

$$F = \frac{2\pi}{\delta\phi} = \frac{\pi\sqrt{xy}}{(1-xy)} \tag{10.44b},$$

The value of T_{min} is zero when $x = y$, or when the condition,

$$cos(\kappa L_c) = (1 - \gamma)^{1/2} exp\left(-\frac{\alpha L}{2}\right)$$

(10.45)

is satisfied.

The FSR is determined by the spacing of two successive resonance peaks occurring at $\phi = 2m\pi$, and $(2m+1)\pi$. Writing the wavenumbers as k and $k+\Delta k$ respectively and assuming $\Delta k << k$, we obtain from Eq.(10.41),

$$\frac{d\beta}{dk}\Delta k = \frac{2\pi}{L}$$

(10.46)

Since $\beta = k\eta$, where η is the effective index, and the frequency shift $\Delta f = (c/2\pi)\Delta k$, we may write

$$\Delta f = \frac{c}{NL}$$

(10.47)

In Eq. (10.47), $N = \eta + k(d\eta/dk)$, is the group index. Using the relation $\delta\phi = \delta(\beta L) = (d\beta/dk)\delta k \cdot L = 2\pi/F$, the FWHM in terms of frequency may be expressed as

$$\delta f = \frac{c}{FNL}$$

(10.48)

Since FSR is inversely proportional to the size of the ring resonator, the ring must be small in order to achieve a high FSR.

The finesse is another key specification of the ring resonator and is dependent on both the internal loss and the coupling (i.e. the external loss) of the resonator. The higher the total losses are, the lower the finesse of the resonator, and it is advantageous to reduce both the internal and external losses in order to obtain higher finesse. However, the external loss due to coupling is necessary and cannot be too small for the resonator to operate as an optical filter. If the external loss is smaller than the internal loss, all the coupled power will be lost inside the cavity and no power will be coupled out. Because of these constraints, the ring resonator must use a strongly guided waveguide to minimize the bending loss for a curved waveguide with a very small radius.

10.5.1   Multiple ring resonators

To increase the finesse, two or more rings could be used as shown in Fig. 10.6. In this case, the combined FSR increases since $FSR = N.FSR_1 = M.FSR_2$ and the total FWHM, in terms of wavelength, $\delta\lambda = \delta\lambda_1\delta\lambda_2/(\delta\lambda_1 + \delta\lambda_2)$, decreases, so there is a net increase of F. The two resonators have to be carefully designed to make sure that both N and M are integers. This way, the N th peak of the resonator 1 located at the same wavelength as the M th peak of the resonator 2 results in a sharper peak (high finesse). All the other peaks are blocked by each other.

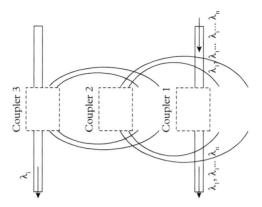

Figure 10.6 *Schematic diagram of a multiple ring resonator.*

Ring resonators possess high spectral selectivity and large free spectral range, and as such find applications in linear passive devices like filters, sensors, and dispersion compensators, as well as active devices like modulators, lasers, and switches. The resonators also find use in fundamental CQED studies. To date, several material systems have been used to fabricate ring resonators. Apart from optical fibres, low index-contrast ($\Delta\eta < 2$) dielectrics, such as, silica, polymers, lithium niobate, silicon nitride, and silicon oxynitrides have been used. High-index contrast materials ($\Delta\eta > 2$), mostly semiconductors, like indium gallium arsenide phosphide (InGaAsP), gallium aluminum arsenide(AlGaAs), GaAs as well as silicon-on-insulators have also been in use. (see Zhou et al 2010 and Deen and Basu 2012 for details and references).

To be useful for planar photonic integration, ring resonators have been reduced in size by using microfabrication technology. In this respect, Photonic Crystal (PhC) ring resonators have provided a very useful solution. The structure of PhC ring resonators and circuits using such resonators have been introduced in the article by Zhou et al (2010).

10.6 Whispering gallery mode resonators

Another class of miniaturized resonator, a dielectric resonator having circular symmetry, is being increasingly used in the study of nanophotonics phenomena. The novelty of these resonators is that they support whispering gallery modes (WGMs) of EM waves circulating within the structure and being confined. The definition of WGMs is due to J W Strutt (Lord Rayleigh). He studied the propagation of sound waves in the interior of the circular gallery of St. Paul's Cathedral in London. A whisper at any point in the circular wall is audible at any other point of the gallery. He correctly interpreted the phenomena as due to total internal reflection in the wall (see Rayleigh 1910). Lord Rayleigh also suggested that such WGMs would exist for EM waves as well in structures having circular symmetry. However, it was not before 1990s that intensive studies on WGMs began to be reported in the literature (see references in Matsko and Ilchenko 2006a, b; for a review see Righini et al 2011).

Optical resonators supporting WGMs have a number of useful characteristics: (1) ultra-high Q factors, (2) small mode volume, (3) small size, (4) operation at optical and telecommunication frequencies, (5) strong electric field leading to strong light-matter interactions, (6) ease of fabrication, and (7) on-chip integration of devices using them. Because of these advantages, the resonators are of great interest for lasers, optical and RF communications, quantum optics, and electrodynamics, and sensing.

WGM resonators (WGMRs) have been fabricated and studied with different shapes. The simplest structures are cylindrical optical fibres and microspheres, and more complex structures include fibrecoils, microdisks, microtoroids, PhC cavities, micropillars, and so on. Structures such as bottle and bubble microresonators have also been fabricated and studied. The most attractive structures are definitely planar structures, like ring and disk resonators, because they are amenable for mass production. The conventional integrated-circuit deposition and etching techniques are used to fabricate these planar microstructures onto wafer substrates. Micropillars have also been extensively used to study important nanophotonics phenomena. However, microspheres, the simplest 3D WGM resonator, are difficult to realize in integrable form.

In the following subsections, we shall develop the working principle of WGM resonators, starting from the simple ray optic theory and then outlining the EM theory. Some important characteristics of the resonators will then be pointed out and different structures investigated so far will be presented. Emphasis will be given to planar structures having cylindrical symmetry.

10.6.1   Ray optic theory

The term WGMs are assigned to EM surface oscillations in optical cavities. The occurrence of different modes is due to almost total internal reflections of light rays from the curved surfaces of the resonator. As noted already, WGMs are supported by cylinders, spheres, toroidal- and spheroidal-shaped dielectric resonators.

The confinement of EM waves in a cylindrical structure may be illustrated by using ray optic theory. Fig. 10.7 shows the propagation of light rays inside the circular cross section of a dielectric cylinder, having a radius a $>>\lambda$, RI η, and the cylinder is surrounded by air. A ray of light, propagating inside, hits the surface with an angle of incidence i. If the angle i exceeds the critical angle $i_c = sin^{-1}(1/\eta)$, then total internal reflection (TIR) occurs. Due to circular symmetry, all subsequent angles of incidence are i, and the ray is trapped within the cross section. If $i \approx \pi/2$, that is, if the ray strikes the surface at near-grazing angle, the ray propagates close to the surface covering a distance $\approx 2\pi a$ in a round trip. A constructive interference occurs when the round-trip length equals an integral multiple of wavelength within the medium.

The resonance condition is therefore

$$2\pi a = m\lambda/\eta m = 1, 2, 3, \ldots . \tag{10.49}$$

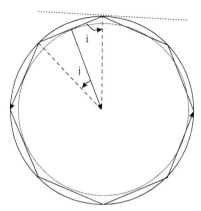

Figure 10.7 *Formation of WGM in a resonator having circular cross section illustrated by total internal reflection of a light ray.*

Note that the same condition has been arrived at in Eq. (10.36) for circular ring resonator.

Example 10.13 *Assume that a = 20 μm, η =1.5 and λ = 1.55 μm. The mode number m ≈ 122.*

10.6.2   Electromagnetic theory

The simple ray optic theory described previously gives a useful insight of the guidance of light wave inside the cavity. However, for more detailed study, particularly to understand the propagation characteristics including the attenuation of waves, more sophisticated theory based on wave equation is necessary. We will now present an outline of the theory, followed by a method to calculate the eigenmodes of the EM waves in a cylindrical cavity.

10.6.2.1   General scalar wave theory

We assume that the wave field, the electric or magnetic field associated with the EM wave, has field amplitude

$$u(\mathbf{r}, t) = \psi_\mu(\mathbf{r})exp(-i\omega_\mu t) \tag{10.50}$$

The field satisfies the wave equation

$$\nabla^2 u - \frac{1}{v^2(\mathbf{r})}\frac{\partial^2 u}{\partial t^2} = f(\mathbf{r}, t) \tag{10.51}$$

where $f(r, t)$ represents the source term and v is the wave velocity. Using the usual separation of variables we obtain the following Helmholtz equation for the source-free case:

$$\nabla^2 \psi_\mu + \left[\frac{\omega_\mu^2}{v^2(\mathbf{r})} \right] \omega_\mu = 0 \tag{10.52}$$

This is an eigenvalue equation for operator $-v^2(\mathbf{r})\nabla^2$ with eigenfunctions and eigenvalues denoted respectively by $\psi_\mu(\mathbf{r})$ and $\omega_\mu^2(\mathbf{r})$. Only discrete values of eigenfrequencies may be obtained for a specified boundary condition. The solutions are called eigenmodes and are denoted by a multivalued index μ, called the mode number. As usual, the general solution is a linear combination of the eigenfunctions and for the source free case, the general solution can be written as

$$u(\mathbf{r}, t) = \sum_\mu A_\mu \psi_\mu(\mathbf{r}) \exp(-i\omega_\mu t) \tag{10.53}$$

10.6.2.2 Scalar whispering gallery modes in a two-dimensional cylinder

As an example of the solution of the eigenvalue equation given by Eq. (10.52), we consider the simplest case of scalar waves in an idealized 2D cylindrical cavity of radius a of a dielectric of RI η_1, surrounded by another material of lower RI η_0. It is assumed that the change in RI occurs abruptly at the interfaces. The Helmholtz equations in the two materials may be written as

$$\nabla^2 \psi_\mu + k_{1,\mu}^2 \psi_\mu = 0, r < a \tag{10.54a}$$

$$\nabla^2 \psi_\mu + k_{0,\mu}^2 \psi_\mu = 0, r > a \tag{10.54b}$$

where the wave numbers are defined as $k_{0(1)\mu} = \omega_\mu / v_{0(1)}$. When the surrounding material is air, $v_0 = c$.

The boundary conditions to be imposed are: (1) the function $\psi_\mu(\mathbf{r})$ should be finite everywhere, and should have no singularities, (2) it should describe only outgoing waves for $r \to \infty$, *that is, there must be* no waves travelling from $r \to \infty$ towards the origin at $r = 0$, (3), and (4) $\psi_\mu(\mathbf{r})$ (and $\nabla \psi_\mu(\mathbf{r})$) should be continuous at the edge of the cavity at $r = a$. (for vector EM waves these two continuity conditions (3) and (4) are slightly modified, depending on the polarization of the eigenmode, to satisfy Maxwell's equations).

We employ cylindrical coordinate system (r, ϕ), assume separation of variables, and apply boundary conditions a) and b). The solutions are written as

$$\psi(r) = \exp(im\phi) \begin{cases} C_\mu \mathcal{J}_{|m|}(\eta k_{0,\mu} r), r < a \\ D_\mu H_{|m|}^{(1)}(k_{0,\mu} r), r > a \end{cases} \tag{10.55}$$

where $J_{|m|}$ and $H_{|m|}^{(1)}$ are respectively the Bessel function and the Hankel function of first kind of order m. The Hankel function ensures that the eigenfunctions decay to zero as

$r \to \infty$. Continuity of $\psi(r)$ at $r = a$ gives the ratio of the amplitude coefficients C and D as

$$\frac{D\mu}{C\mu} = \frac{\mathcal{J}_{|m|}(\eta k_0 \mu a)}{H^{(1)}_{|m|}(k_0 \mu a)} \tag{10.56}$$

Equating the derivatives at $r = a$ yield the following characteristic equation:

$$\frac{\mathcal{J}_{|m|}{}'(\eta k_0 \mu a)}{\mathcal{J}_{|m|}(\eta k_0 \mu a)} = \frac{1}{\eta} \frac{H_{|m|}{}^{(1)'}(k_0 \mu a)}{H_{|m|}{}^{(1)}(k_0 \mu a)} \tag{10.57}$$

The prime in Eq. (10.57) indicates ordinary differentiation. The previous transcendental equation can only be solved numerically and many complex roots of ω_μ are obtained. The mode index is specified by two indices, thus $\mu \equiv (N, m)$. The index N refers to the number of maxima of the field along the radial direction, $N = 1$ being the fundamental mode. A geometrical interpretation of the mode index m can also be given. It indicates that the ray is reflected m times off the edge of the circular cross section, as the ray traverses at grazing angle and makes one full round trip. In this way the ray returns to its origin with the same phase (modulo 2π).

Fig. 10.8 gives the eigenmode profile for $N = 1$ and $m = 4$. There is only one maxima of the field along the radius within the cross section corresponding to $N = 1$. As explained already, there will be 4 such maxima along the azimuthal direction for $m = 4$. It may be possible to obtain the mode profile for higher values of N and m. Modes with $N = 1$ and $m \gg 1$ are called the WGMs of the cavity. Modes with small $N > 1$ are called higher order modes.

Figure. 10.8 *Field distribution for $N =$ and $M = 4$. Light ray is total internally reflected 4 times corresponding to $M = 4$.*

10.6.3   Structures supporting whispering gallery modes

It has been mentioned at the beginning of Section 10.5 that several structures support WGMs. Though a few of them possess higher quality factors, lower mode volumes, etc., it is the planar nature of the cavities that outweighs such advantages. In this subsection, we present schematic diagrams of three representative planar structures: the microdisc, ring resonator, and racetrack resonator, respectively in Figs. 10.9(a),(b), and (c).

The disc resonator has cylindrical shape of reduced height. Another cylindrical resonator of larger height, called a micropillar, has found use in nanophotonics work. Micropillar structures will be described in a later section of this chapter. Ring resonator is a kind of microdisk having a circular hole in the middle. Racetrack resonators are ring resonators elongated along one planar direction. The surface roughness introduced in fabrication process makes the quality factor of these structures quite low as compared to microsphere resonators. However, the planar structures, particularly the ring and racetrack resonators possess very low mode volume.

The three structures mentioned here can be fabricated by using lithography and etching, which are compatible with the complementary metal-oxide semiconductor (CMOS) fabrication technology. Deep ultraviolet (UV), electron beam or nano-imprinting lithography techniques are commonly employed. Each of these lithographic processes has its advantages and disadvantages.

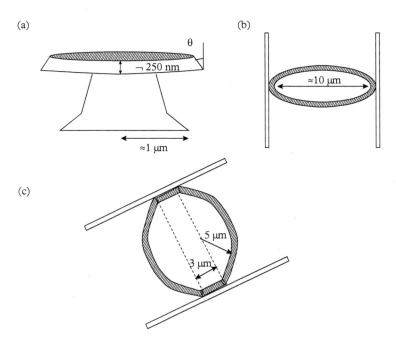

Figure 10.9 *Schematic diagrams of a few structures supporting WGM.* *(a) microdisk, (b) ring resonator, (c) race-track resonator.*

Structures supporting WGMs are made of different materials, including liquids, amorphous materials, and crystals using semiconductors. Details of materials used, methods of fabrication and performance parameters of the resonators may be found in references cited (Matsko and Ilchenko 2006 a, b; Benson et al 2006; Righini et al 2011; Tien et al 2011, for example).

10.7   Wave propagation in periodic structures: Photonic crystals

Propagation of an EM wave in a periodic dielectric structure had been the subject of study since Lord Rayleigh's observation as early as in 1887. He pointed out the existence of a reflection band in a one-dimensional periodic arrangement of two different kinds of dielectrics. The use of Bragg mirrors consisting of multiple layers of two dielectrics as an almost perfect mirror in lasers has been known over many decades. This is an early example of PhCs giving rise to Photonic Band Gaps (PBGs).

A PhC is a medium where periodic variation of dielectric function in one, two, and three dimensions gives rise to pass and stop bands for propagation of EM waves. The Bragg mirrors serve as an example of 1D PhC. However, the pass and stop bands do not appear for any arbitrary direction of wave propagation.

The modern concept of PhC has been introduced by Yablonovitch (1987) and John (1987) and elaborated by Johanapoulous et al (1995: see Reading list at the end of this chapter) amongst others. The structures they proposed are PhCs in the truest sense.

In the following subsections we shall develop the concepts of PhCs and formation of PBGs by first discussing the simplest 1D periodic arrangement of dielectrics, and then extending the concept to higher dimensions. It should be mentioned that the study of EM waves in PhCs requires solution of wave equation in three dimensions, which is normally accomplished by sophisticated numerical methods. We shall illustrate the concept of formation of PBGs by an analytical method first.

Although the name PhCs is used to specify the structures giving rise to PBGs, it is the propagation of EM waves using classical wave theory of radiation that forms the basis for the relevant studies. On the other hand, *photons* refer to quantum particles. Study of PhCs has nothing to do with the particulate nature of EM waves.

10.7.1   One-dimensional periodic array of dielectrics

As mentioned already, this is the simplest and oldest form of PhCs. The formation of bands in the structure can be understood by simple analytical methods.

Consider a periodic arrangement stacking two different dielectrics A and B alternately to form an infinite structure, as shown in Fig. 10.10. The dielectrics have different permittivities ε_1 and ε_2, respectively, and thicknesses a and b, respectively, so that the period of the lattice is $d = a + b$. The permittivities show a step-like variation and is periodic in nature and as such $\varepsilon(x + d) = \varepsilon(x)$. The wave equation for an electric field $E(x, t)$ is

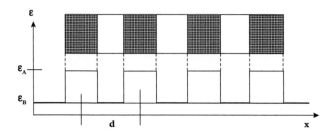

Figure 10.10 *A one-dimensional array of two dielectrics A and B stacked alternately and the corresponding variation of dielectric function.*

expressed as

$$\frac{\partial^2}{\partial x^2}E(x,t) + \frac{\varepsilon(x)}{c^2}\frac{\partial^2}{\partial t^2}E(x,t) = 0 \tag{10.58}$$

We express the electric field in the following form:

$$E(x,t) = E(x)exp(i\omega t) \tag{10.59}$$

to arrive at the Helmholtz equation

$$\frac{d^2}{dx^2}E(x) + \frac{\varepsilon(x)\omega^2}{c^2}E(x) = 0 \tag{10.60}$$

The solutions for the electric field in the periodic structure has the familiar Bloch form given below

$$E(x) = U(x)exp(ikx) \tag{10.61}$$

where k is the Bloch wavenumber, and the Bloch function is periodic with $U(x + d) = U(x)$. The boundary conditions to be satisfied are given by

$$\varepsilon(x) = \begin{cases} \varepsilon_1, nd \leq x < a + nd \\ \varepsilon_2, a + nd \leq x < (n+1)d \end{cases} \tag{10.62}$$

where n is an integer. The eigen functions in the two regions will be

$$E_1(x) = Aexp(ik_1x) + Bexp(-ik_1x)$$
$$E_2(x) = Cexp(ik_2x) + Dexp(-ik_2x) \tag{10.63}$$

where A, B, C, and D are unknown coefficients.

Applying now the boundary conditions that both the electric field and its derivative are continuous at the interfaces of the two dielectrics, one gets the following systems of equations:

$$
\begin{aligned}
A + B &= exp(-ikd)[C\ exp(ik_2d) + D\ exp(-ik_2d)] \\
k_2(A - B) &= k_2\ exp(-ikd)[C\ exp(ik_2d)D\ exp(-ik_2d)] \\
A\ exp(ik_1a) + B\ exp(-ika) &= A\ exp(ik_2a) + DB\ exp(-ik_2a) \\
k_1[A\ exp(ik_1a) - B\ exp(-ik_1a)] &= k_2[A\ exp(ik_2a) - B\ exp(-ik_2a)]
\end{aligned}
\tag{10.64}
$$

Eq. (10.64) may be expressed in the following matrix form:

$$
[M(k_1, k_2, k)] \cdot [C] = 0
\tag{10.65}
$$

The 4 × 4 matrix may be written as

$$
M(k_1, k_2, k) =
\begin{bmatrix}
1 & 1 & -exp[-id(k_2 - k)] & -exp[-id(k_2 + k)] \\
k_1 & -k_1 & -k_2exp[id(k_2 - k)] & k_2exp[-id(k_2 + k)] \\
exp(ik_1a) & exp(-ik_1a) & -exp(ik_2a) & -exp(-ik_2a) \\
k_1exp(ik_1a) & -k_1exp(-ik_1a) & -k_2exp(ik_2a) & k_2exp(-ik_1a)
\end{bmatrix}
\tag{10.66}
$$

The column vector $[C]$ has four elements A, B, C, and D.

In order to have non-trivial solutions, one must have det$[M] = 0$. Expanding the determinant and after some algebra, one arrives at the following implicit equation for $\omega(k)$:

$$
cos(k_1a)cos(k_2b) - \frac{1}{2}\frac{\varepsilon_1 + \varepsilon_2}{\sqrt{\varepsilon_1\varepsilon_2}}sin(k_1a)sin(k_2b) = cos(kd)
\tag{10.67}
$$

Example 10.14 *A plot of $f(\omega)$ versus ω is obtained by calculating $f(\omega)$ with the following values of parameters: $a = 0.2\ \mu m$, $\varepsilon_1 = 2.26$, $b = 0.15\ \mu m$ and $\varepsilon_2 = 11.7$. Since cos (kd) has ± 1, the limiting values are indicated by two horizontal lines.*

Fig. 10.11 shows the plots of $f(\omega)$ and cos (kd) versus ω. One can easily identify allowed and forbidden bands (shaded). The forbidden bands occur whenever the plot of $f(\omega)$ exceeds ± 1. Over this frequency range the EM waves cannot propagate through the crystal. The allowed bands occur when cos $(kd) \leq \pm 1$. Gaps (stop bands) appear between two successive allowed bands. All these features are also present in the Kronig–Penney model for crystalline solids.

The dispersion relation: $\omega(k)$ versus k obtained from Fig. 10.11 is plotted in Fig. 10.12. In Fig. 10.12(a) the dispersion relations for free space and for an isotropic medium, in which $\omega = ck/n$, are shown. In Fig. 10.12(b) the ω-k relation for positive

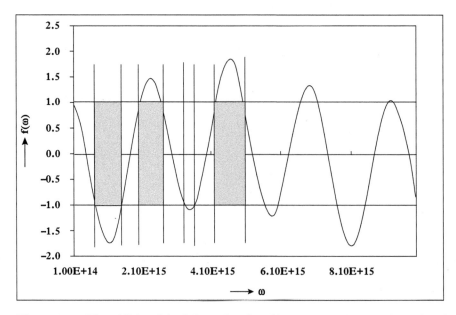

Figure 10.11 *Plots of f(ω) and Cos(kd) as a function of frequency ω to illustrate formation of PBG in 1D PhCs. The range of frequencies over which EM waves do not propagate is shown by shaded rectangles.*

values of k in the range $0 \leq k \leq 2\pi/d$.is shown. As in the case of Kronig–Penney model for electrons in a crystalline solid, there appear breaks in the curves at

$$k_N = \frac{N\pi}{d} \tag{10.68}$$

where N is an integer, both $+$ and $-$. In Fig. 10.12(b), the break appears at $k = \pi/d$. At the break points EM waves in the Bloch form cannot propagate in the medium and remains confined in one of the two layers forming the array.

Since all wave numbers differing by $\frac{2N\pi}{d}$ are equivalent, one can shift the uppermost curve to the left by $-2\pi/d$ and its mirror image in the range $-\frac{2\pi}{d} \leq k \leq 0$ by an amount $\frac{2\pi}{d}$ to the right and obtain the dispersion relation confined to the first Brillouin zone, as shown in Fig. 10.12(c). As usual, the first Brillouin zone is the interval $[-\pi/d, +\pi/d]$ and it contains all the non-equivalent values of k.

The dispersion curve at low values of $k(k \ll \pi/d)$, coincides with the curve for a homogeneous medium. For larger values of k the wave velocity $\frac{d\omega}{dk}$ reduces, and at the zone edge, the velocity reduces to zero. In this situation, the EM field concentrates either in the sublattice of higher RI, occurring at the bottom of the band gap with frequency ω_1, or in the sublattice with lower RI, at the top of the gap having frequency ω_2. The lower and upper bands are termed as dielectric band and air band, respectively.

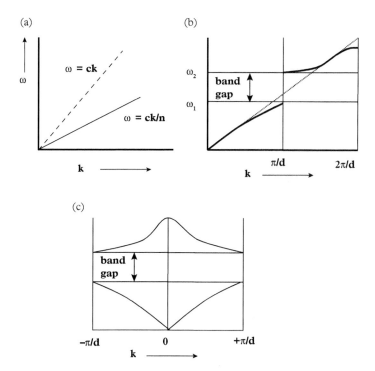

Figure 10.12 *Dispersion (ω Vs k) relation; (a) in free space and in a medium of refractive index n; (b) relation for 1D PhC in the extended zone; and (c) dispersion relation of 1D PhC in the first Brillouin zone.*

The properties of 1D PhC discussed previously is valid when light is incident normal to the layer plane. The treatment for oblique incidence may be found in Yariv and Yeh (2007). It is assumed that the optical wave propagated along the *yz* plane, and the general solution of wave equation can be written as

$$E(y, z, t) = E(z)exp[i(\omega t - k_y y)] \tag{10.69}$$

where k_y is the y-component of wavevector of propagation and is constant as the medium is homogeneous along the y-direction.

Two different modes are of interest: (1) the transverse electric (TE) mode, in which **E** vector is perpendicular to the *yz* plane, and (2) transverse magnetic (TM) mode, in which **E** is parallel to the *yz* plane. The waves are characterized by propagation wavenumbers along the *z*-direction in the two layers as

$$k_{1z} = \sqrt{(n_1\omega/c)^2 - k_y^2} \tag{10.70a}$$

$$k_{2z} = \sqrt{(n_2\omega/c)^2 - k_y^2} \tag{10.70b}$$

The dispersion relationships can be expressed as

$$cos(k_{1z}a)cos(k_{2z}b) - \frac{1}{2}\left(\frac{k_{2z}}{k_{1z}} + \frac{k_{1z}}{k_{2z}}\right) sin(k_{1z}a)sin(k_{2z}b) = cos(kd) \text{ (TE)} \quad (10.71a)$$

$$cos(k_{1z}a)cos(k_{2z}b) - \frac{1}{2}\left(\frac{n_2^2 k_{1z}}{n_1^2 k_{1z}} + \frac{n_1^2 k_{2z}}{n_2^2 k_{1z}}\right) sin(k_{1z}a)sin(k_{2z}b) = cos(kd) \text{ (TM)} \quad (10.71b)$$

10.7.2 Two-dimensional photonic crystals

Two-dimensional PhC microcavities have been and are being used in fundamental studies as well as in device applications. A few examples of practical 2D PhCs will be given
in a later section. In this section, we mention a few of the lattice structures and introduce some highly symmetric points in the reciprocal space. A simple 2D theory is then
presented to illustrate the formation of allowed and forbidden bands.

10.7.2.1 Lattice structure and symmetry points

For 2D PhCs, two lattice configurations, a square lattice, and a hexagonal lattice are
common. Fig. 10.13(a) shows the square lattice and Fig. 10.13(b) shows its reciprocal
lattice, whereas Figs. 10.13(c) and (d) show the hexagonal lattice and its reciprocal lattice, respectively. The high symmetry points in the reciprocal lattices are also indicated
in the figures and the point where $\mathbf{k} = 0$ defines the Γ point for both the lattices. For
square lattice, the points $\mathbf{k} = (\pi/a)\mathbf{x}$ and $\mathbf{k} = (\pi/a)\mathbf{x} + (\pi/a)\mathbf{y}$ are denoted by X and M,
respectively, $\hat{\mathbf{x}}$ and $\hat{\mathbf{y}}$, are the unit vectors in the respective directions. For a hexagonal
lattice, the high symmetry points are $M[\mathbf{k} = (\pi/a)\mathbf{x} + (\pi/a\sqrt{3})\mathbf{y}]$ and K $[\mathbf{k} = (4\pi/3a)\mathbf{x}]$,
where the $\mathbf{k}$-values are given in the brackets.

10.7.2.2 Photonic band gap

As already mentioned, PhCs possess band gap. The simplest PhC, the 1D periodic stack
of two dielectrics like GaAs and aluminum arsenide (AlAs) arranged alternately, has
been shown in Fig. 10.10. The reflectivity spectra shown in Fig. 10.4 show the existence
of a photonic band gap. Light containing wavelengths in the range of high reflectivity
cannot propagate through the crystal. However, the band gap exists only for particular
direction of propagation.

    2D PhCs have a band gap over a wider range. Examples of 2D PhC using cylindrical
nanorods and air holes in a dielectric like Si are given in Fig. 1.6. Just like semiconductors,
the DOS of EM waves is zero in the PBG.

10.7.2.3 Photonic crystal waveguides and resonators

    Defects, like impurities, in semiconductors create localized states, meaning thereby
that the wave function is non-propagating. A similar concept can be introduced in PhCs.

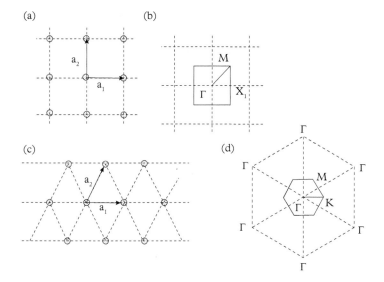

Figure 10.13 *(a) Square lattice and (b) its reciprocal lattice; (c) hexagonal lattice and (d) its reciprocal lattice.*

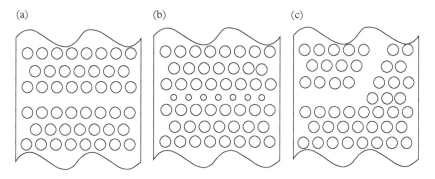

Figure 10.14 *Schematic diagram of 2D PhC Waveguides; (a) simple waveguide having one row of holes missing; (b) linear waveguide with smaller holes; (c) bent waveguide.*

Defects introduce a localized perturbation in the periodic arrangement. Linear perturbation creates a waveguide. Fig. 10.14 illustrates the formation of waveguide. In Fig. 10.14(a) a row of air holes is missing to create a waveguide. One may create smaller holes in a row as shown in Fig. 10.14(b) to have waveguiding. Fig. 10.14(c) shows a waveguide with sharp bend by suitably manipulating the creation of air holes.

Cavities or resonators can be formed by missing holes within the PhC, as illustrated by Fig. 10.15. In (a) a cavity is formed by not creating seven holes in a row. In (b) a

(a)   (b)   (c)

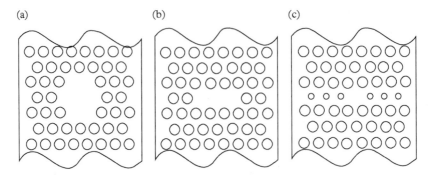

Figure 10.15 *Schematic diagram of 2D PhC Resonators; in (a) a cavity has been formed by removing seven holes; (b) cavity formed by shifting edge holes (c) birefringent cavity by one missing hole and a row of smaller holes.*

cavity is shown to be formed by shifting holes in a row. In (c) a row of smaller holes is created but one hole is missing.

10.7.2.4 *Dispersion relations*

It is quite difficult to find analytical or semi-analytical method to obtain dispersion relation for 2D PhCs. Yariv and Yeh (2007) have presented a semi-analytic method to demonstrate occurrence of PBG in a 2D crystal. In most of the cases, however, numerical technique is employed to obtain the ω-k relationship for different crystallographic directions. The dispersion relation is different for different crystal structure.

The computed dispersion relation for a hexagonal lattice made of Si pillars and air holes has been shown in Fig. 1.7 (Birner et al 2001). An example for a 2D triangular PhC for different crystallographic directions may be found in the work by Boroditsky et al (1999). The crystal is formed by having voids arranged periodically in an InGaAs/InP double heterostructure thin film grown on glass. The film thickness, radius of the circular holes, and spacing between holes are fixed. See also John (1987) and Joannopoulos et al (1995) for further discussions.

10.7.3   Three-dimensional photonic crystals

One-dimensional and 2D PhCs can be fabricated by using epitaxial techniques and lithography and etching. The 2D PhCs illustrated earlier are fabricated by creating air holes in a thin slab of semiconductors. Creation of 3D PhCs by similar techniques is rather cumbersome. An example of 3D PhC, the woodpile structure, is shown schematically in Fig. 10.16. Evidently, each layer is patterned and etched to form the periodic array. The process is followed by growth of another layer and patterning and etching. The number of such layers is however limited. Woodpile structures have been growth using GaAs by Noda et al (2000) and in silicon (Si) by de Dood et al (2003) and the characteristics of band structure are discussed by the authors.

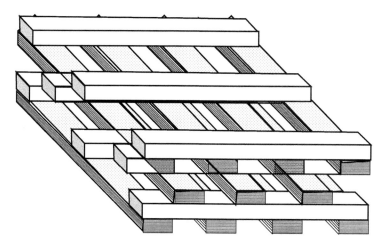

Figure 10.16 *Schematic diagram of a woodpile structure.*

10.8 Micropillar

An important microcavity structure is a micropillar as mentioned in Table 10.1. Several authors (Reitzenstein et al 2007; Reitzenstein and Forchel 2010; Arnold et al 2012; Astratov et al 2007; Jones et al 2010; Gazzano et al 2013) have used this micropillar structure for the study of CQED phenomena and for achieving single photon sources (Gregerson et al 2017).

A schematic diagram of the structure is shown in Fig. 10.17. It is a vertical structure which has two distributed Bragg reflector (DBR) stacks at the upper and lower end. The two DBRs enclose a GaAs cavity in which InAs QWs or in most cases InAs QDs are embedded. The two DBRs confine the optical modes in the vertical direction, whereas almost total internal reflection (ATIR) is responsible to confine the EM waves in the horizontal plane (WGMs). The emitted light is collected by an objective lens placed above the top mirror. It may be mentioned that an almost similar structure with contact layers at the top and below the GaAs substrate is used for VCSELs (see Fig. 11.1 of Basu et al 2015).

The top (bottom) DBR consists of .$N(M)$ pairs of AlAs/GaAs layers. The thickness of DBR layers should be $\lambda_0/(4\eta_{eff})$ and that of the cavity should be λ_0/η_{eff}, where η_{eff} is the effective RI of the fundamental mode, and λ_0 is the chosen wavelength. This choice of thicknesses ensures that the resonance occurs at the DBR stopband for maximum reflection. The number of pairs N or M is quite high to ensure very high reflectivity, particularly for the bottom DBR, as maximum light is to be extracted out of the top mirror. In most cases, N or $M > 20/25$ is common.

In order to obtain maximum transmission efficiency, the reflectivity of the bottom DBR should be maximized by choosing the number of pairs. The choice of pillar

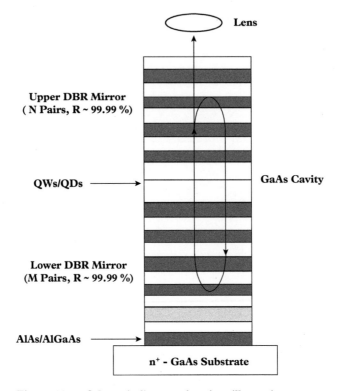

Figure 10.17 *Schematic diagram of a micropillar cavity.*

diameter comes next. It is found that the transmission is low for smaller pillar diameter but reaches almost the ideal unity value for diameter above ~ 2 μm.

Ideally, the structure emits light in the vertical direction. However, in fabricated structures, sidewall imperfections cause scattering of light in the plane. This reduces the quality factor Q. For larger diameter, the overlap between the mode profile and imperfections decreases and Q factor increases thereby, at the cost of higher mode volume, however.

The variation of transmission efficiency with reflectivities of DBR mirrors, effects of pillar diameter on Q values and overall transmission efficiency and optimized design of micropillar have been discussed by Gazzano et al (2013).

10.9   Microcavity

In earlier sections, general characteristics of optical resonators have been introduced followed by a description of few of the resonator structures and principle of light confinement in these structures.

We now discuss some important characteristics of the microcavity resonators used in nanophotonics work. The areas include devices, particularly, lasers, filters, switches,

Table 10.1 *Types of optical resonators and their characteristics.*

Type	Confinement/ dominant mode	Q value	Mode volume	Ease of integration	References
Microsphere	ATIR/WGMs	Ultra-high $Q \sim 10^7 - 10^{10}$;	large mode volume; 3000 μm^3	on-chip integration difficult	Lefevre-Seguin and Haroche (1997)
Microdisk (microring)	ATIR WGMs	High Q $10^4 - 10^5$;	smaller mode volume; $6(\lambda/\eta)^3$	suitable for planar integration	Baba T et al (1997)
Micropillar	DBR in vertical, ATIR in horizontal; FP oscillation	high Q $\sim$ 2.15 x 10^6;	Small mode volume; $5(\lambda/\eta)^3$	easy coupling to fibres	Arnold et al, (2012)
Microtorus	ATIR/WGMs	large Q;	Smaller mode volume than microsphere; $(\lambda/\eta)^3$	suitable for on-chip integration	Armani D K et al (2003)
Racetrack	ATIR/WGMs	Relatively low Q; $\sim$ 850–1500		suitable for on-chip integration	Gmachl et al (1998)
PhC	ATIR in vertical; DBR in horizontal	High Q $\sim$ 10^4	Smallest mode volume 1.2 $(\lambda/\eta)^3$	suitable for on-chip integration	Painter O J et al (1999)

ATIR: almost Total Internal Reflection; Mode volumes are quoted in Vahala (2003).

WDM components, and basic processes like CQED, spontaneous emission control, nonlinear optics, and so on.

We first present a summary of different resonator structures, method of light confinement, order-of-magnitude values of Q, and some sources for the work. Table 10.1 gives the summary, which is principally based on the work by Benson et al (2006) and Vahala (2003).

Several material systems are employed to fabricate microresonators. The key materials are silica, silica on silicon, silicon on insulator, silicon nitride, and oxynitrides, semiconductors such as GaAs, InP, InGaAsP, GaN etc., and crystalline materials like lithium niobate and calcium fluoride are also used. In addition, resonators are also formed by using different polymer materials (see Benson et al 2006 and references quoted for further details).

Problems

10.1 *Prove that the ratio of the transmitted to the incident intensities in a FP resonator filled with a lossy material with loss coefficient α, may be expressed as*

$$\frac{I_t}{I_i} = \frac{(1-R)^2 e^{-\alpha L}}{(1 - Re^{-\alpha L})^2 + 4Re^{-\alpha L}\sin^2\theta}$$

where $\theta = (2\pi f \eta_g/c)L\theta$ *and the reflectivities of the two mirrors are assumed equal.*

10.2 *Express the loss in terms of the ratio of maximum to minimum transmitted intensities.*

10.3 *Show that the expression for transmittivity of FP resonator, Eq. (10.25), reduces to* $T = \dfrac{T_{max}}{1+\left(\frac{4\eta L}{\pi \Delta \lambda}\right)^2 \sin^2\left(\frac{2\pi\eta L}{\lambda}\right)}$ *where* $\Delta\lambda$ *is the FWHM of the transmission at resonance.*

10.4 *Prove that if the transmission of a FP resonator is represented by a Lorentzian, then the response is* $g(f) = [(f-f_0)^2] + (f_0/2Q)^2]^{-1}$, *where f is the frequency,* f_0 *is the resonant frequency and Q is the quality factor of the resonator.*

10.5 *Obtain an expression for photon lifetime in a FP resonator enclosing a medium with loss = α.*

10.6 *Draw different rays as shown in Fig. 10.1 showing the field intensities coming out of mirror 2. The FP resonator however encloses a medium of gain = g and loss coefficient= α. By summing the field intensities, obtain the condition when* $E_t/E_i \rightarrow \infty$. *From this prove that self-sustained laser oscillation condition is* $g = \alpha + (1/2L)\ln(1/R_1 R_2)$

10.7 *Two dielectrics of RIs* η_1 *and* η_2 *are alternately grown. Prove that when each layer is quarter wavelength thick, the stack can be used as a mirror of high reflectivity.*

10.8 *A dielectric mirror using alternate layers of Si and* SiO_2 *are used to form a resonant cavity for a light emitting device at 850 nm. Calculate the thicknesses of each layer.* $[n_{SiO2}= 1.5$ *and* $n_{Si} = 3.4]$.

10.9 *Show that the quality factor for material absorption may be expressed as* $Q_{abs} = 2\pi\eta/\alpha\lambda_0$ *where* η *is the refractive index and* α *is the intensity attenuation coefficient.*

10.10 *Absorption of sapphire due to imperfection of the crystalline structure is* $\alpha = 10.3 \times 10^{-5} cm^{-1}$ *at* $\lambda = 1\mu m$. *Calculate the Q assuming RI = 1.75. Calculate also the Q of crystalline quartz should be better than absorption in fused silica, with* $\alpha \leq 5 \times 10{-6}$ *cm−1 at* $\lambda = 1.55\mu m$. *Assume RI = 1.5.*

10.11 *A FP etalon is to be designed to resolve different longitudinal modes of a laser. The length is 100 cm, and the gain spectrum covers* $10^9 Hz$. *The linewidth of each mode is 0.1 times the longitudinal mode separation. Assume the RI of the laser medium to be 1. Calculate the finesse and reflectivity of the mirrors, assumed to be the same for both.*

References

Armani, D K, Kippenberg T J, Spillane S M, and Vahala, K J (2003) Ultra-high-Q toroid microcavity on a chip. *Nature* 421: 905–908.

Arnold C, Loo V, Lema^ître A, Sagnes I, Krebs O, Voisin P, Senellart P, and Lanco L (2012) Optical bi-stability in a quantum dots/micropillar device with a quality factor exceeding 200 000. *Applied Physics Letters* 100(111111): 1–4.

Astratov V N, Yang S, Lam S, Jones B D, Sanvitto D, Whittaker D M, Fox A M, Skolnick M S, Tahraoui A, Fry P W, and Hopkinson M (2007) Whispering gallery resonances in semiconductor micropillars. *Applied Physics Letters* 91(071115):1–4.

Baba T, Fujita M, Sakai A, Kihara M, and Watanabe R (1997) Lasing characteristics of GaInAsP/InP strained quantum-well microdisk injection lasers with diameter of 2-10 μm, *IEEE Photon. Technological Letters* 9(7): 878–880.

Basu P K, Mukhopadhyay B, and Basu R (2015) *Semiconductor Laser Theory*. Boca Raton, FL, USA: CRC Press (Taylor & Francis).

Benson T M, Boriskina S V, Sewell P, Vukovic A, Greedy S C, and Nosich A I (2006) Micro-optical resonators for microlasers and integrated optoelectronics: Recent advances and future challenges. In: S Janz et al (eds.) *Frontiers in Planar Lightwave Circuit Technology*, pp. 39–70. Berlin: Springer.

Birner A, Wehrspohn R B, Gösele U M, and Busch K (2001) Silicon based photonic crystals. *Advanced Materials* 13(6): 377–388.

Boroditsky M, Vrijen R, Krauss T F, Coccioli R, Bhat R, and Yablonovitch E (1999) Spontaneous emission extraction and Purcell enhancement from thin-film 2-D photonic crystals. *Journal of Lightwave Technology* 17(11): 2096–2112.

Bose S N (1924) Planck's Gesetz and Lichquantumhypothese. *Zeitschrift fur Physik* 26: 178–181.

Chang R K and Campillo A J (eds.) (1996) *Optical Processes in Microcavities*. Singapore: World Scientific.

De Dood M J A, Gralak B, Polman A, and Fleming J G (2003) Superstructures and finite size effects in a Si photonic woodpile structure. *Physical Review B* 67(035322): 1–10.

Deen M J and Basu P K (2012) *Silicon Photonics*. Chichester, UK: John Wiley.

Fabry C and Perot A (1899) Theorie et applications d'une nouvelle methode de spectroscopie interferentielle. *Annals de Chimie et de Physique* 16(7): 115–144.

Gaponenko S V (2010) *Introduction to Nanophotonics*, Chapter 7. Cambridge: Cambridge University Press.

Gazzano O, Michaelis De Vasconcellos S, Arnold C, Nowak A, Galopin E, Sagnes I, Lanco L, Lemaitre A, and Senellart P (2013) Bright solid-state sources of indistinguishable single photons. *Nature Communications* 4: 14–25.

Gmachl C, Cappasso F, Narimanov E, Nockel J U, Stone A D, Faist J, Sivco D L, and Cho A Y (1998) High-power directional emission from microlasers with chaotic resonances. *Science* 280: 1556–1564.

Jeans J H X I (1905) On the partition of energy between matter and Æther. *Philosophical Magazine* 10(91): 91–98.

Joannopoulos J D, Meade R D, and Winn J N (1995) *Photonic Crystals: Molding the Flow of Light.* Princeton: Princeton University Press.

John S (1987) Strong localization of photons in certain disordered dielectric solids. *Physical Review Letters* 58: 2486–2489.

Jones B D, Oxborrow M, Astratov V N, Hopkinson M, Tahraoui A, Skolnick M S, and Fox A M (2010) Splitting and lasing of whispering gallery modes in quantum dot micropillars. *Optics Express* 18(21): 22578–22592.

Lefevre-Seguin,V., and Haroche,S., (1997) Towards cavity-QED experiments with silica microspheres,. *Materials Science and Engineering* B48:53–58.

Matsko A B and Ilchenko V S (2006a) Optical resonators with whispering-gallery modes-part I: Basics. *IEEE Journal of Selected Topics in Quantum Electronics* 12:3–14 .

Matsko A B and Ilchenko V S (2006b) Optical resonators with whispering-gallery modes-part II: Applications. *IEEE Journal of Selected Topics in Quantum Electronics* 12:15–32.

Noda S, Tomoda K, Yamamoto N, and Chutinan A (2000) Full three-dimensional photonic crystals at near-infrared wavelengths. *Science* 289: 604–606.

Painter O J, Husain A, Scherer A, O'Brien J D, Kim I, and Dapkus P D (1999) Room temperature photonic crystal defect lasers at near-infrared wavelengths in InGaAsP. *Journal of Lightwave Technology* 17(11): 2082–2088.

Perot A and Fabry C (1899) On the application of interference phenomena to the solution of various problems of spectroscopy and metrology. *Astrophysical Journal* 9: 87. DOI:10.1086/140557.

Rayleigh J W S (1900) Remarks upon the law of complete radiation. *Philosophical Magazine* 49: 539–540.

Rayleigh L (1910) The problem of the whispering gallery. *Philosophical Magazine* 20: 1001–1004.

Reitzenstein S and Forchel A (2010) Quantum dot micropillars. *Journal of Physics D, Applied Physics* 43(3): 1–25.

Reitzenstein S, Hofmann C, Gorbunov A, Strauß M, Kwon S H, Schneider C, Löffler A, Höfling S, Kamp M, and Forchel A (2007) AlAs/GaAsmicropillar cavities with quality factors exceeding 150,000. *Applied Physical Letters* 90(251109): 1–4.

Righini G C, Dumeige Y, Feron P, Ferrari M, Nunzi Conti G, Ristic D, and Soria S (2011) Whispering gallery mode microresonators: Fundamentals and applications. *Rivita Del Nuovo Cimento* 34(7): 435–488.

Tien M-C, Bauters J F, Martijn Heck J R, Spencer D T, Blumenthal D J, and Bowers J E, (2011) Ultra-high-quality factor planar Si3N4 ring resonators on Si substrates. *Optics Express* 19(14): 13551–13556.

Vahala K J (2003) Optical microcavities. *Nature* 424: 8–39.

Verdeyen J T (1995) *Laser Electronics*, 3rd edn. NJ: Prentice Hall.

Vernooy Ilchenko V S, Mabuchi H, Streed E W, and Kimble H J (1998) High-Q measurements of fused-silica microspheres in the near infrared. *Optics Letters* 23: 247–249.

Yablonovitch E (1987) Inhibited spontaneous emission in solid state physics and electronics. *Physical Review Letters* 58: 2059–2062.

Yamamoto Y and Slusher R E (1993) Optical processes in microcavities. *Physics Today* 46(6): 66–73.

Yariv A and Yeh P (2007) *Photonics: Optical Electronics in Modern Communications.* New York: Oxford University Press.

Yokoyama H (1992) Physics and device applications of optical microcavities. *Science* 2 56: 66–70.

Yokoyama H and Ujihara K (eds.) (1995) *Spontaneous Emission and Laser Oscillation in Microcavities.* New York: CRC Press.

Zhou W, Qiang Z, and Soref R A (2010) Photonic crystal ring resonators and ring resonator circuits. In: *Photonic Microresonator Research and Applications*, I. Chremmos et al. (eds.) Chapter 13, p. 156. Berlin: Springer Series in Optical Sciences. DOI 10.1007/978-1-4419-1744-7_13.

11

Cavity quantum electrodynamics

11.1 Introduction

In Chapter 3, three basic processes of light-matter interaction, as originally proposed by Einstein, were introduced. Spontaneous emission was identified there as a distinct process. It was also indicated that any system at its excited state decays to the ground state spontaneously with a decay time inversely proportional to the Einstein's A-coefficient. Thus, the spontaneous emission represents an irreversible process and there is no way by which the atom in its ground state will occupy an excited state without any stimulus. The rate of spontaneous emission was also derived in Chapter 4 by using time-dependent perturbation theory and considering the density-of-states (DOS) for photons in free space.

The concept that the spontaneous emission of an atom is an *immutable* property of the atom was first challenged by E M Purcell (1946). In his classic paper (essentially an abstract consisting of a few lines) in the *Physical Review*, he pointed out that spontaneous emission rate can be enhanced by several orders in a resonant structure. The abstract is now reproduced.

B 10. *Spontaneous Emission Probabilities at Radio Frequencies.* **E M Purcell, Harvard University Press.**

For nuclear magnetic moment transitions at radio frequencies, the probability of spontaneous emission, computed from

$$A\nu = (8\pi\nu^2/c^3)h\nu(8\pi^3\mu^2/3h^2)\sec.^{-1}$$

is so small that this process is not effective in bringing a spin system into thermal equilibrium with its surroundings. At 300° K, for $\nu = 10^7 \sec.^{-1}$, $\mu = 1$ nuclear magneton, the corresponding relaxation time would be 5×10^{21} seconds! However, for a system coupled to a resonant electrical circuit, the factor $8\pi\nu^2/c^3$ no longer correctly gives the number of radiation oscillators per unit volume, in unit frequency range, there being now *one* oscillator in the frequency range ν/Q associated with the circuit. The spontaneous emission probability is therefore increased, and the relaxation time reduced, by a factor $f = 3Q\lambda^3/4\pi^2 V$, where V is the volume of the resonator. If a is a dimension characteristic of the circuit so that $V \sim a^3$, and if δ is the skin-depth at frequency ν, $f \sim \lambda^3/a\delta$. For a

Semiconductor Nanophotonics. Prasanta Kumar Basu, Bratati Mukhopadhyay, and Rikmantra Basu, Oxford University Press.
© P.K. Basu, B. Mukhopadhyay, R. Basu (2022). DOI: 10.1093/oso/9780198784692.003.0011

non-resonant circuit, and for $a < \delta$, it can be shown that $f \sim \lambda^3/a\delta^2$. If small metallic particles, of diameter 10^{-3} cm are mixed with a nuclear-magnetic medium at room temperature, spontaneous emission should establish thermal equilibrium in a time of the order of minutes, for $\nu = 10^7$ sec.$^{-1}$.

Although Purcell pointed out enhancement of spontaneous emission rate at radio frequencies, the idea is quite general, and hence applies to the optical frequency range also. In this frequency range, the resonator is in the form of an optical cavity. Inhibition of spontaneous emission is also possible, provided the frequency of the oscillator or emitter is detuned from the resonant frequency of the cavity. The enhancement or inhibition of spontaneous emission rate is determined by the number of available photon modes around the emission frequency. If the cavity structure does not allow free propagation of certain modes, then the spontaneous emission decay will get slowed down. If, on the other hand, certain modes around the resonant frequency are accentuated, the decay rate will be enhanced. If no modes are allowed, there will be no spontaneous emission.

The essential element for observing Purcell effect is a resonant cavity. In fact, control of spontaneous emission rate has been observed for Rydberg atoms placed in a parallel plate metallic cavity (see Haroche and Kleppner 1989 and references) and also in semiconductor lasers, the active layer of which is enclosed by a cavity formed by Bragg mirrors (Yamamoto et al 1991). The Vertical Cavity Surface Emitting Laser (VCSEL) is particularly useful in observing control of spontaneous emission (Michalzak and Ebeling 2003; Basu et al 2015).

The use of high Q cavity also facilitates observation of many interesting phenomena of light-matter interaction (Vahala 2003). For example, spontaneous emission, supposed to be an irreversible process, shows reversibility. In a cavity with very high Q factor, the emitted radiation takes a very long time to decay, so that it can be reabsorbed to be promoted again to the excited state. In this way, the spontaneous emission exhibits a reversible nature. This deexcitation followed by reexcitation is referred to as vacuum *Rabi oscillation* (Savona et al 1995; Gibbs et al 2006).

The previously mentioned physical phenomena are the outcome of light-matter interaction. In some cases, the coupling of light and matter is weak so that perturbative approach may be employed to describe the phenomena. In some other cases, the interaction is strong and treatments beyond perturbative approach are needed. In between the weak and strong coupling regimes, there exists an intermediate coupling range showing interesting phenomena. The observed phenomena and their inherent mechanisms form the content of a subject called Cavity Quantum Electrodynamics (CQED).

It is evident that resonant cavities play an important role in this context. In addition to exhibiting interesting phenomena related to light-matter interaction (see review articles and books by Fisher et al 1998; Cao et al 2000; Reitzenstein 2012; Christensen et al 2015), cavities are needed in modern photonic devices. These include Fabry–Perot (FP) resonator in diode lasers, microdisks (Dupuis et al 1999); and Bragg resonators in VCSELs (Basu et al 2015), high quantum cavities in the form of photonic crystals (PhCs) (Costard et al 1998; Boroditsky et al 1999; Deppe et al 2004; Arakawa et al 2012; Huisman et al 2011); and microring resonators (Vahala 2003; Benson et al

2006; Gibbs et al 2006); and other micro/nanocavity systems (Arakawa et al 1992; Albert et al 2011; Arakawa et al 2010; Bayer et al 2001; Pelton et al 2001; Forchel et al 2004). An important benefit accrued from the study of CQED is the realization of very low threshold or even thresholdless lasers (Bjork and Yamamoto 1991; Yamamoto et al 1991), and nano LEDs and lasers with high modulation bandwidth and narrow spectral linewidth (Lau et al 2009; Gregersen et al 2010). A single photon laser is another distinct possibility (Gregersen et al 2017). The cavities have dimensions on the order of few wavelengths to subwavelength and are called microcavities or nanocavities. These cavities provide control of the photon mode densities and dispersion. In conjunction with quantum nanostructures, like Quantum Wells (QW)s or Quantum Dots (QD)s, it is possible to simultaneously control the degrees of freedom of electrons and photons. This allows manipulation of light-matter interaction leading to observation of interesting physical phenomena and to achieve advanced devices.

11.2   Zero-point energy and vacuum field

The Quantum Theory of Radiation, which has been introduced in Chapter 4, views electromagnetic (EM) radiation as an infinite set of harmonic oscillators. Each such oscillator, or a radiation mode, has a frequency ω_k, and wave vector $\mathbf{k}$ and possesses an energy expressed as

$$E(\omega_k) = \hbar\omega_k \left(n_k + \frac{1}{2} \right), n_k = 0, 1, 2, 3, \ldots . \qquad (11.1)$$

where n_k is called the photon number and $\hbar\omega_k$ is the photon energy. Each mode is characterized by its polarization state, and in addition, by the wave vector $\mathbf{k}$.

Eq. (11.1) also indicates that the photon mode has, in addition to a total energy $n_k\hbar\omega_k$ proportional to its number, a background energy $(1/2)\hbar\omega_k$, termed as zero-point energy.

Zero-point energy represents the energy of a state in which no photon exists, that is, $n_k = 0$. This state refers to an *electromagnetic vacuum*. Considering all the modes the total energy associated with an EM vacuum is calculated as

$$E_0 = \frac{1}{2}\hbar \sum_k \omega_k = \frac{1}{2} \int_0^\infty \hbar\omega \frac{\omega^2}{\pi^2 c^3} d\omega = \infty \qquad (11.2)$$

In the previous equation, the summation has been replaced by integration of continuous spectrum of EM waves in free space by including the density-of-states (DOS) function.

It follows from previous statements that, though the energy associated with the lowest state of an oscillator is finite and equals $(1/2)\hbar\omega$, the presence of infinite number of oscillators makes the total energy of EM vacuum infinite. However, the energy density, expressed as the energy per unit volume and per unit frequency interval, is finite and is given by

$$W(\omega) = \frac{1}{2}\hbar\omega\frac{\omega^2}{\pi^2 c^3}$$ (11.3)

The zero-point energy leads to the concept of a *vacuum electric field fluctuation* expressed as

$$|E_{vac}|^2 = \frac{1}{2}\hbar\omega\frac{1}{\varepsilon_r\varepsilon_0 V}$$ (11.4)

Here it is assumed that the EM mode is present in a cavity of volume V formed by a material with relative permittivity ε_r.

The method of calculating the rates of absorption and of emission for a two-level atomic system has been given in Chapter 4 by using Fermi Golden Rule. It was shown (see Eq. (4.71)) that the rate of transition from an upper level (u) to a lower level (l) is expressed as

$$W_{em} = \frac{\pi e^2 \omega_k}{V\varepsilon}|\langle l|\varepsilon_q.r|u\rangle|^2(n_k + 1)\delta(E_u - E_l - \hbar\omega_k)$$ (11.5)

In this expression, the term involving n_k is due to stimulated emission induced by a photon with number n_k, while the term containing 1 represents the rate of spontaneous emission. It is customary to consider spontaneous emission as a special case of stimulated emission, the stimulus coming from the vacuum field fluctuations.

Example 11.1 *The magnitude of E_{vac} is calculated from Eq. (11.4), by assuming $V = (\lambda/\eta)^3$ and taking $\eta^2 = \varepsilon_r$. Eq. (11.4) reduces to $|E_{vac}|^2 = \frac{\pi\hbar c}{\lambda^4\varepsilon_0}$. For $\lambda=1.55$ μm, $E_{vac} \approx 2.5 \times 10^4$ V/m.*

11.3   Control of spontaneous emission

Section 11.1, the introduction of this chapter, has already stated Purcell's original idea that the rate of spontaneous emission can be altered when the emitter is placed in a cavity and certain conditions are fulfilled. The mode and lifetime of the spontaneous emission can be modified by controlling the vacuum field within an optical cavity. This is referred to as Cavity Quantum Electrodynamics (CQED). In this section, we shall provide a derivation of the Purcell factor, and then discuss how the density of modes is changed in a cavity from the free space value and finally mention some experimental findings to observe Purcell enhancement of spontaneous emission rate.

11.3.1   Derivation of Purcell factor

In the original article by Purcell only the expression for the Purcell factor f is given. This factor signifies to what extent the rate of spontaneous emission is altered from its value in free space when the emitter is placed in a cavity. The derivation has been given in the paper by Oraevskii (1994) (see also Gaponenko 2010). In the following, we shall present the derivation given in the book by Basu et al (2015) that follows closely the steps outlined in the book by Numai (2004). A two-level atomic system is considered for this purpose.

The two-level system possesses the ground (g) state and an excited state (x), as shown in Fig. 11.1, with energies E_g and E_x, respectively. The energy of the emitted photon is given by $\hbar\omega_0 = E_x - E_g$, with spectral width $\Delta\omega_0$. The spectral width of a resonant mode of the cavity is $\Delta\omega_c$ and, as shown in Fig. 11.1, $\Delta\omega_0 \ll \Delta\omega_c$ or $\tau_c \ll \tau_0$, where τ denotes the lifetime. Because of its very low lifetime, the emitted light readily comes out of the optical cavity and is not reabsorbed.

The rate of emission due to transition from the excited state to the ground state is calculated by using time-dependent perturbation theory and is expressed as

$$W = \frac{2\pi}{\hbar} \sum_i d_i^2 \left(\frac{\hbar\omega_i}{2V\varepsilon} \right) \sin^2(k_i.r)(n_i + 1)L(E_x - E_g - \hbar\omega_i) \qquad (11.6)$$

where d_i is the electric dipole moment for the ith cavity mode, having angular frequency ω_i and wave vector k_i. The optical cavity has a volume V and the material in the cavity has a permittivity ε, n_i is the photon density, and $L(x)$ is a lineshape function.

The rate for spontaneous transition is expressed by putting $n_i = 0$ (absence of a photon) in Eq. (11.6) as

$$W_{sp} = \frac{2\pi}{\hbar} \sum_i d_i^2 \left(\frac{\hbar\omega}{2V\varepsilon} \right) \sin^2(k_i.r)L(E_x - E_g - \hbar\omega_i) \qquad (11.7)$$

A free space is regarded as an optical cavity with dimensions quite large compared to the wavelength of light and the modes are so close that the dispersion is quasi-continuous (see discussions in Chapter 4). The spontaneous emission rate in free space is obtained

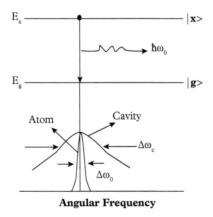

Figure 11.1 *Two level atomic system; emission spectra.*

by replacing the sum over i in Eq. (11.7) by integration with respect to k and it takes the form

$$W_{sp,free} = \frac{\omega_0^3 \eta_r^3}{3\pi c^3 \hbar\varepsilon} \langle d \rangle^2 \tag{11.8}$$

where $\langle d \rangle^2$ is the squared electric dipole moment averaged over all directions of polarization and $sin^2(k_i.r)$ is replaced by its average value ½.

The rate of transition into each cavity mode is proportional to $(1/V)$ and is extremely small. As the number of modes in free space is proportional to V, the spontaneous emission rate in free space does not depend on the volume of the cavity. Furthermore, the spontaneous emission rate does not depend on the position of the atom and is therefore governed by the properties of the atoms themselves.

In a microcavity, the size is of the order of the wavelength giving rise to a small number of discrete resonant modes. Considering that the spontaneous emission couples to the discrete i-th mode in the microcavity, its rate is given by

$$W_{sp,mc} = \frac{\pi}{\hbar} d_i^2 \left(\frac{\hbar\omega_i}{V\varepsilon} \right) \left(\frac{1}{\hbar\Delta\omega_c} \right) sin^2(k_i.r) \tag{11.9}$$

where $\Delta\omega_c$ is the spectral linewidth of the resonant mode of the optical cavity.

Assume now that the emission frequency is resonant with the frequency of the i-th cavity mode, $\omega_0 = \omega_i$. The ratio of the spontaneous emission rates in a microcavity and in free space may easily be expressed from Eqs. (11.8) and (11.9) as

$$\frac{W_{sp,mc}}{W_{sp,free}} = \frac{3}{8\pi} \frac{\lambda_i^3}{\eta_r^3} \frac{1}{V} \left(\frac{\omega_i}{\Delta\omega_c} \right) sin^2(k_i.r) \tag{11.10}$$

The conditions for observing enhanced spontaneous emission rate in a microcavity may now be listed from Eq. (11.10): (1) the size of the cavity should be comparable to the light wavelength, i.e. $V \approx \lambda_i^3/\eta_r^3$; (2) the Q-value of the cavity, given by $Q = \omega_i/\Delta\omega_c$, should be extremely large, and (3) the excited atoms should be placed at the antinode of the standing waves in the optical cavity, so that $sin^2(k_i.r) = 1$.

In contrast, the spontaneous emission will be inhibited if the light frequency and the frequency of the cavity mode are different ($\omega_0 \neq \omega_i$), and the excited atoms are placed at the nodes of the standing wave pattern $[sin^2(k_i.r) = 0]$.

Eq. (11.10) may be rewritten in the following form to give the expression for the Purcell factor:

$$f = \frac{3\lambda^3 Q}{8\pi\eta_r^3 V} \tag{11.11}$$

assuming $sin^2(k_i.r) = 1$. Eq. (11.11) needs a multiplying factor of $2\eta_r^3/\pi$ in order to agree with the expression derived by Purcell.

Example 11.2 *We consider a micropillar having Q = 10^6. Let us take the mode volume V = 5(λ/η)³. The value of f calculated from Eq. (11.11) is 2.3×10^4.*

The expression for rate of spontaneous emission when the atomic transition frequency ω_0 differs from the cavity resonance frequency ω_c has been derived by Bunkin and Oraevskii (1959) and is as follows:

$$W_{sp,mc} = \frac{4\pi d_i^2}{\hbar V} \frac{Q\omega_c^2}{(\omega_0 - \omega_c)^2 Q^2 + \omega_c^2} \tag{11.12}$$

Example 11.3 *At resonance, the spontaneous emission rate decreases Q-fold. Let us estimate the effect of detuning for $\frac{(\omega_0 - \omega_c)}{\omega_c}$ =0.3, 0.5 and 0.7. A very large value of Q is taken. The ratio $W_{sp,mc}/W_{free}$ = 9/Q, 4/Q, and 2/Q, respectively. The rate progressively decreases with increasing detuning.*

11.3.2   Purcell Factor for semiconductor quantum well embedded in two-dimensional photonic crystals

We now consider how the Purcell factor is modified when emission takes place in a semiconductor QW. The treatment is based on the work by Boroditsky et al (1999).

First of all, the classical electric field of the cavity mode, E(**r**) is normalized in the following way:

$$\frac{1}{4\pi} \int \varepsilon(r)|\alpha E(r)|^2 d^3r = \frac{\hbar\omega_0}{2} \tag{11.13}$$

where ε is the permittivity of the material and ω_0 is the resonant frequency of the cavity. The normalization factor is obtained from the following:

$$\alpha^2 = \frac{2\pi\hbar\omega_0}{\int \varepsilon(r)|E(r)|^2 d^3r} \tag{11.14}$$

The integration extends over the quantization volume. The spontaneous emission rate at a given point **r** may be expressed by using Fermi golden rule as

$$W_{sp,c}(r) = \frac{2\pi}{\hbar\langle(d \cdot \alpha E(r))\rangle^{-2}} \frac{g}{\hbar\Delta\omega} \tag{11.15}$$

where **d** is the atomic dipole moment, g is the degeneracy of the cavity mode and $\Delta\omega$ is the linewidth of the mode. The dot product $(d \cdot \alpha E(r))$ is to be averaged over all possible orientations of atomic dipole moments.

In a QW, the main contribution to emission is due to electron heavy-hole (HH) transition. In addition, these transitions are allowed only when the dipole moments of the transition lie in the plane of the QW. Thus, $d_x^2 = d_y^2 = \frac{d^2}{2}$ and $d_z^2 = 0$. Assume also that the mode electric field lies in the QW plane, as for transverse electric(TE) mode. We then obtain

$$\langle (d \cdot \alpha E(r))\rangle^2 = \left(\frac{1}{2}\right) d^2 |\alpha E(r)|^2 \tag{11.16}$$

The prefactor is (1/3) for bulk semiconductors or for interaction with random modes. If the active material is placed at the antinode of the electric field of the mode, the emission rate is given by using Eqs. (11.15) and (11.16) as

$$W_{sp,c} = \frac{2\pi}{\hbar} \frac{d^2}{2} |\alpha E_{max}||^2 \frac{g}{\hbar\Delta\omega}| \tag{11.17}$$

Using Eq. (11.14) for normalization factor in Eq. (11.17) one obtains

$$W_{sp,c} = \frac{\omega_0}{\Delta\omega} \frac{2g\pi^2 d^2}{\hbar} \frac{E_{max}^2}{\int \varepsilon(r)E^2(r)d^3r} \tag{11.18}$$

The spontaneous emission rate for bulk material can be calculated and is given by

$$W_{sp,b} = \frac{4\eta d^2 \omega^3}{3\hbar c^3} = \frac{4\eta d^2 8\pi^3}{3\hbar\lambda^3} \tag{11.19}$$

The Purcell enhancement factor may be written by using Eqs. (11.18) and (11.19) as

$$f_{QW} = \frac{W_{sp,c}}{W_{sp,b}} = Qg \frac{3\lambda^3}{16\pi\eta} \frac{E_{max}^2}{\int \varepsilon(r)E^2(r)d^3r} \tag{11.20}$$

Define mode volume as

$$V_{eff} = \frac{\int \varepsilon(r)E^2(r)d^3r}{|\varepsilon(r)E^2(r)|_{max}} \tag{11.21}$$

The enhancement factor may be finally written as

$$f_{QW} = \frac{3Qg(\lambda/2\eta)^3}{2\pi V_{eff}} \tag{11.22}$$

Comparing this with Eq. (11.11), one notes that f_{QW} is reduced by a factor 2 from the bulk value if the mode volumes are equal and $g = 1$. However, the mode volume V_{eff} may be less in QWs, leading to a higher values of the Purcell factor.

11.4   Mode density in ideal cavities

The spontaneous emission factor given previously for a QW is calculated on the assumption that the spontaneous emission couples to a single mode in the cavity. In real cavities the situation is more complex. The density of available modes in these cavities, though different from the mode density in free space, remains finite. As emission may be coupled to more than one mode, the Purcell factor may be changed accordingly.

As mentioned already, the emitters may be in the form of Rydberg atoms, organic dye, or semiconductor QWs and QDs. We are principally interested in semiconductor lasers. The cavities are formed either by planar mirrors, each of which is formed by Bragg reflectors, or by photonic band gap (PBG) structures, or by using micropillar structures. The counting of mode numbers in each structure is associated with some degrees of complexity.

In this subsection, we illustrate, by using analytical methods, how mode densities alter in two idealized cavities, (1) formed by a pair of plane mirrors and (2) in the shape of an optical wire. The treatment is given by Brorson et al (1990). The mirrors are assumed to be lossless metallic mirrors providing perfect reflection of the EM waves. The optical wire has a rectangular cross section.

11.4.1   Cavity mode density in free space

The authors defined an effective mode density in the following way:

$$g(k) = \frac{1}{V}\sum_k P(k)\rho(k) \tag{11.23}$$

where $P(k)$ denotes the angle dependence of the dipole matrix element and $\rho(k)$ is the mode density of the field in the k direction at frequency $\omega = c|k|$. For a vertically oriented dipole, $P(k) = sin^2\theta$, where θ is the angle between $\mathbf{k}$ and the vertical axis. It may be shown that

$$g(k)dk = \frac{1}{3\pi^2}kdk \tag{11.24}$$

This expression differs from that given in textbooks by a factor $(1/3)$ since an angular dependence of $sin^2\theta$ is considered.

11.4.2   Dipole in between planar mirrors

Now consider that a dipole is placed in the origin of a cylindrical coordinate system as shown in Fig. 11.2. The two mirrors are placed at $\pm L_z/2$. Two possible orientations of the dipole are considered in the work: (1) the vertical electric dipole (VED) with $d\|\hat{z}$ and (2) the horizontal electric dipole (HED) with $d\perp\hat{z}$.

We define the coordinate system by $(\mathbf{r}, z)$. The cavity has no boundaries in the plane, and therefore the wave vectors will lie in the k_r $(= k_x, k_y)$ plane. Since the EM waves

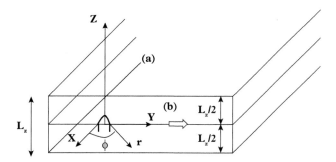

Figure 11.2 *Planar cavity formed by two conducting mirrors. The dipole is located halfway between the mirrors and two possible orientations: VED and HED are shown.*

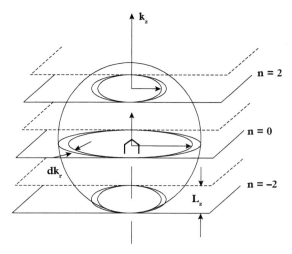

Figure 11.3 *A vertical dipole placed in a planar mirror cavity and the k-space seen by it. The allowed modes are planes perpendicular to the z-axis, which are denoted by even n.*

form standing waves along the z-direction, the propagation vectors along the z-direction will have discrete values given by $k_z = n\pi/L_z$, where $n = \ldots -2, -1, 0, 1, 2, \ldots$. The density of modes along the k_r plane is $(L/2\pi)^2$, where the extensions of the cavity along both x- and y-directions are L. The shape of k-space is shown in Fig. 11.3. It consists of planes intersecting the k_z axis at discrete points separated by π/L_z.

The effective mode density for two-dimensional (2D) propagation is calculated by considering the number of states in an annular ring formed by two circles of radius k_r^n and $k_r^n + dk_r^n$, where

$$k_r^n = \sqrt{k^2 - \left(\frac{n\pi}{L_z}\right)^2} \tag{11.25a}$$

and

$$dk_r^n = \frac{dk}{sin\theta_n} \tag{11.25b}$$

For VED, the angular coupling factor is $P_{ved}(k) = sin^2\theta$. Since the component of electric field $E_z \propto cos[n\pi(z/L_z - 1/2)]$, to match the boundary conditions at the perfectly reflecting mirrors, the VED couples only to even order ™ modes [n = ... −2,0,+2 ....]
  We now write

$$g_{ved}(k)dk = \frac{1}{V} \sum_{n \, even} 2\pi k_r^n dk_r^n sin^2\theta_n \left(\frac{L}{2\pi}\right)^2 \tag{11.26}$$

The sum is taken over all integers $|n| < k/k_c$, where $k_c = \pi/L_z$ denotes the cut-off wave vector. In the present case, $V = L^2 L_z$. Substituting k_r^n, dk_r^n, and noting that $sin\theta_n = k_r^n/k$, the previous equation reduces to

$$g_{ved}(k) = \frac{1}{2\pi^2} k_c k \sum_{n \, even} \left[1 - \left(n\frac{k_c}{k}\right)^2\right] \tag{11.27}$$

This finite sum can be evaluated to give the following for the effective mode density:

$$g_{ved}(k) = \frac{1}{2\pi^2} k_c k \left[1 + 2\left[p - \frac{1}{6}\left(\frac{2k_c}{k}\right)^2 \cdot p(p+1)(2p+1)\right]\right] \tag{11.28}$$

In the previous calculation, $p = [k/2k_c]$ is an integer less than or equal to $k/2k_c$, see Fig. 11.4.
  The expression for mode density for the HED has also been derived and the expression reads

$$g_{hed}(k) = \frac{1}{2\pi^2} k_c k \left[q + \left(\frac{k_c}{k}\right)^2 \left(\frac{4}{3}q^3 - \frac{1}{3}q\right)\right], \quad q = \left[\frac{1}{2}\frac{k}{k_c} + 1\right] \tag{11.29}$$

11.4.3   Rectangular optical wire

In an optical wire, modes are confined along two directions whereas the field propagates along the axis. Fig. 11.5 shows the schematic structure. Here, k_y and k_z components are discrete but k_x is continuous.
  The wire has rectangular cross section with widths L_y and L_z along the respective directions and length L along the x-axis. For a dipole oriented along the z-direction, the allowed modes are odd m and even n due to the perfectly reflecting mirrors. The mode density can be expressed as

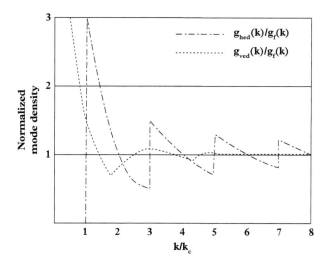

Figure 11.4 *Normalized density of modes in a wire like cavity. The chain dotted and dotted curves correspond to horizontally and vertically emitting dipoles, respectively.*

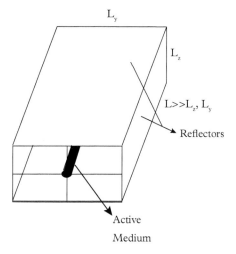

Figure 11.5 *Schematic diagram of a rectangular optical cavity. The dipole sits at the centre of the origin.*

$$g_z(k)dk = \frac{1}{V} \sum_{m \ odd} \sum_{n \ even} 2dx_{n,m} sin^2\theta_{n,m} \left(\frac{L}{2\pi} \right)$$

$$= \frac{1}{2\pi L_y L_z} \sum_{m \ odd} \sum_{n \ even} \frac{2dkk}{\sqrt{k^2 - [(mk_{cy}^2) + (nk_{cz})2]}} \left\{ 1 - \left(n\frac{k_{cz}}{k} \right)^2 \right\} \quad (11.30)$$

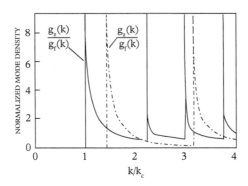

Figure 11.6 *Effective mode density for a planar mirror cavity and an optical wire, normalized by the free space value of mode density.*

The integers m and n should satisfy the condition $k^2 \geq (mk_{cy})^2 + (nk_{cz})^2$, where $k_{cy} = \pi/L_y$, and $k_{cz} = \pi/L_z$ are the cut-off wave vectors.

When the dipole is oriented along the x-direction the effective mode density is expressed as

$$g_x(k) = \frac{1}{2\pi L_y L_z} \sum_{modd} \sum_{neven} \frac{2k}{\sqrt{k^2 - [(mk_{cy}^2 + (nk_{cz})2]}} \left\{ \frac{(mk_{cy})^2 + (nk_{cz})^2}{k} \right\} \tag{11.31}$$

As before, the same condition $k^2 \geq (mk_{cy})^2 + (nk_{cz})^2$ applies for the integers m and n while making the summation.

The effective mode density for planar mirror cavity and rectangular optical waveguide normalized by the same for free space are plotted in Fig. 11.6. It appears that for planar mirrors (1D confinement) there is very little change in the mode density. However, some appreciable change is exhibited by the mode density in an optical wire (2D confinement). Comparing the curves with the DOS of electrons and holes in a QW (1D confinement) and QWR (2D confinement), the effective mode density shows some singularities at the cut off frequencies.

11.4.4   Enhancement factor for broadband emitter

In most cases, the emission from the emitter has a broad spectral width. This is the case for semiconductor lasers. In such situation, the total spontaneous emission rate may be expressed as

$$W_{sp} \propto \int dk R(k) g_{tot}(k) \tag{11.32}$$

R(k) is essentially the transition matrix element, and $g_{tot}(k)$ is the effective mode density taking into consideration all types of orientation of the dipole. An estimate was given by Brorson et al for the emission linewidth $\Delta\lambda/\lambda = 0.02$ for a gallium arsenide (GaAs) emitter taking $L = \lambda/2$ for planar mirror and $L_y = L_z = \lambda/2$ for optical cavity. Finite loss in the reflectors in practical situations and resulting broadening of the mode peaks are totally neglected. Taking the emitted linewidth to be Lorentzian, the emission enhancement factor is only 2.5 for planar mirrors but ~ 12 for optical wires. It may be concluded that reduced dimensionality of mode propagation increases the spontaneous emission rate.

11.5   Experimental observation of Purcell effect

We now present the early experimental results on the inhibition and enhancement of spontaneous emission in semiconductor microcavities. This section begins with a list of conditions to be maintained to observe Purcell effect. A few representative experiments are then described giving the results.

11.5.1   Conditions

Several conditions must be satisfied in order to observe a large Purcell factor.

(1) The emitter's emission frequency must be exactly resonant with the mode frequency.

(2) The emitter must be placed at the antinode of the vacuum field and the dipole should be parallel to the vacuum field.

(3) The cavity must have a small volume so that the vacuum field is enhanced

(4) The cavity Q must be very large but should have an upper limit given by $Q < \lambda_e/\Delta\lambda_e$, where λ_e is the emission wavelength and $\Delta\lambda_e$ is its linewidth.

The first three conditions have been mentioned in Section 11.3.1 in connection with atomic emitters. The fourth condition is also fulfilled there. As shown in Fig. 11.1, the spectral width of the atomic emitter is smaller than the spectral width of the cavity. However, this condition is difficult to maintain for solid state and semiconductor-based emitters. Narrow band (atom-like) emission is possible in quantum dots. We shall see in later sections, the use of microcavities in which QDs are embedded in observing CQED phenomena.

11.5.2   Experimental results

Early CQED experiments showed a 500-fold increase in the spontaneous emission (SE) rate by using Rydberg atoms and high Q microwave cavities. However, the situation is more complicated at optical frequencies. Though planar dielectric cavities using DBRs are well developed, they support a continuum of modes and therefore SE enhancement cannot be expected. With the development of 3D microcavities enhanced SE rate is expected. Silica microsphere, pillar microresonators, photonic disks and wires, and 1D photonic microcavities have the potential for display of enhanced SE rate. However, the large linewidth of commonly used emitters like rare earth atoms in dielectric matrices (~ 20 nm at 300 K) and semiconductor QWs (~1 nm) did not allow the enhancement factor to go beyond unity.

About fivefold enhancement of SE rate was first observed by using self-assembled InAs QBs in a GaAs layer as the emitter. The low linewidth < 0.1 nm of QB emitters allowed use of very high Q cavity. Fig. 11.8 shows the schematic structure of the micropillar, conceptually similar to the same as used by Gerard et al (1998) for the experiment.

The structure consists of 15 period top and 25 period bottom distributed Bragg reflectors (DBR)s made of aluminum arsenide (AlAs)/gallium arsenide (GaAs). The cavity layer is one wavelength thick GaAs into which 5 layers of indium arsenide (InAs) QBs are inserted. Out of these 5 layers, 3 layers are placed at the anti-node, and each of the other two layers are 10 nm above and below the top and bottom DBRs.

The micropillar has 3D confinement of EM modes in the transverse plane defined by sidewalls and along longitudinal direction effected by the two DBRs. A reference sample used for comparison contains similar ensemble of QBs placed in GaAs and no DBRs.

The samples are excited by 1.5 ps pulses from a Ti:sapphire laser and the time resolved PL response was recorded. The photoluminescence (PL) response for reference sample can be fitted by an exponential curve recording a decay time of 1.3 ns. When the 3-layer QBs in the middle are at resonance with the cavity mode, the decay time is 0.25 ns. This clearly indicates enhancement of SE rate. The authors used different sets of micropillars with varying pillar diameters and different Q's. As stated earlier, the maximum value of Purcell factor observed was 5.

The experiments conducted by Gerard et al (1998) did not show suppression of spontaneous emission. In a later experiment, Bayer et al (2001) used similar micropillar structure but having metallic sidewall coating to exclude transverse continuum modes to demonstrate SE suppression.

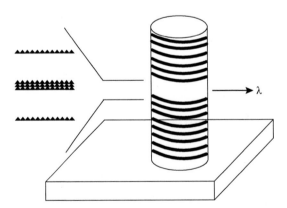

Figure 11.7 *Schematic diagram of a vertical micropillar. The active layer in the centre contains 5 layers having QD assemblies.*

Table 11.1 *Cavity structure, Q, Purcell factor, and reference.*

Cavity Type	Q	Observed Purcell factor	Reference
Micropillar with QD	7,350	36	Reithmaier et al (2004)
Micropillar with metallic sidewall	7×10^3	SE suppression + × 3 times enhancement	Bayer et al (2001)
Microdisk	1.7×10^4	125	Peter et al (2005)
2D photonic crystal	6,000	441	Yoshie et al (2004)
3D PhC-QD	Not quoted	1.5 times enhancement and 10 times inhibition of SE rate	Leistikow et al (2011)

Other works related to Purcell effect have been reported by using a single QD (Ando et al 2001) within a micropost. In addition to micropillars, high Q cavities, like microdisks (Dupuis et al 1999; Peter et al 2005) supporting whispering gallery modes and nanocavities and photonic crystals (Boroditsky et al 1999; Yoshie et al 2004; Leistikow et al 2011; Arakawa et al 2010, 2012), have also been employed. Table 11.1 shows the type of cavity used, typical Q factor, Purcell enhancement factor observed and reference to some of these works.

11.6 Strong light-matter coupling

We now consider the case in which light-matter interaction enters into the strong coupling regime. After giving a brief introduction to the phenomenon, we shall next examine the conditions for strong interaction between EM radiation and quantum matter. The discussion will rely on a two-level system. A consequence of strong light-matter coupling, the Rabi oscillation of a two-level system will then be introduced. The experimental system for observation of Rabi oscillation in semiconductor QDs will then be described and some representative results obtained for the first time for QD nanocavities will be presented.

Although the emphasis in this book is on classical or semiclassical models, there exists a simple model, the Jaynes–Cummins model to describe the interaction. This model will be next introduced and the application of this model to explain the Rabi oscillation will be presented.

11.6.1 Introduction

In the regime of weak light-matter coupling, an emitter, say an atom, makes a transition from the excited state to the ground state, emitting a photon into the continuum

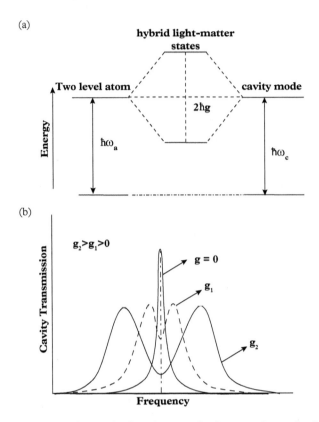

Figure 11.8 *Top: the resonance interaction between a two-level atom and a confined electromagnetic field of a resonator resulting in two new hybrid states separated by Rabi splitting energy 2 ħg; bottom: cavity transmission for different values of coupling. The peaked curve is the relative transmission of a probe through an empty cavity (g = 0). The two double-peaked curves correspond to transmission spectrum due to the splitting in the energy eigen states for two different values of coupling.*

of radiation modes. This is an irreversible process. The opposite regime of strong light-matter coupling can be realized in a microcavity under suitable conditions. In this regime, the emitter or the atom can interact coherently with a cavity mode having a long decay time. This interaction can take place even when the mode lies in its vacuum state (either ground state or state of lowest energy). By using a weak optical probe, it is revealed that the transmission spectrum of the cavity is split into two closely spaced peaks. The atom and the cavity mode become entangled in the microcavity; it means that the entangled states are not factorable into cavity and atom components. If the probe frequency remains equal to the cavity mode's original frequency, then in the presence of the atom, the probe is reflected, or its transmission is blocked. The splitting of the transmission of cavity mode into two eigen frequencies is called the Rabi splitting. In the strong coupling regime, spontaneous emission becomes reversible (Skolnick et al 1998; Vahala 2003; Dovzhenko et al 2018).

11.6.2 Requirements for strong light-matter coupling

The essential requirement for strong coupling to manifest itself is a very high Q cavity. The photons in this cavity survive over a relatively long period before finally escaping out of its walls or being reabsorbed by the system. The long duration of survival of the photon in the cavity allows it to be reabsorbed. In this way a spontaneously emitted photon can again be reabsorbed so that the process becomes reversible. In the strong coupling limit, an 'atom + field' state develops. The two states of the system are denoted by $|E, 0\rangle$ and $|G, 1\rangle$, where E and G denote, respectively, the *Excited* and *Ground* state of the atom and 1 and 0 denote the photon numbers. The state $|E, 0\rangle$ therefore stands for 'excited atom + no photon' and $|G, 1\rangle$ refer to a state with 'de-excited atom (in the ground state) + 1 photon'. The two-level system oscillates back and forth between these two states. This phenomenon is known as vacuum *Rabi oscillation*.

The atom-field interaction is described by a coupling parameter g defined as

$$g = \frac{d|E|}{\hbar} \tag{11.33}$$

where d is the transition dipole moment and $|E|$ is the electric-field amplitude. When a quantum emitter in the form of an atom or a QD is coupled to a radiation mode in a high Q cavity, Rabi oscillation takes place between two states $|E, 0\rangle$ and $|G, 1\rangle$ of the 'atom + cavity' field even in the absence of any external driving field. The oscillation is due to the vacuum field fluctuation associated with the strong cavity mode and a photon emitted by the quantum emitter.

The coupling parameter is defined in a more refined form in terms of the atom's position, $\mathbf{r}$, within the cavity as

$$g(\mathbf{r}) = \left(\frac{d^2 \omega_c}{2\hbar\varepsilon_0 V_M} \right) U(\mathbf{r}) \equiv g_0 U(\mathbf{r}) \tag{11.34}$$

Here V_M is the resonant mode volume and $U(\mathbf{r})$ is the normalized field profile of the mode in the cavity such that $V_M = \int |U(\mathbf{r})|^2 dx$. The coupling coefficient $2g_0$ is termed as the single-photon Rabi frequency which is related to the maximum rate at which a photon is exchanged between the atom and the field.

Example 11.4 *An estimate of the coupling parameter g may be obtained from Eq. (11.33) by noting that in a QD the value of d $\approx$ 10⁻²⁸ C.m and the electric field E $\approx$ 10⁴V/m, which give g $\approx$ 10¹⁰s⁻¹.*

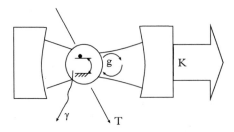

Figure 11.9 *A two-level atom enclosed by a cavity. K is the rate of decay of cavity field, γ denotes the rate at which the atomic dipole radiates into modes different from the modes of cavity field, T is the transit time of an atom through the cavity field and g is the rate of coherent atom-field coupling.*

Fig. 11.9 shows a schematic diagram of a cavity having an emitter (alternatively called an *atom*) (Kimble et al 1992, Kimble 1998, 2008; Birnbaum et al 2005) within it and the environment (free space or material) to which it is connected. The parameter κ denotes the rate of decay of the cavity field which in turn is related to the cavity Q and the width of the cavity mode $\Delta\omega_c$, τ is the rate of decay of the atom from its excited to the ground state, which is related to the linewidth of emission and g is the coupling coefficient. The various parameters are interrelated by the following expressions:

$$\tau = 1/\Delta\omega \text{(i); } \gamma = 1/\tau \text{(ii)}$$

$$Q = \frac{2\pi t_p}{T} = \frac{2\pi c t_p}{L} = \frac{2\pi c}{L\kappa} \text{ (iii); } Q = \frac{\omega_0}{\Delta\omega}; \ \kappa = \frac{2\pi c}{LQ} = \frac{2\pi c}{L}\frac{\Delta\omega}{\omega_0} \text{ (iv)} \qquad (11.35)$$

It is easy to conclude that for strong coupling, the coupling coefficient must dominate dissipation. In other words, the following condition must be satisfied:

$$\frac{g_0}{(\kappa,\tau^{-1})} \gg 1 \qquad (11.36)$$

The previous inequality can be cast in the following alternative form:

$$2g_0 \gg W, \Delta\omega_c \qquad (11.37)$$

Example 11.5 *Let the cavity Q = 10^5 at a wavelength of 1 µm. The linewidth is $\Delta\omega_c \cong 2 \times 10^9$, which is much lower than the value of $2g_0$ obtained in Example 11.4.*

11.6.3   A simple model

We first give a qualitative description of the atom + cavity in the strong coupling regime. As stated in Section 11.4.1, the atom has two states: excited and ground, having energies E_E and E_G, respectively and the transition energy is $\omega_a = (E_E - E_G)/\hbar$. The spontaneous emission rate is denoted by W, which determines the linewidth of transition. As noted before, the cavity resonant frequency is ω_c, the linewidth is $\Delta\omega_c$ so that $Q = \omega_c/\Delta\omega_c$.

In a simple picture, we consider both the atom and the cavity mode as simple oscillators coupled to each other. We use the simple relation $H\psi = \hbar\omega\psi$. Denoting the oscillation frequency and state functions by subscripts a and c, respectively, for atom and cavity oscillation, the coupled equations become

$$\omega_a\psi_a + g\psi_c = \Omega\psi_a \tag{11.38a}$$

$$g\omega_a + \omega_c\psi_c = \Omega\psi_c \tag{11.38b}$$

For nontrivial solution, the determinant formed by the coefficients of the wavefunction must be zero. This leads to

$$(\omega_a - \Omega)(\omega_c - \Omega) - g^2 = 0 \tag{11.39}$$

The solution is

$$\Omega_\pm = \frac{1}{2}(\omega_a + \omega_c) \pm \sqrt{g^2 + \left(\frac{\omega_c - \omega_a}{2}\right)^2} \tag{11.40}$$

Due to coupling there exists a splitting of the resonant frequency and the two modified frequencies of the coupled atom + cavity system takes the form given by Eq. (11.40). When $\omega_c = \omega_a$

$$\Omega_\pm = \omega_a \pm g \tag{11.41}$$

In this situation the splitting value is 2 g, which is termed as the vacuum Rabi frequency. The coupling coefficient, as stated already, is the strength of interaction between an atom and the EM vacuum field contained in the cavity mode expressed by Eqs. (11.33) and (11.34). It is evident from Eq. (11.42) that Rabi splitting increases as the coupling strength increases. This has been illustrated in Fig. 11.9(b).

The atom + field diagram is shown in Fig. 11.10. It is evident that for atom-cavity detuning $\omega_c - \omega_a \ll -2g$, the joint atom-field state $|+\rangle$ is atom-like. It means that the probability that the atom is excited is much greater than the probability of finding a photon in a cavity. The state $|-\rangle$ is cavity-like state. When $\omega_c - \omega_a \gg 2g$, the situation is just the reverse. At resonance, the eigenstates are linear combination of an excited atom, empty cavity product state $|E, 0\rangle$ and a de-excited atom, one photon-cavity mode product state $|G, 1\rangle$.

The first experimental demonstration of vacuum Rabi splitting was made in 1980s. The splitting has been observed in a single atom by Rempe et al (1987).

11.6.4 Experimental observation

Strong coupling phenomena in semiconductor based microcavity structures were first studied by Weisbuch et al (1992). They observed strong exciton-photon coupling in

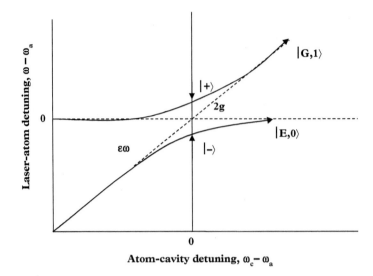

Figure 11.10 *Dispersion diagrams for coupled atom+field system. Dashed lines indicate the uncoupled states. Solid lines denote the dispersion of the two coupled states. At resonance ($\omega_c = \omega_a$), a gap 2g amounting to vacuum Rabi splitting appears between the two branches.*

the reflectivity spectra in microcavities employing GaAs AlGaAs systems. In their work two pronounced peaks were found in the reflectivity spectra indicating vacuum Rabi splitting, which could be controlled by both the cavity and QW exciton properties (Skolnick et al 1998). Since then many other workers have been engaged in studying the phenomena in semiconductor-based systems.

Since the emission lines are quite sharp, semiconductor QDs play a very vital role in the study of strong coupling phenomena in microcavities. The technological challenge lies in achieving high Q cavities with very low mode volume and placing the single QD or an array of QDs in the microcavity. The first successful results in this endeavor have been reported simultaneously by Reithmaier et al (2004) and Yoshie et al (2004) in the same issue of Nature. While the first group used micropillar and $In_{0.3}Ga_{0.7}As$ QDs, the second group used InAs QDs and 2D PhC as the cavity.

In the case of semiconductors, Yoshie and coworkers (2004) reported Rabi oscillation by using a 2D photonic crystal as the cavity and QDs as the emitters. The PhC cavity has a Q factor $\sim 10^4$–10^5 and a very small cavity volume < 0.1 μm^3. This ensured very high vacuum field and strong emitter-field coupling. The 2D PhC made with GaAs and InAs QDs are buried within the structure. The QDs have a very large dipole moment $\sim 10^3$. The QD emission spectrum was changed by using temperature variation, thus providing an efficient way of detuning the atom resonance from the cavity mode frequency. When the pump energy is higher ($\sim$ 770 nm cw-laser, 0.7 mW/μm^2 power density), emission spectra does not show splitting. At middle pump energies (0.8 $\mu W/\mu m^2$), Rabi splitting occurs. Two peaks are further separated when the temperature is changed. The observed

Rabi splitting value is 2 g = 41 GHz = 170 μeV = 0.192 nm. Reithmaier et al (2004) reported similar results by using a semiconductor micropillar as the cavity and embedded QDs as emitters.

Example 11.6 *An estimate of the dipole moment of InAs QD in the PhC cavity may be obtained from the relationship between spontaneous emission lifetime and dipole moment given by $\tau_{sp}^{-1} = \eta\omega_0^3 d^2 / 3\pi\varepsilon_0\hbar c^3$. Using the values τ_{sp}= 2ns, η = 3.4 and emission wavelength = 1.182 μm, one obtains d = 0.926 × 10^{-28} C.m = 28 D, where D = 3.336 × 10^{-30} C.m.*

Example 11.7 *The mode volume of the PhC microcavity is $(\lambda/\eta)^3$. The magnitude of $|E_{vac}|^2 = \frac{\pi\hbar c}{\lambda^4\varepsilon_0}$. This gives E_{vac} = 7.6 × 10^4V/m. Using the value of d obtained in Example 11.6, 2g = 133.5 × $10^9 s^{-1}$.*

Due to strong coupling between cavity modes and excitons, an anticrossing of the SQD exciton and cavity-mode dispersions occur in PL spectra obtained by both the groups as temperature is increased. To understand this, we consider the PL spectra obtained by Reithmaier et al (2004), as shown in Fig. 11.11. At lowest temperature of ~

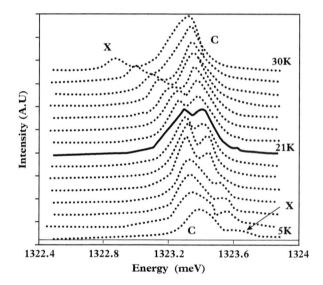

Figure 11.11 *PL spectra showing dot-cavity anticrossing. Temperature is changed in steps from 4 K to 30 K. At 4 K, the X line peak is at a higher energy than C line peak. By changing temperature X and C peaks are aligned at resonance, but a clear splitting is seen around 20 K. With increasing temperature, X lines occur at lower energies.*

Source: Reprinted figure with permission from Reithmaier J P, Sek G, Loffler A, Hofmann C, Kuhn S, Reitzenstein S, Keldysh L V, Kulakovskii V D, Reinecke T L and Forchel A. (2004). Strong coupling in a single quantum dot–semiconductor microcavity system. *Nature*| 432(November): 197–200. Copyright (2004) by the Springer Nature.

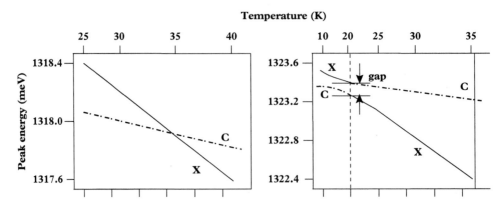

Figure 11.12 *Qualitative variation with temperature of experimental peak energy of QD exciton (X) and cavity mode (C): (a) weak coupling results (b) strong coupling results. In (b) X varies rapidly than C with increase in temperature, but there is no gap between the curves. In (a), There is anticrossing of the two lines. The upper branch corresponds to exciton like dispersion, while the lower branch corresponds to cavity like dispersion. Further there occurs a gap at resonance occurring below 20 K.*

Source: Reprinted figure with permission from Reithmaier J P, Sck G, Loffler A, Hofmann C, Kuhn S, Reitzenstein S, Keldysh L V, Kulakovskii V D, Reinecke T L and Forchel A. (2004). Strong coupling in a single quantum dot–semiconductor microcavity system, *Nature* 432(November): 197–200. Copyright (2004) by the Springer Nature.

5K, the QD exciton X emission occurs at a slightly higher energy than the cavity mode emission energy C. X line has lower intensity and smaller FWHM than the corresponding values for C mode. As temperature is increased, C and X lines coincide. At higher temperatures ~ 30 K, components of emission exchange their properties. Now the lower energy line has a lower intensity and lower FWHM. Therefore, it should be assigned to the QD exciton.

The anticrossing phenomena are further illustrated by plotting and comparing the peak energy, linewidth and intensity of the C and X lines under strong and weak coupling conditions. We show the qualitative variation of only the peak energy for both the coupling conditions in Fig. 11.12. The variation for weak coupling is shown in the left-hand side, and the right-hand side plots the variation for strong coupling. For X, the peak energy decreases rapidly due to strong variation of band gap, but for cavity mode energy the slow decrease is due to slight change of the refractive index (RI) with temperature. Under weak coupling, the two lines cross each other. For strong coupling, the variation of energies of C and X emissions is totally different, as indicated by the two branches in the two sides of the vertical line showing a gap.

It is found that the energies of the higher lying mode decrease strongly with increasing temperature before reaching resonance (at ~ 20 K), and then show a much weaker temperature dependence. The lower energy branch, on the other hand, shows a weak temperature variation, but drops rapidly as the temperature lies above 20 K. The slowly varying sections of the dispersion curves should correspond to the temperature variation of the uncoupled cavity modes. In the upper branch, the anticrossing changes the dispersion from an exciton-like nature to a cavity-mode like one. In contrast, the nature of lower branch changes from cavity-like behaviour to an exciton-like character. As expected, there appears a gap at resonance, which is an indication of coupled exciton-photon behaviour. The gap is about 140 μeV.

11.7 Jaynes—Cummins model

We shall give here an outline of the advanced theory of strong light-matter coupling proposed by Jaynes and Cummins (1963). The theory is based on second quantization treatment, which are presented in detail in advanced text books on many body theory (Mahan 2000; Fetter and Walecka 2003). The model is also treated in books on Quantum Optics (for example Mystere and Sargent 2007). Light-matter interaction using second quantized operators is also treated by Klingshirn 2012).

In the earlier subsections the effects of coupling between light and matter under both weak and strong coupling regimes have been considered. The manifestations in Purcell enhancement of emission and Rabi oscillation in semiconductor have also been discussed. In particular, the Rabi splitting in atomic systems has been introduced without entering into the detailed theory.

A quantitative treatment of light-matter interaction was first presented by Cummings and Jaynes (1964) as early as in 1964. In this section, we introduce this model and arrive at the expression for Rabi splitting in atomic systems.

The model assumes that a coupling exists between cavity modes and a two-level atomic system. Consider first that an EM field of angular frequency ω exists in an optical cavity that coincides with one of the resonant frequencies (modes) in the cavity. The EM field may be considered as composed of harmonic oscillators (photons) of energy $\hbar\omega$. The cavity mode consists of n such photons and its eigenstate are denoted by $|n\rangle$, and $|0\rangle$ is the ground state corresponding to the vacuum (no photons). As before, we use the symbols a^+ and a to denote respectively, the photon creation and destruction operators, as defined by Eq. 4.19. The operational properties of the operators have been described in Section 4.3.

Consider now a two-level atomic systems with ground and excited states eigen functions denoted, respectively, by $|g\rangle$ and $|x\rangle$. It represents a spin-1/2 particle. The state of this two-level atom may be described by a 2-vector as

$$\alpha|x\rangle + \beta|g\rangle = \begin{pmatrix} \alpha \\ \beta \end{pmatrix} \tag{11.42}$$

The Pauli spin matrices are represented by the following expressions:

$$\sigma_x = \begin{pmatrix} 0 & 1 \\ 1 & 0 \end{pmatrix}, \; \sigma_y = \begin{pmatrix} 0 & -i \\ i & 0 \end{pmatrix}, \; \sigma_z = \begin{pmatrix} 1 & 0 \\ 0 & -1 \end{pmatrix}, \; \sigma_+ = \begin{pmatrix} 0 & 1 \\ 0 & 0 \end{pmatrix}, \; \sigma_- = \begin{pmatrix} 0 & 0 \\ 1 & 0 \end{pmatrix} \tag{11.43}$$

The operational properties of the spin matrix operators on the two eigenstates are as follows:

$$\hat{\sigma}_z|x\rangle = |x\rangle; \ \hat{\sigma}_z|g\rangle = -|g\rangle; \ \hat{\sigma}_+|x\rangle = |0\rangle; \ \hat{\sigma}_+|g\rangle = |x\rangle; \ \hat{\sigma}_-|x\rangle = |g\rangle; \ \hat{\sigma}_-|g\rangle = |0\rangle \quad (11.44)$$

The Hamiltonian of the two-level atomic system may be expressed as

$$\hat{H} = \frac{\Delta}{2}\hat{\sigma}_z \quad (11.45)$$

where Δ is the energy difference between the ground and the excited states.

The interaction between the atom and the cavity field may now be introduced. It is assumed that the atom is placed within the cavity and is in resonance with the cavity field. By resonance, it is meant that $\Delta = \hbar\omega$, that is, the photon energy equals the difference in energy between the ground and the excited states.

It is first needed to find the orthonormal basis states of the composite (atom + field) system. The Hilbert space is a (tensor) product of the Hilbert spaces of the atom and the field. The basis for atom space is given by $\{|x\rangle, |g\rangle\}$ and those for cavity field are $\{|n\rangle, n = 0, 1, 2, \ldots\}$. The basis for $|g\rangle$ combined Hilbert space is expressed as

$$\{|x\rangle|n\rangle, |g\rangle|n\rangle, n = 0, 1, 2, \ldots\ldots\} \quad (11.46)$$

The state $|x\rangle|n\rangle$ signifies that the atom is in the excited state and the cavity field is in the state $|n\rangle$ having n photons. The state $|g\rangle|n\rangle$ may be interpreted similarly. These states are product states. Note that superposition of product states, or *entangled* states, are also encountered, an example of which is as follows:

$$|\Psi\rangle = \frac{1}{\sqrt{2}}(|x\rangle|n\rangle + |g\rangle|n + 1\rangle) \quad (11.47)$$

The Hamiltonian of the total field system is now written as

$$\hat{H}_{total} = \overleftarrow{H}_{field} + \hat{H}_{atom} + \hat{H}_{int} \quad (11.48)$$

where $\hat{H}_{int}$ is the *interaction Hamiltonian*.

We now consider some physical processes to find out the nature of the interaction Hamiltonian. One such process is that an atom in the ground state absorbs a photon and moves up to the excited state. The operator describing this process is $\hat{\sigma}_+\hat{a}$, where the $\hat{\sigma}_+$ part takes the atom from the ground to the excited state and the $\hat{a}$ part destroys a photon. The reverse process, in which the atom goes to the ground state by emitting a photon, is associated with the operator $\hat{\sigma}_-\hat{a}^+$. Each process operator is the Hermitian conjugate of the other. Using these two, the interaction Hamiltonian is written as

$$\hat{H}(\hat{\sigma}_+\hat{a} + \hat{\sigma}_-\hat{a}^+)_{int} \quad (11.49)$$

where g is the coupling constant, defined by Eq. (11.41) earlier.

The total Hamiltonian is then written as

$$\hat{H}_{total} = \hbar\omega(\hat{a}^+\hat{a} + 1/2) + (1/2)\hbar\omega\hat{\sigma}_z + \hbar g(\hat{\sigma}_+\hat{a} + \hat{\sigma}_-\hat{a}^+) \tag{11.50}$$

This model is known as the *Jaynes–Cummings model*.

The next task is to identify the eigenvalues and eigenfunctions of the previous Jaynes–Cummins Hamiltonian. To this end, we note that the eigenstates of the harmonic oscillator Hamiltonian $\hat{H}_{int}$ are $\{|n\rangle, n = 0, 1, 2, \ldots\}$ with eigenvalues $\hbar\omega(n + 1/2)$, and $|g\rangle$ and $|x\rangle$ as the eigenstates of the atom Hamiltonian with eigenvalues $\pm\Delta/2$. Both of these bases are non-degenerate, that is, each state in the basis has a distinct eigenvalue.

Consider first that there is no interaction between the atom and the field. The combined Hamiltonian for the field + atom is written as

$$\hat{H}_{free} = \hat{H}_{field} + \hat{H}_{atom} \tag{11.51}$$

The states $|x\rangle|n\rangle$ and $|g\rangle|n\rangle$ are the eigenstates of $\hat{H}_{free}$, the free Hamiltonian. These may be degenerate, however. Consider that the atom is in resonance with the cavity mode, $\Delta = \hbar\omega$. In that case, the states $|x\rangle|n\rangle$ and $|g\rangle|n+1\rangle$ possess the same energy under $\hat{H}_{free}$. They are both eigenstates and have the same energy eigenvalue $\hbar\omega((n + 1) + 1/2)$. Both the states have the same number of energy quanta, either n in the field and 1 in the atom, or (n+1) in the field and 0 in the atom. Thus any superposition of these states in the form

$$\alpha|x\rangle|n\rangle + \beta|g\rangle|n + 1\rangle,$$

with $|\alpha|^2 + |\beta|^2 = 1$, is also an eigenstate with eigenvalue $\hbar\omega((n + 1) + 1/2)$ of the free Hamiltonian, all the eigenstates of which are 2-fold degenerate.

We now switch on the interaction. Since

$$[\hat{H}_{free}, \hat{H}_{int}] = 0 \tag{11.52}$$

the eigenstates of $\hat{H}_{total} = \hat{H}_{free} + \hat{H}_{int}$ are the linear combinations of the degenerate eigenstates of $\hat{H}_{free}$. However, the states $|x\rangle|n\rangle$ and $|g\rangle|n\rangle$ are not the eigenstates of $\hat{H}_{int}$. It may be shown that the eigenstates of $\hat{H}_{int}$ may be expressed by using Eq. (11.47) as

$$|\Psi_n^{\pm}\rangle = \frac{1}{\sqrt{2}}(|x\rangle|n\rangle \pm |g\rangle|n + 1\rangle) \tag{11.53}$$

These states are called *dressed states*. They satisfy the eigenvalue equation

$$\hat{H}_{int}|\Psi_n^{\pm}\rangle = \pm\hbar g\sqrt{n + 1}|\Psi_n^{\pm}\rangle \tag{11.54}$$

The eigenvalue equation for the total Hamiltonian may thus be written as

$$\hat{H}_{total}|\Psi_n^{\pm}\rangle = [\hbar\omega((n + 1) + 1/2) \pm \hbar g\sqrt{n + 1}]|\Psi_n^{\pm}\rangle \tag{11.55}$$

The set of states $\{|\Psi_n^{\pm}\rangle\}$ forms a non-degenerate basis states which may be utilized to determine the evolution of any state.

As the Hamiltonians $\hat{H}_{\text{free}}$ and $\hat{H}_{\text{int}}$ commute, we may resort to *interaction picture* in which the *free* part of the Hamiltonian can be totally ignored (see, Fetter and Walecka (2003); Mahan (2000)). The total Hamiltonian satisfies the following Schrodinger equation:

$$i\hbar \frac{d}{dt}|\Psi(t)\rangle = \left(\hat{H}_{\text{free}} + \hat{H}_{\text{int}} \right) |\Psi(t)\rangle \tag{11.56}$$

Defining a free evolution operator as

$$U_{\text{free}}(t) = \exp\left(-\frac{i}{\hbar}\hat{H}_{\text{free}}t \right) \tag{11.57}$$

and the interaction picture state to be

$$|\Psi(t)\rangle_I = U_{\text{free}}(t)^{-1}|\Psi(t)\rangle \tag{11.58}$$

It is straightforward to show that the interaction picture state satisfies the relation

$$i\hbar \frac{d}{dt}|\Psi(t)\rangle_I = \hat{H}_{\text{int}}|\Psi(t)\rangle_I \tag{11.59}$$

The previous calculation implies that states in the interaction picture evolve only under the interaction Hamiltonian. The interaction picture intuitively views the basis as *rotating with the free evolution of the oscillator*. Only the changes due to the interactions are viewed in terms of this rotating frame.

The following treatment is exclusively in the interaction picture, and the subscript I is no further included.

We now develop the theory of Rabi oscillation in the framework of interaction picture. The basis evolves as

$$|\Psi_n^{\pm}(t)\rangle = \exp\left(-\frac{i}{\hbar}\hat{H}_{\text{int}}t \right) |\Psi_n^{\pm}\rangle = \exp\left(-\frac{i}{\hbar}(\pm\hbar g\sqrt{n+1}t) \right) |\Psi_n^{\pm}\rangle$$

$$= \exp(\mp ig\sqrt{n+1}t)|\Psi_n^{\pm}\rangle \tag{11.60}$$

The initial condition considers an atom in the excited state $|x\rangle$ and the cavity field in the state $|n\rangle$, with n number of photons. The initial state of the combined system may be denoted by $|\psi(t=0)\rangle = |x\rangle|n\rangle$. We expand the initial state in terms of the basis of energy eigenstates as follows:

$$|\psi(t=0)\rangle = \frac{1}{\sqrt{2}}(|\Psi_n^+\rangle + |\Psi_n^-\rangle) \tag{11.61}$$

The time evolution of the state after time t may be described by

$$|\psi(t)\rangle = \frac{1}{\sqrt{2}}(|\psi_n^+(t)\rangle + |\psi_n^-(t)\rangle) = \frac{1}{\sqrt{2}}\left(e^{-ig\sqrt{n+1}t}|\psi_n^+\rangle + (e^{ig\sqrt{n+1}t}|\psi_n^-\rangle)\right)$$

$$= \frac{1}{\sqrt{2}}\left(e^{-ig\sqrt{n+1}t}\frac{1}{\sqrt{2}}(|x\rangle|n\rangle + |g\rangle|n+1\rangle) + e^{ig\sqrt{n+1}t}\frac{1}{\sqrt{2}}(|x\rangle|n\rangle - |g\rangle|n+1\rangle)\right)$$

$$= \frac{1}{2}(e^{-ig\sqrt{n+1}t} + e^{ig\sqrt{n+1}t})|x\rangle|n\rangle + \frac{1}{2}(e^{-ig\sqrt{n+1}t} - e^{ig\sqrt{n+1}t})|g\rangle|n+1\rangle$$

$$= \cos(g\sqrt{n+1}t)|x\rangle|n\rangle - i\sin(g\sqrt{n+1}t)|g\rangle|n+1\rangle \tag{11.62}$$

We aim at measuring the state of the atom at time t, or in other words, we measure the quantum observable $\hat{\sigma}_z$, with eigenstates $|x\rangle$ and $|n\rangle$. The probabilities of occupying the eigenstates are then given by

$$P_x(t) = \cos^2(g\sqrt{n+1}t) \tag{11.63a}$$

$$P_g(t) = \sin^2(g\sqrt{n+1}t) \tag{11.63b}$$

If one puts n = 0 in Eq. (11.63a and b), no photon is involved, and the interaction occurs with the vacuum. In that case,

$$P_x(t) = \cos^2(gt) \tag{11.64}$$

There is now coherent energy exchange between the emitter and the field mode: a phenomenon named as vacuum Rabi oscillation. This is in sharp contrast to the irreversible exponential decay of an excited state into free space. The exchange of energy between the ground (G) and excited (X) states is illustrated in Fig. 11.13.

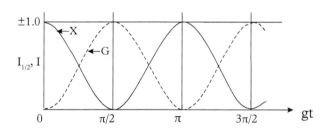

Figure 11.13 *Illustration of vacuum Rabi oscillation between excited (X) and ground (G) states.*

11.8 Microcavities in cavity quantum electrodynamics experiments

Study of light-matter interaction, in both the weak and strong coupling regimes, needs microcavities with high values of Q and low mode volume. In most of the cases, a single QD or an array of QDs is incorporated in some region of the microcavity structure, and the position of the QDs can be fixed with precision by using suitable growth techniques. The fabrication can be done monolithically, and as such, the structures are amenable for large scale integration to yield more complex arrays and circuits. Almost every type of microcavity structure has been considered for the study of light-matter interaction. These include micropillars, microdisks, microspheres, PhC cavities, and even external mirror of very small radius of curvature (Vahala 2003; Benson et al 2006; Khitrova et al 2006).

In this section, the structure, characteristics and representative experimental values of Purcell factor and Rabi splitting of three important microcavities will be discussed. The focus is on QD micropillars, microring resonators, and PhC microcavities. Use of other types will be mentioned very briefly.

Example 11.8 *Experiments on CQED are reported for the following three microcavity structures: (1) InAs QD in PhC, (2) $In_{0.3}Ga_{0.7}As$ QD in micropillar, and GaAs QD in microdisk (Khitrova et al 2006). The emission wavelength in nm, Q and mode volumes in terms of $(\lambda/\eta)^3$ are (1) 1182, 6000,1; (2) 9,377,350,16; (3) 744, 12,000, 8. The Purcell factor is $f = 3Q\lambda^3/4\pi^2\eta^3 V$, and the values are (1) 441, (2) 36, and (3) 125.*

11.8.1 Quantum dot micropillars

A comprehensive review of QD micropillars has been published by Reitzenstein and Forchel (2010). This subsection briefly describes the contents of the article.

A schematic diagram of a micropillar cavity has already been given in Fig. 11.8. The structure consists of a one wavelength thick GaAs cavity, which is sandwiched between a lower and upper AlGaAs/GaAs DBRs. An active layer having InGaAs QDs is placed at the centre of the GaAs cavity and thus the QD layer is placed at the antinode of the cavity. Each layer in the upper and lower DBRs is $\lambda/4$ thick. Different layers are grown by using MBE and then given cylindrical shape by using e-beam lithography. Other shapes of the pillar, as for example elliptical shape, are also achieved. Other III–V and II–IV materials have also been used to fabricate pillar microcavities.

3D light confinement in micropillars is achieved in the vertical direction through reflections from the two Bragg mirrors and waveguiding in the plane of circular cross section. The latter is due to almost total internal reflection (ATIR) as discussed in Section 10.8.

11.8.1.1 *Quality factor and fundamental mode*

To obtain high value of Q for micropillar, the first step is to improve the layout and growth of DBR mirrors. The Q factor of planar microcavity depends on the number and the composition of the mirror pairs in DBR and is expressed as

$$Q = \frac{2L_{eff}}{\lambda} \frac{\pi}{1 - r_l r_u} \tag{11.65}$$

where r_l and r_u are the reflectivities of the lower (l) and upper (u) Bragg mirrors. $L_{eff} = \eta_{cav} d_{cav} + 2\eta_{eff} L_m$ is the effective cavity length, d_{cav} is the cavity layer thickness and $L_m = m_{eff}(d_{GaAs} + d_{AlAs})$ is the effective mirror length. The effective number of mirror pairs is $m_{eff} = (1/2)(\eta_{GaAs} + \eta_{AlAs})/(\eta_{GaAs} - \eta_{AlAs})$. The effective RI of DBR region is $\eta_{eff} = 2\eta_{GaAs}\eta_{AlAs}/(\eta_{GaAs} + \eta_{AlAs})$. The reflectivity of a DBR having m mirror pairs is expressed as

$$r = \frac{\eta_0 - (\eta_{AlAs}/\eta_{GaAs})^{2m}}{\eta_0 + (\eta_{AlAs}/\eta_{GaAs})^{2m}} \tag{11.66}$$

where η_0 refers to RI of air (=1) for top DBR and to RI of GaAs for the bottom mirror.

Example 11.9 *In their work Reitzenstein and Forchel used $\eta_{GaAs} = 3.544$, $\eta_{AlAs} = 2.973$, $d_{GaAs} + d_{AlAs} = 142$ nm, $d_{cav} = 260$ nm, $\lambda = 940$ nm and an effective cavity length of 6.16 μm. For 20/24 mirror pairs in the upper/lower DBRs, the calculated Q of the planar microcavity is 22,000.*

Planar microcavities are suitably etched to yield micropillars to achieve 3D photon confinement. The mode frequency is related to the energy of the resonant frequency E_0 of the planar microcavity by the relation

$$E = \sqrt{E_0^2 + \frac{\hbar^2 c^2}{\epsilon} \frac{x^2_{n\varphi, n_r}}{R^2}} \tag{11.67}$$

where $x_{n\varphi, n_r}$ is the n_r^{th} zero of the Bessel function $\mathcal{J}_{n\varphi}\left(x_{n\varphi, n_r}/R\right)$, and R is the radius of the pillar. Denoting the quantum numbers by $(n_r, n\varphi, 0)$, the lowest modes are $(1,0.0)$, $(1.1,0)$, $(1.2,0)$ and so on. As the pillar diameter decreases, the energy related to the fundamental mode increases due to tighter mode confinement.

The Q factor of micropillar is a strong function of pillar diameter. The Q of the cavity is expressed as

$$Q^{-1} = Q^{-1}_{intrinsic} + Q^{-1}_{edge-scattering} + Q^{-1}_{absorption} \qquad (11.68)$$

The intrinsic Q value of the micropillar decreases with reduced diameter due to radiation loss, shift of photonic band gap etc., and sets an upper limit to the Q value. The value is reduced due to edge scattering loss and absorption loss in the material. These effects have been discussed in Chapter 10. For larger diameter, edge scattering is reduced. The overall Q of the pillar increases with increasing pillar diameter. However, a larger pillar diameter means a larger mode volume, which is not desirable for studying strong coupling phenomena.

Typical values of Purcell factor in the weak coupling regime have been quoted in this chapter.

The necessary conditions to observe strong light-matter interaction will now be illustrated. Eqs. (11.40) and (11.41) have been developed by ignoring loss processes. The following expressions for the two eigen energies have been obtained from a non-perturbative quantum mechanical theory (Andreani et al 1999):

$$E_{1,2} = E_0 - \frac{i(\gamma_c + \gamma_x)}{4} \pm \sqrt{g^2 - (\gamma_c + \gamma_x)^2/16} \qquad (11.69)$$

where E_0 is the energy of the uncoupled X and C modes under zero detuning, and γ_c and γ_x are their respective FWHM linewidths.

The mode splitting, according to Eq. (11.69) is

$$\Delta E_R = \sqrt{g^2 - (\gamma_c + \gamma_x)^2/16} \qquad (11.70)$$

Therefore, to observe mode splitting the following condition is to be satisfied:

$$g^2 - \frac{(\gamma_c + \gamma_x)^2}{16} > 0$$

Usually, $\gamma_x \ll \gamma_c$, even for a high Q microcavity. The threshold condition therefore simplifies to

$$g < \gamma_c/4 \qquad (11.71)$$

Example 11.10 *The threshold value of g is about 0.045 meV in the work of Reithmaier et al (2004). The coupling constant for the micropillar was 0.08 meV. The condition for observing strong coupling is established.*

Strong coupling phenomenon manifests itself in the anticrossing of C and X lines in the PL spectra. The anticrossing has been observed in all the experiments using different microcavity structures and temperature is changed to tune the emission lines. In the case of micropillar, another method, tuning by electric field is employed. The micropillar structure is the same as in Fig. 11.8, but electrical contacts to the structure are made at the top and bottom by adding additional layers of metal (Au). The resulting structure is more or less same as that of VCSEL. The structure is basically a p-i-n diode, the top DBR being the p-type. The QD layer is inserted in the cavity which forms the i-region. A reverse bias to the pin structure shifts the excitonic state by strong Quantum Confined Stark Effect (QCSE). The electro-optical tuning is faster than temperature tuning, and resonance and anticrossing behaviour can be observed more precisely. A number of technical applications, like electro-optical switches or electrically triggered single photon sources, may be envisaged.

11.8.2 Microring resonators

Strong coupling phenomena have been observed in microdisk resonators by Peter et al (2005). The authors formed GaAs QDs in the cavity by thickness fluctuations of a thin GaAs quantum well. Such QDs have a much larger oscillator strength that self-assembled InAs QDs. The microdisks possess high Q and low mode volume and support whispering gallery modes.

Example 11.11 *The coupling constant for exciton-photon interaction is given by* $g = \left(\frac{1}{4\pi\varepsilon_0\varepsilon_r}\frac{\pi e^2 f}{m_0 V}\right)^{1/2}$, *where f is the oscillator strength and m_0 is the free electron mass (Andreani et al 1999). Peter et al (2005) found a value of g = 200 μeV in their experiment using a microdisk for which the mode volume was V = 0.07 μm^3. Converting the value of g, one obtains g = 3.01 s^{-1}. The relative permittivity is taken as 12.5. The oscillator strength is calculated to be 100.*

The emission spectra are obtained by PL study and the usual temperature tuning is employed to change the spectra and to observe the anticrossing of C and X peaks. Figure 11.14 shows the Rabi doublet at resonance. The squares represent experimental data, and the dashed and dotted lines are Lorentzian fits to the two peaks, while the continuous curve is the sum of two Lorentzian lines.

The oscillator strength of QDs formed by thickness fluctuation in microdisks exceeds the value in micropillar (~ 50) and in PhC microcavity (~10) and shows a larger Rabi splitting.

11.8.3 Photonic crystal microcavities

The first experimental demonstration of vacuum Rabi splitting in a photonic crystal microcavity was made by Yoshie et al (2004). They fabricated a two-dimensional triangular lattice PhC using GaAs slab and creating air holes in the slab. Three holes are missing in the slab to form a spacer. Their samples are grown on a (001) GaAs substrate by MBE

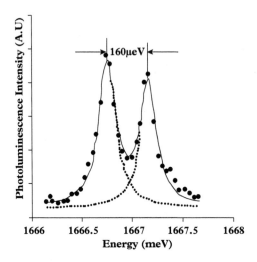

Figure 11.14 *Emission spectrum resonance in the microdisk with GaAs QDs, showing clear Rabi splitting at resonance. Dashed and dotted lines are Lorentzian fit of each peak. The continuous line is obtained by summing the two Lorentzian lines.*

Source: Reprinted figure with permission from Peter E, Senellart P, Martrou D, Lemaı̂tre A, Hours J, Ge´rard J M, and Bloch J. (2005). Exciton-photon strong-coupling regime for a single quantum dot embedded in a microcavity. *Physical Review Letters* 95(067401): 1–4. Copyright (2005) by the American Physical Society.

and a single layer of InAs QDs are grown in the centre of the PhC slab. The spacer containing the missing-holes is surrounded by 14 periods of air holes, in order to ensure good in-plane confinement. The vertical confinement is achieved by total internal reflection at the interface between GaAs slab forming the PhC and air at the top.

The Q value of the cavity is as large as 13,300. However, increased absorption of QD at intermediate power level reduces this value to about 8000. The field strength profile in the cavity has been computed and it is found that the field energy is confined to the defect region with a mode volume of $V \approx (\lambda/\eta)^3 \approx 0.04\,\mu\text{m}^3$. The large Q and small mode volume give a high value of vacuum field strength, facilitating observation of well-resolved Rabi splitting. The evidence for strong coupling comes from the usual PL spectra exhibiting anti crossing and presence of Rabi doublet at resonance. Usual temperature tuning is employed to show anti crossing.

The value of Purcell factor is 441, as quoted by Gibbs et al (2006).

The first experimental study of cavity QED effects using real photonic crystals with a full 3D photonic band gap was made by Leistikow et al (2011). Silicon 3D inverse-woodpile photonic band gap crystals were fabricated by a complementary metal-oxide semiconductor(CMOS)-compatible method. The PhC had a diamond like structure having broad band gaps. First, a 2D array of pores was etched in a silicon wafer by reactive ion etching, and next a second orthogonal set of pores was etched after careful alignment. The 3D PhC so formed was immersed in a dilute suspension of lead sulfide (PbS) colloidal quantum dots in toluene. PbS QDs showed the atomic like energy levels.

The authors observed a 1.5-fold increase in the spontaneous emission rate and under off-resonance conditions, the rate was inhibited by ten times. No results on strong coupling effect were reported.

11.8.4 Other structures

An attractive microcavity structure is a microsphere. It offers a very large value of Q. Also micropillars with elliptical cross section have been used in the study of CQED effects. The reader is referred to Vahala (2003); Gibbs et al (2006); and Reitzenstein and Forchel (2010) for useful references.

11.9 Applications

The concept of control of spontaneous emission in a microwave cavity was introduced by Purcell (1946). Since then, many workers have observed CQED phenomena in atomic systems, solids and semiconductors in optical frequency region also. In all these studies, the prime motivation has been to do fundamental research.

However, as more and more work has accumulated, it becomes apparent that several technological applications of CQED effects are possible. Below is a list of such applications, which is in no way complete, as newer and newer application areas are opening.

(1) Since spontaneous emission rate is enhanced and is more directed, much more efficient lasers and LEDs can be realized. The cavity should be one wavelength thick, which is possible by present-day technology.

(2) As the lifetime is shortened, it is possible to achieve higher modulation bandwidth for lasers and LEDs (Lau et al 2009; Suhr et al 2010).

(3) The fraction of spontaneous emission coupled into laser mode, the β-factor, reduces the threshold of the laser. In the event $\beta \approx 1$, it is possible to get threshold less laser (Brorson et al 1990).

(4) There is transfer of excitation from light to matter and vice versa in the strong coupling regime. This is useful in quantum information processing.

(5) In the strong coupling regime, single photon emitters, and other electro-optic devices based on single photon are feasible.

11.10 Microcavity laser

An important application of enhanced spontaneous emission in a microcavity can be exploited to realize new types of lasers having almost zero or very small threshold current characteristics. In this section, the properties of such lasers will be studied by using a rate equation model developed by Bjork and Yamamoto (1991).

The free carrier density in the active medium is denoted by N and the photon population by p. Their dynamics is described by rate equations, the justification for which is given in the paper. The rate equations are written as

$$\frac{dN}{dt} = \frac{I}{eV} - \left(\frac{1-\beta}{\tau_{sp}} + \frac{\beta}{\tau_{sp}}\right)N - \frac{N}{\tau_{nr}} - \frac{gp}{V}$$

$$\frac{dp}{dt} = -(\gamma - g)p + \frac{\beta NV}{\tau_{sp}} \qquad (11.72)$$

Here, I is the injection current, e is the electron charge, V is the volume of the active material, which may be substantially smaller than the cavity volume, τ_{nr} is the nonradiative recombination lifetime, g is the active material gain in s^{-1} and $\gamma = 1/\tau_p$ is the cavity photon decay rate in s^{-1}. The spontaneous emission lifetime is defined as

$$\tau_{sp} = \frac{1}{\sum_i A_i} \qquad (11.73)$$

where A_i is the spontaneous emission rate of the active material into mode i.

The spontaneous emission coupling rate is written as

$$\beta = \frac{A_0}{\sum_i A_i} \qquad (11.74)$$

The subscript 0 indicates the optical mode associated with the lasing mode.

The optical gain is linearly related to the carrier density by (Basu et al 2015)

$$g = g'(N - N_0) \qquad (11.75)$$

where N_0 is the transparency carrier density at which gain crosses zero and the differential gain g' may be written by using Eq. (11.72) as

$$g' = \frac{\beta V}{\tau_{sp}} \qquad (11.76)$$

Eq. (11.76) is obtained using the fact that when the average photon number in a mode is unity, spontaneous emission equals stimulated emission as known from Einstein's A and B coefficients.

Using Eqs. (11.71)–(11.76), the static and dynamic characteristics of a microlaser can be obtained.

11.10.1   Threshold behaviour

The most promising feature of a microcavity laser is very low threshold or even zero threshold operation. Bjork and Yamamoto introduced three different definitions of the threshold conditions thereby leading to three different expressions for threshold current. These definitions will now be discussed briefly.

The first definition for threshold current is the current at which net stimulated gain equals the loss. From Eq. (11.72) this amounts to

$$\gamma = \frac{\beta V}{\tau_{sp}}(N_{th1} - N_0) \tag{11.77}$$

Subscript 1 indicates the first *definition*. This widely used definition is unphysical, since from Eq. (11.72), the net gain can equal loss only at infinite photon number. However, neglecting stimulated recombination at threshold, one obtains

$$N = \frac{I}{eV\left(\frac{1}{\tau_{sp}} + 1/\tau_{nr}\right)} \tag{11.78}$$

so that

$$N_{th1} = N_0 + \frac{\tau_{sp}\gamma}{\beta V} = N_0\left(1 + \frac{1}{\xi}\right) \tag{11.79}$$

The dimensionless parameter in Eq. (11.79) is defined as

$$\xi = \frac{N_0\beta V}{\gamma\tau_{sp}} \tag{11.80}$$

The second definition of threshold is that the net stimulated emission, that is stimulated emission minus absorption, equals the spontaneous emission. Considering Eqs. (11.72), (11.75), and (11.76), this new definition leads to a threshold photon number expressed as

$$p_{th2} = \frac{N_{th}}{N_{th} - N_0} \tag{11.81}$$

Expressions for photon number, free carrier concentration, and current at threshold are given in the paper. We focus on the third definition of threshold, which is preferred by the authors. It states that the mean photon number in the mode at threshold is unity and thus,

$$p_{th3} = 1 \tag{11.82}$$

From Eq. (11.72), the free carrier density at threshold under steady state is then

$$N_{th3} = \frac{(N_0 + \tau_{sp}\gamma/\beta V)}{2} = \frac{N_0}{2}\left(1 + \frac{1}{\xi}\right) \tag{11.83}$$

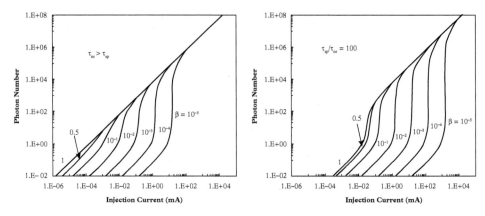

Figure 11.15 *Photon number vs. injection current in a microcavity laser (left) $\tau_{nr} > \tau_{sp}$; (right) $\tau_{sp} = 100\ \tau_{(nr)}$.*

The threshold current, using Eqs. (11.72) and (11.82), may be written as

$$I_{th3} = \frac{e}{2}\left[\frac{\gamma}{\beta}\left(1 + \beta + \frac{\tau_{sp}}{\tau_{nr}}\right) + \frac{N_0 V}{\tau_{sp}}\left(1 - \beta + \frac{\tau_{sp}}{\tau_{nr}}\right)\right] = \frac{e\gamma}{2\beta}\left[1 + \beta + \frac{\tau_{sp}}{\tau_{nr}} + \xi\left(1 - \beta + \frac{\tau_{sp}}{\tau_{nr}}\right)\right]$$

(11.84)

When the pump current is expressed in terms of photon number, one obtains

$$I = \frac{e\gamma}{\beta}\left[\frac{p}{1+p}(1 + \xi)\left(1 + \beta p + \frac{\tau_{sp}}{\tau_{nr}}\right) - \xi\beta p\right]$$

(11.85)

Example 11.12 *As an example of using Eq. (11.85) to calculate current for a given photon number, we assume according to Bjork and Yamamoto, $\gamma = 10^{12}s^{-1}$, $\tau_{sp} = 10\,ns$, $\tau_{sp} > \tau_{sp}$, $N_0 = 10^{18}cm^{-3}$, $V = 10^{-15}\ cm^3$ and $\beta = 10^{-4}$. The vale $\xi = 10^{-5}$ is obtained. We estimate the values of I for $p = 10^1$, 10^2, and 10^3 and obtain $I = 1.46$ mA, 1.60 mA, and 1.76 mA, respectively. The change in I is small over 100-fold increase in the value of p.*

The variation of photon number with injection current for the microcavity described in the previous example is shown in more detail in Fig. 11.15, for different values of β as a parameter and two different conditions: $\tau_{nr} > \tau_{sp}$ and $\tau_{sp} > \tau_{nr}$, shown respectively in the left and right parts of the figure.

From the left panel, when the nonradiative recombination is negligible, it is seen that there is a sharp increase in the photon number as soon as the photon number exceeds unity, for all values of β except 1. This is an argument for the threshold definition 3. When $\beta = 1$, there is smooth transition from below to above the threshold and thus an unambiguous definition of the threshold pump current based on the input-output, or the gain characteristics is difficult to make. All the curves however point out that the threshold current is significantly reduced, almost by three orders, from the values in single mode lasers in which Purcell enhancement is absent. The output power can be easily calculated from the photon number, taking or not taking into account optical losses except mirror losses.

The important role of nonradiative recombination can be understood from the right panel of Fig. 11.15. There is a sharp transition when the photon number crosses 1 for all curves including that for $\beta = 1$. Comparing the two sets of curve, it is easy to conclude that the threshold current increases significantly in this case.

Problems

11.1 *Derive the expressions for the density-of-modes in a planar dielectric (purely 2D) and a 1D wire dielectric.*

11.2 *Calculate the Purcell factor for a cavity having $Q = 10^9$ and mode volume of $5(\lambda/\eta)^3$.*

11.3 *Using Eq. (11.12), calculate the Purcell factor f when the frequency is detuned by 20%.*

11.4 *Determine the value of Q needed to make the spontaneous emission rate 100 times larger than the rate in the free space. The emission wavelength is 1.55 μm, the volume of the cavity is 3 μm in height and 3 μm in diameter, and the RI = 3.5.*

11.5 *Calculate the reflectivity of a DBR consisting of alternate layers of AlAs and GaAs. The top layer is air, and the substrate is GaAs. The number of pairs is 10. The RI of AlAs and GaAs are, respectively, 3.2 and 3.6.*

11.6 *Calculate the approximate value of the bandwidth of GaAs-AlAs Bragg mirror at $\lambda_B = 0.98$ μm, using the expression $\Delta\lambda = 2\lambda_B\Delta\eta/(\pi\eta_{gr})$ (Michalzik and Ebeling: 2003). The RI difference $\Delta\eta$ is 0.56 and the spatially averaged group index η_{gr} is 3.6.*

11.7 *A vertical microcavity is formed by two Bragg mirrors, in which EM mode is confined along the vertical axis, but there is no confinement along the plane. Write the dispersion relation E vs $k_{||}$. Prove that for small $k_{||}$, the dispersion relation is parabolic.*

11.8 *Using the expression for the parabolic E-k relation for photons, obtain a value for photon effective mass in a microcavity.*

11.9 *Considering that a QD represents a two-level atomic system, show that the spontaneous emission lifetime is expressed as $\tau_{sp}^{-1} = \eta\omega_0^3 d^2/3\pi\varepsilon_0\hbar c^3$, where $\hbar\omega_0$ is the energy of the emitted photon, d is the dipole moment, and η is the refractive index of the material.*

11.10 *Calculate the value of dipole moment in an $In_{0.3}Ga_{0.7}As$ QD when the spontaneous emission lifetime is 1 ns. The emission wavelength is 937 nm and RI is 3.51.*

11.11 *Use the analytical expression for spontaneous emission spectrum S(ω) given by*

$$2\pi S(\omega) = \frac{1/2(\kappa + \gamma/2)}{\left(\frac{1}{4}\right)(\kappa + \gamma/2)^2 + (\omega - \omega_0 - g)^2} + \frac{1/2(\kappa + \gamma/2)}{\left(\frac{1}{4}\right)(\kappa + \gamma/2)^2 + (\omega - \omega_0 + g)^2}.$$

Draw a plot of zero-detuning emission using g = 20.6GHz, κ= 42.3GHz = 0.197nm, γ= 21.5GHz =0.1nm. The cavity resonates at 1182.5 nm.

11.12 *Calculate the values of Q for planar microcavity for 21/25, 22/26, and 23/27 mirror pairs in the upper/lower DBRs in a planar microcavity. Use the values of parameters as given in Example 11.9.*

11.13 *Calculate and plot the emission energy of the micropillar fundamental mode for pillar diameter ranging from 1 to 4 μm. The energy for planar microcavity is 1.26 eV.*

11.14 *The intrinsic values of Q for a micropillar are 3.2 × 10⁴(1), 2.5 × 10⁵(2), 4.0 × 10⁵(3), and 5.1 × 10⁵(4), where the numbers in bracket denote the pillar diameter in μm. The Q factor arising out of material absorption is expressed as $Q_{abs} = 4\pi\eta/\lambda_c\alpha_m$, where λ_c is the cavity resonance wavelength and α_m is the absorption coefficient of the material. Take $\lambda_c = 0.94$ μm and $\alpha_m = 260$ m⁻¹. Calculate the values of Q for the micropillar.*

References

Albert F, Braun T, Forchel A, Heindel T, Hofling S, Hopfmann C, Kamp M, Kistner C, Lermer M, Mrowinski P, Reitzenstein S, and Schneider C (2011) Electrically driven quantum dot micropillar light sources. *IEEE Journal of Selected Topics in Quantum Electronics* 17(6): 1670–1680.

Ando H, Gotoh H, Kamada H, Takagahara T, and Temmyo J (2001) Exciton Rabi oscillation in a single quantum dot. *Physical Review Letters* 87(24): (246401): 1–4.

Andreani L, Gerard J M, and Panzarini G (1999) Strong coupling regime for quantum boxes in pillar microcavities: Theory. *Physical Review B* 60:13276–13279.

Arakawa, Y, Ishikawa, A, Nishioka, M, and Weisbuch, C (1992) Observation of the coupled exciton photon mode splitting in a semiconductor quantum microcavity. *Physical Review Letters* 69: 3314–3317.

Arakawa Y, Iwamoto S, Kumagai N, Nomura M, and Ota Y (2010) Laser oscillation in a strongly coupled single-quantum-dot–nanocavity system. *Nature Physics* 6: 279–283.

Arakawa Y, Iwamoto S, Nomura M, Ota Y, and Tandaechanurat A (2012) Cavity quantum electrodynamics and lasing oscillation in single quantum dot-photonic crystal nanocavity coupled systems. *IEEE Journal Selected Top Quantum Electronics* 18(6): 1818–1829.

Basu P K, Mukhopadhyay B, and Basu R (2015) *Semiconductor Laser Theory*. Boca Raton, FL, USA: CRC Press (Taylor & Francis).

Bayer M, Forchel A, Larionov A, McDonald A, Reinecke T L, and Weidner F (2001) Inhibition and enhancement of the spontaneous emission of quantum dots in structured microresonators. *Physical Review Letters* 86(14): 3168–3171.

Benson T M, Boriskina S V, Greedy S C, Nosich A I, Sewell P, and Vukovic A (2006) Micro-optical resonators for microlasers and integrated optoelectronics: Recent advances and future challenges. In: *Frontiers in Planar Lightwave Circuit Technology*, S. Janz et al (eds.) pp. 39–70. Dordrecht NL: Springer

Birnbaum K M, Boca A, Boozer A D, Kimble H J, Miller R, and Northup T E (2005) Trapped atoms in cavity QED: Coupling quantized light and matter. *Journal of Physics B: Advances in Atomic, Molecular and Optical Physics* 38: S551.

Bjork G and Yamamoto Y (1991) Analysis of semiconductor microcavity lasers using rate equations. *IEEE Journal of Quantum Electronics* 27(11): 2386–2396.

Boroditsky M, Vrijen R, Krauss T F, Coccioli R, Bhat R, and Yablonovitch E (1999) Spontaneous emission extraction and Purcell enhancement from thin-film 2-D photonic crystals. *Journal of Lightwave Technology* 17(11): 2096–2112.

Brorson S D, Ippen E P, and Yokoyama H (1990) Spontaneous emission rate alteration in optical waveguide structures. *IEEE Journal of Quantum Electronics* 26(9): 1492–1499.

Bunkin F V and Oraevskii A N (1959) Spontaneous emission in a cavity. *Izvestia Vuzov* 2: 181–188.

Cao H, Tassone F, and Yamamoto Y (2000) *Semiconductor Cavity Quantum Electrodynamics*. New York: Springer.

Christensen P T, Gregersen N, Hueck M, Kaer P, Mork J, and Yu Y (2015) Cavity photonics. In: *Photonics, Scientific Foundations, Technology and Applications*, Vol. II. D J Andrews (ed.) Chapter 2, pp. 21–51. New York, USA: John Wiley & Sons.

Costard E, Gayral B, Gérard J M, Legrand B, Sermage B, and Thierry-Mieg V (1998) Enhanced spontaneous emission by quantum boxes in a monolithic optical microcavity. *Physical Review Letters* 81(5): 1110–1113.

Deppe D G, Ell C, Gibbs H M, Hendrickson J, Khitrova G, Rupper G, Scherer A, Shchekin O B, and Yoshie Y (2004) Vacuum Rabi splitting with a single quantum dot in a photonic crystal nanocavity. *Nature* 432: 200–203.

Dovzhenko D S, Nabiev L R, Rakovich Yu P, and Ryabchuk S V (2018) Light–matter interaction in the strong coupling regime: Configurations, conditions, and applications. *Nanoscale* 10: 3589–3605.

Dupuis C, Gayral B, Gerard J M, Lemaıtre A, Manin L, and Pelouard J P (1999) High Q wet-etched GaAs microdisks containing InAs quantum boxes *Applied Physics Letters* 75(13): 1908–1910

Ebeling K J and Michalzik R (2003) Operating principles of VCSEL. In: *Vertical Cavity Surface Emitting Devices*, Springer series in Photonics, Vol. 6. H Li and K Iga (eds.) pp. 53–98. Berlin: Spinger.

Fetter A L and Walecka J D (2003) *Quantum Theory of Many Particle Physics*. New York: Dover.

Fisher T A, Skolnick M S, and Whittaker D A (1998) Strong Coupling phenomena in quantum microcavity structures. *Semiconductor Science and Technology* 13: 645–669.

Forchel A, Franeck P, Höfling S, Huggenberger A, Kamp M, Kistner C, Münch S, Reitzenstein S, Schneider C, Strauss M, Weinmann P, and Worschech L (2009) Semiconductor cavity quantum electrodynamics with single quantum dots, proceedings of the international school and conference on photonics, PHOTONICA09. *Acta Physica Polonica A* 116(4): 445–450.

Forche lA, Hofmann C, Keldysh L V, Kuhn S, Kulakovskii V D, Loffler A, Reinecke T L, Reithmaier J P, Reitzenstein S, and Sek G (2004) Strong coupling in a single quantum dot–semiconductor microcavity system. *Nature* 432: 197–200.

Gaponenko S V (2010) *Introduction to Nanophotonics*. Cambridge, UK: Cambridge University Press.

Gayral B, Gerardo B D, Hu E L, Imamoğlu A, Kiraz A, Petroff P M, Reese C, Schoenfeld W V, and Zhang L (2003) Cavity-quantum electrodynamics with quantum dots. *Journal of Optics B: Quantum Semiclass* 5: 129–137

Gérard J M, Sermage B, Gayral B, Legrand B, Costard E, and Thierry-Mieg V (1998) Enhanced spontaneous emission by quantum boxes in a monolithic optical microcavity. *Physical Review Letters* 81: 1110 .

Gibbs H M, Khitrova G, Kira M, Koch S W, and Scherer A (2006) Vacuum Rabi splitting in semiconductors. *Nature Physics* 2: 81–90.

Gregersen N, McCutcheon D P S, and Mork J (2017) Single photon sources. In: *Handbook of Optoelectronic Device Modeling & Simulation*, Vol. II. J Piprek (ed.) Chapter 46. Boca Raton, FL: CRC Press.

Gregersen N, Mørk J, Suhr T, and Yvind K (2010)Modulation response of nanoLEDs and nanolasers exploiting Purcell enhanced spontaneous emission. *Optics Express*1 8(11): 11230–11241.

Haroche S and Kleppner D (1989) Cavity quantum electrodynamics. *Physics Today* 42: 24–30.

Huisman S R, Lagendijk A, Leistikow M.D, Mosk A P, Vos W L, and Yeganegi E (2011) Inhibited spontaneous emission of quantum dots observed in a 3D photonic band gap. *Physical Review Letters* 107(193903): 1–5.

Jaynes E T and Cummins F W (1963) Comparison of quantum and semiclassical radiation theories with application to the beam maser. *Proceedings of the IEEE* 51: 89–109.

Khitrova G, Gibbs H M, Kira M, Koch S W, and Scherer A (2006) Vacuum Rabi splitting in semiconductors. *Nature Physics* 2: 81–90.

Kimble H J (1998) Strong interactions of single atoms and photons in cavity QED. *Physica Scripta T* 76: 127–137.

Kimble H J (2008) The quantum internet. *Nature* (London) 453:1023–1030.

Kimble H J, Rempe G, and Thompson R J (1992) Observation of normal mode splitting for an atom in an optical cavity. *Physical Review Letters* 68: 1132–1135.

Klingshirn C F (2012) *Semiconductor Optics*, 4th edn. Heidelberg: Springer.

Lau E K, Lakhani A, Tucker R S, and Wu M C (2009) Enhanced modulation bandwidth of nanocavity light emitting devices. *Optics Express* 17(10): 7790–7799.

Leistikow M D, Mosk A P, Yeganegi E, Huisman S R, Lagendijk A, and Vos W L (2011) Inhibited spontaneous emission of quantum dots observed in a 3D photonic band gap. *Physical Review Letters* 107(193903): 1–5.

Mahan G D (2000) *Many Particle Physics*. New York: Plenum Press.

Mystere P and Sargent M (2007) *Elements of Quantum Optics*, 4th edn. Berlin: Springer.

Numai T (2004) *Fundamentals of Semiconductor Lasers*. New York: Springer Verlag.

Oraevskii A N (1994) Spontaneous emission in a cavity. *Physics-Uspekhi* 37(4): 393–405.

Pelton M, Solomon G S, and Yamamoto Y (2001) Single-mode spontaneous emission from a single quantum dot in a three-dimensional microcavity. *Physical Review Letters* 86: 3903–3906.

Peter E, Senellart P, Martrou D, Lemaître A, Hours J, Gérard J M, and Bloch J (2005) Exciton-photon strong-coupling regime for a single quantum dot embedded in a microcavity, *Physical Review Letters* 95(067401): 1–4.

Purcell E M (1946) Spontaneous emission probabilities at radio frequencies. *Physical Review* 69: 681–681.

Reithmaier J P, Sek G, Loffler A, Hofmann C, Kuhn S, Reitzenstein S, Keldysh L V, Kulakovskii V D, Reinecke T L, and Forchel A (2004) Strong coupling in a single quantum dot–semiconductor microcavity system. *Nature* 432(1): 97–200.

Reitzenstein S (2012) Semiconductor quantum dot–microcavities for quantum optics in solid state. *IEEE Journal of Selected Top Quantum Electronics* 18(6): 1733–1746.

Reitzenstein S and Forchel A (2010) Quantum dot micropiollar. *Journal of Physics D: Applied Physics* 43(033001): 1–25.

Rempe G, Walther H, and Klein N (1987) Observation of quantum collapse and revival in a one-atom maser. *Physical Review Letters* 58: 353.

Savona, V, L C Andereani, P Schwendimann, and A Quatropani (1995) Quantum well excitons in semiconductor microcavities: Unified treatment of weak and strong coupling regimes. *Solid State Communications* 93: 733

Skolnick M S, Fisher T A, and Whittaker D A (1998) Strong coupling phenomena in quantum microcavity structures, *Semiconductor Science and Technology* 13: 645–669.

Suhr T, Gregersen N, Yvind K, and Mørk J (2010) Modulation response of nano LEDs and nanolasers exploiting Purcell enhanced spontaneous emission. *Optics Express* 18(11): 11230–11241.

Vahala K J (2003) Optical microcavities. *Nature* 424: 839–846

Weisbuch C, Nishioka M, Ishikawa A, and Arakawa Y (1992) Observation of the coupled exciton photon mode splitting in a semiconductor quantum microcavity. *Physical Review Letters* 69: 3314–3317.

Yamamoto Y, Machida S, and Björk G (1991) Microcavity semiconductor laser with enhanced spontaneous emission. *Physical Review A* 44(11): 657–668.

Yoshie Y, Scherer A, Hendrickson J, Khitrova G, Gibbs H M, Rupper G, Ell C, Shchekin O B, and Deppe D G (2004) Vacuum Rabi splitting with a single quantum dot in a photonic crystal nanocavity. *Nature* 4 32: 200–203.

12

Bose–Einstein condensation

12.1 Introduction

There are two types of particles in nature: the first type, called fermions, obeys Fermi–Dirac statistics, while the distribution of the second type, called bosons, is governed by Bose statistics. While fermions possess half-integer spins; spins associated on the other hand with bosons have integer numbers.

The property of fermions is determined by the Pauli principle: no two electrons having the same set of quantum numbers can occupy the same state. That means a state may be occupied by two electrons of opposite spins. Such a restriction does not apply to bosons. Bosons are named after the Indian physicist Satyendra Nath Bose, who first gave a complete derivation of Planck's radiation formula (1924). On the basis of this formulation, Einstein predicted in 1924 a state of matter in which separate atoms or subatomic particles coalesce into a single quantum mechanical entity (Masters 2013). This state of matter is termed as Bose–Einstein (BE) condensate and can be described by a wavefunction on a near microscopic scale. In other words, under condensation conditions, all the boson particles can condense into the lowest lying quantum state. The condition to be satisfied to observe BE condensation (BEC) is that the thermal de Broglie wavelength of bosons should be comparable to their average separation. Thus, the temperature and density of particles play an important role. A macroscopic coherence is an attribute to the condensed state.

It took almost 70 years after the prediction by Einstein to demonstrate for the first time BEC for atoms. Eric Cornell and Carl Weiman (2001) cooled a gas of rubidium atoms to 1.7×10^{-7} K to observe BEC. Sophisticated cooling methods were developed to lower the gas temperature down to micro-Kelvin range to demonstrate such type of condensation or phase transition. Almost at the same time, Wolfgang Ketterle created a BEC with sodium atoms (2001). These researchers received the 2001 Nobel Prize for Physics for their ground-breaking work (see Cornell 2019; Ketterle 2019).

The study of BEC in other atomic systems received a great deal of attention after the first successful demonstrations. It is natural to assume that attempts to observe condensation at higher temperatures have been one of the major goals.

The possibility of observing BEC in semiconductors was investigated as early as in 1960s by Moskalenko (1962); Blatt et al (1962); Casella (1963); and Keldysh and Kozlov (1968). Existing theories indicate that the critical temperature below which

Semiconductor Nanophotonics. Prasanta Kumar Basu, Bratati Mukhopadhyay, and Rikmantra Basu, Oxford University Press.
© P.K. Basu, B. Mukhopadhyay, R. Basu (2022). DOI: 10.1093/oso/9780198784692.003.0012

condensation is possible depends inversely on the particle mass. Since excitons have a small mass compared to mass of atoms, the studies first focused on excitons in bulk semiconductors. In particular, the materials of choice were III–V and II–IV semiconductors. A few experiments demonstrating the signature of condensates were reported in 1990s and in the first decade of 2000. However, questions arose whether the experimental data truly demonstrated BEC. Other semiconductor systems in the form of quantum nanostructures were also investigated. These include excitons in coupled Quantum Wells (QW)s, coulomb drag experiments in coupled two-dimensional electron gas (2DEG), and polaritons in microcavities (Butov et al 2002). So far, reports on condensates in bulk and coupled QW structures are not unambiguous. On the other hand, results for microcavity polaritons are most encouraging and a few authors have reported condensation at room temperature even. A new kind of laser: the polariton laser, requiring very low pump power to reach threshold, has been reported. Even room temperature operation of electrically pumped polariton laser has been achieved (Bhattacharyya et al 2013, 2014).

In this chapter, we shall first introduce the basic concept of BEC, and define the relevant parameters like critical temperature, critical density, and so on. Thereafter, attention will be given to the study of BEC in semiconductors. The four basic systems mentioned previously, the underlying principles involved, conditions to be met, lifetime issues and loss mechanisms, the role of ortho and paraexcitons, and other relevant issues will be discussed. Some of the representative experimental demonstrations in meeting the goals will be mentioned. It is beyond the scope of this chapter to give a comprehensive review of all the work. The main emphasis is given on microcavity polaritons.

This chapter will mention some of the important applications of BEC. Finally, a detailed discussion on polariton laser and its modeling will be given.

12.2   Elements of Bose–Einstein condensation

We first illustrate the phenomena of BEC by considering a gas of molecules obeying Bose–Einstein (BE) statistics. After discussing the basic properties, we then point out what is meant by condensation and the temperature and density of particles needed to achieve the condensation (see London 1938). The following treatment has been taken from the book by Agarwal and Meisner (1988).

12.2.1   Bose–Einstein distribution

We consider an ideal Bose gas containing N molecules in a volume V. The most probable number of particles having energy E_i is given by

$$n_i(E_i) = \frac{g_i}{[exp(E_i - \mu)/k_B T] - 1} \tag{12.1}$$

where g_i denotes the degeneracy of the ith level and μ is called the chemical potential. Introducing normalized parameters $\alpha = \mu/k_B T$, $\beta = 1/k_B T$, we may express the total number of particles as

$$N = \sum_{i=0}^{\infty} n_i = \frac{g_0}{exp[\beta(E_0 - \mu)] - 1} + \frac{g_1}{exp[\beta(E_1 - \mu)] - 1} + \dots = n_0 + n_1 + \dots \quad (12.2)$$

The sum is over all levels i. Since population in any level cannot be negative, the difference $(E_i - \mu)$ for all energy levels must be greater than zero, that is $\mu \leq 0$.

The sum in Eq. (12.2) can be replaced by an integral by replacing g_i by the density-of-states(DOS) given as

$$g(E)dE = \frac{V}{4\pi^2 \hbar^3}(2m)^{3/2}E^{1/2}dE \quad (12.3)$$

where m is the mass of the particle. Eq. (12.2) is now written as

$$N = \int_0^{\infty} \frac{g(E)dE}{exp(-\beta\mu)exp(\beta E) - 1} = \frac{V}{4\pi^2 \hbar^3}(2m)^{3/2}\int_0^{\infty} \frac{E^{1/2}dE}{(1/\eta_a)exp(\beta E) - 1} \quad (12.4)$$

The integral in Eq. (12.4) may be written in the following form

$$N = \frac{V}{\lambda^3}F_{3/2}(\eta_a) = Vn_Q F_{3/2}(\eta_a) \quad (12.5)$$

In Eqs. (12.4) and (12.5), $\eta_a = exp(\beta\mu)$ is called the absolute activity, $n_Q = 1/\lambda^3$, is the quantum concentration, that is, the concentration associated with one particle in a cube having each side equal to λ, and $\lambda = (2\pi\hbar^2/mk_B T)^{1/2}$. Using the symbol $x = \beta E$, we may write

$$F_{3/2}(\eta_a) = \frac{2}{\sqrt{\pi}}\int_0^{\infty} \frac{x^{1/2}dx}{(1/\eta_a)exp(x) - 1} = \frac{2}{\sqrt{\pi}}\int_0^{\infty} dx x^{1/2}\eta_a e^{-x}(1 + \eta_a e^{-x} + \eta_a^2 e^{-2x} + \dots)$$

$$(12.6a)$$

Eq. (12.6a) may be written as

$$F_{3/2}(\eta_a) = \eta_a + \frac{\eta_a^2}{2^{3/2}} + \frac{\eta_a^3}{3^{3/2}} + \dots = \sum_{n=1}^{\infty} \frac{\eta_a^n}{n^{3/2}} \quad (12.6b)$$

which is a special case of the general class of the function

$$F_s(\eta_a) = \sum_{n=1}^{\infty} \frac{\eta_a^n}{n^s} \quad (12.6c)$$

The function decreases with increasing value of the argument.

It is of interest to obtain the value of $F_{3/2}$. From Eq. (12.6b)

$$F_{3/2}(\eta_a = 1) = \sum_{n=1}^{\infty} \frac{1}{n^{3/2}} = \zeta\left(\frac{3}{2}\right) = 2.612 \tag{12.7}$$

where ζ is the Riemann zeta function.

Using Eq. (12.5), we may express the total number N in terms of a maximum temperature T_c, called the critical temperature, corresponding to the maximum value of $\eta_a = 1$. Thus

$$N = \frac{V}{\lambda_0^3}.2.612 = V\left(\frac{mk_B T_c}{2\pi\hbar^2}\right)^{3/2} 2.612 \tag{12.8}$$

Therefore

$$T_c = \frac{2\pi\hbar^2}{mk_B}\left(\frac{N/V}{2.612}\right)^{2/3} = 3.312\frac{\hbar^2(N/V)^{2/3}}{mk_B} \tag{12.9}$$

12.2.2   Bose–Einstein condensation

It appears that Eq. (12.5) does not have a solution for $T < T_c$. However, since this difficulty does not arise in the sum given by Eq. (12.2), therefore by improperly changing the summation into an integral in (12.5), large contributions from the first few terms in (12.2) are neglected. This is more serious for low temperatures, $T < T_c$.

For small η_a or large $\exp(-\beta\mu)$, the terms with lowest energies do not contribute much. Therefore, the use of integral introduces little error. However for $\eta_a \to 1$, or small $\exp(-\beta\mu)$, the first few terms in (12.2) become important. So, replacement of summation by an integral is not justified. At sufficiently low temperature, it is true that

$$n_1 = \frac{1}{exp[\beta(E_1 - \mu)] - 1} \ll \frac{1}{exp[\beta(E_0 - \mu)] - 1} = n_0 \tag{12.10}$$

In fact, at $T \to 0$, almost all the particles occupy the ground state, so that

$$\lim_{T \to 0} n_0 = N \cong \frac{1}{exp[\beta(E_0 - \mu)] - 1} \cong \frac{k_B T}{E_0 - \mu} \tag{12.11}$$

Because the wavefunction is symmetric for the Bose–Einstein (BE) case, there is no restriction on the number of particles in a state, unlike only two electrons per state for the Fermi-Dirac (FD) statistics.

Example 12.1 *Let $T = 1K$, and $N = 10^{28}$. This gives $E_0 - \mu \approx 1.38 \times 10^{-45} J$.*

At low temperature, the chemical potential is very close to the energy state E_0, and all particles occupy the lowest energy state. This is called the BEC.

For $T \to 0$, the number of particles in all other states excluding the ground state can be approximated by the following sum

$$N - n_0 = \sum_{i=1}^{\infty} n_i \approx \frac{V}{4\pi^2 \hbar^3} (2m)^{3/2} \int_0^{\infty} \frac{E^{1/2} dE}{(1/\eta_a) exp(\beta E) - 1} = \frac{V}{\lambda^3} F_{3/2}(\eta_a) \qquad (12.12)$$

Using (12.8) to eliminate V, Eq. (12.12) may be cast into

$$N = n_0 + N \left(\frac{T}{T_0} \right)^{3/2} \frac{F_{3/2}(\eta_a)}{2.612} \qquad (12.13)$$

We now consider two limiting cases: $T < T_c$ ad $T > T_c$.

(a) $T < T_c$

Assume that $E_0 = 0$ and $g_0 = 1$. Then $n_0 \approx -\frac{k_B T}{\mu}$.

At low temperatures, μ is very close to zero and $\eta_a \to 1$, signifying quantum region. Therefore, for energy levels above the ground level we put $\eta_a = 1$, and write (12.13) as

$$n_0 \equiv N_0 = N - N' = N \left[1 - \left(\frac{T}{T_c} \right)^{3/2} \right] \qquad (12.14)$$

where $N' = N(T/T_c)^{3/2}$ is the number of particles in the excited state.

Example 12.2 *In the experiment with Rubidium (Rb) atoms, the total number of atoms is 10^4. When $T = 0.9\ T_c$, the number of atoms in the ground state is 1465.*

A plot of N/N_0 as a function of T/T_c is shown in Fig. 12.1(a), in which the number of particles in the excited states equals the total number of particles at the condensation

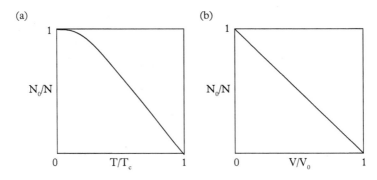

Figure 12.1 *Number of particles in the ground state as a function of (a) temperature and (b) volume.*

temperature $T_0 = T_c$. Below this temperature, more and more particles tend to occupy the ground state and at lowest temperature all the particles occupy the ground state. The BE gas is then degenerate and enters into the quantum regime. The critical temperature T_c is also called the degeneracy temperature.

Instead of expressing the particle number as a function of temperature, one can express it in terms of a critical volume V_0 such that at temperature T

$$N = \frac{V_0}{\lambda_0^3}.F_{3/2}(1) = V_0\left(\frac{mk_BT}{2\pi\hbar^2}\right)^{3/2}2.612 = V_0 n_Q 2.612 \qquad (12.15)$$

Eq. (12.15) is now used to eliminate T in Eq. (12.13) and one obtains

$$N = n_0 + N\left(\frac{V}{V_0}\right)\frac{F_{3/2}(\eta_a)}{2.612} \cong N_0 + N\frac{V}{V_0} \qquad (12.16)$$

Thus

$$N_0 = N\left(1 - \frac{V}{V_0}\right), V < V_0 \qquad (12.17)$$

The variation is shown in Fig. 12.1(b). BE condensation occurs below T_c or V_0.

We can write (12.8) as

$$\lambda_0^3 = \frac{V}{N}.2.612 \qquad (12.18)$$

At critical temperature, de Broglie wavelength is of the order of average inter-particle distance. There is now an overlap of wavefunctions, and the quantum effect sets in.

(b) $T > T_c$

In this regime we have $\eta_a > 1$, signifying classical limit. In Eq. (12.13), the first term n_0 on the right-hand side becomes zero and the second term increases as $T^{3/2}$. Thus, Eq. (12.12) reduces to $N = \frac{V}{\lambda^3}F_{3/2}(\eta_a)$ and from Eq. (12.13)

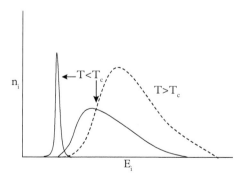

Figure 12.2 *Distribution function for particles for an ideal BE gas below and above critical temperature.*

$$F_{3/2}(\eta_a) = (T_c/T)^{3/2}F_{3/2}(1) = (T_c/T)^{3/2}2.612 = (\lambda/\lambda_0)^{3/2}2.612 \qquad (12.19)$$

For $T \gg 0$, the ground state is practically empty and most of the upper states are occupied. In fact, for $\eta_a \ll 1$, $F_{3/2}(\eta_a) \cong \eta_a$ (see Eq. (12.6b)) and Eq. (12.19) becomes
$$\eta_a = (\lambda/\lambda_0)^{3/2}2.612 = (N/V)\lambda^3 = e^{-\alpha} \text{ (classical limit)}$$
The distribution functions for $T < T_c$ and $T > T_c$ are shown in Fig. 12.2.

Example 12.3 *The de Broglie wavelength of particles is expressed as $\lambda_{dB} = \frac{2\pi\hbar}{\sqrt{2mk_BT}}$. For Rb, $M = 87 \times 1.7 \times 10^{-27}$ kg and atomic radius = 290 pm In order that de Broglie wavelength should equal 2r, $T = 0.32$ K.*

Example 12.4 *In an early experiment the density achieved for atomic hydrogen was $1.8 \cdot 10^{20}$ atoms/m³. Using Eq. (12.9) and 1.7×10^{-27} kg as the mass, the calculated value of $T_C = 50$ μK was in perfect agreement with the experimental value.*

Example 12.5 *In the first observation of the BEC a weakly interacting gas was produced by relatively heavy atoms of ^{87}Rb. 10,000 rubidium-87 atoms were confined within a 'box' with dimensions ~ 10 μm (the density $\sim 10^{19}$ m^{-3}). The expression for the critical temperature, T_c is $T_c = 3.31\hbar^2 M^{-1}n^{2/3}/k_B$ (see Eq. 12.9). The calculated value with $M = 87 \times 1.7 \times 10^{-27}$ kg is $\sim 8 \times 10^{-8}$ K, which is of the same order of the experimental values.*

12.3   BEC in semiconductors

It appears from Eq. (12.9) and the previous example that the BEC criterion may be satisfied at higher temperature by bosons having smaller mass. Excitons in semiconductors have been considered as a promising candidate for BEC, since the first such proposals came from Moskalenko (1962); Blatt et al (1962); and Casella (1963). Their work is followed by Keldysh and Kozlov (1968). The following example given by Keldysh and Kozlov will illustrate the possibility of observing BEC at high temperature using excitonic systems

Example 12.6 *The expression for the critical temperature, T_c is $k_BT_c = 3.31\hbar^2 M^{-1}n^{2/3}$. For excitons in usual semiconductors, $M \sim 10^{-31}$ kg and $N \sim 10^{18}$ cm^{-3}. This gives $T_c \sim 266$K.*

Another system of interest is exciton-polariton. These quasi particles arise due to interaction between photons and excitons. Exciton-polaritons (EP or in short, polaritons) also possess very low effective mass and are thus suitable candidates for observation of BEC in semiconductors.

Experimental demonstration of BEC in semiconductors is beset with several constraints and considerations. First to note is that excitons and EPs are only approximately bosons. At high density and high temperature, excitons for example, dissociate to form electron-hole plasma or even electron-hole liquid. Second, excitons are created by shining light on a semiconductor. Such particles occupy higher energy states and are therefore *hot*, meaning thereby that their mean energy is higher than that of the lattice (Basu and Kundu 1985). These hot excitons quickly relax to the equilibrium states by interactions, mainly with phonons. On the other hand, BEC needs quasi-stable states. This requires that the lifetime of excitons must be quite high compared to electron-hole recombination time. Other loss processes, as for example an Auger recombination, prove to be detrimental. In addition to these requirements, excitons must have a particular state. So called dark excitons are not suitable for BEC (Combescot et al 2007).

There have been numerous attempts to observe BEC in semiconductors. The work may be classified into four groups: (1) excitons in bulk, (2) excitons in coupled quantum wells(QW)s, (3) coulomb drag experiments in couple 2DEG and (4) polaritons in microcavities (Butov et al 2002). Many authors so far have claimed to have observed BECs, which have not been substantiated or rather questioned by other workers. To date, the picture is not clear.

The following sections will describe the work reported for different semiconductor systems. The studies on BEC involving bulk excitons and excitons in coupled QWs will be briefly described, mentioning the important results and basic mechanisms that hinder successful observation of condensation. More promising results have been obtained with polaritons in microcavities. Some representative studies in this connection will be mentioned. Following this, the applications will be listed and, in particular, work on polariton laser operating at room temperature with current injection and the modelling issues related to this kind of laser will be described. The Coulomb drag experiments will not be discussed in this chapter, but the readers are suggested to read the paper by Butov et al (2002) for an introduction.

12.4 Bulk excitons

As already stated, BEC of excitons in bulk semiconductors was predicted independently by Moskalenko (1962); Blatt et al (1962); and Casella (1963). Both electrons and holes possess half integer spins, and when they are bound together by Coulomb interaction, the resulting quasi particle, the exciton, possesses integer spin to act as a boson. A similar situation arises for Cooper pairs.

Two conditions are to be met to observe BEC of excitons: (1) the number of excitons should be conserved; it means that their lifetime should be quite high, higher than the time it takes to thermalize to a well-defined temperature, and (2) the density of excitons should not be too high, since at high density competing phases like electron-hole plasma occur to destroy the binding of the pairs.

Assuming that under the previously mentioned conditions excitons can be treated as a weakly interacting Bose gas, Snoke and Kavoulakis (2014) derived the expression for the critical temperature for BEC. For quantum coherent effects to be important, the thermal de Broglie wavelength of the bosonic particles must be comparable to or greater than the interparticle spacing. The de Broglie wavelength may be expressed as

$$\frac{\hbar^2 k^2}{2m} = \frac{\hbar^2 (2\pi)^2}{2m\lambda_{dB}^2} \sim k_B T \tag{12.20}$$

Thus

$$\lambda_{dB} \sim \frac{2\pi\hbar}{\sqrt{2mk_B T}} \tag{12.21}$$

In three dimensions, the average interparticle distance $r_s \sim n^{-1/3}$, where n is the density of the particles. We now set $\lambda_{dB} \sim r_s$ and obtain

$$n \sim \frac{2^{3/2}}{(2\pi)^3} \frac{(mk_B T)^{3/2}}{\hbar^3} \tag{12.22}$$

Therefore

$$T \sim \frac{(2\pi\hbar)^2}{2mk_B} n^{2/3} \tag{12.23}$$

Standard calculation of statistical mechanics under equilibrium gives 0.17 as a prefactor to the right-hand side of the previous equation.

Experimental work to observe BEC of excitons in bulk semiconductors started with cadmium selenide (CdSe) and copper chloride (CuCl). The most complete study on CuCl was done in early 1980s. However, almost at the same time excitons in semiconductor Cu_2O was identified as a promising and better candidate than excitons or biexcitons in CuCl. Snoke and Kavoulakis (2014) has published a comprehensive review on the work done on excitons in Cu_2O over three decades, giving a large number of useful references.

The direct energy gap between the highest valence and lowest conduction bands is 2.17 eV in Cu_2O (at $T = 10$ K). A yellow-band exciton in this semiconductor is the positronium-like bound state of an electron and a hole with a dipole-forbidden gap of 2.173 eV.

In Cu_2O, holes in the two highest states of the valence band and electrons in the two s-like states of the lowest conduction band form the lowest-energy excitons. These four states are split into a triplet and a singlet state by electron-hole exchange interaction. The separation between the singlet and triplet states is 12 meV. The lowest lying singlet state is called the paraexciton and the higher lying triplet state is called orthoexciton. These para- and ortho-exciton states are the only important ones in the study of BEC.

From the symmetry of the crystal it is known that the lowest paraexciton state is optically inactive, that is, it has zero oscillator strength for interaction with photons. Orthoexcitons however have a symmetry allowed radiative recombination process. Though optically inactive paraexcitons have in principle infinite lifetime making them attractive for observation of BEC, weak phonon-assisted recombination processes as well as impurity assisted recombination yield a typical lifetime ranging from hundreds of nanoseconds to microseconds. It is to be noted that this value is quite long compared to the lifetime in most semiconductors and is order-of-magnitude larger than thermalization times of excitons via electron-phonon interaction, which is tens of picoseconds.

Example 12.7 *An approximate estimate of binding energy of the 1s exciton in Cu_2O is made with $m_e = 0.99\, m_0$, $m_h = 0.58\, m_0$ and a relative permittivity $\varepsilon = 7.2$. The reduced mass is $\mu = 0.37\, m_0$ and the binding energy is 97 meV.*

The 1s exciton exhibits an unusually strong binding energy $E_x = 153$ meV due to the relatively large electron and hole effective masses under dielectric screening (Wolfe and Jang, 2014).

It has been found that biexciton states are non-existent and even if they exist, the binding energy is low ~ 1 *meV*. This fact and the long excitonic lifetime in Cu_2O suggest that the excitons are good candidates for BEC.

A magneto-optical trap has been used to hold cold atoms for BEC investigations. This idea has been applied to Cu_2O crystal. A parabolic potential trap is created by inhomogeneous stress to facilitate exciton densities approach BEC.

Unfortunately however, excitons in high-purity crystals undergo a density-dependent lifetime that opposes BEC. A strong exciton decay process was found to limit their densities to well below the critical density for condensation. The decay is due to an Auger process, which occurs by inelastic scattering of two excitons. In the process, one exciton recombines to create ionization of the other, producing a hot electron–hole (e–h) pair which thermalizes into a single exciton.

The decay rate of excitons by Auger process is described by $dn/dt = -An^2$, where A is the exciton Auger constant, and n is the exciton density. Measurements of Auger constant indicate that experimental values $\sim 10^{-16} \text{cm}^2 \text{ns}^{-1}$ are five orders higher than theoretical values. Wolfe and Jang (2014) proposed that the short lifetimes of excitons are due to their binding into short-lived biexcitons. The e–h pairs in the molecule are in close proximity and as such a much faster Auger decay results than that of two free excitons at the thermodynamic gas density. The shorter exciton lifetime is the result of the rapid thermodynamic binding of excitons into molecules.

Further investigation into other mechanisms for shortening of exciton lifetime is in progress, see for example, Hoang (2019).

In conclusion, in spite of a lot of effort during the past several decades, compelling evidence of BEC in Cu_2O has remained elusive.

12.5   Indirect excitons in coupled quantum wells

Butov and coworkers (1998, 2001, 2002) and High et al (2012) explored creation of degenerate Bose gas of spatially indirect excitons in CQWs. The reasons for selecting such structures are that the excitons have a long lifetime due to spatial separation of electrons and holes, and the cooling time for excitons is quite short. The spatial separation of elec-trons and holes, as illustrated in Fig. 12.3(a), gives rise to a long recombination lifetime, more than three orders-of-magnitude longer than that in direct single QW. The hot photoexcited excitons cool to the temperature of the cold lattice by emission of acoustic phonons. Whereas in bulk semiconductors the ground state mode $E = 0$ couples to a single state of energy $E = 2Mv_s^2$, where v_s is the sound velocity, in QWs the continuum of energy states $E \geq E_0$ are involved increasing the cooling rate via enhanced scattering rate.

It is worth noting that confinement of atomic gases within a potential trap facilitated the condensation of Bose gas. In a QW, an in-plane potential fluctuation, as illustrated in Fig. 12.3(b), exists and excitons drift toward the bottom of the trap and due to the confinement effect T_c increases.

In particular, in the case of confinement by an in-plane two-dimensional (2D) harmonic potential the critical temperature is (Butov et al 2002):

$$T_c = \frac{1}{k_B} \frac{\sqrt{6}}{\pi} \hbar\omega \sqrt{\frac{N}{g}} \tag{12.24}$$

where $\hbar\omega$ is the quantization energy, N is the number of excitons in the trap, and g is the spin degeneracy of excitons.

If the trap is a 2D square potential (2D box) then the expression for critical temperature is

$$T_c = \frac{4\pi\hbar^2 n}{2m_{ex}gk_B} \frac{1}{ln(nS/g)} \tag{12.25}$$

where m_{ex} is the exciton mass, and S is the area of the box.

In the experiments, an n+-i-n$^+$ structure is grown, the i-region of which incorpo-rates gallium arsenide (GaAs)/ gallium aluminum arsenide (AlGaA)s coupled MQW structure. Under normal conditions, the excitons are direct in nature. However, an ex-ternal bias applied to the n$^+$ layers tilts the band, as shown in Fig. 12.3(a) and makes the spatially indirect excitonic state as the ground state. The random potential fluctua-tions at the interface due to interface roughness or due to alloy composition fluctuation (Basu 1997), as shown in Fig. 12.3(c) gives rise to traps as shown in Fig. 12.3(b). The structure is excited by a focused laser pulse of rectangular shape. As indicated in (c), random fluctuation gives rise to many traps in a given sample, out of which the exciton gas will, by itself, select those with optimal parameters for condensation. In such a trap, the degenerate exciton gas will show a strong in-plane photoluminescence (PL) emission peaked at the bottom of the trap, which should be a clear signature of condensation. In the experiments, the spatial distribution of PL spectra could be mapped. The intensity is increased at the bottom of a trap as the temperature is lowered below condensation temperature.

Apart from electrostatic traps, other types of traps including strain-induced types, traps caused by laser induced interdiffusion, laser-induced and magnetic traps are also investigated. However in all these studies, no signature of spontaneous coherence of excitons in these traps have been observed (see High et al 2012 for references).

(a)

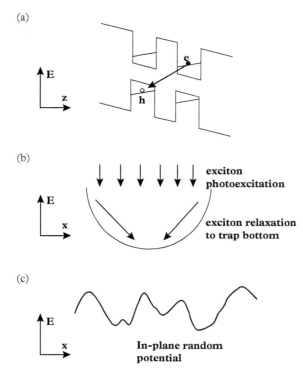

(b)

(c)

Figure 12.3 *(a) Energy band diagram of coupled QW showing the formation of e and h subbands in different wells and spatially indirect recombination process: Z is the direction of growth of QWs; (b) process of exciton relaxation to the bottom of the in-plane trap; (c) in-plane random potential fluctuation. In (b) and (c) X is along the layer plane.*
Source: Reproduced figure with permission from L V Butov, C W Lai, A L Ivanov, A C Gossard, and D S Chemla (2002) *Nature* 417(2): 47–52. Copyright Springer Nature

High et al (2012) claim to have observed BEC in an electrostatic trap for indirect excitons using a diamond-shaped electrode, in which the diamond trap creates a confining potential with the exciton energy gradually reducing toward the trap center.

Example 12.8 *In the case of confinement by an in-plane 2D harmonic potential the critical temperature is $T_c = (\sqrt{6}/\pi k_b)\hbar\omega\sqrt{n/g}$, where $\hbar\omega$ the quantization energy, n is the number of excitons in the trap, and g is the spin degeneracy of excitons.*

In the experiment of High et al (2012), $\omega = \sqrt{\omega_x \omega_y}\omega_x$, $\omega_x = 4.10^9 s^{-1}$ and $\omega_y = 3.10^{10} s^{-1}$, $g = 4$ and $n = 3.10^3$. The calculated critical temperature is 1.79 K, which was close to the experimental value.

The issues and challenges in observing BEC at higher temperatures are discussed in detail by Zimmerman (2008). Combescot et al (2007, 2017) pointed out the role of dark excitons and physical processes related to indirect excitons. The treatments are too involved to be included here. The interested readers are encouraged to study the articles.

12.6   Polariton

The condition to observe BEC in a microcavity is to ensure number conservation of photons and to establish that the photons are under equilibrium. The large Q cavity almost ensures that there is very small leakage of photons out of the cavity. The equilibrium condition can be reached by making the photons interact. This may be accomplished by photon-photon interaction in the presence of non-linearity. Still a better method is to transfer the energy of photons to create electronic excitation. This results in the creation of excitons, the bound electron-hole pair (EHP).

As noted in earlier chapters, a strong electric field develops in the microcavity that leads to coupling between light and matter. Usually the interaction occurs between photons and excitons in a microcavity, and the result is the formation of a quasi-particle named as exciton-polariton, or rather polariton.

Polaritons in optical microcavity have been viewed as a strong candidate to observe BEC at high temperatures. The effective mass of a polariton for zero wave vector is at least four orders of magnitude smaller than the excitonic mass. This implies that the critical temperature needed to observe polaritonic BEC is four orders higher than that in the case of bare excitons for the same particle density. The second advantage offered is that a polariton can easily extend into a phase coherent wavefunction in space even in presence of defects, but an exciton is localized in a fluctuating potential inside a crystal, killing the possibility of BEC. Though polaritons have a very short lifetime as the photon leakage rate from microcavity is large when the cavity photon resonance is detuned at a higher energy than the exciton resonance, the lower polariton has a longer lifetime and shorter cooling time, favouring thermalization.

In the following, we shall first introduce the microcavity polariton. The nature of photon dispersion in a microcavity and the definition of effective mass of photons will then be discussed. The coupling between photons and excitons, giving rise to polaritons will then be introduced and the effective mass of polaritons will be defined. This will be followed by a list of the characteristic features of polaritons. The experimental techniques to observe BEC by microcavity polaritons and the method of analysis will be presented next. Finally, the section will conclude by giving some representative results and the present status of observing BEC.

The reader may consult reviews by Deng et al (2010) and Keeling and Berloff (2011) as well as a popular article by Snoke and Keeling (2017).

12.6.1   Microcavity polaritons

Polaritons are hybrid particles of light and matter, which arise due to strong light-matter interaction in a microcavity. The matter here is a Wannier–Mott exciton formed by a bound pair of an electron and a hole. The light component is a strongly confined photon field in a microcavity, such as a pair of distributed Bragg reflectors (DBRs). As mentioned in Chapters 10 and 11, a strong electric field exists in a microcavity, and it plays a key role in such a strong interaction.

12.6.2 Microcavity dispersion relation

To understand the occurrence of the hybrid mode due to interaction between photons and excitons in a microcavity, let us consider first the dispersion relation of photons in a cavity. In free space, the dispersion relation for photons is given by $E = \hbar\omega = \hbar k(c/\eta)$, where k is the wave number and η is the refractive index.

In a microcavity, such as a Fabry–Perot (FP) resonator, the electromagnetic (EM) waves form a standing wave pattern along the z-direction, the direction perpendicular to the mirror planes. A similar situation arises for microwaves in a parallel plate waveguide, or in a QW for electron waves. Considering the discrete nature of the EM wave along the z-direction, the dispersion relation can now be expressed as

$$\hbar\omega = \hbar(c/\eta)\sqrt{(N\pi/L)^2 + k_{\parallel}^2} \qquad (12.26)$$

where c is the free space velocity of light, η is the RI of the medium forming the cavity, N is an integer, L is the distance between the mirrors and $k_{\parallel}$ is the in-plane wavevector. For small $k_{\parallel}$, the dispersion relation is parabolic, from which the effective mass of photons in a microcavity can be expressed as

$$m_{ph} = \pi\hbar N\eta/Lc \qquad (12.27)$$

Example 12.9 *Assume $N = 1$, RI = 3.5, $L = 0.1 \ \mu m$. The value of photon mass is $1.27 \times 10^{-4}m_0$, where m_0 is the rest mass of electron.*

12.6.3 Quantum well microcavity polariton

As noted in Chapter 10 and Chapter 11, a Quantum well (QW) is placed at the antinode of the resonant field in a semiconductor micrcoacvity to ensure a strong light matter interaction. The $\mathcal{J} = 1$ (heavy-hole) HH exciton doublet then strongly interacts with the photon field. There is exchange of energy between the cavity field and the excitons. If this rate is faster than the decay and decoherence rates for both the cavity field and excitons, the excitation is stored. The elementary excitations are no longer excitons or photons, but a new type of quasi particles known as exciton-polaritons, or polaritons in the present context.

The effects of strong coupling between excitons and cavity photons are described by using quantum mechanical language. It is sometimes more useful to employ a classical model to describe the phenomena (Herzog et al 2019; Novotny 2010; Rodriguez 2016). A classical analogy is two coupled harmonic oscillators (CHO) as shown in Fig. 12.4(a). Under uncoupled conditions, the two oscillators behave independently with their own characteristic oscillation frequencies. When the coupling between them is strong, they begin to exchange energy periodically and the system behaves as a single entity. As in Fig. 12.4(a), the two undamped oscillators are denoted by masses M_A and M_B with respective spring constants k_A and k_B coupled together with spring constant k_C. The equations of motion are written as

$$M_A\ddot{x}_A + k_A x_A + k_C(x_A - x_B) = 0 \tag{12.28a}$$

$$M_B\ddot{x}_B + k_B x_B - k_C(x_A - x_B) = 0 \tag{12.28b}$$

where x_A and x_B are the displacements of the respective balls from their equilibrium positions.

The solution of Eqs. (12.28 a, b) gives

$$\omega_{\pm}^2 = \frac{1}{2}\left[\omega_A^2 + \omega_B^2 \pm \sqrt{\left(\omega_A^2 - \omega_B^2\right)^2 + 4\Gamma^2\omega_A\omega_B}\right] \tag{12.28c}$$

where $\omega_A^2 = \frac{k_A+k_C}{M_A}$, $\omega_B^2 = \frac{k_B+k_C}{M_B}$ and $\Gamma^2\omega_A\omega_B = k_c^2/M_AM_B$

Considering $k_A = k_0$, $k_B = k_0 + \Delta k$, $k_c = \delta k_0$ and $M_A = M_B = M_0$, the previous equation yields

$$\left(\frac{\omega_{\pm}}{\omega_0}\right)^2 = \frac{1}{2}\left[2(1+\delta) + \frac{\Delta k}{k_0} \pm \sqrt{\left(\frac{\Delta k}{k_0}\right)^2 + 4\delta^2}\right]$$

Here δ is taken as 0.08. A plot of frequency spectra as a function of detuning is shown in Fig. 12.4(b).

It is beyond the scope of the present book to present the full quantum mechanical analysis. The total Hamiltonian in the coupled system can be written as a sum of Hamiltonians for cavity photons, excitons and the interaction term. An analogous expression for the coupled system is given in connection with Jaynes–Cummins model in Section 11.7.1 in terms of second quantized notation. We shall not write the full form but ask the readers to consult the review by Deng et al (2010).

The energies of the polaritons are expressed as

$$E_{LP,UP}(k_{\parallel}) = \frac{1}{2}\left[E_{exc}(k_{\parallel}) + E_{cav}(k_{\parallel}) \pm \sqrt{4g_0^2 + \{E_{exc}(k_{\parallel}) - E_{cav}(k_{\parallel})\}^2}\right] \tag{12.29}$$

where E_{exc} and E_{cav} are, respectively, the exciton and cavity resonance energies, g_0 is the coupling strength between exciton and photons, (LP) and (UP) denote, respectively, the lower polariton and upper polariton branches and $+(-)$ signs refer to UP(LP) eigenenergies.

Eq. (12.29) shows that due to strong coupling there appear two polariton branches, upper and lower.

When the energies of uncoupled exciton and photon coincide, that is $E_{exc} = E_{cav}$, the UP and LP energies are separated by $E_{UP} - E_{Lp} = 2g_0$, which is the minimum separation. This splitting is termed as normal mode splitting, in analogy with Rabi splitting in case of atom-cavity interaction. Fig. 12.5 shows the energies of UP and LP branches as a

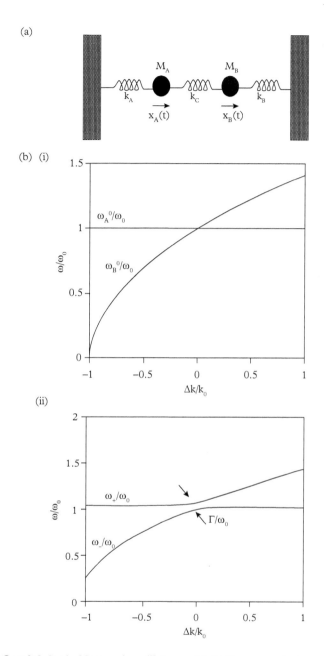

Figure 12.4 *(a) Coupled classical harmonic oscillator model. (b) Frequency spectra as a function of detuning; (I) no coupling, (II) detuning δ = 0.08.*

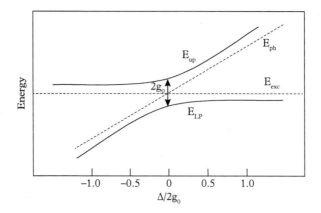

Figure 12.5 *Energies of UP and LP as a function of detuning from the resonance condition. The energies of uncoupled excitons and photons are also shown. At large detuning, the polaritons tend to follow the curves of excitons and photons.*

function of detuning Δ from the resonance condition, normalized by the normal mode splitting. The energies of uncoupled excitons and photons are also plotted in the figure. It is found that for large detuning $|E_{cav} - E_{exc}| \gg 2g_0$, LP and UP branches tend to follow the dispersion of excitons and photons, respectively.

12.6.4 Polariton dispersion and effective mass

It is now of interest to obtain the expression of polariton dispersion, that is, the change of energy with in-plane wave vector $k_{\parallel}$, which may be used to determine the effective mass. Let the detuning between UP and LP at $k_{\parallel} = 0$ be denoted by

$$\Delta_0 = E_{cav}(k_{\parallel} = 0) - E_{exc}(k_{\parallel} = 0) \tag{12.30}$$

The polariton dispersion may now be obtained from Eq. (12.29). It is seen that in the region $\hbar^2 k_{\parallel}^2 / 2m_{cav} \ll 2g_0$, the dispersions are parabolic and

$$E_{\text{LP,UP}}(k_{\parallel}) \cong E_{\text{LP,UP}}(0) + \frac{\hbar^2 k_{\parallel}^2}{2m_{\text{LP,UP}}} \tag{12.31}$$

The effective masses of the polaritons are expressed as

$$\frac{1}{m_{\text{LP}}} = \frac{|X|^2}{m_{exc}} + \frac{|C|^2}{m_{cav}} \tag{12.32a}$$

and

$$\frac{1}{m_{UP}} = \frac{|C|^2}{m_{exc}} + \frac{|X|^2}{m_{cav}} \tag{12.32b}$$

where X and C are Hopfield coefficients, the expressions of which are given in the paper by Deng et al (2010). The exciton effective mass is related to its centre-of-mass motion and the cavity photon mass is already given by Eq. (12.27). Since $m_{exc} \gg m_{cav}$, one may write approximately

$$m_{LP}\left(k_{\parallel} = 0\right) = m_{cav}/|C|^2 \tag{12.33a}$$

$$m_{UP}\left(k_{\parallel} = 0\right) = m_{cav}/|X|^2 \tag{12.33b}$$

Example 12.10 *For $\Delta = 0$, $|X|^2 = |C|^2 = 0.5$. Taking cavity mass to be $10^{-4}m_0$, the polariton mass is $2.0 \times 10^{-4}m_0$.*

Example 12.11 *Using $10^{-4}m_0$ for the polariton mass and a density of 10^{13} m^{-2} the critical temperature turns out to be 293 K from Eq. (12.9).*

The expression for eigen energies Eq. (12.29) is modified in the following form when finite lifetime of excitons and cavity photons are taken into account.

$$E_{LP,UP}(k_{\parallel}) = \frac{1}{2}\left[E_{exc} + E_{cav} + i(\gamma_{cav} + \gamma_{exc}) \pm \sqrt{4g_0^2 + \{E_{exc} - E_{cav} + i(\gamma_{cav} + \gamma_{exc})^2\}}\right] \tag{12.34}$$

Here, γ_{cav} is the rate of decay of cavity photons due to imperfect mirrors and γ_{exc} is the decay rate of excitons due to non-radiative processes.

The dispersion relations for the polaritons: the energy versus in-plane momentum of both the UP and LP branches are shown as solid curves in Fig. 12.6. Also included in the figure are the dispersion relations for bulk photons, photons in a microcavity, and for excitons. The dispersion of uncoupled photons in free space is linear as usual and is shown by a solid line. However, that for photons in a cavity is given by Eq. (12.26). For small momentum the curve is parabolic as may be seen from the dashed curve. The E-k diagram for excitons is almost horizontal due to large excitonic mass as compared to photon mass in a cavity.

In the presence of strong coupling, the quantum superposition of excitons and photons give rise to polaritons. The lower and upper polariton branches are separated at zero momentum. Both the branches follow different dispersions. For higher in-plane momentum, the lower branch merges with the curve for excitons, while the upper branch coalesces with the curve for photons. The parabolic nature of dispersion for both the branches allows the concept of effective masses for polaritons, which are quite low as compared to excitonic effective mass. As mentioned already, the low effective mass of

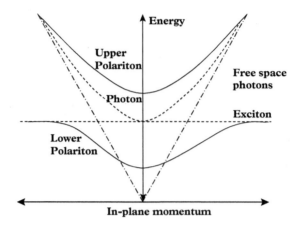

Figure 12.6 *Dispersion diagrams of photons in free space and in a microcavity, of excitons and of polaritons showing its upper and lower branches.*

Table 12.1 *Values of parameters for different BEC systems.*

	Systems		
	Atomic gases	Excitons	Polaritons
Effective mass (in terms of m_0)	10^3	10^{-1}	10^{-5}
Bohr radius (nm)	10^{-2}	10^1	10^1
Particle spacing	10^4	10^3	10^3
Critical Temperature	1 nK- 1 μK	1 mK- 1 K	1-> 300 K
Thermalization time/Lifetime	1ms/1 s ~ 10^{-3}	10 ps/1 ns ~ 10^{-2}	(1–10ps)/(1–10 ps) ~ 0.1–10

LP ensures a higher critical temperature giving rise to the possibility of observing BEC at room temperature.

Comparison of useful parameters of several BEC systems have been made in the form of a Table by Deng et al (2010). We reproduce in Table 12.1 the values quoted by them.

As an application of coupled harmonic oscillator model to obtain polariton dispersion, we use the data for indium gallium nitride (InGaN) multiple quantum well (MQW) enclosed in a GaN microcavity (Wu et al, 2019). The composition and the refractive index(RI) of the alloy are not mentioned. However, we use representative values of the material. The following values are given: $E_{cav} = 3.045$ eV, $E_{exc} = 3.048$ eV, $\Omega = 2g_0 = 45$ meV. The photon effective mass in a cavity of 4 nm width is 6. ×1 $0^{-4}m_0$. RI = 4 is chosen. The exciton is assumed to be dispersionless.

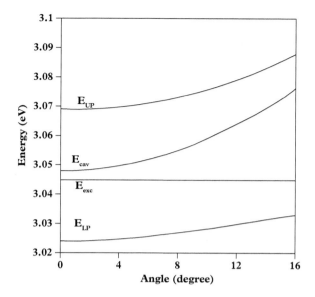

Figure 12.7 *Calculated polariton dispersion as a function of angle for InGaN.*

Example 12.12 *First of all, we give the values of the UP and LP energies at $k_\parallel = \theta = 0$. From Eq, (12.29), we find that $E_{LP} = 3.02\ eV$ and $E_{UP} = 3.074$, the separation being 54 meV.*

The calculated plots of polariton dispersion as a function of angle (proportional to in-plane wave vector: see discussions given later in Section 12.6.4), calculated with parameters given in Example 12.12 are shown in Fig. 12.7. The trend agrees with the variation shown in Fig. 4(b) of Wu at al. (2019).

The dispersion curves for cadmium telluride (CdTe) QWs are shown in Fig. 2 of Keeling et al (2007). This is set as a problem in which the relevant parameters are mentioned.

12.6.5   Experimental observation

For successful observation of BEC of polaritons in a microcavity, the following conditions must be met. (1) The cavity Q must be large to ensure high lifetimes for both the photon and polariton; (2) strong polariton-phonon and polariton-polariton scattering, so that polaritons thermalize very quickly; (3) small exciton Bohr radius and hence large binding energy, so that the saturation density of excitons is high, and (4) strong exciton-photon coupling; this requires large exciton oscillator strength and a high overlap between exciton wavefunction and photon field.

12.6.5.1 Material systems

Experiments have been performed so far by using several material systems. The most preferred choice is to use GaAs/AlAs systems. The growth of GaAs QWs and GaAs/AlAs DBRs by using MBE technique is now standardized. Due to very small lattice mismatch between the two materials strain and defect free QWs can be realized. The interfaces show fluctuations of the order of a monolayer and therefore the inhomogeneous broadening of exciton linewidth is quite low. Values of Q as high as 10^5 or more can be achieved with this system (see references in Chapter 10). Evidence of polariton condensation was obtained in GaAs systems by many authors (Deng et al 2002, 2003, 2007); Balili et al (2007); Lai et al (2007).

CdTe QWs with either CdMnTe or CdMgTe as barriers and CdTe/CdMn(Mg)Te multilayers as DBRs have also been employed for polariton BEC studies. In these systems the lattice mismatch is high; however the larger binding energy and oscillator strength are of advantage. The small Bohr radius makes the saturation density high. The reduced polariton-acoustic phonon scattering does not favour efficient energy relaxation (Richard et al 2005). However, the problems are successfully overcome in the experiments of Kasprzak et al (2006).

Wide band gap materials possess smaller lattice constant, large exciton binding energy, and oscillator strength. These systems are attractive to realize polariton lasers at or above 300 K. Unfortunately however, the growth technique is not as mature as for GaAs systems. It is difficult to find suitable lattice matched layer to grow DBRs. There are several reports on experiments with these material systems, which include zinc selenide (ZnSe), GaN, and zinc oxide (ZnO).

12.6.5.2 Structure design

The primary objective is to achieve maximum coupling between excitons and photons. For this reason the QW is placed at the antinode of the cavity field. For a $\lambda/2$ cavity, the field amplitude becomes maximum at the cavity centre. There are several advantages of using multiple QWs in this regard. Using N_{QW} number of closely spaced multiple thin QWs, the coupling strength $g_0 \propto \sqrt{N_{QW}}$ can be achieved. An increased saturation density of LPs which scales as N_{QW}/a_B^2, where a_B is the Bohr radius, also results with increasing number of QWs. When g_0 becomes comparable with the exciton binding energy, the exciton Bohr radius for LP is further reduced leading to higher saturation density.

12.6.5.3 Pumping methods

Polaritons are created usually by optical pumping. A weak pump signal is highly detuned from the cavity resonance frequency. The particles created by this process give up their energy through various scattering processes and tend to thermalize. In principle, it is possible to create polaritons by electrical injection. However, the method needs very high-quality cavity. The electrical pumping method will be described in a later section.

12.6.5.4   *Detection methods*

Detection of polaritons is made by determining the emissive properties of cavity photons. There is one-to-one correspondence between the in-plane wave vectors of the polaritons and cavity photons. The internal polariton is coupled to external photons by a fixed rate.

The basic properties of polaritons are determined by transmission or reflection measurements. A reflection spectra, for example, give the resonance energy, linewidth of the resonances, and the cavity stop band. Dispersion of the resonances are measured by angle resolved reflection or photoluminescence. By changing the collection angle of the photoluminescence (PL), the energy-momentum distribution, as well as the momentum distribution of polaritons can be measured.

There is a simple relation between the in-plane wave vector $k_\parallel$ and the angle θ between the incident light and the z-direction, the direction perpendicular to the plane of the DBR. The resonance occurs at a wavelength $\lambda_c/\cos\theta$. Noting that

$$E_{cav} = \frac{\hbar c}{\lambda_c}\sqrt{k_\perp^2 + k_\parallel^2} \ and \ k_\parallel = \eta(2\pi/\lambda_c) \tag{12.35a}$$

$$k_\parallel = \eta(2\pi/\lambda_c)\tan\left[\sin^{-1}(\sin\theta/\lambda_c)\right] \approx (2\pi\theta/\lambda_c) \tag{12.35b}$$

By changing the collection angle of the PL, the energy-momentum distribution, as well as the momentum distribution of polaritons can be measured.

12.6.5.5   *Experimental results*

Evidence for phase transition caused by condensation has been first reported by Deng et al (2002) by using GaAs/AlAs structure. Their sample consists of a $\lambda/2$ AlAs cavity sandwiched between two distributed Bragg reflectors (DBRs). The DBRs are made of alternating $Ga_{0.8}Al_{0.2}As$/AlAs $\lambda/4$ layers. Three stacks of QWs, each of which has four 7-nm-thick GaAs QWs separated by 3-nm-thick AlAs barriers, are placed at the antinode position of the photon field. Fig. 12.8 (a) shows the dispersion curves of the QW heavy-hole (HH) exciton, cavity photon, UP, and LP. The cavity photon and QW exciton are on resonance at $k = 0$. The relaxation mechanisms, multiple phonon emission and LP-LP scattering, are also depicted,

The experimental data of integrated PL intensity for different values of pump power are shown schematically in Fig. 12.8 (b). As seen, there are two slowly rising portions in the curve, separated by a sharp, almost steep rise in intensity occurring at the threshold pumping power. There is a sharp rise in polariton number in the lowest LP state at this threshold. It is tempting to accept this as an evidence for BEC. However, the authors indicate that there is a phase transition.

As mentioned already, BEC of polaritons in microcavities are studied by using mainly GaAs/AlAs or CdTe/CdMn(Mg)Te systems. In this connection, the experimental observation of BEC in CdTe/CdMgTe systems by Kasprzak et al (2006) will be described now. The sample used by them consists of 16 QWs with a vacuum field Rabi splitting of 26 meV. The large exciton binding energy of 25 meV and the large number of wells

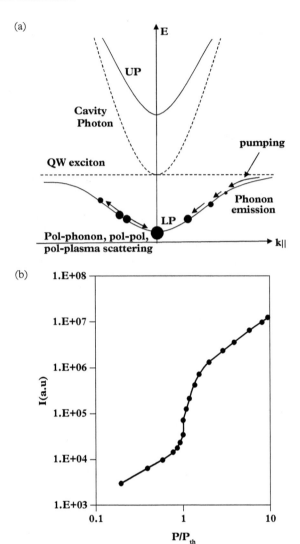

Figure 12.8 *(a) Polariton branches, exciton level, and scattering mechanisms. (b) PL intensity vs. pump power.*

in the DBR ensure strong coupling. Pulses of 1 ms duration with 1% duty cycle derived from a CW Ti:sapphire laser were used to excite the microcavity. The duration of the pulse was four orders-of-magnitude longer than the characteristic times of the system to ensure a steady state condition. The excitation energy of 1.768 eV was well above the cavity resonance energy of 1.671 eV, so that the polaritons created were initially in the incoherent state.

(a)   (b)   (c)

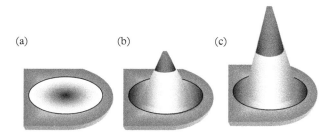

Figure 12.9 *(a): Intensity profile of far-field emission measured by Kasprzak Et Al (2006) at 5 K for three excitation intensities: (a) 0.55 P_{thr}, (b) P_{thr}, (c) 1.14 P_{thr}; where P_{thr}= ¼ 1.67 kW cm^{-2} is the threshold power of condensation. pseudo-3D images of the far-field emission are shown in (a), (b), and (c), the emission intensity is being displayed along the vertical axis (in arbitrary units). In (a), the image is widespread within the angular cone of $\pm 23°$. With increasing excitation power, a sharp and intense peak is formed in the centre of the emission distribution.*

Source: Reproduced figure with permission from Kasprzak J, Richard M, Kundermann S, Baas A, Jeambrun P, Keeling J M J, Marchetti F M, Szyman'ska M H, Andre' R, Staehli J L, Savona V, Littlewood P B, Deveaud B, and Dang L S (2006) Bose–Einstein condensation of exciton polaritons. *Nature* 443: 409–414. Copyright (2006) Springer Nature.

The authors determined the spectral and angular distribution of emission by using a real space imaging set up. The spectral and angular distribution of emission as a function of excitation power may be understood from a 3D pseudo images of the angular distribution of spectrally integrated emission. As shown in Fig. 12.9, the emission is a smooth distribution around $\theta_x = \theta_y = 0$, that is at $k_{\parallel} = 0$ (left). This is expected as the excitation is below threshold. At threshold, the emission intensity at the zero-momentum state becomes predominant as in the middle panel of Fig. 12.9. When the excitation intensity is above threshold, the emission is sharply peaked as in the right-hand part of the same figure.

Fig. 12.10 shows the energy and angle-resolved emission intensities. The width of the momentum distribution shrinks with increasing excitation intensity, and above threshold, the emission mainly comes from the lowest energy state at $k_{\parallel} = 0$.

The emission pattern and the values of radiative lifetime of polaritons yield the values of polariton occupancy. The occupancy of the ground state, its emission energy, and linewidth are shown in Fig. 12.10 as a function of excitation power. The occupancy first increases linearly with excitation power, shows a sharp threshold-like behaviour and then increases exponentially. It should be noted that the occupancy at threshold is unity. The emission blue shift is found to be less than a tenth of the Rabi splitting, at a pump level of ten times the threshold, confirming that the microcavity is still in the strong coupling regime. Significant narrowing of the linewidth, nearly half the value at linear region is found at $k_{\parallel} = 0$. The broadening at higher excitation is due to polariton self-interaction.

Example 12.13 *An idea of polariton lifetime may be obtained from the relation $\tau_{pol}^{-1} = \frac{|X|^2}{\tau_{exc}} + \frac{|C|^2}{\tau_{cav}}$. In the experiment of Sneider et al (2013), $|X|^2 = 0.26$ and $|C|^2 = 0.74$.*

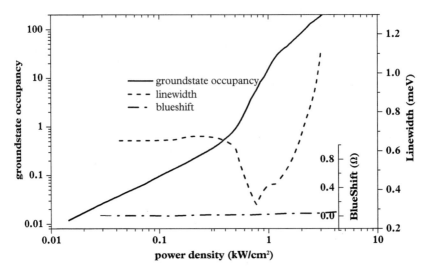

Figure 12.10 *Variation of polariton number, emission energy and linewidth as a function of excitation power.*

Source: Reproduced figure with permission from Kasprzak J, Richard M, Kundermann S, Baas A, Jeambrun P, Keeling J M J, Marchetti F M, Szyman'ska M H, Andre' R, Staehli J L, Savona V, Littlewood P B, Deveaud B, and Dang L S (2006) Bose–Einstein condensation of exciton polaritons. *Nature* 443: 409–414. Copyright (2006) Springer Nature.

$\tau_{exc} = 390\ fs;\ \tau_{cav} = h/4\pi\Delta E = 170\ fs$, *where the linewidth* $\Delta E = 1.54\ meV$ *for* $Q = 1030$; *therefore, a value of* $\tau_{pol} = 199\ fs.is\ obtained.$

The short lifetime of polaritons in earlier experiments raised the question whether polariton condensation is a non-equilibrium effect. As BEC is an equilibrium phenomenon, the observed polaritonic condensate may not truly represent BEC. The issue was resolved in recent experiments by Sun et al (2017), in which a microcavity of very large Q: an order of magnitude higher, has been used. The technique involves growth of double the number of quarter wavelength layers in DBR mirrors. The new microcavity structure has a Q of $\sim 3.2 \times 10^5$. The cavity photon lifetime is ~ 135 ps, leading to a polariton lifetime of 270 ps at resonance. The polaritons are trapped in an annular optical trap. Polaritons within a two-dimensional flat optical trap are seen to unambiguously show thermal Bose–Einstein statistics. This clearly distinguishes polariton condensation from the conventional lasing effect in semiconductor materials.

12.6.6   Room-temperature operations

The obvious way to achieve a room-temperature operation is to design the cavity size and RI of the active layer to yield smaller effective mass, so that the critical temperature exceeds room temperature. However, the problem lies in properly draining out the heat generated. Due to the high temperature, the light-matter coupling is reduced and at the same time the excitons dissociate. These problems arise with GaAs and CdTe, as the exciton binding energies are less than $k_B T$ at room temperature. In these materials, condensates exist typically in the range 4–70 K making them unsuitable for real-world applications. To achieve BEC at room temperature, the exciton binding energy must be 25 meV and the light-matter coupling should be strong enough to ensure LP-UP splitting much higher than 25 meV. The splitting depends on the oscillator strength and on the carrier density.

Wide band gap semiconductors like GaN- and AlGaN-based nanowires (Das et al 2013) and ZnO are attractive for realizing BEC at room temperature. The exciton binding energies in these materials are nearly 50 meV and as such excitons survive even at room temperature. However, the growth technology for these materials is not as mature as for III–V compounds and alloys. The challenges have been overcome and room-temperature polariton lasing has indeed been demonstrated by a few works, as described in Section 12.7.

12.7 Polariton lasers

Exciton-polaritons leading to a condensate form a coherent state and when they ultimately decay photons are emitted coherently. Similar coherent emission of photons occurs in lasers also. Therefore the devices emitting coherent photons from condensate state are called polariton lasers. Such a novel type of coherent source was first proposed by Imamoglu (1996). The first demonstration of polaritonic lasing was given by Deng et al (2003). A comparison between polariton laser and conventional laser is also made in their paper. Kasprzak et al (2006) linked polaritonic condensation to dynamical BEC.

The threshold power (current density) for optically (electrically) pumped Polariton lasers is orders-of-magnitude lower than the values found in photon or conventional laser. The output power in the former laser is therefore low and the application areas are limited. A very high Q microcavity is needed to achieve polaritonic lasing, whereas cavities or microcavities are used in photon lasers. Table 12.2 shows a comparison between the two types of lasers and the distinguishing features are self-explanatory.

Since its first announcement in 2003 by Deng et al, optical pumping has been the only mechanism in polariton lasers reported by different workers. However, for useful practical applications the device must work at room temperature, be pumped electrically and show CW operation. The first report for polariton laser working at room temperature came from Christopoulos et al (2007) and their device was pumped by optical pulses. It took a few more years to achieve the desired feats: room-temperature operation by using electrical pumping.

Das et al (2011) reported room-temperature ultralow threshold GaN Nanowire Polariton Laser which is optically pumped. The room-temperature polariton laser was announced by Bhattacharyya et al (2013); Das et al (2013); and by Schneider et al (2013); in the same year. Bhattacharyya et al (2014) made a further advancement by reporting first electrically pumped polariton laser.

The first observation of a room-temperature low-threshold transition to a coherent polariton state was made by Christopoulos et al (2007) in bulk GaN microcavities in the

Table 12.2 *Comparison between polariton laser and photon laser.*

Polariton Laser	Photon Laser
Emission of photons by decay of polaritons	Emission of photons by recombination of EHPs
(Mainly) occurs under equilibrium condition	Non equilibrium phenomena
Particles obey Bose statistics	Particles obey FD statistics
Strong light matter interaction	Weak coupling between light and matter
Emission from the ground state	Emission involves specified bands in CB and VB
Pumping mechanism: optical and recently electrical	Electrical pumping for devices
Microcavities of very high Q needed to produce strong coupling	Cavities and microcavities are used to provide feedback
Stimulated polariton-polariton scattering	Spontaneous scattering by phonons, defects, etc.
Threshold power (current density) too low: 3–4 orders smaller than in photon laser	Threshold power (current density) high in VCSELs, smaller but still higher in microcavity lasers
Low power	Moderate to high power
Limited applications: in Quantum Cryptography and for the study of fundamental physics	Wide applications (see Fig. 1.2 in Chapter 1)

strong-coupling regime. They used nonresonant pulsed optical pumping to produce rapid thermalization of polaritons A clear emission threshold of 1 MW is observed, and the corresponding absorbed energy density of 29 μJcm^{-2}, is 1 order of magnitude smaller than the best optically pumped InN or GaN QW surface-emitting lasers (VCSELs). Angular and spectrally resolved luminescences show that the polariton emission is beamed in the normal direction with an angular width of $\pm 5°$ and spatial size around 5 μm.

As stated earlier, nitride-based microcavities has the advantage that optical transitions possess a large oscillator strength, and in addition, large exciton-binding energy in nitrides indicate potential high temperature performance. There are confirmed reports that Rabi splittings between upper (UP) and lower (LP) polariton branches in nitrides can exceed 50 meV, an order of magnitude larger than in comparable III–V based microcavities. In III–V microcavities in which QWs are inserted, light-matter coupling increases due to the greater exciton-binding energy and better QW overlap with the cavity standing wave. However, the nitride-based QWs exhibit broad linewidths and quantum-confined Stark effect (QCSE) introduces further limitations. Hence the authors used bulk GaN microcavities in which GaN central spacer is sandwiched between a bottom 35 period $Al_{0.85}In_{0.15}N/Al_{0.2}Ga_{0.8}N$ distributed Bragg reflector (DBR) and a top 10 period dielectric SiO_2/Si_3N_4 DBR.

The structure and important characteristics of the electrically pumped polariton laser operating at room temperature realized by Bhattachayya et al (2014), the first of its kind, will now be described in brief. The structure is made of bulk p-n junction using GaN as the material. The advantage of using GaN has already been mentioned. The schematic diagram of the structure is shown in Fig. 12.11. It consists of a GaN-on-sapphire substrate, on top of which is a layer of n-doped $In_{0.18}Al_{0.82}N$, which in turn supports a lattice matched GaN p-n junction 430 nm thick grown by plasma-assisted MBE. The contact layer is grown on top of the p-layer. The current injection and optical feedback in the microcavity are in orthogonal directions. The large bandgap in InAlN ensures better confinement of photons and reduced substrate leakage of electroluminescence. The microcavity itself consists of six and five pairs of SiO_2/TiO_2 distributed Bragg reflector (DBR) mirrors deposited on opposite sides along the cavity length (690 nm), which essentially acts as an in-plane very short-cavity FP laser.

Usual measurements, like angle resolved PL, and variation of integrated EL intensity, LP emission linewidth, and blueshift of peak emission as a function of injected current density, have been conducted to obtain the important characteristics. The measured curves are qualitatively same as shown in Fig. 12.10.

The device operates as a polariton laser with an injection current threshold of 169 A/cm². With increasing current density, the output characteristics is nonlinear, and linewidth becomes progressively narrower, shows a minimum and then rises. A small blueshift of the emission peak also occurs. The polariton dispersion characteristics have been measured and analyzed from which a Rabi splitting of 32 meV is derived. The polariton condensation is confirmed from measured population redistribution in momentum space and spatial coherence as a function of injection current. The output polarization below and above threshold has also been characterized. Finally, a second threshold in the output at an injection level of 44 kA/cm² has been observed, which indicates onset of photon lasing by population inversion of injected carriers.

Example 12.14 *The LP density at the nonlinear threshold, taking into account all possible sources of uncertainties in the measurement, is calculated to be $(4.7 \pm 1.1) \times 10^{16}$ cm^{-3} with the relation $n_{LP} = \mathcal{J}\tau_{total}/ed$, where the measured total lifetime of excitons ~ 575 ps*

Example 12.15 *The relation between the polariton density N_{3D} at threshold and the blue shift of the resonant excitation energy δE is $\delta E \approx 3.3\pi E_x^B a_B^3 N_{3D}$, where E_x^B is the exciton binding energy and a_B is the exciton Bohr radius. Using measured value $\delta E = 1.9$ meV and reported values $a_B = 3.5$ nm and $E_x^B = 28$ meV, a value $N_{3D} = 1.53 \times 10^{17}$ cm^{-3} is obtained.*

The issue of intensity stability, a defining feature of a coherent state, has been addressed by Kim et al (2016). The observation is that the intensity noise far exceeds shot noise

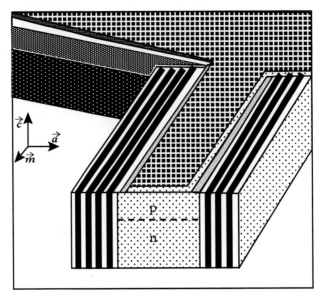

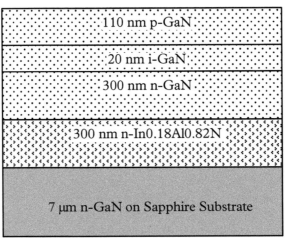

Figure 12.11 *Structure of electrically pumped polariton laser operating at room temperature.*
Source: reprinted figure with permission from Bhattacharya P, Frost T, Deshpande S, Baten M Z, Hazari A, and Das A (2014) Room temperature electrically injected polariton laser. *Physical Review Letters* 112: 236802. Copyright (2014) by the American Physical Society.

thereby limiting the phase coherence. This is detrimental to the use of polariton laser as a coherent source. They reported shot noise limited intensity stability by using an optical cavity of high mode selectivity to enforce single mode lasing and to suppress depletion of condensate.

Recently, Kang et al (2019) used a ZnO nanorod as a core, surrounded by a 5-layer multi QW shell comprising ZnO well and $Zn_{0.9}Mg_{0.1}O$ barrier. They observed better thermal stability and higher oscillator strength for excitons in radial QWs than in bulk excitons. Further, the radial structure gave stronger light matter coupling making the Rabi splitting as large as 370 meV. Persistent polariton lasing at room temperature has been observed by the authors.

12.8   Modelling of electrically driven polariton laser

Nitride based structures due to larger exciton binding energy are more attractive for realization of polariton lasers. A theoretical study of bulk GaN based polariton laser was presented by Solnyshkov et al (2009). Their predicted value of threshold current density $\sim 50 Acm^{-2}$ at room temperature indicates the potential of such devices as low threshold coherent light emitters. Irosh et al (2012) have preferred instead a MQW structure since the binding energy is enhanced in 2D systems and better confinement is ensured. On the other hand, for efficient electrical injection, the number of QWs should be limited. An intracavity pumping geometry has been considered for large number of QWs. A small number of QWs inserted into the i-region of the p-i-n structure is electrically pumped and this subset does not take part in the formation of polariton, but the emission from this subset, which is at an energy larger than the absorption of another subset of QWs, pumps this second subset to create strong coupling regime there.

We shall not present the actual design considered by Irosh et al, but instead focus on their theoretical treatment. They consider two situations: (1) injected electrons and holes are uniformly distributed in the MQWs and (2) intracavity pumping scheme. We shall describe the first one in the following paragraphs.

12.8.1   Formation and scattering processes of excitons

The charge carriers are first injected into the device and after some time either disappear due to nonradiative recombination occurring at dislocations, trapping, or Auger recombination, or they bind into pairs forming excitons. These excitons possess an energy (E)—and in-plane wavevectors (**k**) distributions and their ensemble forms an incoherent reservoir pumped from the electron-hole plasma. This exciton reservoir then feeds the condensate of exciton polaritons, which is a coherent multiparticle state responsible for polariton lasing. The injected electrons are not thermalized, and they are characterized by a Boltzmann carrier distribution with effective temperature exceeding the lattice temperature.

The model assumes that excitons in the reservoir scatter into the condensate state by scattering processes due to acoustic phonon (deformation potential and piezoelectric), LO phonon, exciton-exciton and exciton-free carrier interactions. In addition, decay of excitons from higher lying **k**-state by radiative process, which does not contribute to lasing, is considered by introducing the exciton radiative lifetime. Two processes of depletion of polariton condensate are considered: spontaneous radiative recombination and absorption of acoustic phonons which brings exciton polaritons back to **k** states beyond the inflection point of the lower polariton branch (LPB).

All the previously mentioned processes are described by the following set of rate equations:

$$\frac{dn_k}{dt} = P_k - \frac{n_k}{\tau_k} + \sum_{k' \neq k}[W_{k'k}n_{k'}(n_k + 1) - W_{kk'}n_k(n_{k'} + 1)] \qquad (12.36)$$

Here, n_k is the concentration of polariton with in-plane wave vector k, P_k is the electronic pumping rate, τ_k is the radiative lifetime of polariton, and $W_{kk'}$ is the total scattering rate between states **k** and **k'**. $\mathbf{k} = (k_x, k_y)$, $k_{x,y} = \pm\frac{2\pi j}{L}, j = 0, 1, 2, \ldots$ and $|k| < \omega/c$. L is the lateral size of the system, ω is the frequency of the exciton resonance, and c is the speed of light.

In the direct pumping scheme considered here, we may write

$P_k = 0$, if $E_k - E_X < \Delta$ and $P_k = \frac{Wn_{e-h}}{N}$, if $E_k - E_X \geq \Delta$.

Here E_X is the exciton energy at the center of the first Brillouin zone, $\Delta = 45$ meV is the exciton binding energy deduced from the variational approach, W is the exciton formation rate from the electron-hole plasma, $\tilde{N}$ is the number of states within the light cone which satisfy the condition $E_k - E_X \geq \Delta$ The concentration n_{e-h} of the electron-hole pairs obey the relation

$$\frac{dn_{e-h}}{dt} = \frac{\mathcal{J}}{e} - \frac{n_{e-h}}{\tau_{e-h}} - Wn_{e-h} \qquad (12.37)$$

where $\mathcal{J}$ is the electric pumping rate, and τ_{e-h} is the decay rate of the electron-hole plasma. Assuming that the e-h plasma density = 0 at $t = 0$, one may obtain an analytical solution to Eq. (12.36) as

$$n_{e-h}(t) = \frac{\mathcal{J}}{e}\frac{\tau_{e-h}}{1 + W\tau_{e-h}}\left[1 - exp\left(-Wt - \frac{t}{\tau_{e-h}}\right)\right] \qquad (12.38)$$

The scattering rate $W_{kk'}$ consists of three contributions as

$$W_{kk'} = W_{kk'}^{phon} + W_{kk'}^{el} + W_{kk'}^{pol-pol} \qquad (12.39)$$

where the first term denotes the scattering rate due to phonons, second one is the rate assisted by electrons or holes and the third is the polariton-polariton scattering rate. The rates are evaluated by a formalism available. See also Tasson and Yamamoto (2009) and Deng et al (2010) for scattering mechanisms and the calculation of rates.

The values of parameters chosen correspond to the expected values for state-of-the-art GaN-based microcavities with embedded InGaN/GaN quantum wells.

Here we show the evolution of the ground state polariton occupation number versus pump current density for different temperatures and detunings for the direct injection case. The threshold current density for condensation ($\mathcal{J}_{thr}$) can be clearly identified in each case.

12.8.2   Simplified rate equation modeling: Steady-state solutions

The results obtained from the solution of the full set of semiclassical Boltzmann equations are compared with a simplified semi-analytical rate equation model. Here only the direct pumping scheme will be mentioned. The three rate equations include Eq. (12.37) and the following two

$$
\frac{dn_x}{dt} = -\frac{n_x}{\tau_x} + Wn_{e-h} - an_x(n_p + 1) + aexp(-\beta\nabla_{esc})n_p n_x - bn_x^2(n_p + 1)
$$
$$
- cn_{e-h}n_x(n_p + 1) \tag{12.40}
$$

$$
\frac{dn_p}{dt} = -\frac{n_p}{\tau_p} + an_x(n_p + 1) + aexp(-\beta\nabla_{esc})n_p n_x + bn_x^2(n_p + 1)
$$
$$
+ cn_{e-h}n_x(n_p + 1) \tag{12.41}
$$

In Eqs. (12.40) and (12.41) n_x and n_p are the concentrations of excitons and exciton polaritons, respectively, τ_p is the lifetime of exciton polaritons in the ground state, a is related to the acoustic and optical phonon relaxation rates, $\beta = 1/k_B T$, ∇_{esc} is the characteristic energy splitting between the bottom of the LPB and states beyond the inflection point of the LPB where zero in-plane wave vector are scattered, b is the exciton-exciton scattering rate and c is the rate of exciton relaxation mediated by free carriers.

Approximating $n_p + 1 \approx n_p$ and $n_x + 1 \approx n_x$ above threshold and after some algebra, Irosh et al presented the following steady-state solutions ($t = +\infty$) for the electron-hole pair, exciton, and polariton populations:

$$
n_{e-h\infty} = \frac{\mathcal{J}\tau_{e-h}}{e(1 + \tau_{e-h}W)} \tag{12.42}
$$

$$
n_{x\infty} = \frac{-cn_{e-h\infty} + aexp(-\beta\nabla_{esc} - 1) + \sqrt{[-cn_{e-h\infty} + aexp(-\beta\nabla_{esc} - 1)]^2 + (4b/\tau_p)}}{2b} \tag{12.43}
$$

$$
n_{p\infty} = \tau_p\left(Wn_{e-h\infty} - \frac{n_{x\infty}}{\tau_x}\right) \tag{12.44}
$$

It is to be mentioned that the solution of the three rate equations lead to the same nature of variation of the polariton density $n_{p\infty}$ versus pumping current density as shown in Figs. 12.8, 12.9, and 12.10. By choosing proper values of a, b, and c obtained by proper fitting, a quantitative agreement between the results obtained by the solution of Boltzmann equation and rate equations is also achieved.

Similar sets of rate equations for the intracavity pumping geometry are also derived.

It may be noted that the three rate equations for the steady state are valid above threshold, when $n_p + 1 \approx n_p$ is a valid expression. For the general case, the rate equations at steady state can be written as

$$0 = \frac{\mathcal{J}}{e} - \frac{n_{e-h}}{\tau_{e-h}} - W n_{e-h} \tag{12.45}$$

$$0 = -\frac{n_x}{\tau_x} + W n_{e-h} - a n_x (n_p + 1) + a e^{-\beta \Delta_{esc}} n_p n_x - b n_x^2 (n_p + 1) - c n_{e-h} n_x (n_p + 1) \tag{12.46}$$

$$0 = -\frac{n_p}{\tau_p} + a n_x (n_p + 1) - a e^{-\beta \Delta_{esc}} n_p n_x + b n_x^2 (n_p + 1) + c n_{e-h} n_x (n_p + 1) \tag{12.47}$$

At steady state, equation (12.45) yields

$$n_{e-h} = \frac{\mathcal{J} A \tau_{e-h}}{e(1 + \tau_{e-h} W)} \tag{12.48}$$

At steady state, two different expressions for n_x are as follows

$$n_x = \frac{-[W_x + a(n_p - e^{-\beta \Delta_{esc}} n_p + 1) + c n_{e-h}(n_p + 1)]}{2b(n_p + 1)}$$

$$\frac{\pm \left\{ \left[W_x + a \left(n_p - e^{-\beta \Delta_{esc}} n_p + 1 \right) + c n_{e-h} \left(n_p + 1 \right) \right]^2 + 4b \left(n_p + 1 \right) W n_{e-h} \right\}^{1/2}}{2b \left(n_p + 1 \right)}$$

$$\tag{12.49}$$

$$n_x = \frac{-[a(n_p - e^{-\beta \Delta_{esc}} n_p + 1) + c n_{e-h}(n_p + 1)]}{2b(n_p + 1)}$$

$$\frac{\pm \left\{ \left[a \left(n_p - e^{-\beta \Delta_{esc}} n_p + 1 \right) + c n_{e-h} \left(n_p + 1 \right) \right]^2 + 4b \left(n_p + 1 \right) W_p n_p \right\}^{1/2}}{2b \left(n_p + 1 \right)} \tag{12.50}$$

where $W_x = \frac{1}{\tau_x}$, $W_p = \frac{1}{\tau_p}$ and A is the area.

Eqs. (12.49) and (12.50) are simultaneously solved to have the values of n_p for different values of current density $\mathcal{J}$.

To obtain numerical values, we choose the following values of parameters given in the paper by Irosh et al (2012).

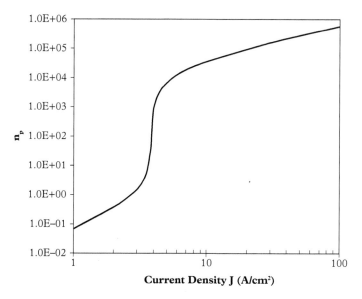

Figure 12.12 *Variation of polariton number with current density in an InGaN/InN MQW structure calculated by modifying the rate equation analysis by Irosh et al (2012) to be valid for any polariton number. The threshold current at 5 Acm² agrees with the value quoted by Irosh et al.*

Area $=A=50\mu m\times 4\mu m$; $W=0.01$ ps^{-1}; $W_x=0.001$ ps^{-1}; $W_p= 0.002$ ps^{-1}; $a=8\times 10^{-11}$ ps^{-1}; $b=9\times 10^{-13}$ ps^{-1}; $c= 1\times 10^{-17}$ ps^{-1}; $\Delta_{esc}=45$ meV; $T=320$K; and $\tau_{e-h}= 5$ ns.

Fig. 12.12 shows the plot of polariton number versus injected current density, which follows the usual variation (see Fig. 12.8(b) and Fig. 12.10).

12.8.3 Modulation response

The rate equations can be employed to obtain the small signal modulation response of the laser (Basu et al 2015). Baten et al (2015) experimentally obtained small signal modulation characteristics of a GaN-based electrically pumped polariton laser operating at room temperature. A maximum −3 dB modulation bandwidth of 1.18 GHz is measured. The experimental results have been analyzed with a theoretical model based on the Boltzmann kinetic equations (equation given by Irosh et al is suitably modified) and the agreement is very good.

Problems

12.1 *Assume that the density of H atoms is 1.8×10^{20} m^{-3} and $T_c = 50\,\mu K$. Obtain a plot of population of the ground state and first excited state as a function of temperature.*

12.2 *Calculate the number of atoms in the first excited state at $T = 0.9\,T_c$, when the total number is 10^4 and the atoms are confined in a cubic box of side $10\,\mu m$. The zero of the energy is placed at the ground state.*

12.3 *Show that $m_{ph} = \pi\hbar N\eta/Lc$ as expressed by Eq. (12.27).*

12.4 *Fig. 12.6 shows the in-plane dispersion of excitons, photons, and polaritons. While other particles show dispersion, excitons are seen to be dispersionless. Explain.*

12.5 *The following parameters are given in the paper of Keeling et al (2007) for CdTe QWs. $M = 0.08\,m_e$, $m = 3\times10^{-5}\,m_e$, $R = 26\,meV$, $\delta = 5.4\,meV$ and $\omega_0 = 1.7\,eV$. Obtain the dispersion curves as a function of angle and in-plane wave vector by using the coupled harmonic oscillator model and compare with the experimental data given in the paper.*

12.6 *Derive the expressions for the eigenfrequencies of the two coupled classical harmonic oscillators.*

12.7 *Using the rate Eqs. (12.37), (12.40), and (12.41) derive the steady state equations valid when the polariton laser works above threshold.*

12.8 *Using the rate equations perform a small signal analysis and obtain the expression for modulation response of the polariton laser. Also derive the expression for resonance frequency of the modulation transfer function.*

References

Abramowitz M and Stegun I A (1975) *Handbook of Mathematical Functions.* NY: Dover Publishing.

Agarwal B K and Eisner M (1988) *Statistical Mechanics.* New Delhi: Wiley Eastern Ltd.

Balili et al (2007) Bose-Einstein condensation of microcavity polaritons in a trap. *Science* 316: 1007–1010.

Basu P K (1997) *Theory of Optical Processes in Semiconductors: Bulk and Microstructures.*, Oxford, UK: Oxford University Press.

Basu P K and Kundu S (1985) Energy loss of two-dimensional electron gas in GaAs-AlGaAs multiple quantum wells by screened electron-polar optic-phonon interaction. *Applied Physics Letters* 47: 264–266.

Basu P K, Mukhopadhyay B, and Basu R (2015) *Semiconductor Laser Theory.* Boca Raton, FL, USA: CRC Press (Taylor & Francis).

Baten M Z, Frost T, Iorsh I, Deshpande S, Kavokin A, and Bhattacharya P (2015) Small-signal modulation characteristics of a polariton laser. *Scientific Reports* 5:(11915): 1–8.

Bhattacharya P, Frost T, Deshpande S, Baten M Z, Hazari A, and Das A (2014) Room temperature electrically injected polariton laser. *Physical Review Letters* 112(236802).

Bhattacharya P, Xiao B, Da A, Bhowmick S, and Heo J (2013) Solid state electrically injected exiton-polariton laser. *Physical Review Letters* 110(206403): 1–5.

Blatt J M, Brandt W, and Boer K W (1962) Bose-Einstein condensation of excitons. *Physical Review* 126(5): 1691–1692.

Bose S N (1924) Plancks gesetz und lichtquantenhypothese. *Zeitschrift für Physik* 26: 178–181.

Butov L V et al (2001) Stimulated scattering of indirect excitons in coupled quantum wells: Signature of a degenerate Bose-gas of excitons. *Physical Review Letters* 86: 5608–5611.

Butov L V and Filin A I (1998) Anomalous transport and luminescence of indirect excitons in AlAs/GaAs coupled quantum wells as evidence for exciton condensation. *Physical Review B* 58: 1980–2000.

Butov L V, Lai, C W, Ivanov A L, Gossard A C, and Chemla D S (2002) Towards Bose-Einstein condensation of excitons in potential traps. *Nature* 417: 47–52.

Butov L V, Mintsev A V, Lozovik Yu E, Campman K L, and Gossard A C (2000) From spatially indirect to momentum-space indirect exciton by in-plane magnetic field. *Physical Review B* 62: 1548–1551.

Casella R C (1963) On the possibility of observing a Bose-Einstein condensation of excitons in CdS and CdSe. *Journal of Physics and Chemistry of Solids* 24(1): 19–26.

Christopoulos S, Von Högersthal G B H, Grundy, A J D, Lagoudakis P G, Kavokin A V, Baumberg J, Christmann G, Butté R, Feltin E, Carlin J, and Grandjean N (2007) Room-Temperature polariton lasing in semiconductor microcavities. *Physical Review Letters* 98(12): 126405.

Combescot M, Betbeder-Matibet O, and Combescot R (2007) Bose-Einstein condensation in semiconductors: The key role of dark excitons. *Physical Review Letters* 99(176403): 1–4.

Combescot M, Combescot R, and Dubin F (2017) Bose-Einstein condensation and indirect excitons: A review. *Reports on Progress in Physics* 80 (066501): 1–44.

Cornell E A (2019) Biographical, Nobel Prize.org. Nobel Media AB. https://www.nobelprize.org/prizes/physics/2001/cornell/biographical/

Cornell E A, and Wieman C (2001) Bose-Einstein Condensation in a Dilute Gas: The First 70 Years and Some Recent Experiments. Nobel Lecture, 8 December, 2001.

Das A, Bhattacharya P, Heoa J, Banerjee A, and Guob W (2013) Polariton Bose-Einstein condensate at room temperature in an Al(Ga)N nanowire-dielectric microcavity with a spatial potential trap. *Proceedings of the National Academy of Science America* 110(8): 2735–2740.

Das A, Heo J, Jankowski M, Guo W, Zhang L, Deng H, and Bhattacharya P (2011) Room temperature ultralow threshold GaN nanowire polariton laser. *Physical Review Letters* 107(066405): 1–5.

Deng H, Gregor G, Santori C, Bloch J, and Yamamoto Y (2002) Condensation of semiconductor microcavity exciton polaritons. *Science* 298: 199–202.

Deng H, Haug H, and Yamamoto Y (2010) Exciton-polariton Bose Einstein condensation, *Review of Modern Physics* 82: 1489–1537.

Deng H, Weihs G, Snoke D, Bloch J, and Yamamoto Y (2003) Polariton lasing vs. photon lasing in a semiconductor microcavity. *Proceedings of the National Academy of Science USA* 100(26): 15318–15323.

Einstein A (1924) Quantum theory of the monatomic ideal gas§ *Sitzungsberichte der preussischen akademie der wissenschaften. Physikalisch-mathematische Klasse* 261–267.

Eisenstein J P and MacDonald A H (2004) Bose-Einstein condensation of excitons in bilayer electronic systems. *Nature* 432: 691–694.

Hertzog M, Wang M, Mony J, and Börjesson K (2019) Strong light–matter interactions: a new direction within chemistry. *Chemical Society Review* 48: 937–961.

High A, Leonard J R, Remeika M, Butov L V, Hanson M, and Gossard A C (2012) Condensation of excitons in a trap. *Nano Letters* 12: 2605–2609.

Hoang C N (2019) Biexciton as a Feshbach resonance and Bose–Einstein condensation of para excitons in Cu2O. *New Journal of Physics* 21:013035.

Imamoglu A, Ram R J, Pau S, and Yamamoto Y (1996) Nonequilibrium condensates and lasers without inversion: Exciton-polariton lasers. *Physical Review A* 53(6): 4250–4253.

Iorsh I, Glauser M, Rossbach G, Levrat J, Cobet M, Butte R, Grandjean N, and Kaliteevski M A, Abram R A, and Kavokin A V (2012) Generic picture of the emission properties of III-nitride polariton laser diodes: Steady state and current modulation response. *Physical Review B* 86(12): 125308.

Kang J-W, Song B, Liu W, Park S-J, Agarwal R, and Cho C-H (2019) Room temperature polariton lasing in quantum heterostructure nanocavities. *Science Advanced* 5(9338): 8.

Kasprzak J, Richard M, Kundermann S, Baas A, Jeambrun P, Keeling J M J, Marchetti F M, Szyman'ska M H, Andre' R, Staehli J L, Savona V, Littlewood P B, Deveaud B, and Dang L S (2006) Bose-Einstein condensation of exciton polaritons. *Nature* 443: 409–414.

Keeling J and Berloff N G (2011) Exciton–polariton condensation. *Contemporary Physics* 52(2): 131–151.

Keeling J, Marchetti F M, Szymanska M H, and Littlewood P B (2007) Collective coherence in planar semiconductor microcavities. *Semiconductor Science and Technology* 22 (2007): R1–R26.

Keldysh L V and Kozlov A N (1968) Collective properties of excitons in semiconductors. *Soviet Physics JEPT* 27(3): 521–528.

Ketterle W (2019) Biographical. *Nobel Prize.org. Nobel Media AB*, https://www.nobelprize.org/prizes/physics/2001/ketterle/biographical/

Kim S, Zhang B, Wang, Z, Fischer J, Brodbeck S, Kamp M, Schneider C, Höfling S, and Deng H (2016) Coherent polariton laser. *Physical Review X* 6(011026): 1–9.

London F (1938) On the Bose-Einstein condensation. *Physical Review* 54: 947–954.

Masters B R (2013) Satyendra Nath Bose and Bose-Einstein statistics. *Optics & Photonics News* April: 40–47.

Moskalenko S A (1962) Inverse optical-hydrodynamic phenomena in a non-ideal excitonic gas. *Soviet Physics Solid State* 4: 199.

Novotny L (2010) Strong couplings, energy splitting and level crossing: A classical perspectives. *American Journal of Physics* 78(11): 1199–1202.

Richard M, Kasprzak J, Baas A, Lagoudakis K, Wounters M, Carusotto I, Andre' R, Deveaud-Ple'dran B, and Dang L S (2010) Exciton-Polariton Bose-Einstein condensation: Advances and issues. *International Journal of Nanotechnology* 7: 668–685.

Richard M, Kasprzak, J, Romestain, R, Andre R, and Dang L S (2005) Spontaneous coherent phase transition of polaritons in CdTe microcavities. *Physical Review Letters* 94:(187401): 4.

Rodriguez S H-K (2016) Classical and quantum distinctions between weak and strong coupling. *European Journal of Physics* 37(025802): 16.

Schneider C, Rahimi-Iman A, Kim N Y, Fischer J, Savenko I G, Amthor M, Lermer M, Wolf A, Worschech L, Kulakovskii V D, Shelykh I A, Kamp M, Reitzenstein S, Forchel A, Yamamoto Y, and Höfling S (2013) An electrically pumped polariton laser. *Nature* 497(7449): 348–352.

Skolnick M S, Fisher T A, and Whittaker D M (1998) Strong coupling phenomena in quantum microcavity structures. *Semiconductor Science and Technology* (13): 645–669.

Snoke D (2002) Spontaneous Bose coherence of excitons and polaritons. *Science* 298: 1368–1372.

Snoke D W and Kavoulakis G M (2014) Bose-Einstein condensation of excitons in: Progress over 30 years. *Reports on Progress in Physics* 77(116501):17.

Snoke D W and Keeling J (2017) The new era of polariton condensates. *Physics Today* 70 (10): 54–60.

Snoke D W, Wolfe J P, and Mysyrowicz A (1990) Evidence for Bose-Einstein condensation of excitons in Cn20. *Physical Review B* 41(16): 11171–11184.

Solnyshkov D, Petrolati E, Di Carlo A, and Malpuech G (2009) Theory of an electrically injected polariton laser. *Applied Physics Letters* 94: 011110.

Sun Y, Wen P, Yoon Y, Liu, G; Steger M, Pfeiffer L N, West K, Snoke D W, and Nelson K A (2017) Bose-Einstein condensation of long-lifetime polaritons in thermal equilibrium. *Physical Review Letters* 118(016602): 1–6.

Tasson F and Yamamoto Y (2009) Exciton-exciton scattering dynamics in a semiconductor microcavity and stimulated scattering into polaritons. *Physical Review B* 59(16): 10830–10842.

Wieman, C E (2019) Nobel Lecture. NobelPrize.org. Nobel Media AB Mon. 30 Sep 2019. https://www.nobelprize.org/prizes/physics/2001/wieman/lecture/.

Wolfe J P and Jang J I (2014) The search for Bose-Einstein condensation of excitons in Cu_2O: Exciton – Auger recombination versus biexciton formation. *New Journal of Physics* 16(123048): 1–18.

Wu J Z, Shi X S, Long H, Chen L, Ying L Y, Zheng Z W, and Zhang B P (2019) Large Rabi splitting in InGaN quantum wells microcavity at room temperature. *Materials Research Express* 6(076204): 1–7.

Zimmermann R (2008) Bose-Einstein condensation of excitons: Promise and disappointment. Oxford Scholarship Online. Published to Oxford Scholarship Online: May 2008 DOI:10.1093/acprof:oso/9780199238873.001.0001

13

Surface plasmon

13.1 Introduction

A very old example of the use of nanoparticles in civilization, as mentioned in Chapter 9, is the Lycurgus cup, in which Au and Ag nanoparticles had been dispersed in glass. It was also mentioned there that tiny particles of Au, Ag, Cu, etc., were used in glass panes in churches to form pictures of different people, flowers, and so on.

It is now known that different colours exhibited by Au nanoparticles are related to a kind of resonance, known as plasmonic resonance. Different colours, and as such, difference in optical properties, are associated with different sizes of the particles, which are different from the optical properties of gold in bulk. The colours also change from material to material.

The plasmonic resonance in metal particles is due to plasma waves restricted in the surface region of the particles. In quantum mechanical language, the plasma waves form quasi-particles named plasmons. As surface waves are involved, the quasi particles are called surface plasmons.

Since the surface plasmons are excited mainly by light waves, there is coupling between surface plasmons, and light quanta termed as photons. This coupling gives rise to a hybrid quasi particle called surface plasmon polariton (SPP).

It has been seen in Chapters 11 and 12 that strong light-matter coupling in semiconductors gives rise to hybrid particles called excitonic polaritons. Together with excitonic polaritons, SPPs cover a wider frequency range of the electromagnetic (EM) spectrum spanning from 200 THz to 1 PHz. The coverage is shown in Fig. 13.1.

The usual structure to support SPPs is an interface between an insulator and a metal (or other conductors including semiconductors). Light incident on the surface of a metal, or in general a conductor, causes fluctuation of the charge carrier density. This generates a surface plasma wave or surface plasmons, which strongly interacts with light to create SPPs.

The strong interaction has two important consequences. First, the local field intensity at the interface can increase by many orders of magnitude. Therefore, minute changes in the local environment are amplified significantly. This has an important application in chemical and biosensing.

Semiconductor Nanophotonics. Prasanta Kumar Basu, Bratati Mukhopadhyay, and Rikmantra Basu, Oxford University Press.
© P.K. Basu, B. Mukhopadhyay, R. Basu (2022). DOI: 10.1093/oso/9780198784692.003.0013

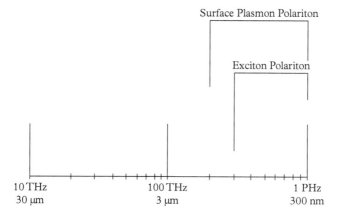

Figure 13.1 *Frequency range covered by Ex-P and SPP in optical region of EM spectra.*

The second important consequence of strong light Plasmon interaction, that is a sub-wavelength operation, is that it has a more striking impact on the field of information processing. As demands for higher speed and bandwidth grow at an unprecedented pace, conventional electronics cannot cope with this demand, in spite of the fact that transistors today have a feature size of less than 10 nm. Optical communication seems to be a viable solution to meet the bandwidth demand. However, the optical components, whose size is diffraction-limited, are bulky and have dimensions of the order of wavelength, which are a few hundred nms. Plasmonics overcome this diffraction limit and are expected to yield smaller optical components, which may be integrated, while leveraging on high speed and the bandwidth of light. The importance of plasmonics in realizing very small dimension devices with very high speed, overcoming the limitations of Electronics and Photonics has been pointed out in Section 1.6 of Chapter 1. The expected superior performance of plasmonics is illustrated by Fig. 1.8, see also Brongersma and Shalaev 2010; Gramotnev and Bozhevolnyi 2010; and Naik et al 2013].

A new field of nanooptics, named as transformation optics, involves manipulation and control of light in the nanoscale. Common materials are not suitable for accomplishing the new functionalities. Artificial or engineered materials, called metamaterials, are needed and have been developed for this purpose. The development of plasmonics has complemented the growth of metamaterials and thereby transformation optics. Metamaterials, with their important physical properties including negative refraction and various applications will be covered in Chapter 15 in this book.

Study of generation, manipulation, and detection of SPPs has led to the development of a new subject, plasmonics. In this chapter, the basic mechanism of plasma wave generation and related theory will first be developed by using a classical Drude model. The concept of plasma frequency and the nature of variation of dielectric function around plasma resonance frequency will then be pointed out. The generation and propagation of plasma waves in the interface of an insulator and a metal will be discussed for two different structures. The dispersion relation of SPPs will next be presented.

Following this, the most important property of surface plasmons, that is, subwavelength propagation and related novel optical properties will be pointed out. After introducing the basic concepts, the use of metal based plasmonics in the important range covered by telecommunication bands and beyond will be examined. Thereafter, the use of alternative materials, in particular semiconductors will be explored. As will be noted, semiconductors and other conductors are the only materials suitable for mid-infrared (IR) and THz regimes. The properties of semiconductors as plasmonic material in these ranges will then be pointed out. Some basic plasmonic devices made of semiconductors will then be introduced. At the end, the envisaged role of semiconductor plasmonics in information processing, and use in communication systems, and data centres will be mentioned. The chapter ends by covering a few applications of plasmonics in basic science and as biosensors.

An emerging research area in surface plasmonics is the stimulated amplification of plasmons. The device envisaged is called surface plasmon amplification by stimulated emission of radiation (SPASER). Its working principle and progress towards achieving the device will be covered in Chapter 13.

13.2   Basic concepts

We first introduce the plasma model for free electron gas that is developed using the standard Drude Lorentz model. The dielectric function of free electron gas is derived by using the model from which definition of the plasma frequency and plasma dispersion relation will be presented. Modifications for real metals and semiconductors will then be pointed out.

13.2.1   The dielectric function of the free electron gas

The plasma model considers free electron gas, the number density of which is n, and the gas moves against a fixed background of positive ion core. The model can explain optical properties of metals over a wide range of frequencies. The simple equation of motion for an electron of the plasma sea can be written as

$$m\frac{d^2x}{dt^2} + m\gamma\frac{dx}{dt} = -e\mathbf{F} \qquad (13.1)$$

In the previous equation, m is assumed to be the effective optical mass of each electron subjected to an external electric field $\mathbf{F}$ and γ is the characteristic collision frequency of the damped oscillation of electrons oscillating in response to the applied electromagnetic field.

Considering that the applied electric field follows a time variation $\mathbf{F}(t) = \mathbf{F}_0\exp(-i\omega t)$, the solution describing the oscillation of the electron can be written as $\mathbf{x}(t) = \mathbf{x}_0\exp(-i\omega t)$. Thus from Eq. (13.1)

$$\mathbf{x}(t) = \frac{e}{m(\omega^2 + i\gamma\omega)}\mathbf{F}(t) \tag{13.2}$$

The macroscopic polarization ($\mathbf{P} = -ne\mathbf{x}$) of the displaced electron can be expressed as

$$\mathbf{P} = -\frac{ne^2}{m(\omega^2 + i\gamma\omega)}\mathbf{F} \tag{13.3}$$

The previous equation can be used to express the dielectric displacement as

$$\mathbf{D} = \epsilon_0 \left(1 - \frac{\omega_p^2}{\omega^2 + i\gamma\omega}\right)\mathbf{E} \tag{13.4a}$$

where

$$\omega_p^2 = \frac{ne^2}{\epsilon_0 m} \tag{13.4b}$$

is the plasma frequency of the electron gas and $\varepsilon(\omega)$ is the dielectric function of the free electron gas given by

$$\varepsilon(\omega) = 1 - \frac{\omega_p^2}{\omega^2 + i\gamma\omega} \tag{13.5}$$

Therefore, the dielectric function can be written as the combination of a real and an imaginary component as $\varepsilon'(\omega) + i\varepsilon''(\omega)$ where

$$\varepsilon'(\omega) = 1 - \frac{\omega_p^2}{\omega^2 + \gamma^2} \tag{13.6a}$$

$$\varepsilon''(\omega) = \frac{\omega_p^2\gamma}{\omega(\omega^2 + \gamma^2)} \tag{13.6b}$$

The characteristic collision frequency γ is related to the relaxation time (τ) of the free electron gas as $\gamma = 1/\tau$. This leads us to reach the following expressions:

$$\varepsilon'(\omega) = 1 - \frac{\omega_p^2\tau^2}{1 + \omega^2\tau^2} \tag{13.7a}$$

$$\varepsilon''(\omega) = \frac{\omega_p^2\tau}{\omega(1 + \omega^2\tau^2)} \tag{13.7b}$$

For high frequency $(\omega \sim \omega_p)$, the damping is negligible leading to $\omega\tau \gg 1$, and $\varepsilon(\omega)$ is predominantly real and is given by

$$\varepsilon(\omega) = 1 - \frac{\omega_p^2}{\omega^2} \tag{13.8}$$

Therefore, equation (13.8) gives the dielectric function of the undamped free electron plasma.

For very low frequency $(\omega \ll \gamma)$, the real and imaginary parts of the complex refractive index are of comparable magnitude given by

$$n \approx \kappa = \sqrt{\frac{\varepsilon''}{2}} = \sqrt{\frac{\omega_p^2}{2\omega\gamma}} \tag{13.9}$$

In this case, the metals will be mainly absorbing with absorption coefficient

$$\alpha = \sqrt{\frac{2\omega_p^2\omega}{\gamma c^2}} \tag{13.10}$$

The absorption coefficient can be related to the dc-conductivity $(\sigma_0 = ne^2/\gamma m = \omega_p^2/\epsilon_0\gamma)$ as

$$\alpha = (2\sigma_0\omega\mu_0)^{1/2} \tag{13.11}$$

According to Beer's law of absorption, the field falls inside the metal as $exp(-z/\delta)$ for low frequency with skin depth

$$\delta = \frac{2}{\alpha} = \frac{c}{\kappa\omega} = \left(\frac{2}{\sigma_0\omega\mu_0}\right)^{1/2} \tag{13.12}$$

This is valid as long as the mean free path of the electron is much lower than the skin depth. At low temperatures the mean free path increases by many orders of magnitude leading to anomalous skin effect and the penetration depth changes.

The previous theory needs some modification, however, for some noble metals. The filled d bands in proximity to the Fermi surface in such metals give rise to an additional polarization for frequencies higher than the plasma frequency. If this additional polarization is expressed as $\mathbf{P}^* = \varepsilon_0(\varepsilon_\infty - 1)\mathbf{F}$, the dielectric function will be of the following form:

$$\varepsilon(\omega) = \varepsilon_\infty - \frac{\omega_p^2}{\omega^2 + i\gamma\omega} \tag{13.13}$$

Usually, ε_∞ is in the range $1 \leq \varepsilon_\infty \leq 10$.

If the dielectric function of the free electron plasma is linked to the classical Drude model, one can get the ac conductivity as

$$\sigma(\omega) = \frac{\sigma_0 \gamma}{\gamma - i\omega} \tag{13.14}$$

Comparing Eqs. (13.5) and (13.14), one can reach the following general expression of dielectric function as

$$\varepsilon(\omega) = 1 + \frac{i\sigma(\omega)}{\varepsilon_0 \omega} \tag{13.15}$$

Example 13.1 *Assume that electron density in a metal is 10^{28} m^{-3}, the mass is the free electron mass, and the permittivity is 1. The plasma frequency is 2.2×10^{15} Hz and the wavelength c/f_p is 134 nm, well below the visible frequency range of 400 nm.*

13.2.2 The dispersion of the free electron gas and volume plasmons

We start with the following well known wave equation derived from Maxwell's curl equations:

$$\nabla \times \nabla \times \mathbf{F} = -\mu_0 \frac{\partial^2 \mathbf{D}}{\partial t^2} \tag{13.16a}$$

Expressing space and time dependencies of all the quantities in the form $exp[i(\mathbf{k}.\mathbf{x} - \omega t)]$, one obtains

$$\mathbf{k}(\mathbf{k}.\mathbf{F}) - k^2 \mathbf{F} = -\varepsilon(\mathbf{k}, \omega) \frac{\omega^2}{c^2} \mathbf{F} \tag{13.16b}$$

where $c \left(= \frac{1}{\sqrt{\mu_0 \varepsilon_0}} \right)$ is the velocity of light. For transverse wave $\mathbf{k}.\mathbf{F} = 0$ and the dispersion relation becomes

$$k^2 = \varepsilon(\mathbf{k}, \omega) \frac{\omega^2}{c^2} \tag{13.17}$$

For longitudinal wave, $\varepsilon(\mathbf{k}, \omega) = 0$ as longitudinal oscillations exist at frequencies corresponding to zeros of $\varepsilon(\omega)$.

Using Eq. (13.8) in Eq. (13.17), the dispersion relation of travelling waves can be written as

$$\omega^2 = \omega_p^2 + k^2 c^2 \tag{13.18}$$

This relation has been plotted in Fig. 13.2 and it is found that for $< \omega_p$, the propagation of the transverse electromagnetic waves is forbidden inside the metal plasma but for $\omega > \omega_p$, the transverse waves propagate with a group velocity $v_g = d\omega/dk < c$.

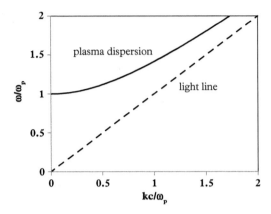

Figure 13.2 *Dispersion relation of free electron plasma. Frequency and wave numbers are normalized. The dispersion for light wave in vacuum is $\omega = Kc$ and is shown by the dashed line.*

In the small damping limit, $\epsilon(\omega_p) = 0$ and the excitation corresponds to a collective longitudinal mode. In this case, $\mathbf{D} = 0$, and the resulting electric field is a pure depolarization field with $\mathbf{F} = -\mathbf{P}/\epsilon_0$.

It is now in order to look into the physical origin of plasma waves. The waves are density waves of charged particles in an electrically neutral medium and the charged particles are usually electrons. At a certain point in space the local density of electrons may exceed the local density of positive charges. Because of the unneutralized electrons accumulated there, an electric field arises due to repulsive forces. This field tries to restore the equilibrium between positive and negative charges, and after some time, the electric field drives away the accumulated electrons charges from the region where they accumulated. Thus, at some point of time, there is overshoot of electron and at a later time there will be a deficit of electrons in the same region. The result is the development of an opposite electrical field which tries to draw back the electrons. This is the usual case of harmonic oscillation.

The physical significance of the excitation at the plasma frequency ω_p can be illustrated schematically by the Fig. 13.3. A collective longitudinal oscillation of the conduction electron gas has been considered against a fixed positive background of the ion cores in a plasma slab. A collective displacement (x') of the electron gas can give rise to a surface charge density $\sigma = \pm nex'$ at the boundaries and a homogeneous electric field $\mathbf{F} = nex'/\epsilon_0$ inside the slab. Therefore, the equation of motion of the displaced electrons can be written as

$$nm\frac{d^2 x'}{dt^2} = -ne\mathbf{F} = -\frac{n^2 e^2 x'}{\epsilon_0} \tag{13.19a}$$

$$\frac{d^2 x'}{dt^2} = -\frac{ne^2 x'}{\epsilon_0 m} \tag{13.19b}$$

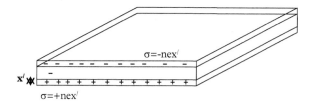

Figure 13.3 *Illustration of the excitation of volume plasmon in a metal arising out of Longitudinal collective oscillations of the conduction electrons.*

$$\frac{d^2 x'}{dt^2} + \omega_p^2 x' = 0 \tag{13.19c}$$

In the previous discussions, the oscillation frequency has been considered in the long wavelength limit, i.e. $k \to 0$. The quanta of these charge oscillations are known as plasmons or volume plasmons. It should be mentioned here that volume plasmons do not couple to transverse electromagnetic waves and their decay takes place only due to Landau damping, i.e. energy transfer to single electrons.

13.2.3   Plasma dispersion

Assuming collisionless ($\omega_0 = 0$) free electron gas, the dielectric function takes the form

$$\varepsilon(\omega) = 1 - \frac{\omega_p^2}{\omega^2} \tag{13.20}$$

It follows easily that when $\omega < \omega_p$, $\varepsilon(\omega) < 1$ and there is no propagation of EM waves. However, for $\omega > \omega_p$, $\varepsilon(\omega) > 1$, and EM waves can propagate through the medium. Using the relation $\omega = ck/\sqrt{\varepsilon}$ and Eq. (13.20) we may write the dispersion relation as

$$\omega^2 = c^2 k^2 + \omega_p^2 \tag{13.21}$$

It is easy to show that in the low frequency limit, $\omega \to \omega_p$, the dispersion law is quadratic as will now be given:

$$\omega = \omega_p + \frac{1}{2}\frac{c^2 k^2}{\omega_p} \tag{13.22}$$

The dispersion with normalized frequency ω/ω_p against k is plotted in Fig. 13.4 as the solid curve. The low frequency curve given by Eq. (13.22) is shown as a dashed curve. Also included in the figure are the dispersion in vacuum $\omega = ck$ and that in a medium of constant RI $\omega = ck/\eta$, both of which are linear.

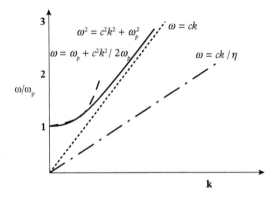

Figure 13.4 *Dispersion relations of plasma (solid and dashed) and EM waves in vacuum (dotted) and in a medium of constant RI (dash-dotted). Solid: plasma; dashed: low-frequency dispersion (Eq. 13.22).*

13.2.4   Real metals and interband transitions

Drude model can describe the dielectric function for optical response of metals only for photon energies below the threshold of transitions between electronic bands. But this model is not sufficient to describe either the real part ε' or the imaginary part ε'' of the dielectric function at high frequencies. As a result, Drude model shows inadequacy in describing the dielectric function of some noble metals like silver and gold. In the case of gold, the validity of the model fails at the boundary between near infrared and the visible light (Maier 2007).

However, the optical properties of gold and silver at visible frequencies can be described by modifying Eq. (13.1) as

$$m\frac{d^2x}{dt^2} + m\gamma\frac{dx}{dt} + m\omega^2 x = -e\mathbf{F} \tag{13.23}$$

The classical model of a bound electron with resonance frequency ω_0 is used to describe the interband transitions and Eq. (13.20) is used to calculate the resulting polarization. It is to be mentioned here that, for modelling the dielectric function more accurately for noble metals, a number of equations have to be solved in order to include the contribution of total polarization. These lead to a Lorentz oscillator term added to dielectric function of the free electrons as given by Eq. (13.5). The Lorentz oscillator term will be of the form $\frac{A_i}{\omega_i^2 - \omega^2 - i\gamma_i\omega}$.

13.3   Surface plasmon polaritons at metal/insulator interfaces

As explained in earlier chapters, exciton polaritons have their origin in the coupling between excitons and EM waves. Surface plasmons arise due to collective electronic oscillation at the metal-insulator interface. SPPs occur due to coupling of the surface plasmon waves with EM waves. In this section, the basic theory of SPPs will be developed first for a metal-insulator interface, and then for a dielectric-metal-dielectric structure.

13.3.1   The wave equation

The physical properties of SPPs can be described with the help of Maxwell's equations of electromagnetic theory. Using the following relations:

$$\left. \begin{array}{l} \nabla \times \nabla \times \mathbf{F} = \nabla(\nabla.\mathbf{F}) - \nabla^2\mathbf{F} \\ \nabla.(\varepsilon\mathbf{F}) = \mathbf{F}.\nabla\varepsilon + \nabla\varepsilon.\mathbf{F} \\ \nabla.\mathbf{D} = 0 \end{array} \right]$$

(13.24)

The curl equation as given in Eq. (13.16a) will be of the form

$$\nabla\left(-\frac{1}{\varepsilon}\mathbf{F}.\nabla\varepsilon\right) - \nabla^2\mathbf{F} = -\mu_0\varepsilon_0\varepsilon\frac{\partial^2\mathbf{F}}{\partial t^2}$$

(13.25)

If the dielectric function $\varepsilon = \varepsilon(r)$ is assumed to be constant over the distances on the order of one optical wavelength, then Eq. (13.25) will take the form as

$$\nabla^2 E - \mu_0\varepsilon_0\varepsilon\frac{\partial^2\mathbf{F}}{\partial t^2} = 0$$

(13.26)

Considering the general harmonic time dependent electric field $\mathbf{F}(\mathbf{r}, t) = \mathbf{F}(\mathbf{r})exp(-i\omega t)$, we get the well-known Helmholtz equation as

$$\nabla^2\mathbf{F} + k_0^2\varepsilon\mathbf{F} = 0$$

(13.27)

where $k_0 = \omega\sqrt{\mu_0\varepsilon_0} = \omega/c$ is the wave vector of the propagating wave in vacuum.

However, for simplicity, the waves are assumed to propagate along the $x$direction and there is no spatial variation in the in-plane y direction, i.e. $\varepsilon = \varepsilon(z)$. As a result, the propagating waves can be described as $\mathbf{F}(x, y, z) = \mathbf{F}(z)exp(i\beta x)$. The complex parameter β is called the propagating constant of the travelling waves. Inserting this equation into Eq. (13.24), one can reach the desired wave equation of the form

$$\frac{\partial^2\mathbf{F}(z)}{\partial z^2} + (k_0^2\varepsilon - \beta^2)\mathbf{F} = 0$$

(13.28)

A similar equation can be obtained for the magnetic field **H**.

Two sets of self-consistent solutions with different polarization properties of the propagating waves exist for harmonic time dependence $\left(\frac{\partial}{\partial t} = -i\omega\right)$ and for propagation along x direction $\left(\frac{\partial}{\partial x} = i\beta, \frac{\partial}{\partial y} = 0\right)$ respectively. The first set gives the transverse magnetic (TM) modes and the second one results in the transverse electric (TE) modes.

The wave equation for TM mode is

$$\frac{\partial^2 H_y}{\partial z^2} + (k_0^2 \varepsilon - \beta^2) H_y = 0 \qquad (13.29)$$

and that for transverse electric(TE) mode is

$$\frac{\partial^2 E_y}{\partial z^2} + (k_0^2 \varepsilon - \beta^2) E_y = 0 \qquad (13.30)$$

13.3.2   Surface plasmon polaritons at a single interface

Consider a single, flat interface as shown in Fig. 13.5 between a dielectric, non-absorbing half space for $z > 0$ with positive real dielectric constant ε_b and an adjacent conducting half space for $z < 0$ with a dielectric constant ε_a. It is to be mentioned here that the frequency is less than the bulk Plasmon frequency (ω_p) to satisfy the metallic properties. To describe the propagating waves, first we look at TM waves for $z > 0$, and the field equations are given by (Maier 2007)

$$H_y(z) = B_2 e^{i\beta x} e^{-k_b z} \qquad (13.31a)$$

$$E_x(z) = iB_2 \frac{1}{\omega \varepsilon_0 \varepsilon_b} k_b e^{i\beta x} e^{-k_b z} \qquad (13.31b)$$

$$E_z(z) = -B_1 \frac{\beta}{\omega \varepsilon_0 \varepsilon_a} e^{i\beta x} e^{k_b z} \qquad (13.31c)$$

and for $z < 0$

$$H_y(z) = B_1 e^{i\beta x} e^{-k_a z} \qquad (13.32a)$$

$$E_x(z) = -iB_1 \frac{1}{\omega \varepsilon_0 \varepsilon_a} k_a e^{i\beta x} e^{k_a z} \qquad (13.32b)$$

$$E_z(z) = -B_1 \frac{\beta}{\omega \varepsilon_0 \varepsilon_a} e^{i\beta x} e^{k_a z} \qquad (13.32c)$$

k_a and k_b are the components of the wave vector perpendicular to the interface in the two media.

Continuity of H_y and $\varepsilon_j E_z (j = a, b)$ at the interface yields $B_1 = B_2$ and

$$\frac{k_b}{k_a} = -\frac{\varepsilon_b}{\varepsilon_a} \tag{13.33}$$

The expression of H_y must satisfy Eq. (13.26) yielding

$$k_a^2 = \beta^2 - k_0^2 \varepsilon_a \tag{13.34a}$$

$$k_b^2 = \beta^2 - k_0^2 \varepsilon_b \tag{13.34b}$$

Combining Eqs. (13.29) and (13.30), one can get the dispersion relation of SPPs propagating at the interface between two half spaces as

$$\beta = k_0 \left(\frac{\varepsilon_a \varepsilon_b}{\varepsilon_a + \varepsilon_b} \right)^{1/2} \tag{13.35}$$

For TE mode of propagation, the field equations are as follows:
for $z > 0$

$$E_y(z) = B_2 e^{i\beta x} e^{-k_b z} \tag{13.36a}$$

$$H_x(z) = -iB_2 \frac{1}{\omega \mu_0} k_b e^{i\beta x} e^{-k_b z} \tag{13.36b}$$

$$H_z(z) = B_2 \frac{\beta}{\omega \mu_0} e^{i\beta x} e^{-k_b z} \tag{13.36c}$$

and for $z < 0$

$$E_y(z) = B_1 e^{i\beta x} e^{k_a z} \tag{13.37a}$$

$$H_x(z) = iB_1 \frac{1}{\omega \mu_0} k_a e^{i\beta x} e^{k_a z} \tag{13.37b}$$

$$H_z(z) = B_1 \frac{\beta}{\omega \mu_0} e^{i\beta x} e^{k_a z} \tag{13.37c}$$

Continuity of H_x and E_y at the interface yields $B_1 (k_a + k_b) = 0$ and the waves to be confined to the surface requires $Re(k_a)$ and $Re(k_b)$ must be positive, i.e. $B_1 = B_2 = 0$. Therefore, SPPs do not exist for TE polarization.

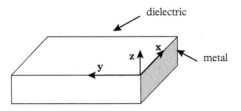

Figure 13.5 *Metal dielectric interface to support propagation of surface plasmon waves. The propagation is along the x-direction, the metal and the dielectric extend along −ve and + ve z directions, respectively.*

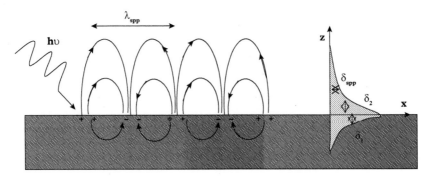

Figure 13.6 *Schematic diagram illustrating excitation of longitudinal SPP at dielectric-metal interface. The formation of alternating +ve and −ve charges, electric field lines and electric field profiles along the z-direction into the two materials are shown.*

The nature of propagating surface plasmon waves at a metal-dielectric interface is illustrated in Fig. 13.6. The wave sticks to the interface and the amplitude decays exponentially in both the dielectric and metallic layers with characteristic field penetration depths denoted by δ. The formation of alternating +ve and−ve charges, as explained in Section 13.2.2, and the corresponding electric field lines in both the layers are also shown. The surface plasmon waves are excited by a light source of photon energy $h\upsilon$, the mechanism of which will be described in Section 13.4. The wavelength of the SP wave is λ_{SPP} and its amplitude decays with a decay length δ_{SPP}. The characteristic length scales are further illustrated in Section 13.6.

13.3.3   Dispersion relation

The dispersion relation for surface plasmons may be obtained from Eq. (13.31). Let us take two dielectrics: air ($\varepsilon = 1$) and SiO_2 ($\varepsilon = 2.25$) and a metal with no damping and dielectric function given by Drude relation. The plots are shown in Fig. 13.7 in which both the frequency and wavevector are normalized. Consider first the range $\omega < \omega_p$. The linear plot for light and the dispersion curve for real part of β for silica-metal are shown as solid lines, while those for air and air-metal are shown as dotted lines.

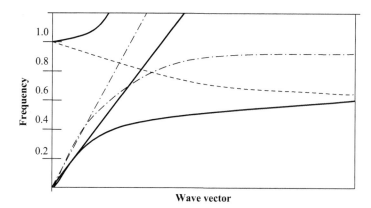

Figure 13.7 *Dispersion relations of SPPs at the air-metal and silica-metal interfaces. The metal obeys Drude dielectric function with negligible damping. Solid curves: silica-metal; dotted curves: air-metal; dashed curve: imaginary part of the normalized wave vector for silica-metal interface.*

The plot for imaginary part of β is shown by dashed line. The plots for the two dielectrics for $\omega > \omega_p$ are as shown in Fig. 13.7 and the two almost overlap each other. Note a frequency gap between two branches, in which β is purely imaginary prohibiting propagation of the waves.

Example 13.2 *Eq. 13.35 can be written as* $kc/\omega_p = (\omega/\omega_p)\sqrt{\left(1 - (\omega_p/\omega)^2\right)/}$
$(2 - (\omega_p/\omega))^2$ *for air-Drude metal interface. Take* $\omega/\omega_p = 0.5$, *or* $\omega_p/\omega = 2$. *The value of* $kc/\omega_p = 0.612$. *For* $\omega/\omega_p = 0.7$, *the normalized wavevector is 3.535.*

For a small wave vector, the SPP curve almost coincides with the light line. When the wave vector is large, the frequency of the SPP approaches the characteristic surface plasmon frequency given by

$$\omega_{sp} = \frac{\omega_p}{\sqrt{1 + \varepsilon_2}}$$

The dispersion curves are modified when damping is included.

13.3.4 Multilayer systems

To focus our discussion to SPPs in multilayer systems consisting of alternating conducting and dielectric thin films, we consider a three-layer system as shown in Fig. 13.8. (Maier 2007) In such a system, if the distance between adjacent interfaces is comparable or less than the decay length of the interface mode, the interactions between bound SPPs sustained in each single interface give rise to coupled mode.

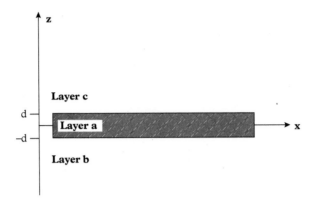

Figure 13.8 *Geometry of a three-layer system consisting of a thin layer a sandwiched between two infinite half spaces consisting layers B and C.*

For $z > d$, the field equations are

$$H_y(z) = A e^{i\beta x} e^{-k_c z} \tag{13.38a}$$

$$E_x(z) = iA \frac{1}{\omega \varepsilon_0 \varepsilon_c} k_c e^{i\beta x} e^{-k_c z} \tag{13.38b}$$

$$E_z(z) = -A \frac{\beta}{\omega \varepsilon_0 \varepsilon_c} e^{i\beta x} e^{-k_c z} \tag{13.38c}$$

For $z < -d$, we have

$$H_y(z) = B e^{i\beta x} e^{k_b z} \tag{13.39a}$$

$$E_x(z) = -iB \frac{1}{\omega \varepsilon_0 \varepsilon_b} k_b e^{i\beta x} e^{k_b z} \tag{13.39b}$$

$$E_z(z) = -B \frac{\beta}{\omega \varepsilon_0 \varepsilon_c} e^{i\beta x} e^{k_b z} \tag{13.39c}$$

For $-d < z < d$, the field components are

$$H_y(z) = C e^{i\beta x} e^{k_a z} + D e^{i\beta x} e^{-k_a z} \tag{13.40a}$$

$$E_x(z) = -iC \frac{1}{\omega \varepsilon_0 \varepsilon_a} k_a e^{i\beta x} e^{k_a z} + iD \frac{1}{\omega \varepsilon_0 \varepsilon_a} k_a e^{i\beta x} e^{-k_a z} \tag{13.40b}$$

$$E_z(z) = C\frac{\beta}{\omega\varepsilon_0\varepsilon_a}e^{i\beta x}e^{k_a z} + D\frac{\beta}{\omega\varepsilon_0\varepsilon_a}e^{i\beta x}e^{-k_a z} \tag{13.40c}$$

The continuity of E_x and H_y yields

$$\left.\begin{aligned}Ae^{-k_c d} &= Ce^{k_a d} + De^{-k_a d}\\ \tfrac{A}{\varepsilon_c}k_c e^{-k_c d} &= -\tfrac{C}{\varepsilon_a}k_a e^{k_a d} + \tfrac{D}{\varepsilon_a}k_a e^{-k_a d}\end{aligned}\right\}at\ z = d \tag{13.41}$$

$$\left.\begin{aligned}Be^{-k_b d} &= Ce^{-k_a d} + De^{k_a d}\\ -\tfrac{B}{\varepsilon_b}k_b e^{-k_b d} &= -\tfrac{C}{\varepsilon_a}k_a e^{-k_a d} + \tfrac{D}{\varepsilon_a}k_a e^{k_a d}\end{aligned}\right\}at\ z = -d \tag{13.42}$$

Again, H_y must satisfy the following wave equation for three regions:

$$k_j^2 = \beta^2 - k_0^2\varepsilon_j,\ j = a, b, c \tag{13.43}$$

Solving the previous system of linear equations, one can reach the expression for the dispersion relation as

$$exp(-4k_a d) = \frac{k_a/\varepsilon_a + (k_a/\varepsilon_a)(k_b/\varepsilon_b) + k_c/\varepsilon_c}{k_a/\varepsilon_a - (k_a/\varepsilon_a)(k_b/\varepsilon_b) - k_c/\varepsilon_c} \tag{13.44}$$

For a special case, if $\varepsilon_b = \varepsilon_c$ and thus $k_b = k_c$ are considered, the dispersion relation given by Eq. (13.41) will be split into a pair of equations:

$$tanh(k_a d) = -\frac{k_b\varepsilon_a}{k_a\varepsilon_b} \tag{13.45a}$$

$$tanh(k_a d) = -\frac{k_a\varepsilon_b}{k_b\varepsilon_a} \tag{13.45b}$$

It can be easily shown that Eq. (13.45a) gives the modes of odd vector parity whereas Eq. (13.45b) gives rise to the modes of even vector parity.

Another commonly used multilayer structure is a metal-insulator-metal (MIM) structure as shown in Fig. (13.9) using Au-Al$_2$O$_3$–Au, in which the typical thicknesses of different layers are indicated. The field patterns may be obtained by extending the procedure outlined previously. We do not include the mathematical details. It may intuitively be concluded that the two decaying field profiles at the two metal-insulator interfaces overlap, in the thin insulator layer lying in the middle, and give rise to the field profile as shown. Recall the pattern of the wavefunction in a thin quantum well(QW) structure.

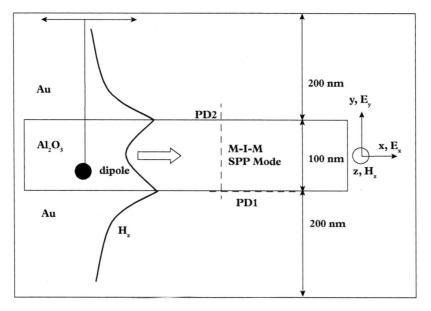

Figure 13.9 *A typical MIM structure and mode profile in the insulator.*

13.3.5 Energy confinement and the effective mode length

Though SPPs are propagating, dispersive electromagnetic waves coupled to the electron plasma of a conductor at a dielectric interface, energy localization below the diffraction limit perpendicular to the interface may also happen and in that case, a significant amount of total electric field energy of the SPP mode resides inside the metal. This energy must be taken into consideration for calculating the spatial distribution of the electric energy density.

Let us consider a one-dimensional (1D) metal/air/metal structure and $u_{eff}(z_0)$ be the electric field energy density at a position z_0 within the air core. In analogy with the light matter interactions in cavity quantum electrodynamics, one can write the following expression relating the effective mode length $L_{eff}(z_0)$ and energy as

$$L_{eff}(z_0)u_{eff}(z_0) = \int u_{eff}(z)dz \qquad (13.46)$$

Therefore, the effective mode length is given by the ratio of the total energy of the SPP mode to the energy density at the position of interest. In a quantized picture for normalized total energy, the inverse of the effective mode length quantifies the field strength per single SPP excitation. The estimation of the effective mode length of metal/air/metal structure gives an insight about the scaling of electric field strength per SPP excitation as a function of the size of the gap in the air.

13.4 Excitation mechanism

It is noted that plasmon waves at the interface of an insulator-metal interface is a 2D wave. As seen from Fig. 13.6, SPP dispersion curve is always to the right of the light dispersion, thus light line never crosses the SPP curve. Excitation of SPP by 3D light is therefore difficult due to this phase mismatch and special techniques must be employed to excite surface waves. In this section, two of these techniques will be introduced.

13.4.1 Prism coupling

Phase matching can be achieved by sandwiching a thin metal film between two different insulators. For simplicity, let us take one of them to be air. Let a photon beam from a higher dielectric material, usually in the form of a prism, be incident at the insulator-metal interface. The reflected beam at the interface will have an in-plane momentum $k_x = k\sqrt{\varepsilon}\sin\theta$, where θ is the angle between the incident photon beam and normal to the interface. The in-plane momentum may excite SPPs at the air-metal interface, but not at the dielectric metal interface.

In the coupling scheme, also known as attenuated total internal reflection, the fields of the excitation beam tunnels to the metal-air interface. Two different schemes for prism coupling are possible, one of which is shown in Fig. 13.10.

The Kretschmann geometry (Kretschmann and Raether 1968) shown in Fig. 13.10(a) has a thin metal film evaporated on a prism. The incident angle at the glass-metal interface must be greater than the critical angle of total internal reflection. The beam tunnels through the metal film and excites the SPP at the air-metal interface. In another configuration, the Otto configuration (Otto 1968), a prism is separated from the metal by a thin air gap. Total internal reflection occurs at the prism-air interface. The field arriving at the air-metal interface via tunnelling excites the SPP. This configuration is desirable when direct contact to the metal surface is to be avoided.

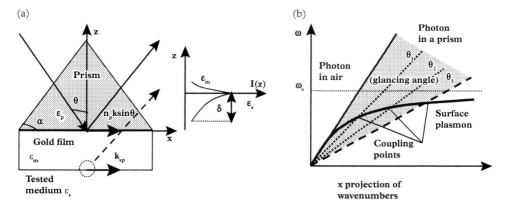

Figure 13.10 *Illustration of prism coupling: (a) structure, (b) illustration of wave vector matching.*

It is easy to find from the dispersion curves in Fig. 13.10(b), that the excited propagation constants lie within a cone defined by air and prism.

The prism coupling method is also suitable for MIM and insulator-metal-insulator(IMI) structures.

13.4.2   Grating coupling

In this scheme the metal surface is patterned in the form of grating of grooves or holes with lattice constant a. A light beam of wave vector $\mathbf{k}$ incident at an angle θ as shown has an in-plane wave vector $k \sin \theta$. The phase matching condition is given by

$$\beta = k \sin \theta \pm m(2\pi/a), \ m = 1, 2, 3, \ldots.$$

The scheme is illustrated in Fig. 13.11. The geometry is shown in (a), and the phase matching condition is shown in (b). In (b) $k_x = k \sin \theta$, is the in-plane wave vector of the photon, which has a deficit of $G = m(2\pi/a)$ needed to satisfy phase matching condition. The grating provides the needed wavevector.

13.5   Materials

SPPs can exist at the interface of two materials. One of them is a positive permittivity material, such as air or glass, or insulators like SiO_2, Si_3N_4, and so on, which are transparent to the frequency of operation. The other material is the plasmonic material, offering negative permittivity, which may be a metal or other material. Knowledge of the range of operating wavelengths, absorption depth, propagation distance along the interface, and dielectric function of the material is essential for the usefulness of the material

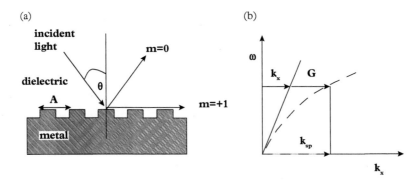

Figure 13.11 *(a) Grating coupling structure (b) illustration of phase matching; K_x and G are respectively the in-plane photon and grating wave vectors.*

and to design some of the related devices. Here we introduce briefly the types of materials used in the study of SPPs and their applications, deferring at present the detailed discussions about the properties of a few of them,

13.5.1 Metals

Metals, in particular noble metals like Au and Ag, are the only materials of choice for visible and near infrared (IR) light. In addition, Cu and Al are also under consideration. Metals have a large number of free electrons, leading to high plasma frequencies and thereby allowing negative permittivity at $\omega < \omega_p$.

Unfortunately, metals suffer from high ohmic losses, degrading the performance of plasmonic devices. The deposition technique also needs careful handling and optimization (McPeak et al 2015). Both the polarizability and loss of a metal are important, and a quality factor defined as $Q_{SSP} = \varepsilon_r^2/\varepsilon_i$, and propagation length L_{SPP} are important figures-of-merit. A list of metals, their wavelength of operation, and Q_{SSP} and L_{SSP} is given in Table 13.1 (McPeak 2015 and Wiki 2019 en.wikipedia.org/wiki/Surface_plasmon_polariton).

The table indicates that Ag has the lowest loss, and highest Q at all regimes excluding the ultraviolet (UV) region, where Al is the only acceptable metal. The performance of Au and Cu is comparable in visible, near IR and telecomm wavelengths, while Cu has a slight advantage over Au at telecomm wavelengths. Of all metals, Au is most stable chemically in natural environments. Al is the best material at UV and is also a complementary metal-oxide semiconductor (CMOS) compatible along with Cu.

Table 13.1 *Properties of some common metals.*

Wavelength range	Metal	Q_{SPP} (x 10^{-3})	L_{SPP} (μm)
UV (280 nm)	Al	0.07	2.5
Visible (650 nm)	Ag	1.2	84
	Cu	0.42	24
	Au	0.4	20
Near-IR(1000 nm)	Ag	2.2	340
	Cu	1.1	190
	Au	1.1	190
Telecom (1550 nm)	Ag	5	1200
	Cu	3.4	820
	Au	3.2	730

13.5.2   Other materials

Excessive loss of noble metals at wavelengths higher that telecomm range makes them unsuitable at mid-IR, far IR, and THz ranges. Furthermore, metal fabrication process is incompatible with CMOS fabrication process. Therefore, alternative plasmonic materials such as conductive oxide like indium tin oxide (ITO), polar materials, highly doped semiconductors, and 2D materials like graphene have attracted attention from many workers. The wavelength range over which each of these materials can work depends on the respective free carrier density and mobility. Semiconductor materials are of extreme importance due to low loss on account of high mobility, tunability of optical properties via doping, carrier injection, or depletion, and full CMOS compatibility. A more detailed discussion about role of semiconductors in SPP study will be presented in a later section.

13.6   Length scales in noble metals

It will be of interest to know the order of magnitude values of different length scales related to SPP waves in the interface between noble metals and insulators. These length scales serve as guidelines for design issues. We have followed the work by Barnes (2006) to estimate the SPP wavelength, the propagation length, and the penetration depths into the metal and the dielectric.

We write the dispersion relation for SPP as (Barnes 2006)

$$k_{SPP} = \frac{\omega}{c}\sqrt{\frac{\varepsilon_m \varepsilon_d}{\varepsilon_m + \varepsilon_d}} \tag{13.47}$$

The relative permittivity of metal contains a real and an imaginary part, and is written as $\varepsilon_m = \varepsilon_{mr} + \varepsilon_{mi}$.

13.6.1   The surface plasmon–polariton wavelength

The SPP wavelength is the period of surface charge oscillation and is expressed by taking the real part of k_{SPP} as

$$k_{SPPr} = \frac{\omega}{c}\sqrt{\frac{\varepsilon_{mr} \varepsilon_d}{\varepsilon_{mr} + \varepsilon_d}} \tag{13.48}$$

From this the wavelength of SPP $\lambda_{SPP} = 2\pi/k_{SPPr}$ may be written as

$$\lambda_{SPP} = \lambda_0 \sqrt{\frac{\varepsilon_{mr} + \varepsilon_d}{\varepsilon_{mr} \varepsilon_d}} \tag{13.49}$$

Example 13.3 *The value $\varepsilon_{mr} = -29.384$ for an Ag film at 780 nm has been quoted by Baburin et al (2018). The wavelength is 766.6 nm for air and 500 nm for glass $(\varepsilon_d = 2.25)$.*

It may be noted that the SPP wavelength is always slightly less than, the free space wavelength, λ_0. The previous example shows that, by using a higher permittivity insulator, the wavelength can be reduced further.

13.6.2 The surface plasmon–polariton propagation length

The imaginary part of the surface plasmon–polariton wavevector is expressed from Eq. (13.47) as

$$k_{SPPi} = k_0 \frac{\varepsilon_{mi}}{(\varepsilon_{mr})^2} \left(\frac{\varepsilon_{mr}\,\varepsilon_d}{\varepsilon_{mr} + \varepsilon_d} \right)^{3/2}$$

The propagation length of SPP L_{SPP} is the distance at which the intensity decreases 1/e times the initial intensity and is given by $L_{SPP} = 1/2k_{SPPi}$. The expression is

$$L_{SPP} = \frac{\lambda_0}{2\pi} \left(\frac{\varepsilon_{mr} + \varepsilon_d}{\varepsilon_{mr}\,\varepsilon_d} \right)^{3/2} \frac{(\varepsilon_{mr})^2}{\varepsilon_{mi}} \approx \frac{\lambda_0}{2\pi} \frac{(\varepsilon_{mr})^2}{\varepsilon_{mi}} \qquad (13.50)$$

The last expression is valid when $\varepsilon_{mr} \gg \varepsilon_d$, and $\varepsilon_d \sim 1$. It follows easily that in order to obtain large propagation length the real part of metallic permittivity should be large and -ve and at the same time the metal loss should be low.

Example 13.4 *The values of Ag quoted by Baburin et al (2018) are $\varepsilon_{mr} = -29.384$ and $\varepsilon_{mi} = 0.365222$ at 780 nm. With air as the dielectric, the calculated value of propagation length is 278.6 μm.*

13.6.3 The surface plasmon polariton field penetration depths

In a material with relative permittivity ε_j the total wavevector is $\varepsilon_j k_0^2$, and considering the x and z components of wavevector, we may write

$$\varepsilon_j k_0^2 = k_{SPP}^2 + k_{z,j}^2 \qquad (13.51)$$

As noted previously, the SPP wavevector always exceeds the photon wavevector freely propagating in the adjacent medium. This means $k_{SPP}^2 > \varepsilon_j k_0^2$, and thus the z-component of the wavevector in both media is imaginary, representing the fields in both the media decay exponentially with distance into the two media. Combining Eqs. (13.47) and (13.51), the penetration depths into the dielectric, δ_d, and metal, δ_m, are

$$\delta_d = \frac{1}{k_0} \left| \frac{\varepsilon_{mr} + \varepsilon_d}{\varepsilon_d^2} \right|^{1/2} \tag{13.52a}$$

$$\delta_m = \frac{1}{k_0} \left| \frac{\varepsilon_{mr} + \varepsilon_d}{\varepsilon_{mr}^2} \right|^{1/2} \tag{13.52b}$$

Example 13.5 *Using the values used in earlier examples $\varepsilon_{mr} = -29.384$ and air as dielectric, the calculated values are: $\delta_d = 661$ nm and $\delta_m = 22.5$ nm.*

13.7   Metal-insulator based plasmonics-Photonics

As pointed out in earlier sections, one of the usual plasmonic structures consists of a single interface between a metal, which acts as the plasmonic material and an insulator possessing positive permittivity. In addition, MIM and IMI structures are also employed. Different metals used and their frequencies of operation are listed in Table 13.1. Amongst the insulators, common dielectrics, such as air, or glass, or insulators like SiO_2, Si_3N_4, and so on, which are transparent to the frequency of operation, are usually employed. In some cases, semiconductors are used as dielectrics.

The structures mentioned previously are employed in realizing passive plasmonic devices like waveguides, branches, tees, etc., some of which will be discussed in Section 13.9.

13.8   All semiconductor plasmonics

As noted already, noble and common metals are employed for SPP studies and applications. However, metals have some limitations in some spectral ranges of immense interest and alternative materials need be investigated for this purpose. Since we aim to focus on semiconductors, we list below the deficiencies of metals and how semiconductors are thought to be the useful replacement.

(1) Noble metals such as Ag or Au have plasmon resonances in blue or deep ultra-violet wavelength ranges. For near and mid-IR ranges (from 1 to 10 µm) there are no suitable metals. This range is extremely important for detection and sensing.

(2) The area integrated plasmonics is receiving much importance and for its development there is a need to explore proper semiconductor based plasmonic devices to be integrated with standard electronic and other signal processing elements. The plasmonic integrated circuits must contain active plasmonic devices or metamaterials either possessing gain or being suitable for coupling with a gain medium. Metals are not suitable for this, whereas semiconductors are.

(3) The quality of metal films developed by normal techniques is far below that of semiconductor layers grown by very high quality epitaxial or other growth processes. As a result, the poor metal quality or poor semiconductor-metal interfaces affect some intrinsic plasmonic properties.

(4) Large metal loss is still a key problem for many plasmonic and metamaterial applications, in particular at mid IR and THz regimes.

(5) In the emerging area of THz systems, no metal is suitable because of the loss issues. Semiconductors play a major role in this regime.

(6) The plasmonic resonance frequency (or wavelength) is fixed for a given metal. Many applications, however, rely on tunable plasmonic properties. Semiconductors offer special advantage in this respect, as resonance frequency and other parameters can be tuned by changing the carrier density by injection and depletion caused by electrical bias, or by changing temperature or by photo illumination.

13.9   Plasmonic properties of semiconductors

In this section, some properties of highly doped semiconductors needed for the study of plasmonic properties and having application potentials will be discussed.

13.9.1   Dielectric function

Three mechanisms, namely the free carrier absorption, lattice vibrations, and background permittivity originating from interband absorption, determine the optical and conductive properties. The resulting permittivity is expressed by (Panah et al 2017)

$$\varepsilon_r = \varepsilon_\infty - \frac{\omega_p^2}{\omega^2 - i\gamma_p\omega} + \sum_j \frac{S_j\omega_{fj}^2}{\omega_{fj}^2 - \omega^2 - i\gamma_j\omega} \tag{13.53}$$

The first term in Eq. (13.53) represents the background permittivity ε_∞, γ_p is the damping constant, the inverse of the scattering time $\tau_p = 1/\gamma_p$.

The second one is the contribution from free carriers, where ω_p is the plasmafrequency, expressed already as $\omega_p = \sqrt{\frac{Ne^2}{\varepsilon_0 m^*}}$

As explained already, N is the carrier concentration, e is the electron charge, ε_0 is the permittivity of free space and m^* is the effective mass of the charge carriers.

The last term arises due to electron-phonon interactions in which a few optical phonons of angular frequency are involved. S_j, ω_{fj} and γ_j are, respectively, the strength, resonance frequency, and damping for the jth Lorentzian oscillator, describing phonon absorption at frequency ω_{fj}.

13.9.2   Carrier-Dependent properties

In all plasmonic studies, heavily doped semiconductors with carrier densities as high 10^{20} cm^{-3} are used. At such high densities, a few of the characteristic parameters are changed from their low doping values.

Band gap narrowing (BGN): it occurs in heavily doped semiconductors, in which the bandgap shrinks with increasing doping level. The narrowing is caused by ionized impurity potentials, by impurity band formation and many-electron interaction (Basu 1997). In heavily doped semiconductors, impurity ions come too close with their neighbours, and this allows the impurity levels, which have been discrete for low doping, to form a band called the impurity band, which merges with the conduction band (CB)or valence band (VB). This is one reason why the band gap is reduced.

A simple model by Jain and Roulston (1991) is used to describe BGN in common semiconductors. The change in gap is expressed as

$$\Delta E_g = \Delta E_x + \Delta E_{corr(min)} + \Delta E_{i(maj)} + \Delta E_{i(min)} \tag{13.54}$$

ΔE_x is the exchange interaction, the shift of the holes is considered in the electron-hole correlation energy, $\Delta E_{corr(min)}$, in n-type in which holes are minority (min) carriers, and the last two terms represent the energy due to carrier-impurity interaction.

Burstein Moss Shift: at high doping density or in presence of heavy injection, semiconductors become degenerate. In particular, semiconductors like indium antimonide (InSb) or indium arsenide (InAs), with small effective mass in conduction band, become degenerate easily at high electron density. As electrons fill up the conduction band, the energy required for a photon to be absorbed across the bandgap increases. This shifts the absorption edge to a higher energy, effectively increasing the bandgap. This increase in the band gap is termed as Burstein–Moss shift (Basu 1997). The absorption coefficient, α, by taking into account of the occupation probabilities of electrons and holes, f_e and f_h, respectively, takes the modified form

$$\alpha = \alpha_0 [1 - f_e - f_h] \tag{13.55}$$

where α_0 is the absorption coefficient in undoped or lightly doped semiconductor. The absorption coefficient is related to the imaginary part of semiconductor dielectric constant by $\varepsilon_i = \frac{\eta_b c \alpha}{\omega}$, where η_b is the background refractive index and c is the speed of light in vacuum.

Example 13.6 *An approximate expression for bandgap shift for InSb is given by Law et al (2014) as $\Delta E_g = (\hbar^2/2m^*)(3n/8\pi)^{2/3}$. The effective mass of electrons with band nonparabolicity is expressed by them as $m^* = m_n \left[1 + \frac{1}{2} \left(\frac{3}{\pi} \right)^{2/3} \left(\frac{\hbar^2}{E_g m_n} \right) n^{2/3} \right]^{1/2}$. Using $m_n = 0.014\, m_0$ and $E_g = 0.17$ eV at 300 K, $m^* = 0.0183\, m_0$ and the shift comes out to be 29.6 meV for $n = 10^{19}$ cm^{-3}.*

Plasma frequency: the usual expression Eq. (13.4b) is still valid, but the effective mass should be modified by considering band non-parabolicity.

13.9.2.1 *Effective mass*

The effective mass of carriers, in particular of electrons, is not constant, but increases from its value at the band edge ($\mathbf{k}$ = 0) due to non-parabolic nature of the E-k diagram. The non-parabolicity is more prominent for lower band gap semiconductors. In addition, the mass is also concentration dependent, as may be seen from the empirical expression given in Example 13.6 for InSb. An empirical formula for gallium arsenide(GaAs) describing the increase of the effective mass due to the band non-parabolicity and increasing Fermi level has been quoted by Cada et al (2015) as

$$m^* = 0.064 + 1.26 \times 10^{-20} n - 4.37 \times 10^{-40} n^2,$$

where the carrier density n is expressed in cm^{-3}.

Dielectric Function: Eq. (13.53) points out a contribution of phonons to the dielectric function. The expression for permittivity will depend on the number of optical phonons involved and therefore this contribution will vary from material to material, and also for a given material on the frequency range of interest. We quote, as an example, the phonon energy, the damping coefficient and other parameters for indium phosphide (InP), as obtained by Panah et al (2017)

$\omega_1 = 9.09$ THz; $\gamma_1 = 0.081$ THz; $S_1 = 1.52$–2.52.

13.9.3 Semiconductors as plasmonic materials

The behaviour of SPPs at the interface between an insulator and a conductor (metal or heavily doped semiconductor) are governed by two wave vectors: k_y and k_z, respectively along the interface and perpendicular to it. The expressions are already given by Eqs. (13.47–13.49) for metals and dielectrics. For the interface between two materials 1 and 2, these are modified as

$$k_y = \frac{\omega}{c} \sqrt{\frac{\varepsilon_1 \varepsilon_2}{\varepsilon_1 + \varepsilon_2}} \tag{13.55}$$

and

$$k_{zj} = \frac{\omega}{c} \sqrt{\frac{\varepsilon_j^2}{\varepsilon_1 + \varepsilon_2}} \tag{13.55}$$

where $j = 1,2$ is the index of the respective media, 1 denotes the insulator which may be air, and 2 represents a conductor. The k_y component must be real and the k_z component imaginary to make SPP propagating. This means that Re $(\varepsilon_2) < $ -ε_1 or the real part of the permittivity of the conductor must be negative and less than the real part of the same in the insulator. In reality, both the permittivities are complex and SPP waves decay along its path of propagation. The propagation length is $L_{SPP} = 1/\text{Im}(k_y)$ and the penetration depth is $L_{1,2} = 1/\text{Im}(k_{z1,2})$.

In the ideal situation, the propagation length should be long but confinement within the metal should be strong. A long propagation length may result if $\varepsilon_2 - \varepsilon_1$ is large; however the penetration depth in the metal is small and most of its energy is carried in the dielectric. The opposite is also true. If the confinement is strong, there will be more absorption in the metal, but the propagation length is shorter.

The plasmonic properties of heavily doped semiconductors, InSb, InP and GaAs have been measured by Chochol et al (2017) for THz applications.

13.9.4   Plasma reflection edge

Eq. (13.7a) valid for free electrons indicates that the real part of the permittivity in a medium is reduced from its static value $\varepsilon_s = 1$ by the contribution from free carriers. Similar analysis for a semiconductor with static permittivity ε_s gives the following expression for the contribution by free carriers to the real (r) and imaginary (i) parts of the permittivity.

$$\varepsilon_r = \varepsilon_s \left(1 - \frac{\omega_p^2}{\omega_0^2 + \omega^2} \right) \tag{13.56a}$$

$$\varepsilon_i = \varepsilon_s \frac{\omega_0}{\omega} \frac{\omega_p^2}{\omega_0^2 + \omega^2} \tag{13.56b}$$

where $\omega_0 = 1/\tau$ is the collision frequency and τ is the mean time between collisions or relaxation time. Eq. (13.56a) indicates that the free carrier contribution is negligible for low carrier density, large collision frequency, and at low frequency. However it becomes important as the light frequency approaches the plasma frequency.

It is known that the reflection coefficient of light for a weakly absorbing medium like semiconductor is $(\eta - 1)^2/(\eta + 1)^2$. Therefore with $\eta = 0$, there will be total reflection from the surface of the material. To arrive at this let us first assume that $\varepsilon_i = 0$ and $\omega_0 = 0$. Then from Eq. (13.56a) ε_r should be zero. This may happen at $\omega = \omega_p$ and total reflection from the surface is possible. On the other hand there will be no reflection when both ε_r and η are zero. Under the same conditions, when $\omega = \omega_p[\varepsilon_s/(\varepsilon_s - 1)]^{1/2}$ there will be no reflection. In semiconductors ε_s is quite large and the two frequencies are very close. The values are known as the *plasma reflection edge*.

13.9.5   Spectral range

We now aim to identify the spectral regions over which known semiconductors, that are grown with state-of-the-art growth technologies, may be suitable and more importantly may replace metals. The following discussion will consider two regions: (1) the standard telecommunication wavelength at 1.55 μm and (2) wavelength regions in the mid, far IR, and THz ranges.

The advantages and disadvantages of metals and semiconductors may be understood by considering two parameters: the crossover wavelength and loss.

13.9.6 Crossover frequency

SPPs are supported when the real part of the permittivity crosses zero. Using the expression for the real part of the permittivity of a semiconductor

$$\varepsilon_r = \varepsilon_s - \frac{\varepsilon_s \omega_p^2}{\omega^2 + \gamma^2}. \tag{13.57}$$

where $\gamma = e/\mu m^*$ and μ is the mobility. Denoting $\omega = \omega_c$ as the crossover frequency at which $\varepsilon_r = 0$, one obtains

$$\omega_p^2 = \varepsilon_s(\omega_c^2 + \gamma^2).$$

The crossover carrier density may be expressed as

$$n_c = (\varepsilon_s \varepsilon_0 m^* / e^2)(\omega_c^2 + \gamma^2) \tag{13.58}$$

13.9.7 Telecommunication wavelength

The values of crossover carrier density and loss at 1.55 μm for n-type GaAs are estimated in the following example.

Example 13.7 *The parameters used are $m^* = 0.067\, m_0$, $\epsilon_s = 10.7$, $\mu = 1000\, cm^2/V.s$, giving a value $\gamma = 2.62 \times 10^{13} s^{-1}$. We obtain $n_c = 3.76 \times 10^{20}\, cm^{-3}$. The imaginary part of permittivity which represents loss is 0.2534. The loss value for Au at this wavelength is > 10 as seen from Fig. 2 of Naik et al (2013).*

It seems that the carrier density required to achieve SPP at 1.55 μm is quite high. This is also true for other materials as calculated by Naik et al (2013) and shown in Table 1 of their work. It is simple to obtain the values of crossover density using the parameter values in that table which is left as a problem. We quote some representative values for a few materials expressed in units of 10^{20} cm^{-3} and given in parentheses against the materials: (n-Si) (16.0), (p-Si) (23.1), (p-GaAs) (24.4), (n-InP) (3.82), (n-GaN) (6.83) and (p-GaN) (42.3).

The problems related to achieving such high levels of carrier density have been discussed in detail by Naik et al for different materials. The density depends on the solid solubility of dopants and doping efficiency. Furthermore the presence of high defect densities and alloying effects degrade the quality of the materials. In summary, none of the common semiconductors that are grown by refined techniques have the prospect of replacing noble metals at the telecommunication wavelength.

On the other hand, transparent conducting oxides like indium tin oxide(ITO), aluminium (Al), or zinc oxide (ZnO), may be heavily doped and they have high mobilities. These materials can serve as metals in the near-IR range including 1.55 μm wavelength.

13.9.8   Mid-infrared and beyond

It is easy to conclude from Eq. (13.58) that if $\omega_c^2 \gg \gamma^2$, the crossover density decreases 16-fold if the operating wavelength increases four times 1.55 μm, that is nearly 6 μm.

While the conclusion that common semiconductors are not suitable at telecommunication wavelengths remains true and is further supported by the work of Khurgin (2017), the previous estimates for plasma frequency and crossover wavelength using the background static permittivity and constant effective mass shows much deviation from the experimentally observed values. The correct estimate should take into account the actual permittivity given by Eq. (13.53) and the changed effective mass due to non-parabolicity and heavy doping effects.

We illustrate this point by taking the following example using parameters for n-InP as given by Panah et al (2016, 2017).

Example 13.8 *The carrier density obtained for Si doped InP by Pannah et al is 3.87 ×*
10^{19} cm^{-3} and the value γ = 3.49 THz is found from the experimental data of the
sample. Using m* = 0.077 m$_0$, and ε$_b$ = 9.44, we obtain ω$_p$ = 4.11×10^{14}Hz and
λ$_c$ = 14.06 μm. Note that the values obtained in the experiment are 50.5 THz (Table
1 and 6 μm (Fig. 5), respectively from the paper by Panah et al 2017.

Using their own experimental results Panah et al (2017) derived the following semi-empirical formula for the plasma frequency of Si doped indium phosphide (InP) as a function of free carrier density n in the range between 0.35 to 4×10^{19} cm^3:

$$\omega_p = \sqrt{An\left(1 - \frac{B}{1.344 - Cn^{1/3}}\right)} \qquad (13.59)$$

where $A = 918.43$ m^{3}s^{-2}, $B = 3.7 \times 10^{14}$ and $C=6.6 \times 10^5$ m and n is in m^{-3} and ω_p is in rad/s.

The term in parenthesis accounts for the changes in the band structure of heavily doped InP.

Example 13.9 *We use n = 3.87×10^{19} cm^{-3}. The calculated value of ω$_p$ comes to be 48.9*
THz, which is close to the experimental value of 50,5 THz.

Table 13.2 *Plasma frequencies of different semiconductors.*

Material	Doping density in 10^{19}cm^{-3}	Plasma wavelength	Wavelength for zero permittivity	Exp/cal	Reference
InSb	0.54–5.11	17.5–5.4		E	Law 2014
InAs	027–7.5	5.6–14.9		E	Law 2012
GaAs	0.24	17.36 µm		E	Cada et al 2015
InP	1.0	11.55 µm		E	Cada et al 2015
InP	0.35–3.87	18.43–50.5 THz 5.94–16.28 µm		E	Panah et al 2017
Si	6–10	6–10 µm		E/C	Shahzad et al 2011
Ge	22	5.4 µm			Prucnal et al 2019
$Ge_{0.95}$ $Sn_{0.05}$	5.10		6.5 (n-type) 14 µm (p-type)	C	Fischer et al 2017

Several authors have measured the plasma wavelengths for different heavily doped semi-conductors by reflectivity measurements. As noted in earlier discussions, it is not simple to obtain expression for plasma wavelengths by using the simple formula involving n, m^* and ε_s. We present instead in Table 13.2, the experimentally obtained values of plasma wavelengths for different semiconductors mentioning the doping range and the sources.

It is clear from Table 13.2 that it is not possible to operate heavily doped semiconductors below a wavelength of approximately 5 µm.

13.9.9   Length scales

The various length scales have been defined in Section 13.6. Expressions given there may be used to calculate the lengths for all semiconductor plasmonic structures. The propagation length has been calculated and illustrated in the following example.

Example 13.10 *The approximate values for permittivity of InP at 10 µm are $\varepsilon_r = -19$ and $\varepsilon_i = 4$. The propagation length calculated from Eq. (13.50) is 132.4 µm.*

A comparison of permittivity values, and different length scales for a number of metals, and semiconductors is presented in the review by Zong et al (2015)

13.10   Components: Source, modulators, waveguides, detector

All applications in computers and communication require circuits in integrated form. Plasmonic integrated circuits, in particular, the circuits monolithically integrated with electronic circuits in silicon platform are still in the development stage. The basic components are of course sources, modulators, waveguides, detectors and signal processing systems. It is beyond the scope of this chapter to introduce even briefly all the components. In the following, we discuss briefly one of the components, plasmonic waveguides. Sources will be introduced in Chapter 14. A few references related to components, integrated circuit (IC)s, and integration, as listed below are given for interested readers. The list includes some representative papers and is not at all complete.

Dionne et al (2010); Krasavin and Zayats (2010); Berini and De Leon (2011); Soref et al (2012); Pickus et al (2013: modulator); Fang and Sun (2015); Subbaraman et al (2015; active and passive interconnects), Taliercio and Biagioni (2019); Zhang et al (2020: THz review: all components), Fukuda et al (2020; integration on Si platform).

See also the review by Carvalho and Mejia-Salazar (2020) for plasmonics for telecommunications applications.

13.10.1   Waveguides

A vast amount of literature exists for plasmonic waveguides covering different metals, semiconductors and platforms. Here we present only a few basic structures which are used by different workers.

The 1D waveguides may have three basic configurations: metal-dielectric, metal film sandwiched between two symmetric dielectric, and a dielectric film clad between two symmetric metal films (Berini and de Leon 2011). The waveguides are usually planar and two dimensional in form. Several 2D waveguides have been investigated, a few of which are illustrated in Fig. 13.12, and may be termed as dielectric loaded, metal (or semiconductor) stripe, gap, low index hybrid, wedge, and channel hybrids. In fact, these structures have already been used in different dielectric waveguides (Deen and Basu 2012).

Waveguides form the building block for other passive components like tee, coupler, Mach–Zehnder interferometer (MZI), tapers, and more complex ones. No attempt is made however to discuss the components in this chapter, but a few references quoted previously mention these components and their characteristics.

13.10.2   Detectors

There are several reports on plasmonic detectors, which find applications at telecommunication wavelength or act as a sensor in biological applications. Here we focus on devices used in communication. The detectors are mostly Schottky barrier photodetectors. Goykman et al (2011) developed an on-chip nanoscale silicon detector suitable

(a)   (b)   (c)

(d)   (e)   (f)

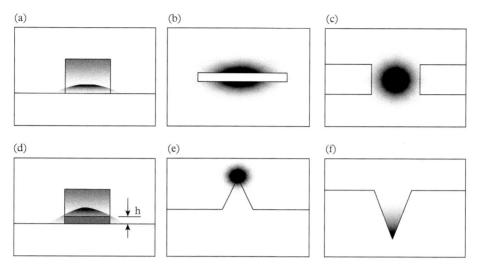

Figure 13.12 *A few 2D SPP waveguide structures; (a) dielectric loaded; (b) metal stripe; (c) gap; (d) low-index hybrid; (e) wedge, and (f) channel or V-groove waveguides.*

Source: Reproduced figure with permission from Berini P and De Leon I (2011) Surface plasmon–polariton amplifiers and lasers. *Nature Photonics* 6(January): 16–24. Copyright (2011) Springer Nature.

for 1.55 and 1.31 µm wavelengths. The device is fabricated on silicon on an insulator, which is compatible with CMOS technology. At the same time, low loss bus photonic waveguide can be fabricated. The silicon surface and the metal layer form the nanoscale Schottky barrier. The internal photoemission gives rise to current which is proportional to the SP wave intensity.

Peale et al (2015) used an Otto coupler to excite SPP in a nearby metal, which serves also as a semi-transparent metal contact to the Schottky diode. The fields of SPP induce photoemission of hot carriers within the contact metal. Otherwise, they penetrate the semiconductor and excite electron-hole pairs. In any case, the resulting current acts as the detector current.

13.10.3 Modulator

Melikyan et al (2011) fabricated an absorption modulator by using a stack of metal/insular/metal oxide/metal layers, which supported a strongly confined asymmetric SPP. The free carrier density in the metal-oxide layer is modulated electrically which in turn changes the absorption and intensity of SPP. The modulator works at 1.55 µm window and operates beyond 100 Gb/s, the operation being only limited by RC time constant.

The Mach–Zehnder Modulator (MZM) has been used by several workers as an intensity modulator as well as a plasmonic switch. The basic principle of an MZM is now presented using the schematic structure shown in Fig. 13.13, which is also applicable to

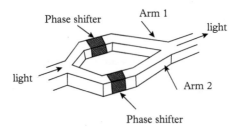

Figure 13.13 *Schematic structure of a Mach–Zehnder modulator used for intensity modulation of light.*

a light modulator (Deen and Basu 2012; Chapters 7 and 11). In Fig. 13.13, the principle illustrated by assuming propagation of light in the two arms is equally applicable for plasmonic waves. As shown, the input optical signal in the input waveguide is split equally by a Y-splitter due to which the input intensities in arm 1 and arm 2 are equal. The propagating electric fields are expressed as

$$E_1 = E_0 sin(\omega t - \beta z) \tag{13.60a}$$

$$E_2 = E_0 sin(\omega t - \beta z) \tag{13.60b}$$

At the input point, $z = 0$, and the two fields are identical. However, the two waves may travel with different propagation constants. Therefore, when the output Y-coupler recombines the two beams, a phase difference appears.

Assume that the two arms have different lengths, L_1 and L_2, and in arm 2 a phase shift is introduced by, say, an electro-optic (EO) effect. The fields at the output terminal are then expressed as

$$E_{01} = E_0 sin(-\beta L_1) \tag{13.61a}$$

$$E_{02} = E_0 sin[-\beta\{L_2 + (\partial\eta/\partial V)lV\}] \tag{13.61b}$$

In Eq. (13.61b), the EO effect is introduced by applying a bias voltage V over a length l.

The output coupler recombines the two fields. The output intensity obtained by using the relation $S = [E{\times}H] = S_0(E_{01} + E_{02})^2$ may be expressed as

$$S = S_0 \left\{ \frac{E_0^2}{2} + \frac{E_0^2}{2} + E_0^2[cos\{\beta(L_2 - L_1) + \Delta\phi_{EO}\}] \right\}$$
$$= S_0[E_0^2\{1 + cos(\beta(L_2 - L_1) + \Delta\phi_{EO})\}] \tag{13.62}$$

where $\Delta\phi_{EO} = \beta lV(\partial\eta/\partial V)$ is the phase shift due to the EO effect. Eq. (13.62) clearly indicates that the output intensity can be changed by changing the phase shift alone induced by the voltage V, giving rise to intensity modulation. As a special case, assume $L_1 = L_2$; the output intensity will be maximum for $\Delta\phi_{EO} = 0$ and minimum (zero) for $\Delta\phi_{EO} = \pi$.

The normalized output intensity therefore varies sinusoidally between 0 and 1 depending on the magnitude of the cosine term. The device can be used as a modulator or a switch to transmit full power or no power to a particular output port. It may be mentioned that the phase difference $\Delta\varphi_{EO}$, which in the present case is introduced via the EO effect, may well be induced by thermo-optic effect also.

Example 13.11 *The voltage-induced change in the phase difference between the two arms of an MZI may be expressed as, $\Delta\phi_{EO} = (2\pi\eta/\lambda_0)L(\partial\eta/\partial V)V$ where V is the applied voltage between two electrodes in one arm, and dn/dV is the change in RI per unit voltage change. Let us assume that for $Ga_{1-x}Al_xAs$, $\partial\eta/\partial V = 5 \times 10^{-5}$, $\eta = 3.6$, and the operating wavelength $\lambda_0 = 850$ nm and the arm length $L = 1$ mm. The required voltage for switching is then $V_\pi = \lambda_0/2\eta L(\partial\eta/\partial V) = 2.36$ volts.*

MZM plasmonic modulators have been fabricated and studied in detail by Leuthold et al (2013), Haffner et al (2014), and Koch et al (2020). In all these studies, Si plasmonic waveguides are used, while organic materials are deposited on Si to introduce phase change by linear EO effect.

13.11 Application of plasmonics in very large-scale integrated data centres and supercomputers

With downscaling of the size of metal-oxide semiconductors (MOS) transistors, the metallic interconnects mainly using copper are also scaled down. Unfortunately, however due to inherent RLC circuit representing the metallic interconnects, very large-scale integration (VLSI) chips show increased delay. At the same time there is excessive heat dissipation due to the metals. It is now well accepted that metallic interconnects cannot meet the increasing demand of huge bandwidth and throughput requirements in data centres and high-performance computing. Optical interconnects have proven to offer a viable solution for high speed and large volume of information transfer and are in use for rack-to-rack and board-to-board transmission links (Deen and Basu 2012).

The power dissipation in supercomputers performing at 10 Peta Flops or more marks an alarming feature. In this respect also, low-energy photonic solutions at board-to-board, chip-to-chip and eventually intra-chip interconnects are expected to drastically reduce energy consumption. The currently used photonic solutions based on the silicon-on-insulator (SOI) photonics platform, offer the advantages of high integration, low cost, and low power consumption at the rate of several Tb/s.

Such photonic solutions cannot reach the compactness of current VLSI circuits due to fundamental limitation imposed by diffraction. The emerging technology relying on plasmonics provides a *beyond photonics* chip-scale platform and is being considered for interconnects due to reduced circuit dimensions and increased energy efficiency (Papaioannou et al 2012).

One of the important components in photonic and plasmonic communication link is waveguide. Various waveguide structures supporting plasma waves have been introduced briefly in Section 13.10.1. The most useful structure for data communication is not yet identified. However, the main obstacle in using the waveguides in data transfer is the very low propagation length of the surface plasmon modes. To circumvent this, an amplifying device named as Surface Plasmon Amplification by Stimulated Emission of Radiation (SPASER), is currently under intense investigations (see Chapter 15).

The aim in practical implementation of active plasmonic circuits is to develop greener and faster network-on-chip (NOC) solutions for data centres and high-performance computers. This entails synergy between plasmonic, electronic and photonic components with very little power consumption, high speed, size reduction and high throughput capabilities. In the aimed NOC solution, the advantages of each constituent, namely, plasmonics, photonics and electronics, are exploited. The most preferred platform is of course silicon. However, plasmonic technology is still in its infancy, as compared to even silicon photonics. The immediate aims are to develop interconnections between silicon and plasmonic waveguide structures in a SOI platform, to develop wavelength division multiplexing systems, and to realize active plasmonic circuits for use in data transmission systems.

Recently Fukuda et al (2020) reported plasmonic circuits merged with silicon Integrated Circuits.

In this context, recent report by Koch et al (2020) merits special mention. Integration of electronic and plasmonic systems has usually been made heterogeneously, that is, these different systems grown on different platform are interconnected by wires. This again leads to power dissipation and reduction of data handling rates. Kohl et al reported a transmitter operating at an active symbol rate beyond 10 GBd, that monolithically integrates a high-speed bi-complementary metal-oxide-semiconductor (BiCMOS) electronic layer and a plasmonic layer on the same Si chip interconnected by vias. The BiCMOS electronic circuit meets the demand of data centres standards and performs 4:1 power multiplexing and delivers on-off keying signals to the plasmonic layer grown above the electronic layer. The plasmonic layer consists of an MZM modulator using hybrid organic-silicon-on-insulator layers, which is highly compact and offers high bandwidth. The ultra-compact modulator has a footprint area of $29 \times 6 \ \mu m^2$.

It may be pointed out that the demand for online services for streaming, storage and computation as well as the advent of artificial intelligence and 5G networks necessitates date dates in the Tb/s region. The current photonics solution provides at most 100 Gb/s, an order of magnitude lower. In addition, the target for computational machines is exascale range ($> 10^{18}$ flops) with a power dissipation of 20 MW and in the fJ/bit range. The available superfast computers could reach less than 100 Peta flops with a power dissipation of > 15 MW (Pleros 2018). It is therefore challenging to develop monolithically integrable plasmonic-photonic-electronic systems to meet the targeted goal.

13.12   Applications of surface plasmons in basic science and characterization

Apart from possible use in communication and information processing, as stated previously, surface plasmon waves, with their enhanced EM field intensity, give rise to a few important phenomena with applications. In this section, we shall briefly mention two of the surface-related physical phenomena.

13.12.1   Surface enhanced Raman scattering

Raman and Krishnan (1928) first experimentally observed in liquids that, in addition to luminescence at the original wavelength, two feeble lines, known as Stokes and anti-Stokes lines appeared, respectively at higher and lower wavelengths than those of the incident wavelength. It was later established that such lines occur due to interaction between incident light wave and vibrational waves of the molecules (Deen and Basu 2012). Since each functional group has its own characteristic vibrational energies, each molecule has a unique Raman spectrum. Raman spectroscopy has therefore been an important analytical tool for identification of molecules.

Typically the Raman signal is several orders of magnitude weaker as compared to the fluorescence emission. As a result, the applicability of Raman scattering was restricted for many years, but with the advent of the laser and improvement in photon detection technology, its utility improved. In 1977, Jeanmaire and Van Duyne (1977) observed a great enhancement of the Raman scattering signal by placing the scatterers on or near a roughened noble-metal substrate. The reason for this enhancement is attributed to interaction between intense electromagnetic field at the surface and the Raman scatterers. The origin of the intensified EM field lies in the generation of localized surface plasmon resonance (LPSR) at the nanoscale sized roughness at the surface of noble metals when excited by visible light. Strong electromagnetic fields are generated when the LSPR of nanoscale roughness features on a silver, gold, or copper substrate is excited by visible light. The term surface-enhanced Raman scattering (SERS) is used for this enhanced scattering process.

SER spectroscopy is now in extensive use in sensitive and selective molecular identification processes. Recently, it has been used as a signal transduction mechanism in biological and chemical sensing. Trace analysis of pesticides, anthrax, prostate-specific antigen, glucose, and nuclear waste are only a few examples in which SERS is used. The other application areas are identification of bacteria, genetic diagnostics, and immunoassay labelling. A miniaturized, inexpensive, and portable SERS instrument is useful for trace analysis in clinics, in the field, and urban settings. Further details may be found in Haynes et al (2005); Stiles et al (2008); and Gaponenko (2010).

13.12.2   Fluorescence enhancement

If an emitter is placed near a metal nanostructure, the fluorescence can be enhanced as well, similar to SERS. In this case, the incident field excites the emitter, and the fluorescence is significantly enhanced due to plasmon resonance in the metal nanoparticle. There is a detrimental effect of quenching the fluorescence due to non-radiative processes and absorption by the surface plasmon resonance. By carefully bypassing these non-radiative processes, the metal nanostructures may be used as nanoantennas (Taliercio and Biagioni 2019; Gaponenko 2010).

An interesting experiment has been performed using nanocrystalline quantum dots(QD)s, which shows in addition to enhancement of decay rate, that the photons from a single emitter can excite a single surface plasmon (SP) in a metallic nanowire(NWR). After propagating through the NWR, these single SPs turn into freely propagating single photons.

13.12.3   Surface plasmon sensors in biology and medicine

The plasmonic interaction between metal nanoparticles is very sensitive to their separation and to the refractive index of the surrounding medium. This high sensitivity is exploited in biological, chemical, and medical sensing, and detection applications. Different chemicals provide a changing dielectric environment and thereby produce a shift in the surface plasmon resonance frequency. This shift is used for identifying different chemicals. In the investigation, a thin metal layer or surface containing a dense package of metal nanoparticles is prepared first with the aim to make the metal particles sensitive to special types of bio-molecules. The chemical to be investigated is then rinsed over the surface. The species then selectively bind to the nanoparticles and change the local dielectric properties. This perturbation of local permittivity introduces a noticeable shift in LSPR. Similarly, injection of functionalized nanoparticles into cells allows binding of the nanoparticles to specific molecules and detecting the location of certain complementary deoxyribonucleic acid (DNA) sequences on chromosomes.

An interesting application in biology is in cancer treatment. Nanoparticles composed of a dielectric core and a metallic shell, when injected into the human body, may be selectively bound to malicious cancer cells. Thereafter, laser irradiation at a precisely engineered plasmon resonance wavelength is used to heat the particles and thereby destroy the cells (Loo et al 2005).

13.13   Intersubband plasmons

Intersubband transitions in QWs and multiple QW (MQW)s have been discussed in Chapter 6. By using a suitable injector, carriers are injected into the first subband which then makes a radiative transition to the ground subband. The collective excitation of 2DEG in the ground state in a doped QW gives rise to an intersubband plasmon. The energy of this particle is determined by the plasma frequency and the difference in energies of the two subbands. By changing the thickness of the QW and the carrier density by different injection level caused by external bias, the whole mid-IR range may be covered.

Recently, a superradiant emission due to a collective excitation has been observed from a structure consisting of InGaAs/InAlAs highly doped quantum wells grown by metal-organic CVD on an InP substrate. The light emission is caused by intersubband plasmons formed by injection of an electrical current into the QW. A dense collection of two-level emitters now oscillates in phase and lead to an increase in the spontaneous emission rate (Laurent et al 2015).

Problems

13.1 *What will be the phase velocity below plasma frequency in presence of free electron plasma by using $v_p = c/\eta$? What will be the group velocity under the same condition? Suggest the use of a proper expression to overcome the paradox.*

13.2 *Derive the expression for normalized ω vs k as shown in Example 13.2. Obtain a plot of the same for silica-Drude metal interface.*

13.3 *Calculate the propagation length and depths of penetration in silica and metal layers at a 1 μm wavelength.*

13.4 *Calculate the plasma reflection edge of GaAs with 10^{18} cm^{-3} electron density.*

13.5 *Calculate the change in absorption coefficient of n-Si at 1.3 μm by using the formula by Soref and Bennet (R A Soref and B R Bennett (1987)), Electrooptical effects in silicon.* IEEE Journal of Quantum Electronics *23: 123–129.*

13.6 *Obtain the expression for the carrier induced change in the permittivity $\Delta \varepsilon$ of a semiconductor in terms of its mobility.*

13.7 *Calculate $\Delta \varepsilon$ of GaAs assuming $m_e = 0.067\ m_0$, carrier density 10^{18} cm^{-3} and a mobility value 5000 cm^2/V.s. Take $\varepsilon_s = 11.8$.*

13.8 *Calculate the Plasmon wavelengths of InAs, InP GaAs, InSb, GaSb, and Si over $n = 10^{19} – 10^{21}$ cm^3. Assume constant permittivity. Calculate the Fermi levels by the Joyce–Dixon formula.*

13.9 *Show that at $T = 0$ K, the increase in the band gap at heavy doping or Burstein–Moss shift is expressed as $\Delta E_g = (\hbar^2/2m^*)(3n/8\pi)^{2/3}$*

13.10 *Barnes (2006) calculated the permittivity of Ag by using Drude formula $\varepsilon(\omega) = 1 - [\omega_p^2/(\omega^2 - i\omega\Gamma)]$, assuming $\omega_p = 1.2 \times 10^{16}$ rad.s^{-1} and damping parameter $\Gamma = 1.45 \times 10^{13}$s^{-1}. The SPP wavelength, propagation length, and penetration depths are calculated using the expressions given. Obtain plots of the wavelength and different lengths over wavelength ranges from 400 to 1,600 nm following Barnes' work.*

13.11 *The relative permittivity of gold at a wavelength of 830 nm is $\approx -29+2.1i$. Calculate the four length parameters assuming air as the dielectric.*

13.12 *Repeat the calculation for n_c and loss for all other semiconductors considered by Naik et al by using parameters in table 1 of their work.*

13.13 *Using the semi-empirical formula Eq 13.59 for InP:Si, calculate and plot the plasma frequency against doping density.*

13.14 *Calculate the real and imaginary parts of the permittivity of InP with carrier density 3.8×10^{19} cm^{-3} (sample 9 of Panah et al 2017) using parameters given in Table 2 of their work.*

13.15 *Write down the equation of motion of free electrons in a semiconductor suffering collisions. Assuming harmonic solutions, obtain the expressions for the real and imaginary parts of the permittivity as given by Eq. (13.56).*

13.16 *Using Eq. (13.56) obtain the expression for the plasma reflection edge.*

13.17 *Using Eq. (13.35) obtain the dispersion relation for the silica-Ag interface, show that the photon dispersion plot does not touch the Plasmon dispersion curve. Using this plot, illustrate the phase matching conditions for both the prism and grating coupling scheme as shown in Figs. 13.10 and 13.11.*

References

Baburin A S, Kalmykov A S, Kiratev R V, Negrov D V, Moskalev D O, Ryzikov I A, Melentie P N, Rodinov I A, and Balykin V I (2018) Toward a theoretically limited SPP propagation length above two hundred microns on an ultra-smooth silver surface [Invited]. *Optical Materials Express* 8(11): 3254–3261.

Barnes W L (2006) Surface plasmon–polariton length scales: A route to sub-wavelength optics. *Pure and Applied Optics: Journal of the European Optical Society Part A* 8: S87–S93.

Basu P K (1997) *Theory of Optical Processes in Semiconductors: Bulk and Microstructures.* Oxford: Oxford University Press.

Berini P and De Leon I (2011) Surface plasmon–polariton amplifiers and lasers. *Nature Photonics* 6: 16–24.

Brongersma M L and Shaleev V M (2010) The case for plasmonics. *Science* 328: 440–444.

Cada M, Blazek D, Pistora J, Postava K, and Siroky P (2015) Theoretical and experimental study of plasmonic effects in heavily doped gallium arsenide and indium phosphide. *Optical Mater Express* 5(2): 340–352. DOI:10.1364/OME.5.000340.

Carvalho W O F and Ricardo Mejía-Salazar J R (2020) Plasmonics for telecommunications applications (Review). *Sensors 2020* 20(2488): 1–21.

Caspers J N, Rotenberg N, and van Driel H M (2010) Ultrafast silicon-based active plasmonics at telecom wavelengths. *Optics Express* 18(19): 19761–19769.

Chochol J, Postava K, Cada M, Vanwolleghem M, Miˇcica M, Halagaˇcka L, Lampin J-F, and Piš-tora J (2017) Plasmonic behavior of III-V semiconductors infra-infrared and terahertz range. *Journal of the European Optical Society-Rapid Publications* 13: 13.

Deen M J and Basu P K (2012) *Silicon Photonics: Fundamentals and Devices.* Chichester, UK: Wiley.

Dionne J A, Sweatlock L A, Matthew T, Sheldon, A, Paul Alivisatos, and Harry A Atwater (2010) Silicon-Based plasmonics for on-chip photonics. *IEEE Journal of Selected Topics in Quantum Electronics* 16(1): 295–306.

Dong Z, Vinnakota R K, Briggs A F, Nordin L, Bank S R, Genov D A, and Wasserman Daniel (2012) Electrical modulation of degenerate semiconductor plasmonic interfaces. *Journal of Applied Physics* 126(043101): 1–6.

Fang Y and Sun M (2015) Nanoplasmonic waveguides: Towards applications in integrated nanophotonic circuits. *Light Science and Applications* 4(6): e294–e294.

Fischer I A, Augel L, Berrier A, Oehme M, and Schulze J (2017) *(Si)GeSn Plasmonics, 2017.* IEEE Photonics Society Summer Topical Meeting Series (SUM), 10–12 July. Puerto Rico: San Juan.

Fukuda M, Tonooka Y, and Ishikawa Y (2020) Feasibility of Plasmonic Circuits Merged with Silicon Integrated Circuits. 2020 Pan Pacific Microelectronics Symposium (Pan Pacific). DOI: 10.23919/PanPacific48324.2020.9059350.

Gaponenko S V (2010) *Introduction to Nanophotonics.* Cambridge, UK: Cambridge University Press.

Goykhman I, Desiatov B, Khurgin J, Shappir J, and Levy U (2011) Locally oxidized silicon surface-plasmon Schottky detector for telecom regime. *Nano Letters* 11: 2219–2224.

Goykhman I, Desiatov B, and Levy U (2013) Silicon plasmonics. In: *Plasmonics: Theory and Applications, 149 Challenges and Advances in Computational Chemistry and Physics.* T V Shahbazyan and M I Stockman (eds.) Chapter 4, pp. 149–166. Dordrecht: Springer Science + Business Mediam.

Gramotnev D K and Bozhevolnyi S I (2010) Plasmonics beyond the diffraction limit. *Nature Photonics* 4(2): 83–91.

Haffner C, Heni W, Fedoryshyn Y, Niegemann J, Melikyan A, D. L. Elder et al (2015) All-plasmonic Mach-Zehnder modulator enabling optical high-speed communication at the microscale. *Nature Photonics* 9: 525–528.

Haynes C L, McFarlandA D, and Van Duyne R P (2005) Surface enhanced Raman spectroscopy. *Analytical Chemistry A* 338(20): 338A–346A.

Jeanmaire D L and Van Duyne R P (1977) Surface Raman spectro-electrochemistry: Part I. Heterocyclic, aromatic, and aliphatic amines adsorbed on the anodized silver electrode. *Journal of Electroanalytical Chemistry* 84: 1–20.

Hill M T, Marell M, Leong E S P, Smalbrugge B, ZhuY, Sun M, van Veldhoven P J, Geluk E J, Karouta F, Oei Y-S, Nötzel R, Ning C-Z, and Smit M K (2009) Lasing in metal-insulator-metal sub-wavelength plasmonic waveguides. *Optics Express* 17(13): 11107–11112.

Jain S C and Roulston D J (1991)A simple expression for band gap narrowing (BGN) in heavily doped Si, Ge, GaAs and Ge_xSi_{1-x} strained layers. *Solid-State Electronics* 34(5): 453.

Krasavin A V and Zayats A V (2010) Silicon-based plasmonic waveguides. *Optics Express* 18:(11): 11791–11799.

Kretschmann E and Raether H (1968) Radiative decay of non-radiative surface plasmons excited by light. *Z Naturforschung* 23A: 2135–2136.

Khurgin J B (2017) Replacing noble metals with alternative materials in plasmonics and metamaterials: How good an idea? *Philosophical Transactions of the Royal Society A* 375(20160068): 10.

Koch U, Uhl C, Hettrich H, Fedoryshyn Y et al (2020) A monolithic bipolar CMOS electronic-plasmonic high-speed transmitter. *Nature Electronics* 3: 338–345.

Laurent T, Todorov Y, Vasanelli A et al (2015) Super radiant emission from a collective excitation in a semiconductor. *Physical Review Letters*115(187402): 4.

Law S, Adams D C, Taylor A M, and Wasserman D (2012) Mid-infrared designer metals. *Optics Express* 20(11): 12155–12165.

Law S, Liu R, and Wasserman D (2014) Doped semiconductors with band-edge plasma frequencies. *Journal of Vacuum Science & Technology B* 32(052601): 7.

Leuthold J, Koos C, Freude W, Alloatti L et al (2013) Silicon-Organic hybrid electro-optical devices. *IEEE Journal of Selected Topics in Quantum Electronics* 19(6): 1–13.

Li D and Ning C -Z (2011) All-semiconductor active plasmonic system in mid-infrared wavelengths. *Optics* 19(15): 14594–14603.

Loo C, Lowery A, Halas N, West J, and Drezek R (2005) Immuno-targeted nanoshells for integrated cancer imaging and therapy. *Nano Letters* 5(4): 709–711.

Maier S A (2007) *Plasmonics: Fundamentals and Applications.* Boston, MA: Springer.

Mao X and Cada M (2013) Optical surface plasmon in semiconductors. *Integrated Photonics Research, Silicon and Nanophotonics 2013OSA Technical Digest (online) (Optical Society of America, 2013)* paper IM1B.4, Rio Grande, Puerto Rico United States. https://doi.org/10.1364/IPRSN.2013.IM1B.4.

Mao X (2013) Optical Surface Plasmon in Semiconductors. MASc thesis, Halifax, Canada: Dalhousie University Press.

Melikyan A, Lindenmann N, Walheim S, Leufke P M, Ulrich S, Ye J, Vincze P, Hahn H, Schimmel Th, Koos C, Freude W, and Leuthold J (2011) Surface plasmon polariton absorption modulator. *Optics Express* 19(9): 8855–8869.

McPeak K M, Jayanti S V, Kress S J P, Meyer S, Iotti S, Rossinelli A, and Norris D J (2015) Plasmonic films can easily be better: Rules and recipes. *ACS Photonics* 2: 326–333.

Naik G V, Shalaev V M, and Boltasseva (2013) Alternative plasmonic materials: Beyond gold and silver, *Advanced Materials* 25: 3264–3294.

Otto A (1968) Excitation of non-radiative surface plasma waves in silver by the method of frustrated total reflection. *Z Physik* 216: 398–410.

Panah M E A., Han L, NorrmanK, Pryd N, Nadtochi, A, Zhukov A E, Lavrinenko A, and Semenova E (2017) Mid-IR optical properties of silicon doped InP. *Optical Materials Express* 7(7): 2260–2271.

Panah M E A, Takayama O, Morozov S V, Kudryavtsev K E, Semenova E, and Lavrinenko A (2016) Highly doped InP as a low loss plasmonic material for mid-IR region. *Optics Express* 24(25): 29078–29089.

Papaioannou S, Vyrsokinos K, Kalavrouziotis D, Giannoulis G, Apostolopoulos D, Avramopoulos H, Zacharatos F, Hassan K, Weeber J -C, Markey L, Dereux A, Kumar A, Bozhevolnyi S I, Suna A, de Villasante O G, Tekin T, Waldow M, Tsilipakos O, Pitilakis A, Emmanouil E, Kriezis E, and Pleros N (2012) Merging plasmonics and silicon photonics towards greener and faster 'network-on-chip' solutions for data centers and high-performance computing systems. In: *Plasmonics – Principles and Applications*, Papaioannou et al (eds.) Chapter 21. InTech open. UK https://www.intechopen.com/chapters/40348

Peale R E, Smith E, Smith C W, Khalilzadeh-Rezaie F, Ishigami M, Nader N, Vangala S, and Cleary J W (2016) Electronic detection of surface plasmon polaritons by metal-oxide-silicon capacitor. *APL Photonics* 1(066103): 8.

Pickus S K, Khan S, Ye C, Li Z, and Sorger V J (2013) Silicon plasmon modulators: Breaking photonic limits, *IEEE Photonics Society Newsletter* December: 4–10.

Pleros N (2018) Silicon photonics and plasmonics towards network-on-chip functionalities for dis aggregated computing. Optical Fibre Communication Conference. Pages Tu3F.4. 2018/3/11. New York: Optical Society of America.

Prucnal S, Liu F, Voelskow M, Vines L, Rebohle L, Lang D, Berencén Y, Andric S, Boettger R, Helm M, Zhou S, and Skorupa W (2019) Ultra-doped n-type germanium thin films for sensing in the mid-infrared, *Scientific Reports* 6(27643): 1–8.

Raman C V and Krishnan R S (1928) A new type of secondary radiation. *Nature* 121: 501–501.

Rosenberg A, Surya J, Liu R, Streyer W, Law S L, Leslie L S, Bhargava R, and Wasserman D (2014) Flat mid-infrared composite plasmonic materials using lateral doping-patterned semiconductors. *Journal of Optics* 16(094012): 9.

Shahzad M, Medhi G, Peale R E, Buchwald W R, Cleary J W, Soref R, Boreman G D, and Edwards O (2011) Infrared surface plasmons on heavily doped silicon, *Journal of Applied Physics* 110(123105): 6.

Soref R A and Bennett B R (1987) Electrooptical effects in silicon. *IEEE Journal of Quantum Electronics* 23: 123–129.

Soref R, Hendrickson J, Justin W, and Cleary J W (2012) Mid- to long-wavelength infrared plasmonic photonics using heavily doped n-Ge/Ge and nGeSn/GeSn heterostructures. *Optics Express* 20(4): 3814–3824.

Stiles P L, Dieringer J A, Shah N C, and van Duyne R P (2008) Surface enhanced Raman spectroscopy. *Annual Review of Analytical Chemistry* 1: 601–626.

Taliercio T and Biagioni P (2019) Semiconductor infrared plasmonics. *Nanophotonics* 8(6): 949–990.

Wikipedia (2021) Surface Plasmon. polariton.en.wikipedia.org/wiki/Surface_plasmon_polariton.

Yu H, Peng Y, Yong Yang Y, and Li Z -Y (2019) Plasmon-enhanced light–matter interactions and application. *NPJ Computational Materials* 5: 45.

Zhang X, Xu Q, Xia L, Li Y, Gu J, Tian Z, Ouyang C, Han J, and Zhang W E (2020) Terahertz surface plasmonic waves: A review. *Advanced Photonics* 2(1): 1–19.

Zhong Y, Malagari S D, Hamilton T, and Wasserman D (2015) Review of mid-infrared plasmonic materials. *Journal of Nanophotonics* 9(093791): 1–21.

14

Spasers, and plasmonic nanolasers

14.1 Introduction

Surface plasmon polaritons (SPPs) can store light energy in electron oscillations at a subwavelength scale and can thus overcome the Rayleigh diffraction limit. The problem lies however in the very short propagation length of the wave. Some kind of amplification is therefore needed to enable propagation of the waves over a large distance. The possibility of achieving such amplification as well as a self-sustained oscillation was first proposed by Bergman and Stockman (2003). An acronym SPASER for surface plasmon amplification by stimulated emission of radiation, similar to light amplification by stimulated emissions of radiation (LASER), is then introduced in the literature.

Development of such devices was first announced in 2009 by three groups almost simultaneously. Noginov et al (2009) used an assembly of nanoparticles with gold core surrounded by a dyed silica gain medium. Oulton et al (2009) employed a nanowire on a silver screen, and Hill et al (2009) demonstrated amplification in an electrically pumped semiconductor layer of 90 nm width surrounded by silver.

Some similarities and differences between spasers and lasers have been listed by Stockmann (2008). While lasers directly emit photons, their nano plasmonic counterpart, the spaser, does not. In spasers, surface plasmons, instead of photons, are amplified. The resonant cavity in a laser is replaced by a nanoparticle in a spaser which supports the plasmonic mode. Just like in a laser, the energy source for the spasing mechanism is an externally excited gain medium and usually the excitation field is optical, the frequency of which differs from the frequency of operation of the spaser. There are other similarities also. Both photons in lasers and SPPs in spasers are bosons with spin 1. SPPs are electrically neutral excitations, a characteristic of photons also. In addition, SPPs are the most collective material oscillations found in nature. They therefore are the most harmonic, interact very weakly with one another, and undergo stimulated emission just like in a laser.

After the first proposal, followed by successful experimental demonstrations, the field *spaser* has become an important area of theoretical and experimental investigations. Its various applications include nanoscale lithography, fabrication of ultra-fast photonic nano circuits, single-molecule biochemical sensing, and microscopy, which merits special mention. There are a few good reviews of the basic physics and progress in the

Semiconductor Nanophotonics. Prasanta Kumar Basu, Bratati Mukhopadhyay, and Rikmantra Basu, Oxford University Press.
© P.K. Basu, B. Mukhopadhyay, R. Basu (2022). DOI: 10.1093/oso/9780198784692.003.0014

development of SPASERs from time to time, such as Oulton (2012); Gwo and Shih (2016); Deeb and Pelouard (2017); Wang et al (2018); Xu et al (2019); and Azzam et al (2020). Applications of spasers are discussed by Ma and Oulton (2019); and Ning (2019).

This chapter begins with the first theoretical proposal of spasers by Bergman and Stockman (2003), discussing the principle of operation, proposed structure, and a brief outline of the underlying theory. Further work and alternate structures proposed by them are then introduced. The first experimental demonstrations, almost concurrently reported by the three groups, as mentioned earlier, are then described. A short description of the model (Stockman 2008) explaining the work by Noginov et al (2009) then follows. We then begin discussions on semiconductor based spasers and the underlying theory. Following this, the models used by Khurgin and Sun (2012a,b,c, 2014) and Khurgin (2015) are discussed in detail. Though these models underline scepticisms about better performances of spasers, having subwavelength dimensions in all the directions, than conventional photonic lasers, the models provide analysis of spasers in terms of rate equation models similar to those used in conventional lasers. The works also draw attention to devices called surface plasmon emitting diodes(SPED)s, describing their characteristics. Scepticisms are removed by Wang et al (2017) by reporting low-threshold power spasers with subwavelength dimensions and their superiority. The work and unusual scaling laws proposed by Wang et al, valid for such types of spasers, are described.

The chapter ends with a summary of recent developments, following reviews by different authors.

Example 14.1 *We make a realistic calculation of propagation length in Ag-semiconductor interface by using the expression (13.50)* $L_{SPP} = \frac{\lambda_0}{2\pi} \left(\frac{\epsilon_{mr}+\epsilon_d}{\epsilon_{mr}\epsilon_d} \right)^{3/2} \frac{(\epsilon_{mr})^2}{\epsilon_{mi}} \approx \frac{\lambda_0}{2\pi} \frac{(\epsilon_{mr})^2}{\epsilon_{mi}}$ *at photon energy of 2.25 eV; the free space wavelength is 551 nm. The values* $\epsilon_{mr} = -12.65$ *and* $\epsilon_{mi} = 0.3866$ *are given for silver (Ag) at this wavelength in the paper by Johnson and Christy (1972). We use* $\epsilon_d = 12$ *for a typical semiconductor used as the dielectric. The calculated propagation length is now 10.1 nm only.*

Note that plasmonic waveguides and other elements like modulators in a Mach–Zehnder configuration (Berini and de Leon 2011; Deen and Basu 2012) usually have lengths of the order of wavelength or more. Since the propagation length in the previous example is too low in comparison with plasmon wavelength, the need for amplification of surface plasmon (SP) waves is easily understandable.

14.2   Early investigations on surface plasmon amplification

We present in this introductory section, the first theoretical proposal by Bergman and Stockman (2003) pointing out the possibility of amplification of SP waves. We briefly mention the ingredients of their theory and main results, including their proposed structure. Further elucidation of the basic principle using energy level diagrams and energy transfer mechanisms then follows. This section ends with a brief description of the three experimental works establishing the phenomena of amplification by stimulated emission of SP modes.

14.2.1   First proposal

As mentioned earlier, Bergman and Stockman (2003) first considered the amplification of SP by stimulated emission and studied the possibility of achieving a quantum nano-generator. The spaser radiation consists of bosons but is highly localized. The scheme is similar to a conventional laser. The first element, the emitter, is an assembly of Quantum Dots (QD)s which transfer their emitted optical energy non-radiatively to the SPs existing in a metallic nano system. Stimulated transitions taking place between quantized SP levels in the metallic nanoparticles lead to the amplification of SP waves as usual. The nanoparticle itself acts as a nanocavity which provides proper feedback to the amplified wave.

Bergman and Stockman expressed the electric field in quantized form in terms of the eigenmodes of the SP wave and SP creation and annihilation operators as

$$\phi(\mathbf{r}, t) = \sum_n C_n \varphi_n(\mathbf{r}) e^{-\gamma_n t} [a_n e^{-i\omega_n t} + a_n^\dagger e^{i\omega_n t}] \qquad (14.1)$$

Here, the electric field is $\mathbf{E}(\mathbf{r}, t) = -\nabla \phi(\mathbf{r}, t)$, C_n is the normalization constant, $\varphi_n(\mathbf{r})'$ s are eigen solutions for SP mode equations, ω_n and γ_n are, respectively, the frequency and damping constant of the n-th SP mode, and a_n and $a_n^\dagger$ are annihilation and creation operators of the SP in the state φ_n. Eq. (14.1) and other equations in the paper lead to the quantized Hamiltonian expression in terms of harmonic oscillators in the form $H = \sum_n \hbar \omega_n (a_n^\dagger a_n + 1/2)$.

The active host medium is assumed to contain a number of dipolar emitters, having two levels with respective population of $\rho_0(\mathbf{r})$ and $\rho_1(\mathbf{r})$ in the ground and excited states. These emitters are placed at the points $\mathbf{r}_a$, $a = 1, 2, \ldots$ and have dipole moments $\mathbf{d}^{(a)}$ with a transition matrix element d_{10}. The perturbation Hamiltonian for interaction between the active medium with the SP field is now described by $H' = \sum_a \mathbf{d}^{(a)} \cdot \nabla \phi(\mathbf{r}_a)$. Applying Fermi's golden rule and taking into account Eq. (14.1) and other equations, the rate equation for the number N_n of SPs in the n-th mode is expressed as

$$\frac{dN_n}{dt} = (B_n - \gamma_n)N_n + A_n \qquad (14.2)$$

Here, B_n and A_n are the Einstein coefficients for stimulated and spontaneous emission of SP modes. Expression for B_n is derived in the paper by Bergman and Stockman.

The dimensionless gain is defined as $\alpha_n = (B_n - \gamma_n)/\gamma_n$ and when $\alpha_n > 1$, that is, $B_n > \gamma_n$ amplification of SP wave of the n-th mode occurs, and the number of SPs in a single eigenmode grows exponentially as $N_n \propto exp(\gamma_n \alpha_n t)$.

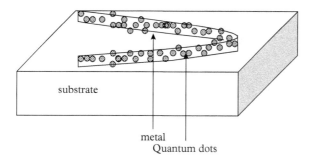

substrate

metal
Quantum dots

Figure 14.1 *Schematic of the spaser conceptually similar to the originally proposed structure by Bergman and Stockman (2003). The resonator of the spaser is a metal nanoparticle shown as a gold V-shape. It is covered by the gain medium depicted as nanocrystal QDs. This active medium is supported by a neutral substrate.*

The structure originally proposed for a spaser is a V-shaped gold layer in the form of nanoparticles. The metal inclusions are covered by QDs, which act as the emitters. A schematic diagram, similar to the structure proposed by Bergman and Stockman (2003) is given in Fig. 14.1.

By considering lead sulfide (PbS) and lead selenide (PbSe), QDs, using their dielectric functions, to calculate values of d_{10} etc., the authors could show that gain¬12 could be achieved for a few SP modes in the energy range 1–2 eV. Their work is the first report on the possibility of amplification of SP modes, which leads to build-up of coherent SP modes.

14.2.2 Spasers explained

As mentioned earlier, Bergman and Stockman (2003) considered a two-level emitter in the form of QDs to transfer energy to the SP modes. In a paper titled *spasers explained* Stockman (2008) explained more clearly the operation of spasers, their characteristics, and suggested a simpler and more attractive structure to observe SP amplification. The operation was further elaborated by Stockman (2010, 2013).

Similarities and differences between a spasers and a lasers have already been pointed out in Section 14.1. It is noted that both lasers and spasers have a common physical basis: stimulated emission and coherence. In fact, as will be made clear in a later section, the quasi particles involved in both the cases are polaritons: SPPs in a spaser and exciton-polariton in a laser.

Stockman (2008) considered a metal-dielectric core-shell structure for the possible observation of SPASER phenomena. The structure is shown in Fig. 14.2(a) in which a thin layer of Ag is deposited surrounding a dielectric core rod and works as the nanoshell. On top of this metal layer a few monolayers of QDs are allowed to form. Alternatively, the nanocrystal quantum dots (NQD)s themselves forming the gain medium can be incorporated in the core matrix which is surrounded by the metallic shell.

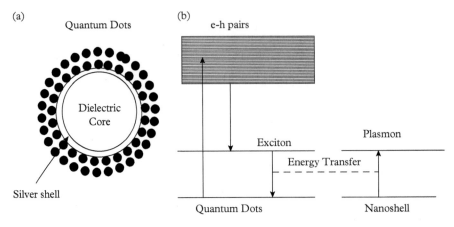

(a) Quantum Dots

(b) e-h pairs

Dielectric Core

Silver shell

Exciton

Plasmon

Energy Transfer

Quantum Dots

Nanoshell

Figure 14.2 *(a) The core-shell structure to observe possible spaser action. The thin nanoshell containing Ag nanoparticles surrounds the dielectric core. A few monolayers of Nanoparticle QDs sit on the nonoshell and act as the gain medium. (b) Energy level diagrams in NQDs and in nanoshell showing various transitions and energy transfer processes between NQDs and Ag nanoparticles. The levels in nanoshell correspond to SP states.*

Source: Reproduced figure with permission from Stockman M I (2008) Spasers explained. *Nature Photonics* 2(June): 327. Copyright (2011) Springer Nature.

The energy levels and transitions are indicated in Fig. 14.2(b). An external source first creates electron-hole pairs (EHP)s in the QDs. The EHPs then relax to form excitons. The exciton energy level is above ground level. This two-level system provides the energy for SP emission into the spasing mode. In a free environment, the excitons recombine and emit photons. However, in the spaser structure, the emitted photons transfer their energy to the SP modes as indicated by the dashed line in Fig. 14.2(b). The non-radiative energy transfer to the SPs has a higher probability than that of the radiative decay (photon emission), and their ratio is given by the Purcell factor

$$F = \frac{3}{4\pi^2} \left(\frac{Q}{V}\right) \left(\frac{\lambda}{2\eta}\right)^3 \tag{14.3}$$

where V is a characteristic size of the spaser metal core, and Q is the plasmonic quality factor, λ is the wavelength, and η is the refractive index(RI). A value of $Q = 100$ may easily be obtained for good plasmonic metal like Ag.

Example 14.2 *An estimate of the Purcell enhancement factor is made by using $\lambda= 0.5\ \mu m$, radius R=14 nm of the nanoparticle, $\eta = 1$ and $Q = 15$. The enhancement factor is 1.5×10^3. The value in most of the structures is in the range below 100.*

The SP modes already existing create a strong electric field to excite the gain medium, which in turn, creates more SP modes signalling a feedback mechanism. When this feedback is strong and the lifetime of SP modes is quite long, self-sustained generation of coherent SP modes occurs. The spaser action is the outcome of a non-equilibrium phase transition just like in a laser.

14.2.3 Experimental demonstrations

As stated already, three different groups within a period of 5–6 months reported spaser action by using different structures and different material combinations. We will now discuss the experimental observations of these groups one by one.

Noginov et al (2009) reported the first nanoparticle spaser, using a spherical core-shell structure, much like Fig. 14.2(a). Their medium was a chemically synthesized gold nanosphere forming the core of the structure. It was surrounded by a silica dielectric shell which contained the gain medium, the organic dye Oregon Green 488 (OG-488). The immobilized dye molecules were optically pumped by nanosecond optical pulses, the frequency of which coincided with the absorption band of the dye molecules. The shell medium provided adequate gain to sustain SP oscillations in the gold nanosphere. The SP oscillations were converted into visible laser light of wavelength 531 nm.

The spectra of the emitted light were measured by varying the pump pulse power. For low power the spectra were somewhat broader but narrowed down with increasing pump power and became very narrow with intense illumination. The onset of a threshold as pump power increased was clearly established, confirming spaser action.

In their experiment, Hill et al (2009) used metal insulator semiconductor insula-tor metal (MISIM) waveguides containing electrically pumped semiconductor gain medium cores. This work extended the authors' previous work of electrically pumped metallic nano-lasers (Hill et al 2007) and was another step to realizing the eventual plas-mon mode lasers. The core sizes of the waveguide were a few tens of nanometres in two dimensions (2D). Fig. 14.3 shows a schematic cross-sectional diagram of the structure, more or less similar to that used by Hill et al (2009).

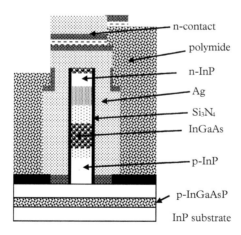

Figure 14.3 *Cross-sectional view of the MISIM structure used by Hill et al (2009).*

The MISIM waveguide consisted of a double-heterojunction bipolar transistors with step-graded indium gallium arsenide phosphide (InGaAsP) collector (InP/InGaAs/InP) pillar having rectangular cross-section grown on p-InGaAsP pedestal grown on the indium phosphide (InP) substrate. The pillar was surrounded by a 20 nm thick insulating silicon nitride (SiN) layer. A thin layer of silver encapsulated the pillar and insulator. The middle InGaAs layer in the double heterostructure was the gain medium, and due to its larger RI than in InP, light is confined vertically within it. Injection of electrons took place via the top of the pillar, and a large area lateral contact (not shown in the diagram) made on the p-InGaAsP layer injected the holes. The length and width of the pillar were below the diffraction limit.

The intensity and linewidth of laser light coming out of the substrate were measured and analyzed. The light-current characteristics showed the usual threshold behaviour and usual narrowing of spectral width above the threshold was observed.

Oulton et al (2009) experimentally demonstrated the laser action of SPPs with mode areas as small as $\lambda^2/400$, signifying operation at subwavelength scale. A cross-sectional view of their sample is shown in Fig. 14.4. The sample consisted of a cadmium sulfide (CdS)nanowire, separated from the Ag film by an insulating magnesium fluoride (MgF$_2$) layer of nm thickness. Due to the close proximity of the nanowire and metal film, light was confined within an extremely small area, more than 100 times smaller than a diffraction-limited spot. They used a CdS nanowire grown on quartz substrate, which was a typical nanowire (NWR) laser, for comparison with the CdS-Ag laser, which they called as the plasmonic laser.

The laser devices were optically pumped by 405 nm sources and emission from the dominant CdS exciton line at 489 nm took place. At moderate pump intensities, amplified spontaneous emission peaks, corresponding to plasmonic modes resonating between two reflective end-facets of the NWR, were observed. For these modes, propagation losses were compensated by gain. The spectra of emission were compared with the spectra of photonic NWR laser and gross differences were observed. All these differences, in particular, the nanometre-sized optical modes supported by the plasmonic nanolaser, which cannot be supported by conventional lasers, gave evidence for subwavelength operation, with gain surmounting the high losses in metals.

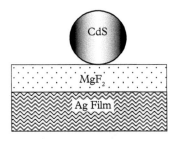

Figure 14.4 *Schematic cross-sectional diagram of the CdS cylindrical nanowire used in the experiment by Oulton et al (2009).*

14.3   Models for the Noginov et al experiment

An analytical rate equation model was developed by Zhong and Li (2013) to understand the behaviour of a spaser using a gold (Au) core, silica shell, and gain medium containing organic dye molecules. They used the parameters given by Noginov et al to obtain plots and contour maps of field, power output, and so on. In that sense, this work attempts to explain the experimental results by an analytical method. A more refined theory has later been presented by Kewes et al (2017). A detailed quantum mechanical theory of spasers is given in several publications, see for example, Stockman (2013) and Premaratne and Stockman (2017) and references cited therein.

In the model by Zhong and Li (2013), the chromophores, that is, the organic dyes have four states 0, 1, 2, and 3. A pump induces an upward transition $0\rightarrow3$, and by non-radiative process, the molecules reach the state 2, which is the upper lasing level. The energy released via $2\rightarrow1$ transition to the lower lasing level is nonradiatively transferred to stimulate a SP mode, effecting a $1\rightarrow2$ upward transition. Some energy is lost by non-radiative processes by downward $1\rightarrow0$ and $3\rightarrow0$ transitions. The four rate equations involving population in different states are solved to obtain the relationship between the characteristics of the spaser (the output power, saturation, and threshold) and the nanocavity parameters (quality factor, mode volume, loss, and spontaneous emission efficiency), atomic parameters (number density, linewidth, and resonant frequency), and external parameters (pumping rate). The theory shows that spaser operation is very difficult to achieve using a single gold nanoparticle plasmonic nanocavity as the pump power to reach threshold is quite high. The authors claim that their theory can commonly be used in understanding and designing all novel microlaser, nanolaser, and spaser systems.

The work by Noginov et al (2009) has been subjected to criticism by a few workers (see Premaratne and Stockman 2017 for details). The experiment could not be replicated. Both semiclassical and quantum theories questioned the validity of different aspects of this work. The threshold pump powers calculated by a few workers were too high. The degree of coherence of emission was not unambiguously determined. The dye molecules used were fragile and suffered photodegradation.

We make no attempt to describe further developments using dye molecules as the gain medium, as the work is not related to semiconductor-based systems, which forms the main focus of this book.

Example 14.3 *The spontaneous decay rate γ_2 of the gain medium excitation into the LSP mode is given by Azzam et al as $\gamma_2 = \frac{2|d|^2}{4\pi\varepsilon_0\hbar\varepsilon_d\varepsilon_{mr}\varepsilon_{mi}R^3}$, where R is the nanoshell radius. Assume $d = 10^{-28}$ Cm (12D; D = dipole moment of H-atom), $\varepsilon_d = 12$, $\varepsilon_{mr} = 12.65$, $\varepsilon_{mi} = 0.3866$, and R = 5 nm. The value of lifetime becomes 4.3 ps.*

14.4   Semiconductor spasers and plasmonic nanolasers

We will now devote the following sections to discussing spasers and plasmonic nanolasers using semiconductors, both bulk and nanostructures. In the literature, several comprehensive reviews and tutorials have appeared from time to time discussing the developments of the devices and the basic principles involved. In this connection, mention may be made of the papers by Ning (2010); Oulton (2012); Berini and de Leon (2011); Leosson (2012); Stockman (2013); Gwo and Shih (2016); Premaratne and Stockman (2017); Wang et al (2017); Deeb and Pelouard (2017); and Azzam et al (2020). Some of these papers also discuss the work utilizing gain media other than semiconductors, e.g. organic dyes.

Gwo and Shih (2016) have mentioned the difference between the terms *spasers* and *plasmonic nanolasers*. The former involves localized SP (LSP) modes, whereas the operation of the latter relies on propagating SPPs. For example, LSP modes are involved in the experiment of Noginov et al (2009). The organic dyes can be replaced by metal nanoparticles or nanoshells. Both Hill et al (2009) and Oulton et al (2009) used SPP propagation-based approaches.

In both cases, the plasmon modes are excited and confined at metal surfaces lying in close proximity to the semiconducting gain medium, resulting in a dramatic reduction of the laser mode volume. Ideally, all the 3Ds of the devices should remain in the sub-wavelength ranges. In the experiments by Hill et al, and Oulton et al, the SPP modes, which are termed as plasmonic gap modes, propagate along the length of the device. These subwavelength modes are reflected at the two ends of the semiconductor acting as a Fabry–Perot resonator. The lengths of the devices are in the micrometre ranges. Some workers prefer to call these devices as SPP microlasers, or rather SPP nanolasers. In later sections, these lasers are termed as photonic nanolasers.

Early SPP-based plasmonic nanolasers showed very high threshold current. In addition, these devices did not have all the 3Ds in the subwavelength range and at least 1D was larger than the diffraction limit. It was also doubtful if true laser action was exhibited, or the nonlinear power dependence and line narrowing were the outcome of amplified spontaneous emission (Gwo and Shih 2016).

Khurgin and Sun (2012a,b,c, 2014) made a detailed study of the loss mechanisms in metals, and of the comparison of gain, threshold, and linewidth of spasers and SPP nanolasers. Their estimates were based on the reported values of dielectric functions of noble metals. It was concluded that, though spasers show low threshold current, the threshold current density is more important, which is enormously high for spasers. They also considered surface plasmon emitting diodes (SPEDs) and pointed out their advantages. The steps to come out of the discouraging situation were also suggested by them. Khurgin (2015) discussed the issue of loss in plasmonics and metamaterials.

The first demonstration of continuous wave operations of a semiconductor plasmonic laser having subdiffractional dimensions in all the three directions came from Lu et al

(2012). The basic device structure was a MIS nanocavity structure used by Oulton et al. However, Lu et al in their sample used an atomically smooth, epitaxially-grown Ag film as the plasmonic material platform and obtaining high gain and indium gallium nitride (InGaN) semiconductor nanorods. Later, these authors used gallium nitride and continuously tunable $In_xGa_{1-x}N$ alloy gain media to obtain the operation at ultraviolet (UV) and for all colour emissions (Lu et al 2014).

In principle, the SPP nanolasers are not much different from spasers. The only difference is that in spasers localized plasmon modes are involved whereas propagating SPPs are amplified or processed in SPP nanolasers.

14.5 Theoretical models by Khurgin and Sun

A detailed analysis of semiconductor based spasers and plasmonic nanolasers has been provided by Khurgin and Sun (2012a,b,c, 2014). Their theory is capable of predicting most of the characteristics of these devices like power output versus pump power, coherence, linewidth, and small-signal modulation characteristics. They considered electrically injected operation. The most attractive feature of their analysis is the inherent simplicity, and their models show that the basic analysis including the rate equation models is common for semiconductor lasers including vertical cavity surface emitting lasers (VCSEL)s, microcavity lasers with an enhanced Purcell factor, spasers, and plasmonic nanolasers. Their unified theory is able to make a useful comparison of spasers, vertical-cavity surface-emitting lasers, and SPEDs. The theory also provides the answer to the question 'How small can "Nano" be in a "Nanolaser"?' (Khurgin and Sun 2012c). This investigation is really essential for proper development of photonic integrated circuits (IC)s, which may compete with the current electronic ICs in increasing the speed and reducing the power consumption of future information processing systems.

14.5.1 Semiconductor gain versus metallic loss

The answer to the question of whether metallic losses can be overcome by a suitable gain medium and advantages and limitations of such a combination are discussed by Khurgin and Sun in a number of papers (2012a,b,c, 2014, abbreviated later as KS). We first discuss their paper (2012a) in which two planar structures are considered as plasmonic waveguide.

A schematic of the first structure is given in Fig. 14.5. The electric field associated with SPP mode propagates along the x-direction in the interface between metal and semiconductor with in-plane wave vector k_x and is expressed as $E_z(z)exp[i(k_xx-\omega t)]$. The depth ($z$)-dependent electric field decays in the semiconductor as $E_z(z) \sim \exp(-q_sz)$ and in the metal as $E_{zm}(z) \sim \exp(q_mz)$, and the dispersion relation is

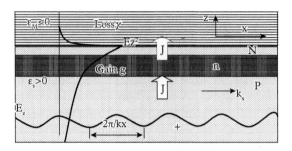

Figure 14.5 *Schematic of a metal-semiconductor heterostructure studied by Khurgin and Sun (2012a).*

$$k_x^2 - q_{s,m}^2 = k_s^2 = \frac{\epsilon_{s,m}\omega^2}{c^2} \tag{14.4}$$

where the symbols used are standard. The dispersion relation for the in-plane wave vector is

$$k_x^2(\omega) = \frac{k_s^2 \epsilon_m}{\epsilon_s + \epsilon_m} \tag{14.5}$$

The permittivity of the metal is negative and is expressed by the Drude model in terms of plasma frequency ω_p, as

$$\epsilon_m = 1 - \frac{\omega_p^2}{\omega^2 + i\omega\gamma} \tag{14.6}$$

When the light frequency ω approaches the SP frequency $\omega_{sp} = \omega_p/(\epsilon_s + 1)^{1/2}$, both k_x and q_s become large. Large k_x signifies that the effective wavelength of light is short so that subwavelength dimension of the plasmonic waveguide in the form of narrow strip line along x-direction is allowed. An effective RI may be introduced such that $\eta_{eff} = k_x/k_s > 1$. A large q_s implies a tight confinement within the semiconductor defining an effective thickness $d_{eff} = \ln(10)k_s/2\pi q_s$, as shown in Fig. 14.5. The semiconductor contains a very large fraction of energy ¬ 90%. It is shown that $\eta_{eff} > 3$ and $d_{eff} < 0.1$ may be attainable.

Unfortunately however, high absorption in metals, even in noble metals like Au and Ag make the coefficient γ quite large, so that the SPPs cannot propagate over a large distance along x. Compensation of loss by using a semiconductor gain medium has been addressed by Khurgin and Sun (2012a) by considering noble metals like Ag and Au and their properties reported in the literature. They considered two different configurations: a metal–semiconductor (heterojunction) structure and a metal-semiconductor-metal (M-S-M) structure. We have shown in Fig. 14.5 the schematic structure of the first.

The essence of the theoretical model using the structure in Fig. 14.5 will now be described. The top metal is either Ag or Au. The semiconductor heterostructure is an N-n-P type, in which the lower band gap n layer is sandwiched between two wider gap layers (N and P). Injection of carriers is made via the top metal contact and a contact made to the P-type layer. KS (2012a) considered two material systems: (1) Au and $In_xGa_{1-x}As_yP_{1-y}$ suitable for the longer wavelengths, and (2) Ag and $In_x Ga_{1-x}N_yAs_{1-y}$ for the shorter wavelength ranges. It was assumed that both the quaternary semiconductors can be grown to cover wide ranges of band gap.

It may be shown by using Maxwell's equations that the ratio

$$\frac{\mu_0 H^2}{\epsilon_0 \epsilon_s E^2} \sim \frac{1}{\eta_{eff}^2} < 1 \tag{14.7}$$

This is the ratio of energy stored in the magnetic field to that in the electric field, so that the remaining energy is in the form of kinetic energy. The effective loss coefficient in the metal is therefore

$$\gamma_{eff} = \gamma(1 - \eta_{eff}^{-2}) \tag{14.8}$$

It is shown that when the value of η_{eff} reaches the moderate value 1.5, the effective loss factor γ_{eff} becomes commensurate with metal loss γ.

The gain confinement factor is defined as

$$\Gamma = 2q_s \int_0^{d_a} exp(-2q_s z) dz \tag{14.9}$$

where d_a is the thickness of the active n-type material and q_s is the decay length of the field inside it; $d_a = 1/2q_s$ may be a good choice to make $\Gamma = 0.63$. The modal gain per second is expressed as

$$g(\omega) = B\sqrt{\hbar\omega - E_{gap}}[f_c(\omega) - f_v(\omega)]\Gamma \tag{14.10}$$

where B is the Einstein coefficient for stimulated emission and f's denote the Fermi-occupation probabilities. Putting $g(\omega) = \gamma_{eff}(\omega)$ for transparency condition, the transparency carrier density is 10^{18}–10^{19} cm^{-3}, which seems reasonable. The transparency current density is expressed as

$$\mathcal{J}_{trans} = \frac{4ed_a\epsilon_s^{3/2}}{c\lambda^2}BF_p \int_{E_{gap}}^{\infty} \sqrt{\hbar\omega - E_{gap}}[f_c(\omega) - f_v(\omega)]d\omega \tag{14.11}$$

The previous expression for the transparency current density involves a B coefficient, instead of an A coefficient which is inversely proportional to the spontaneous recombination lifetime (Basu et al 2015). B is however related to the A coefficient (see Chapter 4).

Equation (14.11) includes, in addition to usual factors used in laser theory, the Purcell factor given by

$$F_p = 1 + \frac{\pi \Gamma q_s k_x \omega (dk_x/d\omega)}{k_s^3} \tag{14.12}$$

Purcell factor can be large as values of q_s and k_x are large due to confinement and especially due to a reduced in-plane group velocity $(dk_x/d\omega)$. It can be shown that this factor scales roughly as (η_{eff}^5), confirmed by exact calculations.

The stimulated emission coefficient B depends on frequency, effective masses, and the oscillator strength. For InGaAs at 1500 nm a value 6.5×10^{14} $s^{-1}eV^{-1/2}$and for InGaN at 400 nm a value $13.5 \times 10^{14}s^{-1}eV^{-1/2}$ is quoted by KS. The value of γ for Au are taken as 12.3×10^{13} s^{-1} and that for Ag as 3.6×10^{13} s^{-1}.

The Purcell's factor normally enhances the fluorescence and Raman scattering. However, it plays a detrimental role here by increasing the transparency current densities. The KS's calculation shows that even in modestly confined plasmonic waveguides with $d_{eff}<0.25$ (half of the diffraction limit) the value is as high as 100 kA/cm^2 for silver waveguides and 300 kA/cm^2 for gold waveguides. These numbers are at least two orders of magnitude larger than the threshold current density in typical high-power double-heterostructure semiconductor lasers.

Example 14.4 *The transparency current density given by Eq. (14.11) may be calculated if one assumes T = 0 K, for which the Fermi functions becomes the unity, and the integral can be evaluated analytically. We consider InGaAs for which a value of B = 6.5×10^{14} eV$^{-1/2}$s^{-1} is used. The value of d = 0.2 µm, Purcell factor F = 10, λ = 1550 nm, ε$_s$ = 11 and ħω − E$_{gap}$ = 0.1 eV are taken. The calculated transparency current density is 4.26×10^4 A/cm^2. Compare this value with the plot given in Fig. 4(e) by KS (2012a).*

14.5.2   Threshold, linewidth, and coherence

Khurgin and Sun (2012b) presented a more detailed theory of SPP generators and expressions for different characteristics of interest for a laser (spaser) using semiconducting gain medium.

Structure: with the aim of reducing the electric field inside the metal, KS considered nanoparticles of spherical or prolate spheroid shape. It was found that for spherical shell, the electric field had three degenerate modes that would increase the threshold current 3-fold. On the other hand, a prolate spheroid, the structure of which is shown in Fig. 14.6, would support the lowest mode along the long axis at the chosen wavelength of 1550 nm. The gain medium was In$_{0.53}$Ga$_{0.47}$As lattice matched to InP. The authors found the ratio b/a, where b and a are respectively, the half widths along short and long axes, for which the SP resonance has its dipole moment directed along the long axis. The ratio b/a was ~0.425 in case of Au and 0.385 in case of Ag. According to KS, the SP resonance occurred at 1320 nm and the corresponding energy was about 200 meV above the 740 meV bandgap of InGaAs. This ensured that the density-of-states (DOS)

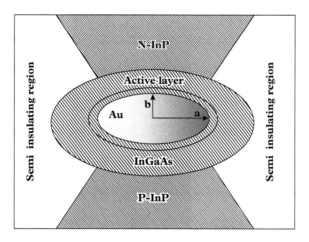

Figure 14.6 *Schematic diagram of injection pumped nanolaser studied by Khurgin and Sun (2012b). The gain medium is InGaAs (shown in figure) or GaAs. The nanoparticle is made of noble metal. Electric connection is made to the top and bottom InP layers (not shown).*

at the SP frequency in $In_{0.53}Ga_{0.47}As$ was sufficient to obtain required gain. The next two dipole modes with dipole moment normal to axis a had much shorter wavelength and thus can be excluded from consideration.

4.5.2.1   *Analysis*

The effective volume V_{eff} of a mode is defined as

$$U_t = \frac{1}{2}\epsilon_0\epsilon_s E_{max}^2 V_{eff} \tag{14.13}$$

where E_{max} is the maximum field in the mode along the long axis, $\epsilon_s = \eta^2$, and the total energy of the mode is expressed as

$$U_t = \frac{1}{2}\epsilon_0 \int_{metal} E^2(r)dV + \frac{1}{2}\epsilon_0\epsilon_s \int_{semi} E^2(r)dV \tag{14.14}$$

The two integrals are taken over the volumes of metal and semiconductor, respectively. The magnetic energy is expressed by

$$U_{mag} = \frac{1}{4}\mu_0 \int_{metal+semi} H^2(r)dV \tag{14.15}$$

The potential (electrostatic) energy in the metal is $U_{pot} = \epsilon_0 E^2/4$ and in semiconductor is $U_{pot} = \epsilon_0\epsilon_s E^2/4$. From the energy conservation argument, the potential energy should equal the sum of the magnetic energy and the kinetic energy of the electrons in a metal.

Thus a factor ½ in Eq. (14.13) instead of ¼ appears. Calculations showed that the effective volume is about 2 orders of magnitude smaller than the volume of the nanoparticle, since the field would concentrate near the poles of the spheroid.

The kinetic energy of the electron is thus given by $E_{kin} = U_t/2 - U_{mag}$. This leads to the following expression for the effective non-radiative relaxation rate in the metal as

$$\gamma_{nrad} = 2\gamma \frac{U_{kin}}{U_t} = \gamma \left(1 - 2\frac{U_{mag}}{U_t}\right) \tag{14.16}$$

where γ is the intrinsic momentum relaxation rate of the metal. As the magnetic energy decreases with the decrease in the nanoparticle volume, the kinetic energy becomes $\frac{1}{2}U_t$ and the non-radiative rate approaches γ. The rate of decrease in the energy via dipole radiation is

$$\gamma_{rad} = \frac{2\eta\omega}{9}\left(1 + \frac{|\epsilon_m|}{\epsilon_s}\right)\left(\frac{2\pi}{\lambda}\right)^3 ab^2 \tag{14.17}$$

and thus it increases with the particle size. The total mode decay rate is $\gamma_{m0} = \gamma_{rad} + \gamma_{nrad}$, and it is more or less constant for any shape of nanoparticles, for $\omega \gg \gamma$, which is satisfied over the optical and mid-IR regions.

4.5.2.2 Gain and mode confinement

The expression for the gain coefficient in a semiconductor, according to the Fermi Golden Rule, reads as

$$\frac{dn_c(r)}{dt} = \frac{2\pi}{\hbar}\frac{1}{3}\frac{e^2 P^2}{m_0^2\omega^2}\frac{1}{2\pi^2}\left(\frac{2\mu_r}{\hbar^2}\right)^{3/2}(\hbar\omega - E_g)^{1/2}(f_c - f_v)E^2(r) \tag{14.18}$$

where all the symbols have been defined earlier in Chapter 5. The electron-hole pair density n_c can be expressed by the usual expressions involving the Fermi functions f_c and f_v involving the respective quasi-Fermi levels for electrons and holes.

The gain coefficient is evaluated by first integrating Eq. (14.18) over the mode volume defined by Eq. (15.13). The total energy is then expressed as $U = \hbar\omega N_{sp}$, where N_{sp} is the total number of SPPs. The rate equation for SPPs and carriers then becomes

$$\frac{dN_{sp}}{dt} = -\frac{dN_c}{dt} = \frac{2\pi}{\hbar}\frac{1}{3}\frac{e^2 P^2}{m_0^2\omega^2}\frac{1}{2\pi^2}\left(\frac{2\mu_r}{\hbar^2}\right)^{3/2}(\hbar\omega - E_g)^{1/2}(f_c - f_v)\frac{2\hbar\omega N_{sp}}{\epsilon_0\epsilon_s}\Gamma = g(\omega)N_{sp} \tag{14.19}$$

Here, $g(\omega)$ is the gain per unit time and the gain confinement factor is defined as

$$\Gamma = \frac{\int E^2(r)dV}{E_{max}^2 V_{eff}} \tag{14.20}$$

The integration in the numerator is over the entire volume of the gain medium.

The values of confinement factor have been estimated by KS (2012b) for different values of the spacing layer thicknesses and nanoparticle size.

4.5.2.3 *Approximate expression for gain*

A simplified expression for gain from Eq. (14.19) may be obtained by introducing the approximation $m_c^{-1} = m_0^{-1} + (2P^2/m_0 E_g)$ and $E_g \sim \hbar\omega$ and introducing the fine structure constant $\alpha_0 = e^2/4\pi\epsilon_0\hbar c$. The expression is

$$g(\omega) = \frac{8\alpha_0 c}{3\epsilon_s} R_p \left[\frac{2\mu_r}{\hbar^2}(\hbar\omega - E_g)\right]^{1/2} (f_c - f_v)\Gamma = Ak\omega(f_c - f_v)\Gamma \qquad (14.21)$$

The material related coefficient $R_p = \frac{2P^2\mu_r}{m_0^2 E_g}$ is nearly unity for most of the III–V compounds. The wave vector of electron-hole pair (EHP) resonant with frequency ω is given by

$$k_\omega = \left[\frac{2\mu_r}{\hbar^2}(\hbar\omega - E_g)\right]^{1/2} \qquad (14.22)$$

The factor $A = 8\alpha_0 c R_p/3\epsilon_s$ is 5.73×10^7 cm/s for GaAs and 5.65×10^7 cm/s for InAs and is considered to be material independent.

Example 14.5 *Let us consider* T = *0 K and modal loss* $\gamma_{m0} = 10^{14} s^{-1}$. *At T = 0 K* $k_\omega^3 = 3\pi^2 n_c$ *and from Eq. (14.21)* $g(\omega) = 1.75\times10^8 n^{1/3}\Gamma$. *Assuming* $\Gamma = 0.5$, *the carrier density needed to provide gain* $g(\omega) = \gamma_{m0} = 10^{14} s^{-1}$ *is* 10^{18} cm^{-3}.

4.5.2.4 *Modal gain*

The calculation of gain needed to overcome total modal loss is however complicated at room temperature since the lineshape functions for modal gain need be included. The evaluation of modal gain should take into account the finite linewidth of the SP resonant frequency ω_0 given by

$$\gamma_m = \gamma_{mo} - g_m \qquad (14.23)$$

The modal gain is evaluated by convolving the gain $g(\omega)$ in Eq. (15.21) with the Lorentzian line shape function of width γ_m, and thus

$$g_m = \int_0^\infty g(\omega)\frac{(\gamma_{m0} - g_m)/2\pi}{(\omega - \omega_0)^2 + (\gamma_{m0} - g_m)^2/4}d\omega \qquad (14.24)$$

4.5.2.5 *Rate of spontaneous emission*

The spontaneous emission in SPASERs has two components, the first is the recombination into the resonant SP mode and second is into the free space modes. The latter is affected by the resonant mode, which is linearly polarized. The free space modes are pulled by the polarization, and considering only one component of polarization, the density of free space modes for the spheroid is given by

$$\rho_\omega = \frac{2\eta^3 \omega^2}{3\pi^2 c^3}$$

(14.25)

The effective field density of vacuum fluctuations in the energy interval $d(\hbar\omega)$ is expressed as

$$dE_p^2 = \frac{4\eta\omega^3}{3\pi\epsilon_0 c^3} d(\hbar\omega)$$

(14.26)

We substitute Eq. (14.24) in Eq. (14.18) and follow all the steps used for stimulated recombination, and obtain

$$\tau_{sp,\omega}^{-1} = \frac{2\pi}{\hbar} \frac{1}{3} \frac{e^2 P^2}{m_0^2 \omega^2} \frac{4\eta\omega^3}{3\pi^2 \epsilon_0 c^3} f_c(1 - f_v) = \frac{16}{9} \alpha_0 R_p \frac{\hbar}{\mu_r} \frac{\eta\omega^3}{c^2} f_c(1 - f_v)$$

(14.27)

where $E_g \sim \hbar\omega$ is used. The total spontaneous emission rate is obtained by integrating Eq. (14.27) over semiconductor DOS and the mode volume V_a and is given by

$$r_{sp,fs} = \frac{V_a}{2\pi^2} \left(\frac{2\mu_r}{\hbar^2} \right)^{\frac{3}{2}} \int \frac{1}{\tau_{fs,\omega}} (\hbar\omega - E_g)^{\frac{1}{2}} d(\hbar\omega) = \frac{64}{9} \alpha_0 R_p \frac{V_a \eta}{\hbar\lambda^2} \int f_c(1 - f_v) k_m d(\hbar\omega)$$

(14.28)

4.5.2.6   *Rate equations and threshold*

The rate equations for the SPP in mode, N_{sp}, and carriers in the gain region, N_c, are now expressed as

$$\frac{dN_{sp}}{dt} = (g_m - \gamma_{m0})N_{sp} + r_{sp,m}$$

(14.29)

$$\frac{dN_c}{dt} = R_{ex} - g_m N_{sp} - r_{sp,m} - r_{sp,fs} - r_A$$

(14.30)

where different rates are already defined. In Eq. (14.30), R_{ex} is the pumping or injection rate of the carriers. The dominant non-radiative recombination rate r_A is expressed as $r_A = CV_a N_c^2$. When injection carrier density is high, this rate is due to the Auger recombination. Eqs. (14.29) and (14.30) are identical to standard Statz-de Mars equations (1960) for a laser.

In addition to two intrinsic parameters N_{sp} and N_c, KS introduced an intrinsic output parameter, R_{sp}, the SP generation rate, as

$$R_{sp} = \gamma_{m0} N_{sp} \tag{14.31}$$

which corresponds to the steady state situation, signifying that the SP generation rate equals total SP decay rate, both radiative and non-radiative.

The extrinsic output parameter, the photon generation rate, depends on the product of SP generation rate and radiative decay rate and may be expressed as

$$R_{out} = \gamma_{rad} N_{sp} = [\gamma_{rad}/(\gamma_{rad} + \gamma_{nrad})]R_{sp} \tag{14.32}$$

For small mode volumes, the last two terms in Eq. (14.30) may be neglected, yielding the steady state solution as

$$N_{sp} = \frac{R_{ex}}{\gamma_{m0}} \tag{14.33}$$

Thus, using Eq. (14.29), the SP generation rate equals the excitation rate and

$$R_{sp} = R_{ex} \tag{14.34}$$

In other words, the spaser is thresholdless.

In the opinion of KS, the increase in coherence leading to narrowing of linewidth signifies more effectively the onset of lasing. Following KS, at a critical condition, the effective linewidth is reduced to half the value of the linewidth of the cavity without gain and therefore

$$g_m = \frac{\gamma_{m0}}{2} \tag{14.35}$$

Eqs. (14.29, 30), and (14.35) yield the expression for approximate number of SPs at threshold as

$$N_{sp,th} = \frac{r_{sp,m}}{\gamma_{m0} - g_m} \approx \bar{n}_{sp,m} \approx 1 \tag{14.36}$$

This indicates that at threshold there is only one SP mode, the usual definition for laser threshold introduced by Bjork et al (1994). There are, however, other definitions of threshold, but Eq. (14.35) seems to be the most appropriate.

Eq. (14.36) is now substituted in Eq. (14.30) in which the last two terms are neglected. This gives the threshold excitation rate $R_{ex,th} \approx \gamma_{m0}$ and the threshold current is given by $I_{ex,th} \approx e\gamma_{m0}$.

Example 14.6 *The values of γ_{m0} at 1320 nm are 1.3×10^{14} s^{-1} and 3.4×10^{13} s^{-1} for Au and Ag, respectively. Using the approximate expression $I_{ex,th} \approx e\gamma_{m0}$, the threshold current values are 16 μA for Au and 6 μA for Ag-based spasers.*

4.5.2.7 *Results*

KS made detailed calculations of different parameters and predicted the characteristics of single mode spasers. A summary of their findings is given in the following points.

(1) A gold elliptical nanoparticle with half-axis $a = 20$ nm, with no spacing layer was considered. Assuming confinement factor of 0.8 they calculated the SP generation rates, and linewidth on input current. The calculated transparency current was 7.56 μA and threshold current was 22 μA, not too far from the values quoted in Example 14.6.

(2) The carrier density was found to increase linearly with current and then rather slowly with current near threshold and got clamped to a value near 1×10^{19} cm^{-3}. The behaviour of the single mode spaser was found to follow the same pattern of lasers, but the knee for the spaser is not as sharply defined as in the laser.

(3) The plots of material gain profile and modal gain spectral density were obtained for different pumping conditions. Below transparency, material gain profile showed the inverted U shape, the profiles being wider and with higher peak as the pump was at threshold, above and well above the threshold. The modal gain spectra progressively became more peaked and narrowest, almost Lorentzian-shaped, well above the threshold. All these were the usual characteristics of a single mode operation.

(4) While the threshold currents calculated for Au and Ag nanoparticles were not very high, the threshold current density for an Au device of 120 nm length, $\lambda/3$ for InGaAs, was about 220 kAcm^{-2} and for Ag device it was 120 kA.cm^{-2}. A size of 120 nm may hardly be called truly sub-wavelength. Calculated values for lower sizes exceeded MA.cm^{-2}, which no semiconductor-based device can sustain.

(5) The authors considered injection-pumped generation of SPs through spontaneous emission of SP modes, rather than the stimulated emission as in spasers, to realize a powerful source in a small volume. The large enhancement of the Purcell factor can be of advantage for the device. They called the device as Surface Plasmon Emitting Diodes (SPEDs) and predicted nearly 90% intrinsic efficiency and also that at about 100 A.cm^{-2} current density level, high power and efficiency would be achievable.

14.6 Current theoretical models and experiments

The limitation of spasers is that their enormously large threshold current density prevents them superseding nanoscale semiconductor lasers, even though their advantage of having subwavelength dimensions has been the main outcome of the series of papers by KS. However, their claim has been reexamined in some later investigations, both theoretical and experimental. The inherent weakness of the models by KS has been discussed by Wang (2017) and in depth by Azzam et al (2020). An important attribute of

M-S-M plasmonic waveguides pointed out by Li and Ning (2009, 2010), not considered by KS, led to a detailed experimental study of spasers and conventional semiconductor nanolasers by Wang et al (2017). This section constitutes description of these works and conclusions drawn from the studies.

14.6.1   Modal gain and loss in metal-semi-conductor-metal plasmonic waveguides

The typical value of semiconductor gain is in the range 10^3–10^4 cm^{-1}, far below the absorption loss $\sim 10^6$cm^{-1} of noble metals. It seems therefore impossible to overcome the loss incurred by SPP propagating mode in typical semiconductor-metal interface, since such a mode exists in both the layers. Li and Ning's work (2009, 2010) has established for the first time that a dramatic slowing down of the energy flux of the SPP mode in an M-S-M structure should lead to a large optical gain, the modal gain. The modal confinement factor in such waveguides becomes several orders of magnitude larger than unity when metals are used, whereas the ideal value of confinement factor in dielectric waveguides is unity. Since any mode penetrates much less into the metal, the effective modal loss is a few orders of magnitude lower than the effective modal gain in the semiconductor, resulting in a giant optical gain for SP propagating modes. In the following, the outline of the theory developed by Li and Ning (LN) is presented.

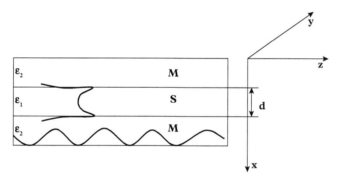

Figure 14.7 *MSM structure considered by Li and Ning (2010). The SP wave propagates along the z-direction. The profile of power distribution of an SP mode along the depth (x-direction) is shown.*

Example 14.7 *The values of dielectric constant at 3.99 eV for copper (Cu), given in the paper by Johnson and Christy (1972), yield for the real part $\eta_r = 1.38$ and the imaginary part $\eta_i = 1.729$ of the refractive index η. Using the relationship between η_i and the loss $\alpha = 2\omega\eta_i/c$, the value of the metallic loss comes as 7×10^5 cm^{-1}.*

The MSM structure considered by LN is shown in Fig. 14.7. The semiconductor layer in between the two metal layers has a thickness d. The modes are confined along the x-direction, are uniform along the y-direction, and the propagation direction is in the z-direction.

The dispersion relation for the symmetric transverse magnetic™ mode has been expressed by Eq. (13.45) and is reproduced here as

$$\epsilon_2 k_1 \tanh(k_1 d/2) = -\epsilon_1 k_2 \qquad (14.37)$$

The propagation wave vector is defined as usual by

$$k_z^2 = \epsilon_{1,2} k_0^2 + k_{1,2}^2 \qquad (14.38)$$

The subscripts 1 and 2 refer, respectively, to the semiconductor and the metal, as indicated in Fig. 14.7.

Putting Eq. (14.38) in Eq. (14.37) gives the eigenvalue equation in terms of real and imaginary parts of the propagation wave vector k_z. Separating real and imaginary parts, two transcendental equations are obtained. Solving these two and inserting the values of parameters for metal and semiconductor, one may obtain the values of k_z. We quote the results for a particular frequency in the following examples.

Example 14.8 *Li and Ning (2010) first considered a semiconductor with real part of permittivity $\epsilon_{1r} = 12$ and different values of its imaginary part ϵ_{1i}. We shall use one such value $\epsilon_{1i} = -0.4$ in this example. The permittivity values of different noble metals are given by Johnson and Christy. We choose following LN, Ag as the metal and photon energy $= 2.25$ eV. The corresponding values for permittivity are $\epsilon_{2r} = -12.6506$ and $\epsilon_{2i} = 0.3866$. The calculated values of real and imaginary parts as found are $k_{zr} = 1.8 \times 10^7/m$ and $k_{zi} = -8.79 \times 10^8/m$. Since the modal gain is defined by $G_m = -2k_{zi}$, the modal gain is substantially large.*

Example 14.9 *It is of interest to obtain the value of material gain corresponding to $\epsilon_{1i} = -0.4$, as used in previous example. Using the relation $G_0 = -\epsilon_{1i}\omega/(\eta_r c)$, where η_r is the real part of the semiconductor RI, and with its value 3.464, the material gain is 1.36×10^4 cm^{-1}. The modal gain calculated by LN is 2×10^7 cm^{-1} at plasma frequency corresponding to 2.32 eV of energy.*

This example shows that the modal gain may be almost three orders of magnitude larger than the material gain.

LN presented plots of real and imaginary parts of k_z against photon energy for four values of ϵ_{1i}, 0, −0.3, −0.4, and −0.5 using $d = 100$ and 200 nms and for a bilayer. The metal was Ag and the semiconductor was assumed to have $\epsilon_{1r} = 12$. In all these plots, both the real and imaginary parts, show a monotonic rise of values with photon energy, a broad peak for $\epsilon_{1i} = 0$, and sharper peaks near plasma frequency for higher values of ϵ_{1i}, a sharp decrease at plasma frequency, monotonous decrease for k_{zr}, but rise from a minima for k_{zi}. The plots for different values of d are barely distinguishable below plasma frequency but deviate more at higher energies above the plasma edge. The plots for k_{zi} show, to some extent, the same nature of the atomic susceptibility versus frequency plot. The reader is urged to examine further details from the original plots.

The origin for giant modal gain as compared to material gain, as illustrated by Examples 14.8 and 14.9, has been investigated by the authors. They defined an energy velocity for the propagating waves and calculated its value considering that energy flows in opposite directions in semiconductors and metals. The large modal gain, according to their calculation, was due to too low a value of energy velocity, as low as 200 m/s. A low energy velocity enhances more exchange of energy between the waves and the media with gain or loss. The result is that more absorption and emission occur, respectively for lossy and gain medium in the case of slow wave propagation. This was believed to be the reason for giant modal gain in MSM structure.

LN (2009) considered a more realistic semiconductor $Zn_{0.8}Cd_{0.2}Se$ and calculated the real and imaginary parts of k_z at 77 K. They found a pronounced negative value for k_{zi} near the plasma resonance ensuring a huge modal gain.

It is pointed out in a later work (Azzam et al 2020), another advantage of metals in small lasers, in spite of huge metal loss. The overall cavity Q is defined as $1/Q = 1/Q_a + 1/Q_r$, where Q_a and Q_r, are, respectively, determined by the internal absorption and far-field radiation. In a nanolaser, far-field radiation is reduced so that the overall cavity Q is increased leading to a smaller threshold.

14.6.2 Unusual scaling laws for sub-wavelength spasers

The crux of KS's theories is that the Purcel effect has a detrimental effect on the threshold current density of an SP-based nanolaser and in that respect the Spasers or SP nanolasers are no better than conventional semiconductors with reduced dimensions. In their work, however, the dimensions hardly reached subwavelength scale in all three dimensions. Further, the operating wavelength is far away from the SP resonance to limit the loss.

In this respect, the work by Li and Ning (2009, 2010) presents an important observation leading to giant optical gain near the SPP resonance.

To answer the questions related to intrinsically high threshold and to explore the possibility of plasmonic nanolasers surpassing the performance of photonic nanolasers, Wang et al (2017) performed detailed experimental investigations using 170 plasmonic and photonic nanolasers using cadmium selenide (CdSe) films of different thicknesses. Both of these two types have the same gain material and cavity feedback mechanisms. They could find an extremely low threshold for plasmonic nanolasers, 10 kW/cm^{-2} or so, having pump densities corresponding to current photonic lasers. Further comments about the findings by Wang et al (2017) may be found in the review by Azzam et al (2020).

In brief, Wang et al. systematically studied several key parameters like physical size, threshold, power consumption and lifetime. They presented measured data for many samples as a function of volume of the samples for two different ranges of thickness d of the nonanolasers, (1) 100 nm $< d < 150$ nm and (2) 250 nm $< d < 350$ nm. They found wide variations in the measured values and obtained an exponential fit to the variation.

We discuss the salient features of the nature of variation of two parameters, threshold pump power and power consumption for the lower range of the thickness.

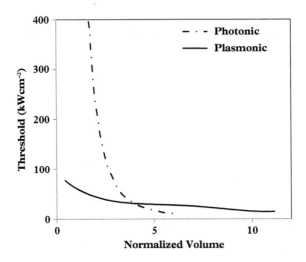

Figure 14.8 *Qualitative variation and comparison of threshold current as a function of volume of spasers and photonic nanolasers.*

Source: redrawn from Fig. 4(a) of Wang et al (1889) *Nature Communication* 8.

The qualitative variations of threshold current density with volume normalized by λ^3 for plasmonic and photonic nanolasers are shown in Fig. 14.8.

It appears that for both the types of nanolaser the threshold is more or less constant for larger volumes, but in both cases, the values rise rapidly for smaller volumes. The rise is quite rapid for photonic nanolasers reaching the volume $\sim 5\lambda^3$ and decreasing further. This means that photonic nanolasers having dimensions near to the diffraction limit will have enormous threshold unreachable in practical systems. The threshold for plasmonic nanolasers also increases with decreasing volume, but the rise is at a smaller rate. The threshold for plasmonic nanolasers with thickness below the diffraction limit $d < 100$ nm is less than 100 kWcm^{-2}. At this range no photonic lasing was observed for about 30 devices studied.

Qualitative variation of the power consumption with normalized volume of plasmonic and photonic nanolasers is depicted in the plots of Fig. 14.9. The plots are obtained by best fit to the data for $150 < d < 250$ nm. It is clear from the plots that while the power consumption decreases monotonically with decreasing volume for plasmonic nanolasers, the trend is different for photonic nanolasers. Here, the power decreases first with decreasing volume, reaches a minimum around cavity size $\sim 5\lambda^3$ and then rises rapidly with further decrease of volume, giving rise to an almost V-shaped curve. This trend implies once more that photonic nanolasers are not suitable for sizes below the diffraction limit.

Similar results were obtained for nanolaser thickness t in the range 250 nm $< t < 350$ nm. However, very little differences between the values of threshold and power consumption were found between the two types.

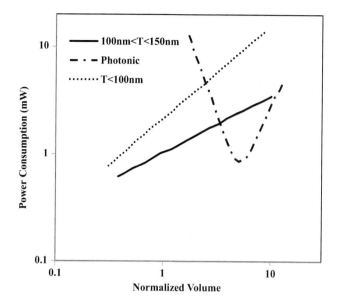

Figure 14.9 *Qualitative variation and comparison of power dissipation as a function of volume of spasers and photonic nanolasers. Redrawn from Fig. 4(c) of Wang et al (1889) Nature Communication 8.*

    Wang et al presented a set of scaling laws, which advocate of compactness, higher speed, lower power consumption of plasmonic nanolasers when the cavity size approaches or goes below the diffraction limit.

14.6.2.1   Rate equation analysis

The rate equations for population and photon number are expressed as

$$\frac{dN_2}{dt} = \eta P_{pump} - R_{nr}N_2 - RN_2 - \Gamma R\beta(N_2 - N_0)N_{ph} \tag{14.39}$$

$$\frac{dN_{ph}}{dt} = -\gamma N_{ph} + \beta RN_2 - \Gamma R\beta(N_2 - N_0)N_{ph} \tag{14.40}$$

In the previous equations, N_2 is the excited carrier density, N_{ph} is the photon number in a single mode laser, P_{pump} is the pump rate, and η is the conversion efficiency of pump photons into EHPs, N_0 is the excited state population at transparency, R_{nr} is the non-radiative recombination rate, R is the rate of spontaneous emission ($\sim 1/\tau$), Γ is the mode confinement factor, β is the spontaneous emission coupling factor, and γ is the rate for total cavity loss.

Putting $\frac{d}{dt} = 0$ under steady state and solving the previous two equations, one obtains the following quadratic equation for N_{ph}:

$$\gamma N_{ph}^2 - \left[\eta P_{pump} - \frac{R_{nr} + R}{\Gamma R \beta} - (R_{nr} + R)N_0 + \beta R N_0 \right] N_{ph} - \frac{\eta P_{pump}}{\Gamma} = 0 \qquad (14.41)$$

The two solutions to the quadratic equation show linear variation at low and high values of the pump power and the usual kink in the light output vs. pump power curve (see Fig. 11.16). The threshold is defined as the point of intersection of the two linear portions. The pump power density is related to pump power by $P = P_{pump}\hbar\omega/A$, where $\hbar\omega$ is the photon energy and A is the area of the square-shaped sample. The expression for threshold pump power becomes

$$P_{th} = \frac{\hbar\omega}{\eta} \left[\frac{1 - \eta_i\beta}{\eta_i\Gamma\beta A}\gamma + R_{nr}n_0 d + (1 - \beta)n_0 d\frac{1}{\tau} \right] \qquad (14.42)$$

The internal quantum efficiency of the gain medium, $\eta_i = R/(R_{nr} + R)$ is approximately 1, so that $R_{nr} \ll R$, and $1/\tau \approx R$, $n_0 = N_0/V$ is the excited state density at the transparency, and $V = Ad$ is the volume of the sample.

The relative contributions to the threshold power by different parameters may now be assessed. The first term within square bracket of Eq. (14.42), proportional to cavity loss γ, is inversely related to Q of the cavity. The second term takes care of contribution from carrier loss due to non-radiative processes, and the third term proportional to n_0/τ represents power density needed to achieve population inversion of carriers.

Eq. (14.42) also elucidates the debated role of Purcell effect, which increases β, reducing the first and last terms. However, as the spontaneous emission rate increases, τ decreases and makes the last term larger. Assuming that the second term, the non-radiative contribution is negligible, the threshold power is determined by the first and the third terms.

Wang et al. made an empirical fit for the measured lifetime by $\tau \propto P_{th}^{\alpha_1}$ and obtained the values of the exponent α_1 for both the plasmonic and photonic nanolasers for different ranges of thickness of the nanofilm. Their conclusion was that, for photonic lasers of smaller thickness approaching diffraction limit, the first term becomes large making the threshold values quite large. On the other hand, for plasmonic nanolasers, the smaller cavity size leads to higher β and thus both shorter lifetime and lower thresholds are achieved simultaneously. A room temperature, plasmonic nanolaser was found with threshold as low as 10 kW cm^{-2}, the pumping density of which was comparable to the same for modern nanolaser diodes. They demonstrated that the threshold power increased with reduced dimension for both types, but the increase was more gradual for plasmonic type. This was attributed to better confinement and Purcell effect.

Their results demonstrated unambiguously that plasmonic lasers would show superiority over photonic nanolasers below the diffraction limit. In summary, their work clarified the long-standing debate over the viability of plasmonic nanolasers having metallic layers in laser technology.

In the opinion of Wang et al., both sides of the long-standing debate are valid and there is a need to make detailed studies on actual physics.

Example 14.10 *An estimate of the threshold power of plasmonic nanolaser is made by using Eq. (14.42). The parameter values are taken from the work of Wang et al: wavelength $\lambda = 700$ nm, area $A = 2$ $\mu m \times 1.6$ μm, $d = 140$ nm, $\beta = 0.09$, $\tau = 1.3$ ns. We assume $\Gamma = 0.3$, $\gamma = 10^{14}s^{-1}$ and $n_0 = 10^{18}$ cm^{-3} and $\eta = \eta_i = 1$. The second term, the non-radiative contribution within square bracket in Eq. (14.42) is neglected. The first and the third terms within bracket become 10.53×10^{26} and $0.98 \times 10^{26} m^{-2}s^{-1}$, respectively. The threshold power is now $32.8 \times 10^7 Wm^{-2} = 32.8$ kW.cm^{-2}. This value comes closer to the value reported by Wang et al.*

14.7 Further developments

The first plasmonic nanolaser or spaser was announced only in 2009. Since then, many workers contributed to the development of spasers by theoretical predictions, modeling, critically examining the practicability of spasers in terms of threshold and power consumption, relative merits and demerits of plasmonic and photonic nanolasers, different material systems, novel physical properties, and future predictions (Gwo and Shih 2016; Wang et al 2018; Xu et al 2019; Ma and Oulton 2019; Ning 2019; Li and Gu 2019). In this section, we shall present very briefly some developments, relying mostly on the historical review *Ten Years of Spasers and Photonic Nanolasers* by Azzam et al (2020). Mention is also made of some suggestions about improved metal deposition techniques.

14.7.1 Review of work done in the last 10 years

Azzam et al (2020) discussed many aspects of spasers and plasmonic nanolasers. In this subsection we shall briefly outline their reports under a few heads only.

14.7.1.1 Different types of nanolasers

Different types of nanolasers studied since 2009 and the corresponding structural diagrams are given in Fig. 2 of Azzam et al. Here is the list: (1) a cavity combining DBRs with metal; (2) a double patch metal cavity; (3) a core-shell NWR with InGaN-GaN on a single crystal Ag film; (4) a CdSe NWR coupled to a Ag wire; (5) a semiconductor-metal coaxial cavity; (6) QWs in a metal pan; (7) a microdisk laser with a plasmonic disk; (8) a metamaterial laser; (9) a perovskite based plasmonic nanolaser; (10) a metallic trench FPO resonator plasmonic nanolaser; (11) 1D plasmonic crystal laser; (12) 2D plasmonic crystal nanolaser.

14.7.1.2 Electrically injected nanolasers

The spasers based on localized surface Plasmon modes and plasmonic nanolasers relying on propagating SPP waves reported so far are mostly pumped by optical sources.

However, electrically injected plasmonic nanolasers certainly are of need for the development of integrated photonic circuits, as mentioned in Chapter 13. In this respect, the work reported by Hill et al (2007, 2009) makes use of electrical pumping. The same metal has been used both for plasmonic confinement and providing contacts for electrical injection. The studies by Hill et al were conducted at low temperature. There are however several challenges to overcome. It is highly difficult to fabricate high-quality metallic structures with small dimensions. At the same time, the contact resistance of the metal increases substantially with smaller size. It is also challenging to reduce the RC delay for high-speed applications and careful design is needed to reduce capacitance due to the plasmonic or contact metals.

14.7.1.3   *Room-temperature operation of electrically injected devices*

Ma et al (2011) reported first room temperature operation of plasmonic nanolasers by using optical pumping. Achieving room-temperature operation of any device by electrical injection is the ultimate goal in device research. Room temperature operation is definitely needed for achieving photonic integrated circuits leading to electronics-photonics integration, for interconnects in data centres and high-performance computers.

The progress in the development of electrically injected plasmonic and metallic cavity nanolasers is depicted in Fig. 5 in the paper by Azzam et al (2020). As already stated, the work by Hill et al (2007, 2009) pertains to low temperatures. Reports for higher operating temperatures came subsequently. Lee et al (2011) obtained operation at 140 K using InGaAs. Ding et al (2011) reported 260 K operation with InGaAs and then obtained room temperature operations (Ding et al 2012, 2013). Flynn et al (2011) reported a room-temperature semiconductor spaser operating near 1.5 μm. Room-temperature plasmonic lasing in a continuous wave operation mode from an InGaN/GaN single nanorod with a low threshold was found by Hou et al (2014).

It is to be noted that in all the works reported previously, only the work Hill et al (2009) involved plasmonic gap mode. In all other studies operation has been demonstrated for dielectric mode. In that sense, plasmonic nanolasers at room temperature under electrical injection have not yet been realized.

14.7.1.4   *Interconnect*

The interconnect bottleneck in very large-scale integrated (VLSI) circuits has been introduced in Chapter 1 and further discussed in Chapter 13. The current solution is to resort to optical interconnects. The next step would be to realize spaser based interconnects that will further increase the speed of operation and at the same time introduce plasmonic systems having size compatibility with current electronic systems. It is known that the coupling between transistors in a processing chip is electrostatic. During switching, a transistor charges or discharges the capacitor related to the metallic interconnect and the process needs long time of operation and dissipates heat.

An interesting scheme using spasers has been proposed by Stockman (2018) in a patent application, which has not yet been translated into practice. In the proposal, a transmitting transistor electrically pumps a spaser of about 10 nm size, the same size of the transistor. The signal from the spaser is propagated through a copper waveguide, and in the receiver side; the SPP pulse is converted into charge by a Ge nanocrystal, which is fed into the receiving transistor. It is shown that a single transistor can produce sufficient drive current to provide electrical pumping for the spaser.

14.7.2 Better plasmonic film

In addition to using noble metals for plasmonics, UV or complementary metal-oxide semiconductor (CMOS)-compatible metals like aluminum (Al) or Cu are now being investigated. The structures are also used to investigate quantum effects at the limit of single surface plasmons. McPeak et al (2015) noted that in most cases metals are deposited under wrong conditions, thereby compromising with the performance. They present simple prescriptions for deposition of high-quality plasmonic films of Al, Cu, Ag, and Au by using commonly available equipment like thermal evaporator. They also provide recipes to obtain films with optimum optical properties.

Problems

14.1 *Calculate the value of dipole moment d_{10}, expressed as $d_{10} = e\sqrt{fK/(2m_0\omega_n^2)}$ by Bergman and Stockman (2003). Use $f = 1$, $K = 3\ eV$, and $\hbar\omega_n = 1.6\ eV$.*

14.2 *Using the values of permittivity of Ag at 551 nm wavelength given in Example 14.1, determine the minimum value of gain coefficient needed for the spaser.*

14.3 *Consider the expression for the Einstein's coefficient for stimulated emission, denoted by A_n and given by Eq. (5) in the paper by Bergman and Stockman (2003). Compare this expression with the similar expression for B_n for photonic lasers given in Chapter 5 and make a list of differences in the two equations.*

14.4 *Show that the transparency current density for an MSM structure as given by Eq. (14.11) increases with injected carrier density at 0 K.*

14.5 *Using the approximation given by Eq. (5.150) in Eq. (14.21), obtain a plot of gain versus frequency for GaAs spaser at 300 K. Assume that $A = 5.65 \times 10^7$ cm/s and confinement factor $\Gamma = 0.8$.*

14.6 *Using Maxwell equations and proper boundary conditions, derive the eigenvalue equation (14.37) for an MSM structure.*

14.7 *Show that when the thickness d of the semiconductor in Fig. 14.7 tends to infinity so that the MSM structure reduces to a single interface M-S structure, the dispersion relation $\epsilon_2 k_1 = -\epsilon_1 k_2$, becomes valid.*

14.8 *Calculate and plot the real and imaginary parts of k_z in the MSM waveguide, as in Example 14.6 and 14.7 in semiconductor and metals for d = 100 nm, 200 nm, and single interface, as a function of SP energy. Use $\epsilon_1" = 0, -0.2, -0.4$ for semiconductor and $\epsilon_1" = 12$. Optical constants for Ag are to be taken from Johnson and Christy.*

14.9 *Repeat the problem stated previously by using Au as the metal.*

14.10 *Solve Eq. (14.41) and then derive the expression Eq. (14.42) for threshold pump power.*

14.11 *Consider the curves of output power vs. input power given by Fig. 2c in the paper by Wang et al (2017). Estimate the values β for each curve.*

References

Azzam S I, Kildishe A V, Ma R M, Ning C-Z, Oulton R, Shalaev V M, Stockman M I, Xu J-Land Zhang X (2020) Ten years of spasers and plasmonic nanolasers. *Light: Science & Applications* 9(90): 1–21.

Basu P K, Mukhopadhyay Bratati, and Basu Rikmantra (2015) *Semiconductor Laser Theory.* Boca Raton, FL, USA: CRC Press.

Bergman D J and Stockman M I (2003) Surface plasmon amplification by stimulated emission of radiation: Quantum generation of coherent surface plasmons in nanosystems. *Physical Review Letters* 90(2): 027402.

Berini P and De Leon I (2011) Surface plasmon–polariton amplifiers and lasers. *Nature Photonics* 6(January): 16–24.

Bjork G, Karlsson A, and Yamamoto Y (1994) Definition of a laser threshold. *Physical Review A* 50(2): 1675–1680.

Deeb C and Pelouard J-L (2017) Plasmon lasers: Coherent nanoscopic light sources. *Physical Chemistry Chemical Physics* 19: 29731–29741. DOI: 10.1039/C7CP06780A.

Deen M J and Basu P K (2012) *Silicon Photonics: Fundamentals and Devices.* Chichester, UK: Wiley.

Ding K et al (2011) Electrical injection, continuous wave operation of subwavelength metallic-cavity lasers at 260 K. *Applied Physics Letters* 98(231108): 4.

Ding K et al (2012) Room-temperature continuous wave lasing in deep subwavelength metallic cavities under electrical injection. *Physical Review B* 85: 041301.

Ding K et al (2013) Record performance of electrical injection sub-wavelength metallic-cavity semiconductor lasers at room temperature. *Optics Express* 21: 4728–4733.

Flynn R A, Kim C S, Vurgaftman I, Kim M, Meyer J R, Mäkinen A J, Bussmann K, Cheng L, Choa F-S, and Long J P (2011) A room-temperature semiconductor spaser operating near 1.5 μm. *Optics Express* 19(9): 8954–8961.

Gwo S and Shih C-K (2016) Semiconductor plasmonic nanolasers: Current status and perspectives. *Reports on Progress in Physics* 79(086501): 34.

Hill Martin, Marell Milan, Leong Eunice, Smalbrugge Barry, Zhu Youcai, Sun Minghua, Van Veldhoven Peter J, Geluk Erik J, Karouta Fouad, OeiYok, Noetzel Richard, Ning Cun Zheng, and Smi Meint K (2009) Lasing in metal-insulator-metal sub-wavelength plasmonic waveguides. *Optics Express* 17: 11107–11112.

Hill M T, Oei Y-S, Smalbrugge B, ZhuY, de Vries T, Van Veldhoven P J, Van Otten F W M, Eijkemans T J, Turkiewicz J P, de Waardt H, Geluk E J, Kwon S-H, Lee Y-H, Nötzel R, and Smit M K (2007) Lasing in metallic-coated nanocavities. *Nature Photonics* 1(10): 589–594.

Hou Y, Renwick P, Liu B, Bai J, and Wang T (2014) Room temperature plasmonic lasing in a continuous wave operation mode from an InGaN/GaN single nanorod with a low threshold. *Scientific Reports* 4(5014): 1–6.

Johnson P B and Christy R W (1972) Optical constants of the noble metals. *Physical Review B* 6(12): 4370–4379.

Kewes G, Herrmann K, Rodríguez-Oliveros R, Kuhlicke A Benson, O, and Busch K (2017) Limitations of particle-based spasers. *Physical Review Letters* 118: 237402.

Khurgin J B (2015) How to deal with the loss in plasmonics and metamaterials. *Nature Nanotechnology* 10(January): 2–6.

Khurgin J B and Sun G (2014) Comparative analysis of spasers, vertical-cavity surface-emitting lasers and surface-plasmon-emitting diodes. *Nature Photonics* 8: 468–473.

Khurgin J B and Sun G (2012a) Practicality of compensating the loss in the plasmonic waveguides using semiconductor gain medium. *Applied Physics Letters* 100: 011105.

Khurgin J B and Sun G (2012b) Injection pumped single mode surface plasmon generators: threshold, linewidth, and coherence. *Optics Express* 20: 15309–15325.

Khurgin J B and Sun G (2012c) How small can 'Nano' be in a 'Nanolaser'? *Nanophotonics* 1(2012): 3–8.

Lee J H et al (2011) Electrically pumped sub-wavelength metallo-dielectric pedestal pillar lasers. *Optics Express* 19: 21524–21531.

Leosson K (2012) Optical amplification of surface plasmon polaritons: Review. *Journal of Nanophotonics* 6(061801): 9.

Li D B and Ning C Z (2010) Peculiar features of confinement factors in a metal-semiconductor waveguide. *Applied Physics Letters* 96(181109): 4.

Li D Band Ning C Z (2009) Giant modal gain, amplified surface plasmon-polariton propagation, and slowing down of energy velocity in a metal-semiconductor-metal structure. *Physical Review B* 80(153304): 1–4.

Li X and Gu Q (2019) High-speed on-chip light sources at the nanoscale. *Advances in Physics* 4(1):1658541.

Lu Y J et al (2012) Plasmonic nanolaser using epitaxially grown silver film. *Science* 337: 450–453.

Lu Y J et al (2014) All-color plasmonic nanolasers with ultralow thresholds: Autotuning mechanism for single-mode lasing. *Nano Letters* 14: 4381–4388.

Ma. Ren-Min, Oulton R F, Sorger V J, Bartal G, and Zhang X (2011) Room temperature sub-diffraction-limited plasmon laser by total internal reflection. *Nature Materials* 10(2): 110–113.

Ma R M and Oulton R F (2019) Applications of nanolasers. *National Nanotechnology* 14: 12–22.

McPeak K M, Jayanti S V, Kress S J P, Meyer S, Iotti S, Rossinelli A, and Norris D J (2015) Plasmonic films can easily be better: Rules and recipes. *ACS Photonics* 2(2): 326–333.

Ning C Z (2010) Semiconductor nanolasers. *Physics Status Solidi B* 247:774–788.

Ning C Z (2019) Semiconductor nanolasers and the size-energy-efficiency challenge: A review. *Advanced Photonics* 1: 014002.

Noginov M A, Zhu G, Belgrave A M, Bakker R, Shalaev V M, Narimanov E E, Stout S, Herz E, Suteewong T, and Wiesner U (2009) Demonstration of a spaser-based nanolaser. *Nature* 460(7259): 1110–1112.

Oulton R F (2012) Surface plasmon lasers: Sources of nanoscopic light. *Materials Today* 15(1–2): 26–34.

Oulton R F, Sorger V J, Zentgraf T, Ma R-M, Gladden C, Dai L, Bartal G, and Zhang X (2009) Plasmon lasers at deep subwavelength scale. *Nature* 461: 629–632.

Premaratne M and Stockman M I (2017) Theory and technology of SPASERs. *Advances in Optics and Photonics* 9(1): 79–128.

Statz H and de Mars G (1960) In: *Quantum Electronics*, C H Townes (ed.) p. 530. Columbia, USA: Columbia University Press.

Stockman M I (2018) Spasers to speed up CMOS processors. US patent: 10,096,675.

Stockman M I (2008) Spasers explained. *Nature Photonics* 327–329:

Stockman M I (2010) Spaser as nanoscale quantum generator and ultrafast amplifier. *Journal of Optics* 12: 024004.

Stockman M I (2013) Spaser, plasmonic amplification, and loss compensation. In: *Active Plasmonics and Tuneable Plasmonic Metamaterials*, 1st edn. Anatoly V Zayats and Stefan A Maier (eds.) Chapter 1, pp. 1–45. New York: John Wiley & Sons, Inc.

Wang D, Wang W, Knudson M P, Schatz G C, and Odom T W (2018)Structural engineering in plasmon nanolasers. *Chemical Reviews* (118): 2865–2881.

Wang S, Wang X-Y, Li B, Chen Hua-Zhou, Wang Yi-Lun, Dai Lun, Oulton Rupert F, and Ma R-M (2017) Unusual scaling laws for plasmonic nanolasers beyond the diffraction limit. *Nature Communications* 8(1889): 1–8.

Xu L, Li F, Liu Y, Yao F, and Liu S (2019) Surface plasmon nanolaser: Principle, structure, characteristics and applications. *Applied Science* 9(861): 15.

Zhong X-L and Li Z-Y (2013) All-analytical semiclassical theory of spaser performance in a plasmonic nanocavity. *Physical Review B* 88: 085101-1-10.

15

Optical metamaterials

15.1 Introduction

Metamaterials are engineered materials and have properties that are not exhibited by naturally occurring materials. The name derives from the Greek work *meta*, meaning *beyond*. These artificially structured materials have assumed scientific and technological importance over the last two decades.

Though the name *metamaterial* is a recently coined one, the idea of creating artificial materials to perform in the desired fashion is quite old. In 1894, Lippmann (1894) was the first to produce such material. He created a refractive index variation in a photographic film and was able to produce good colour photographs. Four years later, J C Bose (1898) proposed twisted jute as an artificial material and demonstrated that it could rotate the polarization of electromagnetic (EM) waves. Such materials are now-a-days called artificial chiral materials. Almost a half century later, an artificial dielectric having higher refractive index (RI) than in the then available natural dielectrics, was made by inserting metallic pieces into a dielectric of lower permittivity. These were used as radomes of radar. Works on artificial dielectrics were continued by Engineer et al (1967). Walsh (1973) proposed an artificial semiconductor structure made with alternate layers of metals and insulators.

An important milestone in the path of metamaterials is laid down by the Russian physicist Victor Veselago in 1967 (see Veselago 1968: English translation). He considered the propagation of electromagnetic (EM) waves in a medium having negative permittivity and permeability. He showed that in such materials phase velocity may be anti-parallel to the direction of the Poynting vector. Since the usual right-handed rule for propagation of EM waves is not followed, the materials were later called left-handed metamaterial. A practical way to realize such metamaterials was suggested by John Pendry (1999, 2000). He proposed that an array of metallic wires aligned in the direction of the wave propagation could provide negative permittivity. He also proved that a slip ring in the shape of letter C could provide negative permeability. An array of slip rings and metallic wires could give rise to negative RI, according to Pendry and his coworkers (Pendry et al 1999; Pendry 2000; Pendry 2004; Pendry and Smith 2004; Pendry 2005).

Semiconductor Nanophotonics. Prasanta Kumar Basu, Bratati Mukhopadhyay, and Rikmantra Basu, Oxford University Press.
© P.K. Basu, B. Mukhopadhyay, R. Basu (2022). DOI: 10.1093/oso/9780198784692.003.0015

The experimental demonstration of a practical EM metamaterial was first given by Smith et al (2000a,b, and c). They used a periodic array of split-ring resonators and thin metallic wires. Using artificial lump-element loaded transmission lines, a negative index metamaterial was realized in microstrip technology (2000a,b,c,. In the next year, complex negative refractive index (RI) and imaging by a flat lens using left-handed metamaterials were demonstrated. Over the following few years, there was an explosion of metamaterials research. One of the interesting properties of negative RI materials is complete reflection of EM waves. This makes practical realization of *The Invisible Man*, the lead character of eighteenth-century science fiction written by H G Wells, alluring. The first invisibility cloak, though imperfect, was realized at microwave frequencies in 2006 by several workers (see references in en.wikipedia).

Metamaterials can be classified into different categories. Apart from EM metamaterials, elastic, acoustic, structural, Hall, thermo-electric, and non-linear metamaterials are also present. Depending on the wavelength range in the EM spectrum the materials cover, metamaterials are classified into microwave, millimetre (mm) wave, terahertz radiation (THz), and optical, including infrared (IR) and visible ranges. Plasmonic metamaterials are also available. The materials used include metals, semiconductors: both bulk and nanostructured, and insulators. Metamaterials are realized by employing different structures like metal-insulator-metal (MIM), microstrip, photonic band gaps(PBG)s, and semiconductor nanostructures like Quantum Wells (QW)s, superlattices (SL)s, nanowire (NWR)s and Quantum Dots (QD)s.

Metamaterials show the potential for applications in diverse fields. These include optical filters, remote aerospace applications, sensor detection and infrastructure monitoring, smart solar power management, crowd control, radomes, high-frequency battlefield communication, lenses for high-gain antennas, and medical devices. They are used in improving ultrasonic sensors and they even shield structures from earthquakes. An interesting property is to allow imaging below the diffraction limit suffered by conventional glass lenses. Because of that these lenses are called superlenses. Though metamaterials are expected to produce an invisibility cloak, such a feat has not yet been achieved (see Metamaterials: Wikipedia: The free Encyclopedia; link given in reference list; 2021).

The aim of the present chapter is to cover, apart from introducing the subject and mentioning basic properties, semiconductor-based metamaterials at optical frequencies. Search for such semiconductor metamaterials has only recently started and the subject is therefore not too mature. In order to describe related work, it is first necessary to prepare the background. The effect of negative permittivity and permeability and negative RI on propagation of EM waves will first be introduced. The resonator structures used to obtain negative ε and μ will then be introduced. Initial work on metamaterials relates to artificial structures using periodic arrangement of metallic resonator elements like wires and slip ring capacitors. The problems of using metallic components at higher frequencies will be mentioned and how semiconductor-based structures can eliminate the problems will then be pointed out. The proposals made in the literature to realize optical metamaterials and related growth processes, structures used, and available experimental results will finally be presented.

15.2 Left-handed material with negative refractive index

As stated in the Introduction, Victor Veselago first considered the electrodynamics of substances with simultaneously negative values of ε and μ in his 1967 paper (translated version published in 1968). It is known that the RI may be written as

$$\eta = (\varepsilon_r \mu_r)^{1/2} \qquad (15.1)$$

For dielectrics at optical frequencies, μ_r, the relative permeability is unity. If both μ_r and the relative permittivity ε_r are negative, the RI is still positive. However, there might be significant changes in propagation as Veselago noticed. It is well known from Maxwell equations, that for a plane propagating wave in the form $\exp[-i(\omega t - \mathbf{k}.\mathbf{r})]$ in a medium with material constants ε and μ, the following relations connecting $\mathbf{k}$ and electric field $\mathbf{E}$ and magnetic field $\mathbf{H}$ are valid,

$$\mathbf{k} \times \mathbf{H} = -i\omega\varepsilon\mathbf{E} \text{ and } \mathbf{k} \times \mathbf{E} = i\omega\mu\mathbf{H}$$

The three vectors, $\mathbf{k}$, $\mathbf{E}$, and $\mathbf{H}$ thus constitute a right-handed set: a well-known result. On the other hand, when both ε and μ are negative, the three sets of vectors give rise to a left-handed system, justifying the name left-handed material, or metamaterial (LHM). It is also known that $\mathbf{k}$ coincides with the direction of phase velocity, whereas the Poynting vector, $\mathbf{S} = \frac{1}{2}(\mathbf{E} \times \mathbf{H})$, and the group velocity are in the same direction. For usual right-handed materials (RHM)s, phase and group velocities are in the same direction, whereas for LHM, $\mathbf{k}$ and $\mathbf{S}$ are in opposite directions. This implies that in LHM the energy flows along forward direction, but wave propagates along the backward direction.

Since the RI is the square root of a positive real quantity, it is more general to write $\eta = \pm(\varepsilon_r\mu_r)^{1/2}$. The +ve sign applies to RH materials, but it is expedient to use the −ve sign for LHM. The use of the −ve sign has the role to play in refraction, as shown in Fig. 15.1(b). Whereas Snell's law $\eta_1 \sin\theta_1 = \eta_2 \sin\theta_2$ tells that ray 12 refracted into medium 2 follows the path 23 in RHM, negative η_2 directs the refracted ray to follow path 24.

Propagation in a lossy medium and when RI is negative needs more attention. The wave vector $\mathbf{k}$ is complex when loss is present. Considering that both the permittivity and permittivity are complex numbers written as

$$\varepsilon = \varepsilon' + i\varepsilon'' \text{ and } \mu = \mu' + i\mu''$$

where single and double dashes over ε and μ denote respectively the real and imaginary parts of the permittivity and the permeability. One may hence write the wave number, assuming small dispersion, as

$$k = k' + ik'' = \frac{\omega}{c}\sqrt{(\varepsilon' + i\varepsilon'')(\mu' + i\mu'')} = \frac{\omega}{c}\sqrt{\varepsilon'\mu'}\left[1 + \frac{i}{2}\left(\frac{\varepsilon''}{\varepsilon'} + \frac{\mu''}{\mu'}\right)\right] \qquad (15.2)$$

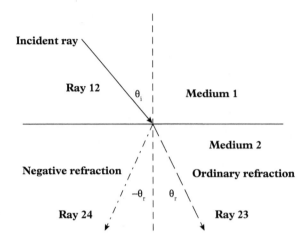

Figure 15.1 *Propagation of light rays in two media 1 and 2 and illustration of Snell's Law in right-hand and left-hand metamaterials. Incident ray 12 at the interface follows the path 23 in right-hand side, which is also the direction of both the phase and group velocities. In left-hand side, the refracted ray is along 24, which is also the direction of Poynting vctor, but the phase velocity is along 23.*

It follows that the change of the sign of the real part of **k** does not automatically change the sign of the imaginary part of **k**. Change in the sign of the imaginary part of the wave vector may only be accomplished by changing the sign of the imaginary parts of ε and μ. This necessitates introduction of negative absorption as in an amplifier. This is in no way connected to the possible transition from usual materials to the materials with negative refraction.

It may easily be surmised that in a material in which either ε or μ is $-$ve (single negative (SNG)), but not both of them (double negative (DNG)), the wave cannot propagate but has a decaying nature. Based on these results, the materials can be classified as double positive, ε negative, double negative and μ -ve. These materials occupy different quadrants in the μ vs. ε plot as shown in Fig. 15.2. The double positive materials are usual dielectrics, whereas the DNG materials are metamaterials or negative index metamaterial (NIM). Examples of ε negative materials are electrical plasmas, metals, and thin wire structures, while gyromagnetic materials or yet-to-be found magnetic plasmas belong to μ negative material.

Example 15.1 *The expression for permittivity for a plasma as given in Eq. (13.8) is $\varepsilon(\omega) = 1 - \frac{\omega_p^2}{\omega^2}$. This clearly indicates that when $\omega < \omega_p$, the permittivity is $-$ve.*

15.3   Structures for microwaves

NIMs or metamaterials cannot be found in nature. They are engineered structures. Whereas in natural crystals, the properties are derived from the constituent atoms, the properties in NIMs are derived from the constituent units that can be engineered at will.

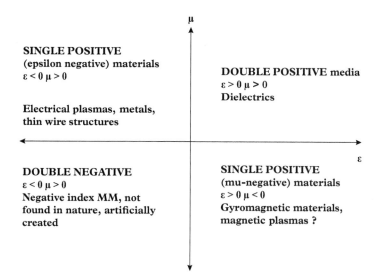

Figure 15.2 *μ vs. ε plot. Different quadrants are occupied by four different types having +ve and -ve values of permeability and permittivity.*

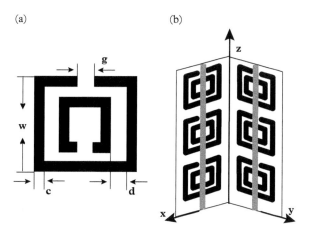

Figure 15.3 *(a) A representative split ring capacitor; (b) a periodic array of wires and SRRs make a NIM.*

A single cell of metamaterial consists of two regions that store electrical and magnetic energy. A combination of an inductor and a capacitor may serve as such a unit cell. Pendry and coworkers (1999, 2000) first proposed that an array of metal wires can act as an inductor and an array of split ring resonators (SRRs) can form the combination of a unit cell. The schematic structures of the wires and SRRs are shown in Fig. 15.3.

The wire mesh formed with a proper geometrical choice can provide a diluted metal showing negative ε according to the relation

$$\varepsilon(\omega) = 1 - \frac{2\pi c^2}{\omega^2 a^2 \ln(a/r)} \tag{15.3}$$

The authors also proposed that a SRR will provide strong coupling to the magnetic field leading to a negative μ following the relation:

$$\mu(\omega) = 1 - \frac{(\pi r \omega)^2 / a^2}{\omega_0^2 - \omega^2} \tag{15.4}$$

The first structure used to demonstrate negative refraction at microwave frequencies was realized by Smith et al (2000) by using a standard circuit board and depositing lithographically metal strips and SRRs on its opposite sides. These unit cells are repeated on the same board.

15.4   Perfect lens

It is well known that the best resolution offered by conventional lens is about half the wavelength. There have been many attempts during the last decades to overcome this diffraction limit. The real breakthrough is made by Pendry (2000) who realized that Veselago's flat lens may work wonders. He showed that a point source may be focused on another point in the focal point. The material with $\epsilon_r = \mu_r = -1$ can reproduce complete Fourier spectrum of the point object, including both the propagating and evanescent components of the wave.

Let us assume that the source is located at z = 0, and a flat lens is at z = d/2, and a plane wave has the form

$$F = F_0 exp(ik_x x)exp(-k_z'' z) \tag{15.5}$$

where k_z'' is the imaginary component of the wavevector in the z-direction and the wave will have an amplitude of $exp(-k_z'' d)$. Let the thickness of the perfect lens be d. The wave is amplified at the same rate, so that one obtains

$$exp(-k_z' d/2)exp(k_z' d)exp(-k_z' d/2) = 1 \tag{15.6}$$

The original amplitude of the evanescent wave is restored by the negative index material.

Further useful discussion is available in Solymar and Walsh (2010), and Pendry (2005).

15.5   Negative index of refraction with positive permittivity and permeability

Negative index of refraction (NIR) discussed so far is obtained for materials and structures in which both ε and μ are negative. However, NIR can be exhibited by materials and structures with positive ε and μ also. Pendry, in his 2004 paper, pointed out that structures showing chirality can be used to obtain NIR. Almost at the same time, it was theoretically proposed by a few workers, e.g. Podolskiy and Narimanov (2005), that NIR can be shown by anisotropic structures, either naturally occurring or engineered. In the following, we shall present the basic theory for propagation of EM waves in these two structures and mention the conditions for observing NIR.

15.5.1 Chiral metamaterials

The phenomenon of double refraction or birefringence in calcite and sodium chloride crystals and the control of polarization states of light by exploiting this have been known for quite some time. Similarly, optical activity, that is, the rotation of a linearly polarized light by placing a quartz crystal between crossed polarizers has been studied over many decades. The need to control both linear and circular polarization states with better precision and size reduction exists even today in various disciples in science.

Optical activity in a chiral material, as for example, quartz, amino acids, and sugar, occurs due to coupling between electric and magnetic fields. The electromagnetic coupling in these materials is rather weak as compared to dielectric polarization. Optical activity and dielectric polarization are characterized by a parameter, called chirality parameter, and the refractive index.

A new type of artificially structured metamaterial, called a chiral metamaterial has recently been developed, in which the chirality parameter is dramatically enhanced, so as to make it comparable to the refractive index. These chiral metamaterials can even show a negative refractive index without requiring any negative permittivity or negative permeability. Furthermore, their optical properties can be tuned by changing geometric or material parameters. In that way, the RI can be made negative or very large and the optical properties can be controlled dynamically. The reader is referred to the papers by Oh et al (2015); Wang et al (2016); and Ashalley et al (2021) for review, materials, and applications.

In a chiral material, there exists coupling between electric and magnetic fields, as a result of which the Maxwell equations are modified in the following way:

$$\mathbf{D} = \varepsilon\mathbf{E} + i\chi\mathbf{B} \tag{15.7a}$$

$$\mathbf{H} = \frac{\mathbf{B}}{\mu} + i\chi\mathbf{E} \tag{15.7b}$$

Both the electric displacement vector **D** and magnetic induction **B** are related to both the electric field **E** and magnetic field **H**. The parameter χ, named as the chiral parameter represents the coupling strength between the electric and magnetic fields. When $\chi = 0$, the constitutive equation has the same form as in normal isotropic materials.

Using Maxwells equation $\nabla \times \mathbf{E} = -\partial\mathbf{B}/\partial t$, expressing all fields and other vectors in the form of $exp(-i\omega t)$, and using Eq. (15.7a), one obtains

$$\nabla \times \mathbf{E} = i\omega\mathbf{B} = \frac{\omega(\mathbf{D} - \varepsilon\mathbf{E})}{\chi} = i\nabla \times \mathbf{H}/\chi - \omega\varepsilon\mathbf{E}/\chi \tag{15.8}$$

where the relation $\nabla \times \mathbf{H} = \partial \mathbf{D}/\partial t$ has been used. This equation, using (15.7b) may be written as

$$\nabla \times \mathbf{E} = -\nabla \times \mathbf{E} + \frac{1}{\omega\mu\chi}\nabla \times \nabla \times \mathbf{E} - \frac{\omega\varepsilon\mathbf{E}}{\chi} \qquad (15.9)$$

Rearranging terms, one obtains

$$\nabla \times \nabla \times \mathbf{E} - 2\omega\mu\chi\nabla \times \mathbf{E} - \omega^2\mu\varepsilon\mathbf{E} = 0 \qquad (15.10)$$

Using the relation

$$\nabla \times \nabla \times \mathbf{E} = \nabla(\nabla \cdot \mathbf{E}) - \nabla^2\mathbf{E} = -\nabla^2\mathbf{E} \qquad (15.11)$$

The wave equation for a chiral material may be expressed as

$$\nabla^2\mathbf{E} + 2\omega\mu\chi\nabla \times \mathbf{E} + \omega^2\mu\varepsilon\mathbf{E} = 0 \qquad (15.12)$$

and similarly

$$\nabla^2\mathbf{H} + 2\omega\mu\chi\nabla \times \mathbf{H} + \omega^2\mu\varepsilon\mathbf{H} = 0 \qquad (15.13)$$

is obtained for the wave equation of the magnetic field. In terms of the wave vector $\mathbf{k}$, the wave equation for the electric field can be expressed as

$$-\mathbf{k} \times \mathbf{k} \times \mathbf{E} - i\omega\mu\chi\mathbf{k} \times \mathbf{E} - \omega^2\mu\varepsilon\mathbf{E} = 0 \qquad (15.14)$$

The wave number k satisfies the following equation:

$$k^2 = \left(\frac{\omega^2\mu\varepsilon - k^2}{2\omega\mu\chi}\right)^2 \qquad (15.15)$$

Solution of the quadratic equation gives the following expressions for the two eigensolutions for circular polarizations:

$$k_R = \omega\mu\chi + \omega\sqrt{\mu\varepsilon + \mu^2\chi^2} \qquad (15.16a)$$

$$k_L = -\omega\mu\chi + \omega\sqrt{\mu\varepsilon + \mu^2\chi^2} \qquad (15.16b)$$

The subscripts R and L refer respectively to right-handed polarized and left-handed polarized waves.

The difference in wave numbers for RCP and LCP waves leads to two different RIs given by

$$\eta_\pm = \sqrt{\varepsilon\mu} \pm \chi \tag{15.17}$$

where the subscript + is valid for RCP and − for LCP. It is interesting to note that when the chiral parameter is quite large the RI can be negative even though both ε and μ are positive quantities.

Example 15.2 *Let us assume ε = 4 and μ= 1. Then χ should be > 2 to make refractive index negative as follows easily from Eq. (15.17).*

15.5.2 Indefinite or hyperbolic metamaterial

Indefinite metamaterial (MM) systems that possess negative permittivity only along one direction show negative permittivity as reported by Smith et al (2004). Podolski et al (2005) developed the theory of optical processes in such MMs. All angle negative refraction using metallic nanowires representing such indefinite MMs was demonstrated by Liu et al (2008). In this subsection, we shall discuss the properties of hyperbolic MMs (HMMs), following two comprehensive articles by Shekhar et al (2014) and Guo et al (2020).

The relative permittivity tensor of anisotropic dielectric takes the form

$$\varepsilon = \begin{bmatrix} \varepsilon_{xx} & 0 & 0 \\ 0 & \varepsilon_{yy} & 0 \\ 0 & 0 & \varepsilon_{zz} \end{bmatrix} \tag{15.18}$$

A similar tensorial form for the relative permeability may be written.

The components for electric biaxial material satisfy the condition $\varepsilon_{xx} \neq \varepsilon_{yy} \neq \varepsilon_{zz}$, whereas for uniaxial material, one writes $\varepsilon_{xx} = \varepsilon_{yy} = \varepsilon_\|, \varepsilon_{zz} = \varepsilon_\perp$. Again similar relations and symbols exist for magnetic biaxial and uniaxial materials.

Putting Eq. (15.18) in Maxwell's curl equations, and writing the electric and magnetic fields in terms of frequency ω and wave vector $\mathbf{k}(\equiv k_x, k_y, k_z)$ as

$$\mathbf{E} = \mathbf{E}_0 \exp i(\omega t - \mathbf{k.r}) \text{ and } \mathbf{H} = \mathbf{H}_0 \exp i(\omega t - \mathbf{k.r}),$$

and assuming $\mu_\| = \mu_\perp = 1$, the wave equations for the electric field in a uniaxial electric medium may be written as

$$k_x (k_x E_x + k_y E_y + k_z E_z) = (k^2 - k_0^2 \varepsilon_\|) E_x \tag{15.19a}$$

$$k_y (k_x E_x + k_y E_y + k_z E_z) = (k^2 - k_0^2 \varepsilon_\|) E_y \tag{15.19b}$$

$$k_z \left(k_x E_x + k_y E_y + k_z E_z \right) = \left(k^2 - k_0^2 \varepsilon_\perp \right) E_z \qquad (15.19c)$$

where k_0 is the wave vector in free space. The following dispersion relation is obtained by solving the system of equation in Eq. (15.19):

$$\left(k_x^2 + k_y^2 + k_z^2 - \varepsilon_\parallel k_0^2 \right) \left(\frac{k_x^2 + k_y^2}{\varepsilon_\perp} + \frac{k_z^2}{\varepsilon_\parallel} \right) = 0 \qquad (15.20)$$

The first term within bracket in Eq. (15.20) indicates that the electric field is in the x-y plane and perpendicular to the z-axis, the optic axis ($\mathbf{z} \cdot \mathbf{E} = 0$). This describes a transverse electric (TE) wave, corresponding to an ordinary wave. The second term describes a transverse magnetic (TM) wave; the magnetic field is now in the x-y plane and $\mathbf{z} \cdot \mathbf{H} = 0$, corresponding to an extra-ordinary wave.

15.5.2.1   *Types I and II hyperbolic metamaterials and dispersion relations*

Consider first the situation when $\varepsilon_\parallel > 0$ and $\varepsilon_\perp < 0$. In this case both the TE and TM waves can coexist. The isofrequency (IFC) contour for the TE mode is then a sphere as shown in the left panel of Fig. 15.4. The IFC of the TM mode is a two-fold hyperboloid as shown in the middle panel of Fig. 15.4. The two hyperboloid surfaces are also oriented along the optic axis, as may be seen from the figure. The corresponding material is called a Type I or a dielectric-type HMM. In another situation, when $\varepsilon_\parallel < 0$ and $\varepsilon_\perp > 0$, the TE mode is absent, and the IFC of the TM mode is a one-fold hyperboloid, as shown in the right panel of Fig. 15.4. This type of material having this characteristic is called a Type II or a metal-type HMM.

Structures: two commonly used structures for HMM will now be introduced, representing one-dimensional (1D) and two-dimensional (2D) structures.

1D HMM: a multilayer (superlattice) structure in which a metal and a dielectric form alternate layers, as shown in Fig. 15.5, leads to a 1D HMM. The layer thicknesses should

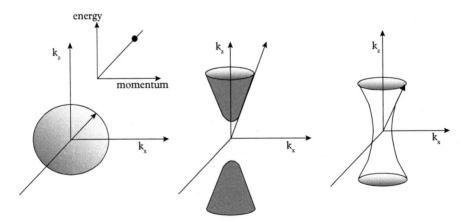

Figure 15.4 *Isofrequency surfaces of (left) isotropic MM, (middle) type I HMM, and (right) type II HMM.*

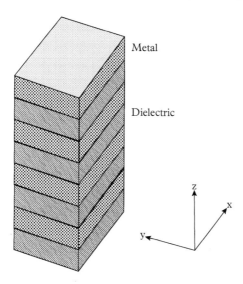

Metal

Dielectric

Figure 15.5 *A common 1D HMM structure using a multilayer combination of a metal and a dielectric forming alternate layers.*

be small compared to the operating wavelength. From effective medium theory, the components of the permittivity tensor in Eq. (15.18) can be expressed from 11 as

$$\varepsilon_{\parallel} = f\varepsilon_1 + (1-f)\varepsilon_2 \qquad (15.21a)$$

$$\frac{1}{\varepsilon_{\perp}} = \frac{f}{\varepsilon_1} + \frac{1-f}{\varepsilon_2} \qquad (15.21b)$$

where the fill fraction is given in terms of thickness t of the layers by

$$f = \frac{t_1}{t_1 + t_2}$$

A number of material combinations may be used to realize a 1D HMM with a multilayer structures operative at different wavelengths. A few examples of the material combination are shown in Table 15.1.

A 2D HMM is shown in Fig. 15.6. A 2D HMM may be realized by embedding metallic nanowires, typically silver (Ag) or gold (Au), in a dielectric host like a nanoporous

Table 15.1 *Choice of materials in type I superlattices at different wavelength regimes.*

	UV	Visible	Near IR	Mid IR
Layer 1: metal or other conductors	Au or Ag	Ag or Ag	transition metal nitrides or transparent conducting oxides	III–V degenerately doped semiconductors
Layer 2: dielectric	Alumina	TiO$_2$ or SiN	dielectrics	As above

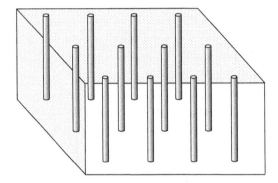

Figure 15.6 *Two-dimensional HMM using an array of metallic nanowires embedded in a dielectric.*

alumina template. The fill fraction needed in this structure to achieve type I behaviour is quite low compared to that in multilayer structure. The nanowire array exhibits low losses, broad bandwidth, and high transmission.

The effective parallel and perpendicular components of permittivity in the NWR system can also be expressed by employing effective medium theory (Shekhar et al 2014) and the following expressions are obtained:

$$\varepsilon_{\parallel} = \frac{(1+f)\varepsilon_m\varepsilon_d + (1-f)\varepsilon_d^2}{(1+f)\varepsilon_d + (1-f)\varepsilon_m} \qquad (15.22a)$$

$$\varepsilon_{\perp} = f\varepsilon_m + (1-f)\varepsilon_d \qquad (15.22b)$$

15.5.2.2 *Novel properties of hyperbolic metamaterials*

The general theory of all angle negative refraction by HMMs is given by Liu et al (2008). A plot of angle of refraction vs. angle of incidence for different geometrical parameters used for an array of metallic NWRs clearly shows the possibility of negative refraction for all angles. The basic theory is used to illustrate the negative refraction using dielectrics only by Sayem et al (2016), though at a somewhat limited range. This work with proper illustration will be described in Section 15.5.3.

HMMs show a few novel characteristics, namely, enhanced rates for spontaneous emission, photoluminescence and absorption, enhanced Purcell factor, lifetime shortening and abnormal refraction, reflection, and scattering. HMMs are used for density-of-states (DOS) engineering. Detailed discussions of these characteristics using useful structures are available in the papers by Shekhar et al (2014) and Guo et al (2020), and interested readers may obtain useful information of these novel characteristics.

15.5.3 Lossless all dielectric/semiconductor asymmetric anisotropic metamaterial

It is not always necessary to use DNG materials to observe negative refraction. The use of chiral metamaterials and hyperbolic metamaterials (HMM) or so-called indefinite structures has already been discussed in this connection. HMMs are made from multilayers of metals and dielectrics and therefore possess loss. Dielectric–dielectric/semiconductor subwavelength multilayers may work as anisotropic elliptical metamaterials with limited loss in some frequency regions providing better waveguiding. Though natural birefringent or uniaxial materials show negative refraction, very narrow angles of incident, and refracted rays limit their practical applications.

Recently, Sayem et al (2016) studied an asymmetric anisotropic metamaterial (AAM) made from lossless dielectric or semiconductor materials at the telecommunication wavelength band. They are able to show that broad angle negative refraction is possible by using multilayered structures made of several pairs of semiconductors and or dielectrics. This negative refraction does not require any negative permittivity like others and there are no resonance-like phenomena. So, there is no possibility of loss as long as the dielectrics or semiconductors used are lossless in the respected frequency range of interest. By changing the fill fraction and the rotation angle, tunability can also be achieved.

The schematic of the proposed anisotropic structure is shown in Fig. 15.7. The interfaces between the two dielectrics lies along the x' axis. In the (x', y', z') coordinate system the relative permittivity tensor is expressed as

$$\varepsilon' = \begin{bmatrix} \varepsilon_{\parallel} & 0 & 0 \\ 0 & \varepsilon_{\parallel} & 0 \\ 0 & 0 & \varepsilon_{\perp} \end{bmatrix} \tag{15.23}$$

The dashed axes are rotated through an angle ϕ around a y-axis to coincide with the (x, y, z) axes. The permittivity tensor in the new coordinate system is obtained as

$$[\varepsilon_{new}] = R_y [\varepsilon'] R_y^T \tag{15.24}$$

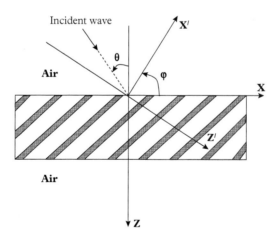

Figure 15.7 *Schematic of the proposed asymmetric anisotropic multilayer metamaterial. Individual slab interface lies along the x axis. The structure is rotated around y axis at an angle φ with respect to the x axis. A TM (p) polarized electromagnetic wave (magnetic field polarized along y axis) is incident at angle θ with respect to the z-axis where d is the thickness of the structure.*

where R_y is the rotation matrix around y-axis and R_y^T is the transposed matrix. R_y is given by

$$R_y = \begin{bmatrix} \cos\phi & 0 & -\sin\phi \\ 0 & 1 & 0 \\ \sin\phi & 0 & \cos\phi \end{bmatrix} \tag{15.25}$$

Using Eq. (15.18) one obtains

$$[\varepsilon_{new}] = \begin{bmatrix} \varepsilon_{xx} & 0 & \varepsilon_{xz} \\ 0 & \varepsilon_{yy} & 0 \\ \varepsilon_{zx} & 0 & \varepsilon_{zz} \end{bmatrix} = \begin{bmatrix} \varepsilon_{\parallel}\cos^2\phi + \varepsilon_{\perp}\sin^2\phi & 0 & (\varepsilon_{\parallel} - \varepsilon_{\perp})\sin\phi\cos\phi \\ 0 & \varepsilon_{yy} & 0 \\ (\varepsilon_{\parallel} - \varepsilon_{\perp})\sin\phi\cos\phi & 0 & \varepsilon_{\parallel}\cos^2\phi + \varepsilon_{\perp}\sin^2\phi \end{bmatrix} \tag{15.26}$$

As shown in Fig. 15.7, a TM polarized EM wave (magnetic field polarized along y axis) is incident from the ambient medium (air) to the structure and the wave propagates in z–x plane forming an angle θ with the z axis. The EM field is assumed to vary as $exp - i[\omega t - k_z z - k_x x]$.

Assuming that the material is non-magnetic, the normal component of the propagation constant may be expressed as

$$k_z = \frac{\varepsilon_{xz}k_x \pm \sqrt{(\varepsilon_{xz}^2 - \varepsilon_{xx}\varepsilon_{zz})k_x^2 - k_0^2\varepsilon_{zz}(\varepsilon_{zx}\varepsilon_{xz} - \varepsilon_{zz}\varepsilon_{xx})}}{\varepsilon_{zz}} \tag{15.27}$$

where $k_x = k_0 \sin\theta$ is the tangential component of wave vector and $k_0 = 2\pi/\lambda$, is the free space wave number. For a symmetric case, $\varepsilon_{zx} = \varepsilon_{xz}$. In Eq. (15.27), + and– signs refer to components of wave number propagating downward and upward, respectively.

For TM wave, the H field is along the y-direction. The time averaged Poynting vector is $\mathbf{S} = \frac{1}{2}\text{Re}\,[\mathbf{E} \times \mathbf{H}^*]$. The x and z components of the time-averaged Poynting vector may be expressed as

$$\langle S_x \rangle = \frac{1}{2}\frac{|H_y|^2}{\omega\varepsilon_0}\text{Re}\left(\frac{\varepsilon_{xx}k_x - \varepsilon_{zx}k_z}{\varepsilon_{zz}\varepsilon_{xx} - \varepsilon_{zx}\varepsilon_{xz}}\right) \tag{15.28}$$

$$\langle S_z \rangle = \frac{1}{2}\frac{|H_y|^2}{\omega\varepsilon_0}\text{Re}\left(\frac{\varepsilon_{zz}k_z - \varepsilon_{xz}k_x}{\varepsilon_{zz}\varepsilon_{xx} - \varepsilon_{zx}\varepsilon_{xz}}\right) \tag{15.29}$$

The angle of refraction for wave vector can be written as

$$\tan\theta_{r,k} = \frac{\text{Re}(k_x)}{\text{Re}(k_z)} \tag{15.30}$$

and the angle of refraction for the Poynting vector is expressed as

$$\tan\theta_{r,S} = \frac{\text{Re}(\varepsilon_{xx}k_x - \varepsilon_{zx}k_z)}{\text{Re}(\varepsilon_{zz}k_z - \varepsilon_{xz}k_x)} \tag{15.31}$$

When $\phi = 0$, $\varepsilon_{zx} = \varepsilon_{xz} = 0$, and Eq. (15.31) reduces to

$$\tan\theta_{r,S} = \frac{\mathrm{Re}(\varepsilon_{xx}k_x)}{\mathrm{Re}(\varepsilon_{zz}k_z)} \tag{15.32}$$

reproducing the result for a uniaxial crystal.

As the dielectric thicknesses are fraction of wavelengths, all the tensor components in Eq. (15.23) are positive. Also for the case when $\phi = 0$, $\varepsilon_{xx} = \varepsilon_{\parallel}$, $\varepsilon_{zz} = \varepsilon_{\perp}$, $\varepsilon_{zx} = \varepsilon_{xz} = 0$, the right-hand side of Eq. (15.32) is always positive, meaning that there is no negative refraction. This same conclusion applies when $\phi = \pm 90°$. However, when the structure is rotated, depending on the values of $\varepsilon_{xx}, \varepsilon_{zz}$, and $\varepsilon_{zx} = \varepsilon_{xz}$, the right-hand side of Eq. (15.29) may be negative, leading to negative refraction.

In the present situation, the ISF contour in the wave vector space is elliptic. A graphical method to establish occurrence of negative refraction has been presented in the work of Yonghua et al (2005). In the following, we shall present a numerical calculation showing negative angle of refraction in the multilayer dielectric using silicon (Si) and silica (SiO$_2$) as examples.

15.5.4 Calculation of permittivities by using effective medium theory

The multilayer structure considered by Sayem et al (2016) consists of dielectric 1 with permittivity ϵ_1 and thickness t_1 and dielectric 2 with permittivity ϵ_2 and thickness t_2. The Bloch equation for the unrotated case can be written as

$$\cos(k_z t) = \cos(k_{z1}t_1)\cos(k_{z2}t_2) - \frac{1}{2}\left(\frac{\epsilon_2}{k_{z2}}\frac{k_{y1}}{\epsilon_2} + \frac{\epsilon_1}{k_{z1}}\frac{k_{z2}}{\epsilon_1}\right)\sin(k_{z2}t_2)\sin(k_{z1}t_1) \tag{15.33}$$

where $k_{zi}^2 = \epsilon_i k_0^2 - k_x^2$, $i = 1, 2$ *and* $t = t_1 + t_2$ is the thickness of the unit cell. In deeply sub-wavelength limit $k_0 t \ll 1$, the components of the permittivity tensor in Eq. (15.33) can be expressed as

$$\varepsilon_{\parallel} = f\tilde\epsilon_1 + (1-f)\varepsilon_2 \tag{15.34a}$$

$$\frac{1}{\varepsilon_{\perp}} = \frac{f}{\varepsilon_1} + \frac{1-f}{\varepsilon_2} \tag{15.34b}$$

where the fill fraction is given by $f = \frac{t_1}{t_1+t_2}$. These expressions are obtained by using *effective medium theory.* The values obtained from Eq. (15.34) may be used to calculate the components in the rotated case from Eq. (15.26).

15.5.5 Numerical example

We now present calculated results for Si-SiO$_2$ multilayer dielectrics showing occurrence of negative refraction by using the theory presented by Sayem et al (2016). The chosen

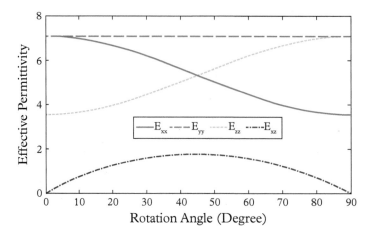

Figure 15.8 *Effective permittivity values of the asymmetric anisotropic metamaterial as a function of rotation angle around the y axis.*

wavelength is 1550 nm and permittivity values of 2.085 and 12.099 are chosen for SiO_2 and Si, respectively.

The tensor components as a function of rotation angle are displayed in Fig. 15.8.

Calculated values of angle of refraction as a function of incident angle with a fill factor of 0.5 are shown in Fig. 15.9. It is clear that negative refraction is possible upto a value of 38 degrees for incident angle and the maximum value of negative angle is nearly 18 which occurs for $\phi = 0$ degrees.

15.6   Low-loss plasmonic metamaterial

Metals, even Ag and Au that have highest conductivities, are not suitable for use as optical MM due to excessive resistive loss. This demands new materials for devices at optical frequencies and in particular for telecommunication bands. Boltasseva and Atwater (2011) suggested ways to exploit plasmonics which allow routing and manipulation of light at the nanoscale. However, loss is also a detrimental factor in metal-based plasmonics. The real part of the dielectric permittivity is also an important consideration, because it determines the optical performance of the system.

The optical properties of conducting materials are determined by their carrier concentration and carrier mobility. The former must be high enough to ensure a high value of negative permittivity, while the latter must be high to reduce loss. Loss due to interband transition must be low, that means the working frequency must be less than the value corresponding to fundamental absorption edge. As metals cannot satisfy these conditions, potential materials to replace silver or gold are alkali metals, intermetallics, various alloys, transparent conducting oxides (TCOs), and graphene. Till now Ag,Au, and their alloys with slightly improved properties are the best materials in the visible wavelength

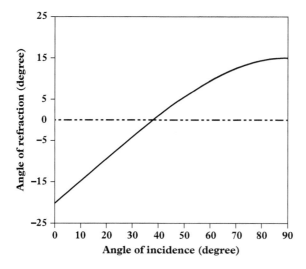

Figure 15.9 *Angle of refraction of the AAM as a function of incident angle for SiO₂–silicon.*

range. Although alkali metals have attractive properties, their extreme chemical reactivity limits their use. At near-infrared (NIR) heavily doped (10^{21} carriers cm^{-3}) TCOs such as indium tin oxide, or zinc oxide doped with aluminum or gallium are suitable plasmonic MM. Useful semiconductors in the mid-infrared (MIR) region are silicon carbide, gallium arsenide, and nanostructured semiconductors. Since the losses associated with metals partly arise from too large values of free electron densities, reduction in carrier density is possible in intermetallics, a mixture of metals with other metal, and nonmetals resulting in metal silicides, germanides, borides, nitrides, oxides, and metallic alloys. For example, it is found that titanium nitride may replace Au as a plasmonic MM above 550 nm. Similarly, in the NIR region, the optical properties of silicides and germanides of tungsten and tantalum could be optimized to produce low-loss plasmonic materials. Nonstoichiometric oxides such as vanadium, titanium, and aluminum oxides are suitable in the NIR and visible ranges and show switchable optical properties.

Boltasseva and Atwater (2011) have given a *metamaterial map* showing carrier density, carrier mobility, and interband losses of various plasmonic MM and also indicating the regions of EM spectrum in which these materials may find applications.

Use of oxides and nitrides as alternative plasmonic materials in the optical range has been further discussed by Naik and Boltasseva (2010); and Naik et al (2011, 2012).

Heavily doped semiconductors, and TCOs and semiconductor nanostructures are useful in THz region, MIR, NIR, and even visible regions. The plasmonic properties, as well as other optical properties of these materials, are exploited to produce MM. An interesting property of semiconductors is that by changing the carrier concentration by illumination or by electric field, the permittivity can be modulated, tuned and even switched from positive to negative values and vice versa. The following sections will deal with semiconductor-based MM.

15.7 Semiconductor metamaterials

We now focus our attention to metmaterials using semiconductors. The metamaterials discussed so far are metal based and their frequency range of operation cannot be extended to near and far infrared regions. Semiconductor-based MMs are useful in these ranges. We cover the advantages of semiconductor-based MMs, the structures used and the use of quantum nanostructures in long wavelength ranges.

15.7.1 Introduction

The wire mesh and SRRs used to create negative permittivity and permeability are made of metals. Initially these structures worked at the microwave region. In order to extend the operation at shorter wavelengths, the structures need to be scaled down. Scaling near 100 nm has been achieved using present-day lithography.

However, the LH metamaterials designed for the visible near-infrared IR spectrum based on metal inclusions suffer from several disadvantages, as listed below:

(1) Metal-based LMMs rely on resonance phenomena to achieve $-$ve ε and μ and therefore entail substantial losses.

(2) Since electrical and magnetic resonances peak at different frequencies, the band over which LMMs show simultaneous $-$ve ε and μ has a limited width, thus restricting the applicability of the structures.

(3) Compensation of the heavy loss is not possible by inclusion of gain media, since the gain obtained does not compensate the loss. Further, it is difficult to combine the lossy structures with gain media.

(4) The fabrication processes are not compatible with present nanofabrication processes.

Semiconductor based NMs can overcome these difficulties in the following ways:

(1) Semiconductors are non-magnetic and therefore resonances for both $\varepsilon(\omega)$ and $\mu(\omega)$ are not possible. The structures relying on resonance of $\varepsilon(\omega)$ only can be designed to have low loss. The loss can even be compensated by gain within the medium.

(2) Non resonant configurations possessing either chirality or anisotropic permittivity can be used, minimizing loss.

(3) The useful bandwidth can be increased by proper design.

(4) Use of proper material in bulk or nanostructured form can cover different wavelength regimes.

(5) Tunability of operation can be accomplished by doping, injection, alloying or properly designed nanostructures.

(6) HMMs made of alternating material and semiconductor layers can be used at visible and higher frequencies.

(7) Subwavelength structures can be realized by exploiting plasmonics in doped semiconductors.

(8) Standard nanofabrication processes can be employed to fabricate structures that have small size, can be integrated with other devices, and systems and batch-fabricated.

Though all the advantages are only anticipated or predicted, realization of actual structures showing all the advantages is still far away.

In the following subsections, a few of the representative works reported so far in connection with semiconductor optical metamaterials will be reported.

15.7.2   Demonstration of negative refraction in all-semiconductor structures

The first demonstration of negative refraction using an all-semiconductor approach came from Hoffman et al (2007) who used a superlattice made of intrinsic (i) indium aluminum arsenide (i–InAlAs) and doped indium gallium arsenide (InGaAs) grown on indium phosphide (InP) substrate. The structure had an anisotropic dielectric function and an electrical resonance and NIR was exhibited in the long wavelength range around 11 μm. They extended their work to significantly thicker structures and by using also a QW in which intersubband transition was the underlying mechanism. Molecular beam epitaxy(MBE) and metal organic chemical vapor deposition (MOCVD) growth processes were employed to fabricate inherently three-dimensional (3D) and planar structures. As will be pointed out, the range of operational wavelength is determined by the free carrier density and layer thicknesses, both easily controlled parameters in the material growth.

The schematic of the multilayer structure is shown in Fig. 15.10 which also indicates that the structure shows anisotropic permittivity, that is, the permittivity along the plane, $\epsilon_\parallel$, is different from the permittivity, $\epsilon_\perp$, along the growth direction.

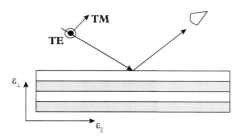

Figure 15.10 *Schematic of the superlattice structure made of i–InAlAs and n+-InGaAs. The orientation of the electric field and the components of the permittivity are shown.*

Since the electric field vector, **E**, and the electric displacement vector, **D**, are not usually parallel in anisotropic materials, as a consequence, the Poynting vector, **S**, pointing in the direction of energy flow and wave vector, **k**, directed along the wave front normal, need not be parallel. Noting also that only the tangential component of **k** is conserved at the heterointerface, the refracted beam may exhibit normal positive refraction with respect to **k** but negative refraction with respect to **S**. The directions of the vectors **k** and **S** in the refracted beam will be related to the angle of incidence, θ_i.

A uniaxial anisotropic material with $\epsilon_{\parallel} > 0$ and $\epsilon_{\perp} < 0$ will exhibit negative refraction with respect to **S** for the transverse magnetic (TM) polarization for all incidence angles. Other anisotropic materials exhibit negative refraction for only a small range of angle of incidence. In an isotropic NIM, both ϵ and μ are negative, and both **S** and **k** refract negatively. It should be noted that the transverse electric (TE) polarization does not experience anisotropy, and for such waves both **k** and **S** refract normally.

The two layers are thick enough to prohibit quantization, but at the same time thin compared to the wavelength. The i–InAlAs layer has constant permittivity, while the n$^+$ InGaAs is suitably doped so that negative permittivity may be obtained around plasma resonance. The expressions for components of permittivity tensor are given by

$$\epsilon_{\parallel} = \frac{\epsilon_{InAlAs} + \epsilon_{InGaAs}}{2}; \; \epsilon_{\perp} = \frac{2\epsilon_{InAlAs}.\epsilon_{InGaAs}}{\epsilon_{InAlAs} + \epsilon_{InGaAs}} \tag{15.35}$$

The i–InAlAs has a constant permittivity given by $\epsilon_{InAlAs} = \epsilon_{\infty InAlAs} = 10.23$, the high-frequency permittivity and the permittivity of doped InGaAs is expressed from Drude model as

$$\epsilon_{InGaAs}(\omega) = \epsilon_{\infty InGaAs} \left[1 - \frac{\omega_p^2}{\omega^2 - i/\gamma} \right] \tag{15.36}$$

$\epsilon_{\infty InGaAs} = 12.15$ is the high frequency permittivity, ω_p is the plasma frequency and $\gamma = 0.1 \times 10^{-12} s^{-1}$ is the damping parameter.

Example 15.3 *We assume that the doping density is 7.5×10^{18} cm^{-3}, the plasma frequency $\omega_p^2 = ne^2/m_e\epsilon\epsilon_0$. Taking $m_e = 0.041m_0$, $\omega_p^2 = 4.8 \times 10^{28}$, $\omega = 1.89 \times 10^{14}$ at 10 μm wavelength. Using other values of parameters as given above the permittivity $\epsilon_{\perp} = -14.55$ and $\epsilon_{\parallel} = 2.99$ are obtained.*

Fig. 15.11 shows a plot of the calculated values of $\epsilon_{\parallel}$ and $\epsilon_{\perp}$ as a function of wavelength. The solid lines correspond to the real parts of the permittivity. The spectral interval over which the metamaterial exhibits negative refraction is indicated by the shaded region.

15.7.3   Metamaterials with low-dimensional semiconductors

Ginzberg and Orenstein (2008) gave a concept of MM assembly where metallic parts are replaced by semiconductor based low-dimensional quantum structures. They showed

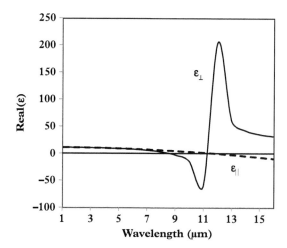

Figure 15.11 *Calculated values of $\varepsilon_\parallel$ and $\varepsilon_\perp$ as a function of wavelength. The doping density in n+-InGaAs layer is 7.5×10^{18} cm^{-3}. In the shaded spectral region, the anisotropy results in negative refraction for all angles of incidence.*

that such structures exhibit significantly lower losses and even may have positive gain by injection of carriers. The proposed structures can be grown by established epitaxial growth processes. The inherent mechanism relies on achieving negative permittivity due to resonant transitions between subbands in Quantum Well (QW) or Quantum Dot (QD) structures and does not use free carrier plasma effect in semiconductors.

15.7.4 Left-handed material by quantum system anisotropy

The study considers an exceptionally high anisotropic waveguide structure proposed by Podolskiy and Narimanov (2005) and developed by the Ginzberg and Orenstein (2008), in which some of the propagating TM optical modes are expected to be effectively left-handed. Two possible structures consisting of either QWs or coupled QDs are examined to achieve the required anisotropy.

The origin of the anisotropy lies in the difference between the susceptibilities along the plane and vertical growth directions. The polarization dependent matrix element in the intersubband transitions in a QW, and polarization dependent tunnelling efficiency for coupled QDs contribute to this anisotropy. Though the effect is pronounced in the vicinity of transition resonances, the loss is relatively low as compared to that in metal-based materials. By applying voltage the structure can be converted from LHM to RHM. Using quantum cascade laser (QCL) configuration, gain can be introduced to compensate the losses (Ramovic et al 2011).

According to Ginzberg and Orenstein the materials suitable at visible and IR regions would be nitride alloy families, such as, indium nitride (InN)/gallium nitride (GaN)/aluminum nitride (AlN) QWs/QDs, due to a large gap, large conduction band offsets, and low background permittivity to facilitate realization of negative permittivity.

The structures considered are MQWs and vertically stacked QD arrays enclosed between upper and lower cladding layers.

The EM wave polarized in the plane of the QWs growth strongly interacts with the resonant CB electron causing intersubband transition between subbands 1 and 2, while the polarization perpendicular to the growth direction is ineffective according to selection rules (Basu 1997). The permittivity along the growth direction is expressed as

$$\varepsilon_\perp = \varepsilon_{back} - ine\frac{|\mu_{12}|^2(f_1 - f_2)}{\varepsilon_0\hbar}\frac{1}{i(\omega - \omega_{21}) + \gamma}\Gamma \qquad (15.37)$$

where n is the carrier density, μ is the dipole moment, f's denote occupation probabilities, ω_{21} is the transition angular frequency, γ is dephasing rate and Γ is the mode confinement factor.

Due to sharply peaked nature of absorption or population inversion as dictated by Eq. (15.37), the permittivity will show the usual nature, that is, positive peak below resonance, zero-crossing at ω_{21}, followed by negatively valued region.

Example 15.4 *As an illustration of occurrence of negative permittivity, they considered GaAs/Al$_{0.3}$Ga$_{0.7}$As QWs. The parameters taken were well depth ΔE_c = 230 meV, well width L=10 nm, and the volume density of charge of 1 × 10^{18} cm^{-3}. Dephasing rate of the dipolar transitions was chosen to be γ_{12} = 3 meV, ε_{back} = 13. The waveguide core is designed to contain about 30% QW layers.*

For 13 μm wavelength, $\varepsilon_\perp$ = −1.56+ i5.22 was obtained and the −ve permittivity was found to cover the range 13–14 μm.

By considering pairs of vertically coupled InN/AlN QDs, it has been found that LHM character may be obtained over the range 965.3 to 965.45 nms.

The effect of inclusion of gain has also been studied in the paper. In this connection, the work by Ramović et al (2011) may be mentioned. They consider structural layers as in a GaAs/AlGaAs QCL. The same model as given by Eq. (15.35) is employed. When the structure is passive, there is no −ve permittivity. However, when population inversion takes place, under application of strong magnetic field, negative permittivity values are obtained: even the sheet carrier density is as low as 10^9 cm^{-3} because of narrow absorption linewidth and large matrix element.

Vuković et al (2015) considered the possibilities of achieving negative refraction in QCL-based semiconductor metamaterials in the THz spectral range lower than 2 THz. Their objective is to assess the degree of population inversion needed for LHM.

Liu et al (2016) published a report for all dielectric metamaterials using III–V semiconductors, consisting of arrays of high index Mie resonators having subwavelength-size. Both the permittivity and permeability of the structure can be controlled exhibiting lower loss than metals at optical frequencies. By selectively oxidizing the semiconductors, a high RI contrast between the resonators and their surroundings were achieved.

The GaAs/AlGaAs based dielectric metamaterials exhibited strong magnetic and electric dipole resonances that could be widely tuned at NIR wavelengths. The MMs are fully integrable with optical functionalities.

15.8   Metasurfaces

Metasurfaces, as the word implies, is the planar or 2D version of metamaterials having subwavelength thickness. Due to their light weight and ease of fabrication, metasurfaces attract attention from workers for various applications. They are useful for blocking, absorbing, concentrating, dispersing, and guiding waves. A recent review on metasurfaces and its applications has been published by Li et al (2018). These include impedance metasurfaces, metasurface absorbers, both active and passive, wavefront engineering comprising waveguiding, beam shaping, leaky antenna, holography, modulators and polarizers, cloaking, lensing, and imaging. In all these applications, fundamental limitations including narrow bandwidth, linear response, fixed functionality, etc., come into play. By applying active electronic devices like transistors, diodes, and varactors to traditional passive elements, metasurfaces with nonlinearity, power dependency, tenability, and tuning ability have been realized.

Since it is beyond the scope of this book to introduce the basic principles of the application areas of metasurfaces, we shall limit our scope to the discussion of hybrid structures integrating metasurfaces with conventional semiconductor devices and ICs. These hybrid structures find important applications in the THz regime. The hybrid structures and their applications, particularly in communication areas, will form the contents of the following sections.

15.8.1   Terahertz range

Terahertz (THz) frequencies cover the range from 0.1 to 10 THz in the EM spectrum, and the corresponding wavelengths range from 3 mm to 30 μm. The band lies between the microwave and IR regions. Astrophysicists use THz systems to detect THz radiation: one of the components of the BIG BANG. It passes through clothing but is reflected by metals and explosives, making it potentially useful for security scanning. It is a better alternative to X-ray scanning of bodies, as it does not produce ionizing of cells, unlike X-rays. The high frequency of the waves means they could also transfer wireless data very rapidly. Therefore, ultrafast signal processing and massive data transmission using THz waves are important areas of application in information technology.

Other areas of application of THz systems are as follows: Space Science: cosmology, planetary cosmology and cosmo chemistry; environmental science: atmospheric sensing, pollution detection; defence: chemical agent detector, biological agent detector, digital RADAR; communication: space to space, short battle field; local area network (LAN); materials processing. Of course, scientific investigation using THz is also a challenging and rewarding field.

THz radiation can penetrate most dielectric materials and non-polar liquids. It also produces less harm to human tissues than X-rays. Because of these features, THz imaging has assumed importance in bio-detection, security screening in airports, illicit drug detection, and skin cancer detection. THz imaging, due to the availability of phase-spectroscopic images, can be used for material identification. THz systems can image dry dielectric substances like paper, plastics, and ceramics.

The demand for high-speed wireless access is increasing at an exponential rate. Individual subscribers, the consumer market, and other communities are now demanding 100 Gb/s data for super hi-vision and ultra-high-definition TV, 5G, or even 6G mobile networks. THz wireless communication, although it incurs high atmospheric loss, has been identified to provide useful solutions for highly demanded services.

As noted previously, high-speed imaging and wireless communication for 5G technology are two potential research areas that need immediate attention for expanding the horizon of information technology. It should be noted in this connection that usual electronic devices are unable to enter into the THz range due to inherent limitations. Photonic devices also cannot go below the MIR range and cannot invade FIR and THz ranges. Semiconductor-based devices like QCLs and uni-travelling carrier photodetectors (UCT-PD) are being introduced for emitters and detectors, respectively, in this region. In addition, a breakthrough is needed in the area of THz modulation to develop long-distance communication and high-speed and high-resolution THz imaging with compressive sensing. However, it is difficult to design a high-speed modulation scheme for THz waves and thus, during the past 10 years, THz modulators have captured worldwide attention (Wang et al 2016).

15.8.2   THz modulator

The usually employed scheme integrates a conventional MM grown on a substrate which also supports an active device like a Schottky Barrier (SB) diode or a transistor like high-electron mobility transistor (HEMT). The planar form of the MM allows illumination to fall vertically on the surface. We mention first the basic principle involved in realizing an amplitude modulator.

15.8.2.1   A. Principle

Fig. 15.12 shows a schematic diagram of the MM, the tuning of which is electrically controlled. The equivalent circuit of this device is shown in Fig. 15.12b, which includes an inductor L, a capacitor C, and a resistance R_l which represents the loss in the circuit. The permittivity of the MM can be expressed as

$$\varepsilon = \varepsilon_b \left[1 - \frac{F^2 f^2}{f^2 - f_0^2 + i\gamma f} \right], f_0 = \frac{1}{\sqrt{LC}} \tag{15.38}$$

Here ε_b is the background permittivity, f_0 is the resonant frequency of the MM, γ is associated with loss R_l, and F is the filling factor of the geometry of the unit cell.

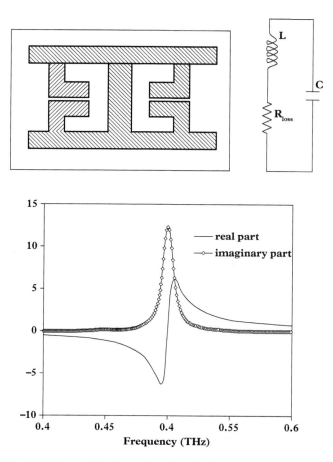

Figure 15.12 *Unit cell of planar SRR. Equivalent circuit. Calculated real and imaginary parts of permittivity.*

A plot of the real and imaginary part of the change in permittivity due to carriers, normalized by ϵ_b as a function of frequency is given in Fig. 15.13 with $f_0 = 0.5$ THz, $\gamma = 0.01$ THz, and F = 0.5. The plots are of the same nature of the plots of atomic susceptibility.

The principle of amplitude modulation lies on dynamic shifting of the resonant frequency so as to change the permittivity of the structure at the chosen frequency. The scheme is illustrated by the equivalent circuit shown in Fig. 15.13, in which the capacitance of the MM is shunted by another resistive element R_{sh}. The effect of this shunt resistor is to change the value of γ. Suppose the value changes from 0.1 to 0.5 THz. The plots of real and imaginary parts of susceptibility for this changed value are shown in Fig. 15.13 by thicker lines. Since the imaginary part of the permittivity is related to absorption coefficient, use of shunt loss element changes the absorption or transmission of the incident THz radiation and provides amplitude modulation.

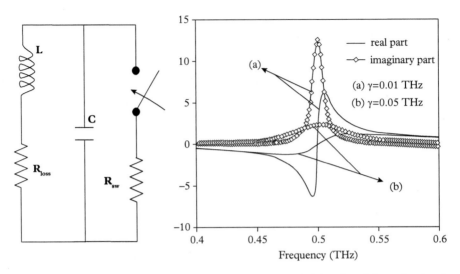

Figure 15.13 *(left) A conceptual circuit equivalent of a metamaterial-based modulator showing the use of shunt resistor. (right) Calculated real and imaginary parts of permittivity.*

Example 15.5 *Consider the values of normalized absorption at resonant frequency 0.5 THz with two values of damping coefficients $\gamma = 0.1$ THz and 0.05 THz, $F = 0.5$ in both cases. The values of absorption are 25 and 5, respectively. The change causes a corresponding change in the refractive index, which is utilized in modulation*

15.8.2.2   B. Structure

The previously mentioned principle of modulation was first utilized by Chen et al (2006). On the two sides of the planar MM structure are deposited two metal pads, one making ohmic and another making Schottky contacts with the substrate. A voltage applied between these contacts controls the substrate charge density near the split gate. This changes the value of the shunt resistance and alters the tuning frequency. This results in the change of absorption and hence alters the intensity of the THz radiation.

A scheme proposed by Rout and Sonkusale (2016) using a MM element based on electric-inductance (L) capacitance (C) (ELC) resonator, patterned using the top 2.1 μm thick gold metal is now described. A pseudomorphic high-electron mobility transistor (pHEMT) grown on SI GaAs substrate is placed underneath each split gap. The source and drain of the HEMT are connected to each side of the split gap. The gate-to-source/drain voltage (V_{GS}) controls the 2-dimensional electron gas (2DEG) channel charge between the split gap, thus electronically controlling the damping frequency. The channel charge offers the shunt resistance needed to alter the resonant frequency of the electrically controlled LC circuit (see also Rout 2016).

The average permittivity of the material may be expressed as

$$\varepsilon = \varepsilon_\infty \left[1 - \frac{f_p^2}{f^2 - f_0^2 + i\gamma_e f} \right]$$ (15.39)

where ε_∞ is the permittivity of the substrate material and f_p is the plasma frequency.

The operation of the modulator can be understood with the help of the equivalent circuit shown in lower part of Fig. 15.14. The HEMT offers an inductance and resistance combination in parallel to the capacitor.

When the pHEMT is 'off' ($V_{GS} = -1V$), the values for the left high-electron mobility transistor (LHEMT) and right high-electron mobility transistor (RHEMT) are negligible and the resonant frequency f_0 can be expressed in terms of the equivalent circuit parameters shown in Fig. 15.14 as

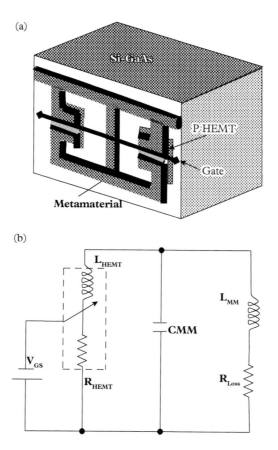

Figure 15.14 *Equivalent circuit of THz modulator using HEMT.*

$$f_0 = \frac{1}{2\pi\sqrt{L_{MM}C_{MM}}}$$

where C_{MM} is the capacitance associated with the split gap of the metamaterial and L_{MM} is the inductance associated with the current loop in each half of the metamaterial loop. The circuit elements may be calculated from the geometry of the unit. The damping factor is expressed as

$$\gamma_e = \frac{1}{2\pi}\frac{R_{loss}}{L_{MM}} \tag{15.40}$$

When the p channel high-electron mobility transistor(p-HEMT) is ON ($V_{GS} = 0$ V), a high mobility 2DEG channel is formed between source and drain and its conductivity is described by Drude model as

$$\sigma_{2D}(\omega) = \frac{\sigma_0}{1 + i\omega t} \tag{15.41}$$

where $\sigma_0 = n_{2D}e^2\tau/m_e$ is the standard expression for the dc conductivity σ_0. The 2DEG channel is a series combination of resistance $R_{HEMT} = 1/\sigma_0$ and an inductance $L_{HEMT} = \tau/\sigma_0$,

The equivalent circuit may be converted into a series parallel combination of R_E, L_E, and C_{MM}. The new resonant frequency and damping constants are expressed as

$$f_1 = \frac{1}{2\pi\sqrt{L_E C_{MM}}} \tag{15.42}$$

and $\gamma_{e1} = 2\pi(f_1)^2\frac{L_E}{R_E}$

In the structure used by Rout and Sonkusale, $f_1 \sim f_0$ and the damping is dominated by p-HEMT, so that $\gamma_{e1} \gg \gamma_e$,

The imaginary part of the permittivity is directly proportional to the absorption of the electromagnetic wave transmitting through the metamaterial sample.

Example 15.6 *We choose some realistic values given by Rout and Sonkusale. For $V_{GS} = 0V$: $f_0 = 0.55$ THz $\gamma_e = 0.15$ THz and for $V_{GS} = -1V$: $f_0 = 0.5$ THz, $\gamma_{e1} = 0.025$ THz.*

The plots of absorption for the two bias voltages are shown in Fig. 15.13 and they indicate changes in absorption or transmission through the MM. The primary reason for the modulation is due to the increase in the damping factor or loss in the ELC resonator due to the conductance in the p-HEMT that shunts the metamaterial capacitor C_{MM}.

The device offers element level control, so that higher switching speed and future opportunity of creating more exotic devices can be achieved. Such amplitude modulator can be put into an array to implement a THz spatial light modulator (SLM) without any moving parts. Arrays of such an MM-based SLM working at different frequencies can offer hyper-spectral imaging.

Other methods of realizing THz modulators include dynamically controlling the capacitance or the inductance of the split gap capacitance. For example, optically pumping the substrate may dynamically control the capacitance of the split gap. See Rout (2016) for useful references.

Different THz amplitude and phase modulators with type such as photoinjected or electrically controlled, materials like semiconductors, graphene, MoS_2, superconductors etc., and basic performance like modulation speed, depth of modulation, and references to the workers are given in Tables 1 and 2 in the review by Wang et al (2019).

15.9 Beam steering

Steering of beam by electronic method is well known in connection with radio detection and ranging (RADAR) systems and is attractive as it is quite fast, and avoids the heavy parts used by steering the beam mechanically. In phased array RADAR, a 1D or 2D array of antennas are used and electrical signals with progressive phase changes are introduced to each of the antenna elements. Due to constructive interference, the beam is directed along a specified direction in accordance with the phase difference introduced between the successive elements.

The same concept is applied to optical systems, an example of which is light detection and ranging (LIDAR). Development of compact and lightweight solid state beam steering system has assumed importance in recent years in view of push towards self-driving cars. The LIDAR systems used in these cars need rotating mechanisms to scan the surroundings and other objects. Electrically tunable metamaterials or metasurfaces can provide an inexpensive solution. The metamaterials are arranged in a phased array and by applying bias light can be directed at any desired angle. So far, highly doped semiconductors, transparent conducting oxides (TCO), and 2D materials like graphene or transition metal dichalcogenides have been used to realize the electrically tunable systems. Ultrathin metasurfaces, due to their exceedingly short interaction length with light, are attractive. However, resonant antenna structures made of Au or Ag lack CMOS compatibility.

An example of solid-state tunable metamaterial as developed by Morea et al (2018) will now be presented. They have used an array of nanopillars as the metamaterial, a 3D design for increased interaction length with light, composed of a highly doped semiconductor, an insulator, and a TCO to form a field effect structure. The structure is fabricated in this way. First, n-type germanium(Ge) with carrier density of 1.5×10^{19} cm^{-3} is epitaxially grown on Si wafers. With electron-beam lithography, the Ge layer is next etched with e-beam lithography with an Al mask to form nanopillars of diameter around 40 to 50 nm, periodicity of 110 nm, and height of approximately 500 nm. The Ge nanopillar array is then coated with a gate oxide and n-type al doped zinc oxide (AZO), which is used instead of a metal as the gate, because it offers both high optical transparency and electrical conductivity. The structure is thus essentially a semiconductor-oxide-semiconductor (SOS) capacitor that can accomplish the desired field effect.

The applied bias controls the charge carrier density of both Ge and AZO with the oxide layer in between them. A positive bias accumulates carriers in n-Ge, while depletes the AZO. A negative bias creates the opposite effect. The control is more effective as it is

3D in a gate-all-around structure in nanopillars, as compared to planar geometry. The changes in carrier density alter the plasma frequency and the permittivity. In terms of electrical equivalent circuit, the capacitor and resistor values are changed to change the resonant frequency. Using effective medium theory, it may be shown that the periodic array shows anisotropic permittivity characteristics of a MM.

The applied bias not only changes the resonant frequency but modulates the phase. The reflectivity of light wave and its phase are changed. Application of slightly different biases to the elements of the arrays of nanopillars, can make a phased array and steer the direction of light. The devices require low power and by decreasing total device area, high frequency operation typical of transistor speed can be achieved. At mid IR (2.5 and 11 μms), wavelength shift up to 240 nm and a high differential reflectance of 40% are measured by covering the voltage range from −4 V to 4 V; simulation indicates that a phase of 270° is possible. These results make the structure suitable for beam steering and other applications. In addition, as the fabrication process is CMOS compatible, the structures suit well for mass production using standard semiconductor foundries.

Problems

15.1 *Derive the expressions for real and imaginary parts of permittivity in an electrical plasma subject to damping. Give the plots of the two components as a function of frequency. Show that for vanishing damping, the expression for permittivity takes the form as given in Example 15.1.*

15.2 *Derive the expressions, Eqs. (15.22a) and (15.22b) for the effective parallel permittivity $\epsilon_\parallel$ and effective perpendicular permittivity $\epsilon_\perp$ for multilayers HMM by using the principles of continuity of electric field and electric displacement at the interface of two dielectrics.*

15.3 *Repeat the exercise as given in Problem 15.2 for an array of metallic NWs embedded in a dielectric.*

15.4 *Follow the general theory of all-angle negative refraction by indefinite media as given by Liu et al (2008). Derive the angles of refraction for the wave vector and the Poynting vector. Using $\varepsilon_x = 4.515$ and $\varepsilon_z = -2.530$ (corresponding to real parts of ε_x and ε_z for nanowires with the filling ratio $p = 0.227$ at 632.8nm wavelength), plot the two angles as the function of incidence angle. Show that negative refraction occurs for all incidence angles.*

15.5 *Give the detailed derivation of the permittivity tensor given by Eq. (15.26).*

15.6 *Derive the expressions for parallel and perpendicular components of permittivity for the all-dielectric/semiconductor asymmetric anisotropic metamaterial as given by Eqs. (15.34(a) and (b)).*

15.7 *Show that, for normal incidence, the angle of negative refraction becomes maximum for AAM and for normal incidence, $\theta = 0°$ and the angle of negative refraction reaches maximum value $\tan^{-1}\left[\dfrac{\epsilon_\parallel - \epsilon_\perp}{2\sqrt{\epsilon_\parallel \epsilon_\perp}}\right]$.*

15.8 *Optical parameters of materials used (at $\lambda = 1.55\ \mu m$): $\varepsilon_{air} = 1$, $\varepsilon_{SiO2} = 2.085$, $\varepsilon_{silicon} = 12.09$ $\varepsilon_{GaAs} = 11.38$, $\varepsilon_{germanium} = 18.28+0.05i$. Obtain plots of the angle of refraction vs. the angle of incidence for $30°$ orientation for air-Si, air-Ge, SiO_2-Si, and SiO_2-Ge SiO_2-GaAs.*

15.9 *Show that the maximum angle of $-ve$ refraction depends on the difference $\varepsilon_{\parallel} - \varepsilon_{\perp}$ in AAM.*

15.10 *Obtain a plot of the refraction angle of a pointing vector against angle of incidence for Si-SiO_2 multilayer for 30 degrees of an orientation angle.*

15.11 *Draw plots of $\varepsilon_{\parallel}$ and $\varepsilon_{\perp}$ versus wavelength using the data given in Example 15.3 and indicate the region over which $\varepsilon_{\perp}$ is negative.*

15.12 *Repeat the exercise in Problem 15.3 when the thicknesses of the InGaAs layer are 0.3 and 0.7 times the total thickness of the two layers.*

15.13 *Obtain plots for real and imaginary parts of perpendicular permittivity using the $GaAs/Al_{0.3}Ga_{0.7}As$ QW structure considered by Ginzberg and Orenstein. The parameters are well depth $\Delta E_c=230\ meV$, well width $L=10\ nm$, and the volume density of charge of $1\times10^{18}\ cm^{-3}$, dephasing rate of the dipolar transitions was $\gamma_{12} = 3\ meV$, and $\varepsilon_{back} =13$. The waveguide core is designed to contain about 30% QW layers.*

15.14 *Obtain plots of real and imaginary parts of refractive index using equivalent circuits of resonators with and without the shunt resistor. Use $f_0 = 0.5$ THz, $\gamma = 0.01$ and 0.05 THz, $F = 0.5$ and background permittivity $\varepsilon_b = 12$.*

15.15 *In the work of Rout and Sonkusale (2016), the following dimensions were used, 2.1 μm thick gold (Fig. 15.2(a)). Dimension of each element is 42 μm wide by 30 μm in height and they are repeated with a period of 55 μm × 40 μm. The line width of the metamaterial is 4 μm and the split gap is 3 μm.*
Use expressions for L, C, and R given in the paper and obtain the plots of absorption for HEMT in the off and on states.

15.16 *Calculate the plasma frequency and permittivity in Ge-oxide-AZO SOS capacitor used in the experiment by Morea et al (2018).*

References

Ashalley E, Ma C-P, Zhu Y S, Xu H-X, Yu P, and Wang Z-M (2021) Recent progress in chiral absorptive metamaterials. *Journal of Electronic Science and Technology* 19(100098): 25.

Basu P K (1997) *Theory of Optical Processes in Semiconductors: Bulk and Microstructures*, pp. 298–302. Oxford: Clarendon Press.

Boltasseva A and Atwater H A (2011) Low-loss plasmonic metamaterials. *Science* 331: 290.

Bose J C (1898) Twisted jute used for rotating the polarization of an electromagnetic wave. *Proceedings of the Royal Society* 63: 146.

Chen H-T, Padilla W J, Joshua M, O Zide O, Gossard A C, Taylor A J, and Averitt R D (2006) Active terahertz metamaterial devices. *Nature* 444(7119): 597–600.

Desouky M, Mahmoud A M, and Swillam Md A (2017) Tunable mid IR focusing in InAs based semiconductor hyperbolic metamaterial. *Scientific Reports* 7(15312): 7.

Engineer M H, Datta A N, and Nag B R (1967) Microwave faraday rotation in nickel-powder artificial dielectric. *Journal of Applied Physics* 38(2): 884–885.

Ginzburg P and Orenstein M (2008) Non-metallic LHM based on - + anisotropy in low D quantum structures. *Journal of Applied Physics* 103(083105).

Guo Z, Jiang H, and Chena H (2020) Hyperbolic metamaterials: From dispersion manipulation to applications. *Journal of Applied Physics* 127:(071101): 1–28.

Hoffman A J, Alekseyev L, Howard S, Franz K J, Wasserman D, Podolskiy V A, Narimanov E, Sivco D L, and Gmachl C (2007) Negative refraction in semiconductor metamaterials. *Nature Materials* 6: 946–950.

Hoffman A J, SridharA, Braun, P X, Alekseyev L, Howard S, Franz K J, Cheng L, Choa F-S, Sivco D L, Podolskiy V A, Narimanov E, and Gmachl C (2009) Midinfrared semiconductor optical metamaterials. *Journal of Applied Physics* 105(12): 2411.

Klar T A, Kildishev A V, Drachev V P, and Vladimir M. Shalaev V M (2006) Negative-Index metamaterials: Going optical? *IEEE Journal of Selected Top Quantum Electronics* 12(6): 1106–1115.

Li A, Singh S, and Sievenpiper D (2018) Metasurfaces and their applications. *Nanophotonics* 7(6): 989–1011.

Li K, Simmons E, Briggs A F, Bank S R, Wasserman D, Podolskiy V A, and Narimanov E (2020) Ballistic Metamaterials. *Optica* 7: 173–1780.

Lippmann G (1894) *Lippmann's and Gabor's Revolutionary Approach to Imaging.* Klaus Biedermann (ed.) Nobelprize.org. (accessed 6 December 2010).

Liu S, Sinclair M B, Keeler G A, Reno J L, and Brener I (2016) All-Dielectric metamaterials using III–V semiconductors. *Optics and Photonics News* 57(December).

Liu Y, Bartal G, and Zhang X (2008) All-angle negative refraction and imaging in a bulk medium made of metallic nanowires in the visible region. *Optics Express* 16(20): 15439–15448.

Metamaterial (2021) *Wikipedia, The Free Encyclopedia,* https://en.wikipedia.org/w/index.php?title=Metamaterial&oldid=1003301499 (accessed February 16, 2021).

Min L and Lirong Huanga L (2015) All-semiconductor metamaterial-based optical circuit board at the microscale. *Journal of Applied Physics* 118:013104.

Min L, Wang W, Huang L, Ling Y, Liu T, Liu J, Luo C, and Zeng Q (2019) Direct-tuning methods for semiconductor metamaterials. *Scientific Reports* 9: 17622.

Morea M, Zang, K, Kamins T I, Brongersma M L, and Harris J S (2018) Electrically tunable, CMOS-compatible metamaterial based on semiconductor nanopillars. *American Chemical Society Photonics* 5: 4702–4709.

Naik G and Boltasseva A (2010) Semiconductors for plasmonics and metamaterials. *Rapid Research Letters* 4: 295–297.

Naik G V, Kim J, and Boltasseva A (2011) Oxides and nitrides as alternative plasmonic materials in the optical range. *Optical Materials Express* 1(6): 1090–1099.

Naik G V, Liu J, Kildishev A V, Shalaev V M, and Boltasseva A (2012) All-semiconductor meta-material with negative refraction in the near-infrared QTh1A. *CLEO Technical Digest* ©OSA.

Oh S and Hess O (2015) Chiral metamaterials: Enhancement and control of optical activity and circular dichroism. *Nano Convergence* 2(24): 1–14. Available at: https://doi.org/10.1186/s40580-015-0058-2.

Padilla W J, Basov D N, and David R, Smith D R (2006) Negative refractive index metamaterials. *Materials Today* 9(7–8): 28–35.

Pendry J B (2000) Negative refraction makes a perfect lens. *Physical Review Letters* 85: 3966.

Pendry J B (2004) Negative refraction. *Contemporary Physics* 45: 191–202.

Pendry J B (2005) Negative refraction & the perfect lens. Available at: http://esc.u-strasbg.fr/docs/2010/lectures/StrasbourgTalk.pdf

Pendry J B, Holden A J, Robbins D J, Stewart W J (1999) Magnetism from conductors and enhanced non-linear phenomena. *IEEE Transactions on Microwave Theory and Techniques* 47(17): 2075–2084.

Pendry J B and Smith D R (June 2004) Reversing light with negative refraction. *Physics Today* 57(6): 37.

Podolskiy V A and Narimanov E (2005) Strongly anisotropic waveguide as a nonmagnetic left-handed system. *Physical Review B* 71:201101R.

Ramović S, Radovanović J, and Milanović V (2011) Tunable semiconductor metamaterials based on quantum cascade laser layout assisted by strong magnetic field. *Journal of Applied Physics* 110: 123704.

Rout S (2016) Active Metamaterials for Terahertz Communication and Imaging. Ph.D. Thesis, USA: Tufts University.

Rout S and Sonkusale S (2016) Wireless multi-level terahertz amplitude modulator using active metamaterial-based spatial light modulation. *Optics Express* 24(13): 14618–14631.

Sayem A A, Mahdy M R C, and Rahman Md S (2016) Broad angle negative refraction in lossless all dielectric or semiconductor based asymmetric anisotropic metamaterial. *Journal of Optics* 18(015101): 8.

Shcherbakov M R, Liu S, Varvara S, Zubyuk V, VaskinA, Polina P, Vabishchevich P, Keeler G, Pertsch T, Dolgova T V, Staude I A, Brener I, and Fedyanin A (2017) Ultrafast all-optical tuning of direct-gap semiconductor metasurfaces. *Nature Communications* 8(17): 1–6.

Shekhar P, Atkinson J, and Jacob Z (2014) Hyperbolic metamaterials: Fundamentals and applications. *Nano Convergence* 1(14): 1–17.

Shelby R A, Smith D R, and Schultz S (2001) Experimental verification of a negative index of refraction. *Science* 292: 77–79.

Smith D R et al (2000c) Transmission through a set of split-ring resonators exhibiting a stop band in the region where the permeability is negative. *Physical Review Letters* 84: 4184–4186.

Smith D R, Kolinko P, and Schurig D (2004) Negative refraction in indefinite media 1032. *Journal of the Optical Society of America B* 21(5): 1032–1043.

Smith D R, Padilla W J et al (2000a) The first negative index material, meta or otherwise!. *Physical Review Letters* 84(14): 23–26.

Smith D R, Padilla W J, Vier D C, Nemat-Nasser S C, and Schultz S (2000b) Composite medium with simultaneously negative permeability and permittivity. *Physical Review Letters* 84(18): 4184–4187.

Solymar L and Walsh D (2010) *Electrical Properties of Materials*, 8th edn. Chapter 15. Oxford, UK: Oxford University Press.

Sonkusale S R, Xu W, and Rout S (2014) Active metamaterials for modulation and detection. *Computers, Materials & Continua* 39(3): 301–315.

Taliercio T and Biagioni P (2019) Semiconductor infrared plasmonics. *Nanophotonics* 8(6): 949–990.

Veselago V, Braginsky L, Shklover V, and Hafner C (2006) Negative refractive index materials. *Journal of Computational and Theoretical Nanoscience* 3: 1–30.

Veselago V G (1968) The electrodynamics of substances with simultaneously negative values of ε and μ. *Soviet Physics Uspekhi* 10(4): 509–514.

Vuković N, Danicic A, Radovanovi, J et al (2015) Possibilities of achieving negative refraction in QCL-based semiconductor metamaterials in the THz spectral range. *Optical and Quantum Electronics* 47(4): 883–891.

Walsh D (1973) Artificial semiconductors. *Nature* 243: 33–35.

Wang Z, Cheng F, Winsor T, and Liu Y (2016) Optical chiral metamaterials: A review of the fundamentals, fabrication methods and applications. *Nanotechnology* 27(412001): 20.

Yonghua L, Pei W, Peijun Y, Jianping J, and Hai M (2005) Negative refraction at the interface of uniaxial anisotropic media, *Optics Communications* 246: 429–435.

16

Nanolasers

16.1 Introduction

It was mentioned in earlier chapters, laser actions in semiconductor were first reported almost simultaneously by four different groups. Continuous wave and room temperature operation of lasers were achieved by using double heterojunction structures. Subsequently, the performance of semiconductor lasers was improved in terms of lower threshold current and current density, higher modulation bandwidth, sharper linewidth, better temperature insensitivity, etc., by using improved growth technology, and quantum structures, both strained and unstrained. Different materials systems were also employed to cover a wide spectrum of electromagnetic (EM) spectra including ultraviolet (UV), visible, near and far infrared, and even the terahertz (THz) region. Table 16.1 in Basu et al (2015) mentions important milestones in the journey of semiconductor lasers.

Two approaches are followed in the development of lasers and other semiconductor-based photonic devices, namely, the top-down and the bottom-up methods. Some of the lasers developed by both the methods have been introduced in Chapters 7 and 8 of the present book. In addition to improving individual devices, another burning issue is integration with other photonic devices like modulators, waveguides, detectors, and above all with processing systems based on complementary metal-oxide semiconductor (CMOS) circuits. Integration of electronics and photonics on the same chip entirely grown on silicon (Si) substrate, or alternatively, on different platforms leading to hybrid systems, has been the subject of current research and development activities of both academia and industry. The Si-based electronic systems have shown remarkable progress in terms of this tiny footprint, very low power dissipation and the ultimate limit of high speed. The targets for integrated photonic systems are the same, but in this case, it is possible to achieve higher speed. The application areas of photonic integrated circuits (IC)s consist of course of optical fibre communication and the development of optical interconnects for high-performance computers and other equipment in data centres (Miller 2000, 2009; Deen and Basu 2012).

Though photonic devices cannot reach the size of today's electronic devices, due to fundamental limitations by diffraction, reduction of size of devices, in particular of lasers, brings forth advantages in power consumption and dissipation. The use of microcavities

Semiconductor Nanophotonics. Prasanta Kumar Basu, Bratati Mukhopadhyay, and Rikmantra Basu, Oxford University Press.
© P.K. Basu, B. Mukhopadhyay, R. Basu (2022). DOI: 10.1093/oso/9780198784692.003.0016

in nanostructured lasers in the form of vertical cavity surface emitting lasers (VCSELs) and other structures has proved to be effective. An important phenomenon observed in microcavities is the inhibition and enhancement of spontaneous emissions in solid states first predicted by Yablonovitch (1987) and then demonstrated by Yablonovitch et al (1988). The change in spontaneous emission in cavities was predicted by Purcell (1946). The effect can lead to the thresholdless operation of lasers, predicted by Yamamoto et al (1991). The strong light matter interaction in microcavities gives rise to interesting phenomena including Bose–Einstein condensation. A new kind of laser, known as polariton laser, has been developed as a consequence (Schneider et al 2013; Bhattacharyya et al 2014).

Important theoretical and experimental works related to the development of semiconductor photonic lasers starting from the proposal of VCSEL by Iga (1977) to announcement of a room-temperature polariton laser (Bhattacharyya et al 2014) have been listed in Table 16.1. Several materials and their combinations have been investigated by different workers in this connection. We have not included such systems in the table but mentioned only pioneering activities. As an exception, we have mentioned two reports on germanium tin (GeSn) alloy-based lasers. This group IV alloy shows direct band gap nature for x > 0.08 or under strain and can be grown on Si platforms, thereby opening the prospect for electronic-photonic integration. Further details of Group IV lasers may be found in chapter 14 of Basu et al (2015).

The ultimate limit for length in photonic lasers, as governed by diffraction, is $\lambda/2\eta$, where η is the refractive index. This means that semiconductor lasers can never reach the below 10 nm size of field-effect transistors (FET)s. In this connection, spasers and plasmonic lasers exploiting localized or propagating surface plasmon hold promise for sub-wavelength operation and such lasers with lengths near 100 nm have already been reported. The basic working principle, structures, and materials used, and their characteristics are already discussed in Chapter 14.

In the following sections, progress in achieving small lasers, both photonic and plasmonic nanolasers will be discussed. First, the parameters of lasers will be summarized and the problems faced by reduction of size will be pointed out. The progress in nanolasers in a chronological order is then outlined. The expressions for threshold by including Purcell effect will then be mentioned. Plots of normalized threshold against the ratio of cavity and material loss with spontaneous emission factor as a parameter are then given to make a comparative analysis of all important types of nanolasers. Some figures-of-merit and performance parameters of nanolasers, namely, influence of laser volume on power and threshold, dynamics, spectral linewidth, and energy density are then introduced. Some applications of nanolasers as integrated optical interconnects, the performance of such interconnects against current, and future demand of speed, energy per bit, and electrically injected device are then discussed. Applications in some other areas like sensing and in biological areas are finally mentioned. The following sections closely follow the reviews by Ma and Oulton (2019) and by Ning (2019).

Table 16.1 *Outline of development of a semiconductor photonic laser.*

Year	Work	Reference
1977	VCSEL	Iga (1977)
1982	Concept of QWR and QD Lasers	Arakawa and Sakaki (1982)
1987	Inhibited spontaneous emission	Yablonovitch (1987)
1988	Demonstration of inhibited and enhanced spontaneous emission in double heterostructures	Yablanovitch et al (1988)
1991	Proposal for thresholdless laser	Yamamoto et al (1991)
1991	VCSEL	Jewel et al 1991
1992	Microdisk lasers	McCall et al 1992
1994	QD lasers	Kirstaedter et al (1994)
1994	Quantum Cascade Laser	Faist et al (1994)
1999	Photonic crystal laser	Painter (1999)
2001	Single nanowire laser	Johnson et al (2001)
1990–96	Blue Laser	Akasaki, Amano and Nakamura
2013	Polariton laser	Schneider et al (2013)
2014	Polariton laser at room temperature	Bhattacharyya et al (2014)
2019	GeSn VCSEL	Huang et al (2019)

16.2   Parameters of lasers

In all the treatments of determining the ultimate limits on laser size and their effects on the device characteristics of small lasers (Ning 2010; Chang et al 2010; Hill and Gather 2014), a generalized model with a Fabry–Perot (FP) resonator as shown in Fig. 16.1 is employed. The active material enclosed by the resonator mirrors has a length L, thickness d, and volume V_a, refractive index η, and material loss coefficient α_i. The resonator cavity is resonant at a free space wavelength λ_0, and the end mirrors have reflectivities R_1 and R_2.

For self-sustained oscillation, the electric field amplitude E_0 at any point retains the same value after a round trip through the resonator. In other words, one may write

$$E_0 \sqrt{R_1 R_2}\, exp[(g_m - \alpha_i)L]exp(i2\eta L/\lambda_0) = 1 \qquad (16.1)$$

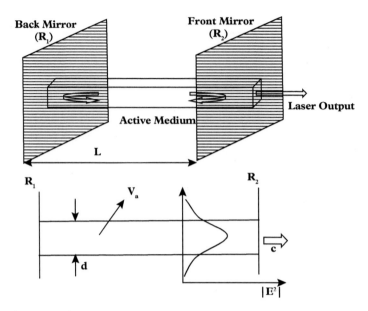

Figure 16.1 *Schematic diagram of a FP resonator enclosing a gain medium. The length, width, active volume of the medium, mode volume, and electric field profile across the width of the medium are shown.*

where g_m is the modal gain index. The second exponential dictates phase condition and it should equal unity, giving

$$L = \frac{\lambda_0}{2\eta} m, \quad m = 1, 2, 3, \dots. \tag{16.2}$$

The shortest length of the cavity is thus half a wavelength with mode number $m = 1$.
The amplitude requirement in Eq. (16.1) leads to the condition

$$g_m = \alpha_i - \frac{1}{2L} \ln(R_1 R_2) \tag{16.3}$$

from which the length may be expressed as

$$L = \frac{-\ln(R_1 R_2)}{2(g_m - \alpha_i)} \tag{16.4}$$

The modal gain g is related to the material gain g_a and the mode confinement factor Γ by

$$g_m = \Gamma g_a \tag{16.5}$$

A typical gain spectra for a double heterostructure (DH) laser, as displayed in Fig. 5.9, shows that for a fixed injected carrier density, *n*, the gain increases first with photon energy, attains a maximum value g_{max}, and then decreases. The maximum gain increases with injected carrier density and assuming linear variation, it may be expressed as

$$g_{max} = a(n - n_{tr}) \tag{16.6}$$

where n_{tr} is the transparency carrier density for which $g_{max} = 0$, $a = \partial g_{max}/\partial n$ is called the differential gain.

At steady state, the current density is related to the spontaneous emission rate R_{sp}, total recombination lifetime τ_r, and spontaneous radiative emission lifetime τ_s by

$$\mathcal{J} = edR_{sp} = \frac{end}{\tau_r} = \frac{end}{\eta_i \tau_s} \tag{16.7}$$

where $\eta_i = \tau_r/\tau_s$ is the internal quantum efficiency. Using Eqs. (16.6) and (16.7)

$$g_{max} = \frac{a\eta_i \tau_s}{ed}(\mathcal{J} - \mathcal{J}_{tr}) \tag{16.8}$$

The transparency current density is therefore $\mathcal{J}_{tr} = edn_{tr}/\eta_i \tau_s$.

At threshold, the modal gain $g_m = \Gamma g_{max} = \alpha_i + \alpha_{mir}$, that is, maximum gain should equal the total loss comprising the material and mirror losses. Using Eq. (16.9), the threshold current density of semiconductor laser is expressed as

$$\mathcal{J}_{th} = \mathcal{J}_{tr} + \frac{ed}{\Gamma \eta_i \tau_s a}\left[\alpha_i + \frac{1}{2L}\ln\left(\frac{1}{R_1 R_2}\right)\right] \tag{16.9}$$

Example 16.1 *We consider a gallium arsenide (GaAs) DH laser having the following values of parameters: $L = 200~\mu m$, $d = 0.2~\mu m$, $R_1 = R_2 = 0.32$, $\alpha_i = 10^3~m^{-1}$, $n_{tr} = 1.8 \times 10^{24}~m^{-3}$, $\Gamma = 0.02$, $\tau_s = 1~ns$, $\eta_i = 0.8$, and $a = 7\times10^{-20}~m^2$. The transparency current density is $0.72 \times 10^8~Am^{-2}$. The mirror loss is $0.57\times10^4~m^{-1}$. The second term in the right-hand side of Eq. (16.9) is $16.75 \times 10^7~Am^{-2}$. Therefore, the threshold current density $\mathcal{J}_{th} = 23.95 \times 10^7 Am^{-2}$. If the width of the laser is $w = 10~\mu m$, so that the area is $10 \times 200~\mu m^2$, the threshold current is $0.48~A$.*

The photon lifetime, τ_p, in the absence of lasing is related to the total loss in the cavity and the quality factor Q of the cavity as

$$\frac{1}{\tau_p} = c\frac{\alpha_i + (1/2L)\ln(R_1 R_2)}{\eta} = \frac{2\pi c}{Q\lambda_0} \tag{16.10}$$

At threshold the modal gain is related to photon lifetime by

$$\frac{1}{\tau_p} \sim \frac{c\Gamma g_a}{\eta} \tag{16.11}$$

One may note that Eq. (16.2), signifying phase condition for self-sustained oscillation with smallest mode number $m = 1$, cannot be used for minimization of laser length, as the calculated laser length is too small. This may be clear if we look into Eq. (16.4) for the amplitude requirement. The minimum length is limited by the gain coefficient g_m as well as the reflectivities which determine the mirror loss. Increasing L reduces the mirror loss. It is the gain limit which puts a constraint on the length. To achieve lower length, mirror losses are to be reduced, Q factor of the cavity should be increased, and modal gain should be maximized by increasing the mode confinement factor Γ and material gain g_a.

As the waveguides in semiconductor lasers are dielectric type, the small difference between refractive indices of active and cladding layers causes leakage of the optical field into the cladding layers. For small thickness d of the active layer, the mode confinement factor is drastically reduced due to this leakage, as illustrated by the following example for a DH laser.

Example 16.2 *We show the approximate variation of mode confinement factor Γ with thickness d of the active (a) layer InGaAs with indium phosfide (InP) as the cladding (cl) layer in a InP-InGaAs-InP double heterostructure. An approximate formula $\Gamma = \frac{\beta^2}{2+\beta^2}$, where $\beta = \frac{2\pi d}{\lambda_0}\sqrt{\eta_a^2 - \eta_{cl}^2}$ is used. Taking the values $\eta_a = 3.77$, $\eta_{cl} = 3.544$ and $\lambda_0 = 1.55\ \mu m$, we find $\Gamma = 0.69$, 0.12 and 0.033, respectively, for d = 0.4, 0.1, and 0.05 μm. The calculation clearly indicates that mode confinement factor decreases with the decreasing transverse dimension.*

The photon lifetime defined by Eqs. (16.10) and (16.11) is an important device parameter for optimizing the speed of operation of the laser (Hill and Gather 2014). Higher speed may be attained by reducing the length of the laser. In communication, on-off keying is employed, that is, the modulating pulse turns the laser on or off. There is always some turn on and turn off delay between the times when the pulse is applied or switched off and the laser emission is turned on and off. The turn-off delay is more severe, and both the delays are determined by the photon lifetime τ_p. However, for direct current modulation like on off keying (OOK) in communication, factors like non-linear gain compression is responsible for limiting the modulation bandwidth, or bit-rate, before the ultimate limit by τ_p is reached. Shorter photon lifetime is, however, related to increased material and mirror losses, thus increasing the demand for higher modal gain.

In addition, small cavity significantly changes the rate of spontaneous emission which in turn increases the modulation bandwidth when the device is used as a light-emitting diode (LED).

A few other desirable laser characteristics are discussed now.

Low power operation: the power output is given by

$$P_{out} = (h\nu/e)(\mathcal{J} - \mathcal{J}_{th})$$
(16.12)

It is therefore desirable to make the threshold current density small for low power operation. From Eq. (16.9) and (16.10) high values of photon lifetime and mode confinement factor are therefore needed. With a small active volume this will ensure a small pump energy reaches the threshold. This will of course reduce the power output, but in some cases, low power output with high conversion efficiency is needed (high slope efficiency). There must be a trade-off between internal losses and mirror losses to balance the threshold and slope efficiency.

Spectral purity: for certain applications, spectral purity, that is very narrow linewidth of emission, is a prime requirement. In such cases, a high value of Q is desirable, as the linewidth varies inversely with Q.

16.3   Progress in nanolasers

A chronological listing of different kinds of nanolasers, showing year of announcement, type, feedback mechanism employed, device size, gain medium, operating wavelength, pumping mechanism, temperature of operation, and threshold pump power has been given by Ma and Oulton (2019). We reproduce in Table 16.1 such a coverage of important milestones in the development of nanolasers, starting from the typical vertical cavity surface emitting lasers (VCSEL) and ending with the report in 2018 on hyperbolic metacavity.

A few recent addition to the entries in Table 16.1 is given in Table 16.2, which is by no means exhaustive.

16.4   Threshold pump power of nanolasers: Purcell effect

We now discuss some useful characteristics of nanolasers, following the review by Ma and Oulton (2019), which is quite exhaustive. In our discussions, however, we focus on fewer points. The intrinsic merit of nanolasers lies in their unique property to localize EM fields simultaneously in frequency, time, and space.

The expressions given in the previous section apply to semiconductor lasers in which the active medium is either a bulk (DH configuration) or a quantum nanostructure. However, the use of microcavity structure modifies the spontaneous emission rate via the Purcell effect and the equations are modified as indicated in Section 11.10. An unambiguous definition of threshold emerges from the studies, which states that rates of spontaneous and stimulated emission into the laser mode should be equal at the threshold condition. The following general expression for the threshold power of a laser has been derived by Wang et al (2017).

$$P_{th} = \frac{\hbar\omega}{\eta_a A} \frac{(1+\beta)}{2} \left[\frac{\gamma}{\beta\Gamma} + F\frac{2n_{tr}V_{phy}}{\tau_0} \right] \tag{16.13}$$

Table 16.2 *Progress in nanolasers.*

Year	Type	Feed back	Size (μm)	Gain Medium	λ (μm)	Pump	T(K)	Threshold	Ref
–	Typical VCSEL	F-P	D = 10 L = 1	InGaAs/ GaAs	0.98	El-CW	Room Temp	~ 1 kW cm^{-2}	Coldren et al (2012)
2007	Metallic-coated pillar	F-P	D = 0.21 L = 2	InGaAs	1.4	El-CW	77 K	~ 26 kW cm^{-2}	Hill et al, (2007)
2009	Spaser	L-SPR	D = 0.044	Dye	0.53	Opt-P	Room Temp	–	Noginov et al (2009)
2009	Plasmonic nanowire	F-P	D = 0.1 L = 10	Cds	0.49	Opt-P	10 K	~ 100 MW cm^{-2}	Oulton et al (2009)
2009	Gap plasmon	F-P	t = 0.09 L = 6	InGaAs	1.5	El-CW	10 K	~ 20 kW cm^{-2}	Hill et al (2009)
2010	Plasmonic nanodisk	W-G	t = 0.235 D = 0.9–1,2	InAsP MQWs	1.3	Opt-P	8 K	~ 120 kW cm^{-2}	Kwon et al (2010)
2010	Metallic-coated disk	W-G	t = 0.48 D = 0.49	InGaAsP	1.43	Opt-P	Room Temp	~ 70 kW cm^{-2}	Nezha et al (2010)

Year	Device	Cavity	Dimensions (µm)	Material		Pumping	Temperature	Threshold	Reference
2010	MIM nanopatch	D-R	$t = 0.23$ $D = 0.4–0.62$	InGaAsP	1.4	Opt-P	78 K	~ 80 kW cm^{-2}	Yu et al (2010)
2011	Plasmonic nanosquare	WG	$T = 0.045$ $L = 1$	CdS	0.5	Opt-P	Room Temp	~ 20 GW cm^{-2}	Ma et al (2011)
2012	Plasmonic coaxial	FP	$D = 0.2$ $H = 0.21$	InGaAsP MQWs	1.4	Opt-CW	4.5 K	–	Khajavikhan et al (2012)
2012	Plasmonic nanowire	FP	$D = 0.06$ $L = 0.48$	InGaN	0.51	Opt-CW	78 K	~ 3 kW cm^{-2}	Lu et al (2012)
2013	Metallic-coated pillar	FP	$L = 1.39$ $W = 1.15$ $H = 1.7$	InGaAs	1.59	El-CW	Room Temp	~ 100 kW cm^{-2}	Ding et al (2013)
2014	Plasmonic nanowire	FP	$D = 0.05$ $L = 0.2$	InGaN	0.468–0.642	Opt-CW	7 K	~ 100 kW cm^{-2}	Lu et al (2014)
2014	Metallic-coated disk	WG	Cavity volume: 3 µm^3	InGaAs	0.015	El-CW	77K	~ 300 kW cm^{-2}	Gu et al (2014)
2014	Plasmonic nanowire	FP	$D = 0.1$ $L = 15$	GaN	0.375	Opt-P	Room Temp	~ 3 MW cm^{-2}	Zhang et al (2014)

Continued

Table 16.2 *Continued*

Year	Type	Feed back	Size (μm)	Gain Medium	λ (μm)	Pump	T(K)	Threshold	Ref
2015	Plasmonic nanowire	FP	D = 0.15 L = 5	GaAs-AlGaAs	0.8	Opt-P	8K	~1 kW cm^{-2}	Ho et al (2015)
2016	Plasmonic nanowire	FP	D = 0.022 L = 1-4	ZnO	0.38	Opt-P	Room Temp	~100 MW cm^{-2}	Chou et al (2016)
2017	Spaser	L-SPR	D = 0.022	Dye	0.528	Opt-P	Room Temp	~3 MW cm^{-2}	Galanzha et al (2017)
2018	Plasmonic pseudo-wedge	FP	Wedge Length = 0.08	ZnO	0.37	Opt-P	77K	~55 MW cm^{-2}	Chou et al (2018)
2018	Hyperbolic metacavity	W-G	L = 0.2	AlGaN	0.29	Opt-P	Room Temp	~90 kW cm^{-2}	Shen et al (2018)

Reproduced with permission.

Source: Table 1 in Ma and Oulton (2019) Applications of nanolasers. *Nature Nanotechnology 14: 12–22.* copyright Springer Nature 2019.

Here, η_a is the fraction of pump power absorbed by the gain medium, $\hbar\omega$ is the energy of the emitted photons, A is the area of the pump beam, γ is the cavity loss rate, excluding the intrinsic loss in the gain medium, Γ is the overlap factor between the optical mode and the gain region, V_{phy} is the volume of the gain medium; n_{tr} is the transparency carrier density, and $2n_{tr}$ is the density to create population inversion n_{inv}, τ_0 is the natural spontaneous emission lifetime, F is the Purcell factor, and β is the spontaneous emission factor.

The first term within square brackets in Eq. (16.13) is due to cavity mode loss, while the second term accounts for the intrinsic loss of the gain medium. Defining a term ζ for the ratio of the two loss terms as

$$\zeta = \frac{\text{cavity loss}}{\text{material loss}} = \frac{\gamma\tau_0}{\beta F n_{inv} V_{phy}}$$

and

$$R_{th} = \frac{\eta_a A P_{th}}{\hbar\omega}$$

where R_{th} is the rate of photon generation at the threshold and $\Gamma R_{th}/\gamma$ being the normalized threshold pump rate, Eq. (16.12) may be written as

$$\frac{\Gamma R_{th}}{\gamma} = \frac{(1 + \beta^{-1})(1 + \zeta^{-1})}{2} \tag{16.14}$$

At the minimum pump rate $R_{th,min} = \gamma/\Gamma$, the photons must be supplied to the cavity at the same rate as the rate by which they are lost from the cavity.

The normalized threshold pump rate for any laser may now be determined by the factors β and ζ. The threshold can be minimized by making β approaching 1, that is, by making Purcell effect stronger. It can also be minimized by maximizing ζ, so that cavity loss far exceeds material loss. The factor $(1 - \beta)$ measures the loss in energy due to coupling to non-lasing modes and ζ accounts for the energy spent in creating population inversion in the medium.

A parameter $\overline{N} = \Gamma n_{inv} V_{phy}$ is defined to enable ζ to account for various types of gain media. This parameter $\overline{N}$ is the number of excited centres needed to make the gain medium transparent to the cavity mode. Table 16.3 gives the values of different parameters for different types of lasers. A value $\tau_0 = 10^{-9}$ s is chosen, a typical value for most inorganic gain media.

The normalized threshold as a function of inverse loss ratio ζ^{-1} are plotted in Fig. 16.2 for different values of spontaneous emission coupling factor β as a parameter. As may be seen from Table 16.3, different lasers possess parameter values covering different ranges, and therefore they will have some areal distribution in the diagram. The areas covered by spasers and plasmonic nanolasers in the diagram are shown. The threshold energy consumption of a laser is determined by $\zeta^{-1} \propto V_{phy}$. Reducing the volume of the laser always has a benefit provided that $\zeta < 1$.

Table 16.3 *A few examples of recent work in a nanolaser cum spaser.*

Year	Type	Feed back	Size (μm)	Gain Medium	λ (μm)	Pump	T(K)	Threshold	Ref
2021	Plasmonic nanowire	FP		ZnO		Opt	300		Wang et al (2021)
2021	Photonic crystal	FP				Opt	300		Parkhomenko et al (2021)
2021	Microdisk	WGM	300 nm	InP	870	Opt-P	300	2pJ/pulse	Tiwari et al (2021)

FP: Febry–Perot; **L-SPR**: Localized SPR; **WG**: Whispering gallery; **DR**: Dipole Resonance
D: Diameter; L: Length; t: Thickness; W: Width; H: Height
El-CW: Electrical, continuous wave; **Opt-CW**: Optical, continuous wave; **Opt-P**: Optical, pulsed
λ: Emission Wavelength

Table 16.4 *Parameters for typical small laser systems used to estimate the ratio of cavity and material losses,* ζ.

Type	γ (s−1)	βF	$\overline{N}$	ζ
Spaser	10^{14}	10^2	10^3	1
Plasmonic nanolaser	10^{13}–10^{14}	1–10	10^5	10^{-2}–1
Nanowire laser	10^{12}–10^{13}	10^{-2}–10^{-1}	10^6	10^{-2}–1
Photonic crystal laser	10^{11}–10^{12}	10–10^2	10^3	10^{-3}–10^{-1}
VCSEL	10^{11}–10^{12}	10^{-4}–10^{-3}	10^5	1–100

Reproduced with permission from, Applications of Nanolasers. *Nature Nanotechnology* 14: 12–22. Copyright Springer Nature 2019.

Example 16.3 *We use the parameter values for VCSEL from Table 16.3:* γ *(s⁻¹)* $= 10^{12}$, $\beta = 10^{-4}$, *and* $\zeta = 10$. *The normalized threshold from Eq. (16.13) is* 5.5×10^{3}. *This value falls in the area covered by VCSELs around the curve in Fig. 16.2 (not shown for VCSELs, but shown for spasers and plasmonic nanolasers).*

It may be noticed from the table that $\zeta < 1$ for semiconductor lasers and material loss predominantly determines the threshold. Lowering of threshold may be accomplished by maximizing ζ. When $\zeta \rightarrow 1$, maximum advantage is achieved. By using Quantum Well (QW) or Quantum Dots (QD) as the gain medium, ζ may be increased beyond 1. As $\zeta^{-1} \propto \overline{N}$, the energy consumption by a laser at threshold may be reduced by reducing V_{phy}, n_{inv} or overlap factor Γ. It may also be noted from Fig. 16.2, that spasers operate under near-optimal conditions for energy consumption at threshold, that is, with

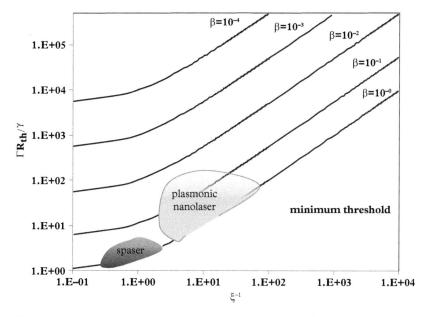

Figure 16.2 *Plots of normalized threshold as a function of the ζ for different values of β as a parameter. The areas covered by spacers and plasmonic nanolasers indicate the ranges of threshold energy, β and ζ.*

$\zeta \approx 1$ and $\beta \approx 1$. However, the cavity loss rate is quite high, $\gamma \approx 10^{14}$ s^{-1} and spasers need to be operated under pulsed condition to avoid excess heating.

16.5   Intrinsic merits of nanolasers

In this section, a few important characteristics of nanolasers that include power output and threshold, laser dynamics, spectral linewidth, and energy density will be discussed.

16.5.1   Influence of laser volume on power consumption and threshold

The fact that a reduction of laser volume leads to reduced power consumption and laser threshold and at the same time leads to faster operation has been known for a long time. Small physical volume reduces the number of cavity modes and reduces the coupling to non-lasing modes, thereby increasing β. The larger the value of β is, the smaller is the pump power needed to reach threshold, and for the ideal situation $\beta = 1$, thresholdless operation may be achieved. The normalized threshold pump powers calculated with different values of β and ζ are shown in Fig. 16.2. The advantages of spasers and plasmonic nanolasers having sub-wavelength dimensions are clearly indicated there.

16.5.2 Nanolaser dynamics

Due to strong confinement of modes in a nanocavity interaction between gain medium and cavity modes gets stronger. The role of Purcell effect is to enhance the spontaneous emission rate by the Purcell factor F, so that the rate is modified as $\tau = \tau_0/F$, where τ_0 is the natural lifetime. In addition, the emission lifetime into a particular mode is changed and becomes τ_0/F_m, which in turn modifies β. Two simple expressions for the spontaneous emission factor β for two different limits have been developed by van Exeter et al (1996). In one such, the cavity resonance is broader (low Q) than the linewidth of the emitter. In this case, $F_m \propto Q/V_m$, where V_m is the effective optical volume of the laser mode. In the other extreme, the cavity Q is quite high, so that the emission linewidth is broader than the cavity linewidth. In that case, F_m is determined by emission linewidth and V_m. For semiconductor lasers, linewidth is usually narrower than resonance linewidth of cavities with Q < 100.

Example 16.4 *We assume Q = 100 and since $Q = \frac{\nu}{\Delta\nu} = \frac{\lambda}{\Delta\lambda}$, the linewidth at 1550 nm wavelength. The linewidth of earlier VCSELs is about 2 nm (Michalzik and Ebeling 2003), and very narrow linewidth VCSELs in kHz–MHz ranges have been reported recently. Even the 2 nm linewidth is less than 15 nm linewidth of the cavity. The Purcell factor $F_m \propto Q/V_m$ may therefore be used.*

It has been shown earlier that enhanced spontaneous emission into the lasing modes increases the speed of response of nanolasers and nanoLEDs (Altug et al 2006; Lau et al 2009; Ni and Chuang 2012), which has also been demonstrated experimentally. However, a high Q cavity and resulting narrow cavity resonance are not advantageous, as a long photon lifetime lowers the modulation bandwidth of a laser.

Example 16.5 *The relationship between the Q of a cavity and photon lifetime τ_{ph} is $Q = \omega\tau_{ph}$ and from this the linewidth (FWHM) is $\Delta\nu_{1/2} = \frac{\nu}{Q} = 1/2\pi\tau_{ph}$ (Yariv and Yeh 2007).*

16.5.3 Spectral linewidth

A very narrow linewidth of about 0.3 nm (100 GHz) at 700 nm has been observed for a metal-based nanolaser (Wang et al 2018). Though this value is comparable to the linewidths for recent VCSELs, it should be noted that small gain volume and high spontaneous emission noise due to small number of modes in nanolasers act as a deterrent for achieving narrower linewidth. Even then, nanolasers are more coherent than incoherent light sources, and with their capability of localizing strong light field in a narrow space are suitable for a variety of sensing applications.

16.5.4 Nanolaser energy density

The optical mode volume V_m is related to the local optical density-of-states (DOS) and determines to what extent the light-matter interaction is enhanced. Optical energy is delivered within this volume by the light generated within the physical gain volume V_{phy}, and the ratio V_{phy}/V_m assumes importance.

In a semiconductor laser operating in continuous wave (CW) mode the output power may be expressed as $P = \eta n h \nu V_{phy}/\tau_s$, where η is the external quantum efficiency, n is the internal carrier density, $h\nu$ is the energy of emitted photon, and τ_s is the stimulated recombination time. Above threshold, n is clamped to the threshold carrier density and τ_s depends on the pumping condition.

Example 16.6 *The emission power is expressed as $P = \frac{\eta n h \nu}{\tau_s} V_{phy}$. We use $\eta = 1$, $h\nu = 0.8\ eV$, corresponding to $\lambda = 1550\ nm$, $n = 5.10^{18}\ cm^{-3}$ and $\tau_s = 100\ ps$ and $V_{phy} = 0.1\lambda^3$. The calculated power is 2.38 mW.*

The variation of output power of a nanolaser as a function of gain physical volume using the parameters given in Example 16.6 is shown in Fig. 16.3 for different values of τ_s as parameters. As the physical volume is reduced, the lifetime is also reduced due to Purcell factor F_m and the output power is increased.

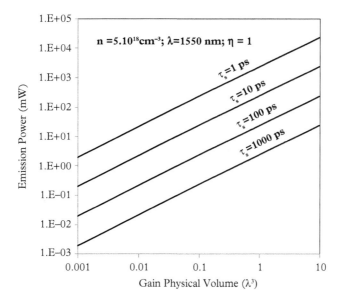

Figure 16.3 *The dependence of the emission power of a laser on its gain physical volume for different stimulated recombination lifetime, τ_s. Operation at 1550 nm, 100% quantum efficiency, and a threshold carrier concentration of $5 \times 10^{18}\ cm^{-3}$ is considered.*

When the nanolaser is pumped by ultrashort optical pulses of duration less than τ_s, gain switching phenomena lead to ultrashort optical pulses. The intracavity energy density for a single pulse may be estimated by using $U = nh\nu V_{phy}/V_m$. As plasmonic nanolasers have $V_{phy}/V_m \gg 1$, while the ratio is about unity in conventional lasers, a pulse with very high optical energy density over a period τ_s may be obtained. Energy densities $U > 50\,\text{J.cm}^{-3}$ and peak pulsed intensities $I > 10^{11}\text{W.cm}^{-2}$ may be obtained.

This high local field intensity in nanolasers makes feasible applications like data storage, sensing, imaging, optical probing, and spectroscopy in a near field.

16.6   Optical Interconnect

There is an exponentially rising demand for global internet traffic, the current rate of which is about 1 Zb (10^{21} bytes) per year (Cisco 2017). Associated with this is the growth of energy consumption, which must be restricted to sustain future data networks. At present, VCSELs provide best solution for short distance optical interconnects. VCSELs have the largest dimension to a few μms with energy consumption lowered to fJ/bit and are the favoured replacement for electrical interconnects in data centres and supercomputers. Nanolasers in on-chip integrated optical interconnects must have low power consumption, sufficient output to maintain large signal-to-noise ratio (SNR), high modulation rate, efficient coupling to waveguides, continuous wave (CW) room temperature operation by electrical injection.

The following paragraphs will describe briefly how the challenge of meeting all these stringent requirements are simultaneously met.

16.6.1   Energy per bit for a nanolaser

Optical interconnects aiming to replace mature on-chip electrical interconnects, must have total system energy loss less than 1 pJ/bit, with drive laser's energy consumption nearing 10 fJ/bit. The energy consumption by a laser consists of the energy needed to operate it above threshold (see Eq. (16.12)) and for modulation, due to the modulation format and to link budget (Senior 1993).

Example 16.7 *In order to have an idea of energy/bit in a nanolaser, consider that the threshold power is 10 kW.cm^{-2}, a diffraction limited area of $\lambda^2/4$, refractive index 3.26, and operating wavelength 1550 nm. The threshold power consumption is 5.6 μW. For 10 Gb.s^{-1} data rate, the energy consumption is 0.56 fJ/bit.*

With further reduction in device volume, the threshold is first reduced, but for a smaller volume there is an increase (see Fig. 14.8). This may be due to rising metallic loss. A compromise between the device volume and cavity loss is therefore needed.

The output power of a communication laser is simply the energy per bit multiplied by the modulation speed or bit rate. At a specified bit rate, the output power must be less than the prescribed limit 1 fJ/bit and above the thermal noise limit.

An estimate of thermal noise limit, that is, the minimum number of photons need to exceed the thermal noise, is presented in the following subsection.

16.6.2   Thermal noise limit

In simple digital communication systems, known as on-off keying, analog signals are converted into digital form containing bit 1 (full power) and bit 0 (no power). In fibre optic communication systems, the bits represent optical pulses (1) and no pulse (0). The bit stream is received by the photodetector in the receiver and then converted into corresponding electrical pulses. Errors usually occur in receiving the pulses due to poor signal power and noise in the receiving system. A simple analytical theory exists to estimate the bit error rate (BER) in terms of SNR, which in turn gives the idea of the number of photons to be incident on the photodetector.

Assuming that the probability distribution functions of receiving the bits are Gaussian in nature and that 1s and 0s are transmitted in equal numbers, the BER can be expressed in terms of SNR as (Senior 1993)

$$BER = \frac{1}{2} erfc \left[\frac{(S/N)^{1/2}}{2\sqrt{2}} \right] \tag{16.15}$$

where *erfc* is the complimentary error function, S and N represent respectively the signal and noise power.

Example 16.8 *Assuming BER = 10^{-9}, the value of the argument in erfc is 4.24. This yields SNR = 144.*

16.6.3   Estimate for an avalache photodetector

The signal current in an Avalache Photodetector (APD) is multiplied by the multiplication factor M and when I_p is the primary photocurrent caused by the average number of photons, p_{av} incident on the APD. The signal power is therefore

$$S = (MI_p)^2 \tag{16.16}$$

Asuming that shot noise exceeds thermal noise in the detector, the mean square shot noise current is expressed as

$$\overline{i_n^2} = 2eBI_p M^2 F(M) \tag{16.17}$$

where B is the bandwidth of the detector-amplifier combination, and F(M) is called the excess noise factor caused by random multiplication process in an APD. The SNR may thus be expressed as

$$\frac{S}{N} = \frac{(MI_p)^2}{2eBI_pM^2F(M)} = \frac{I_p}{2eBF(M)} \tag{16.18}$$

The signal current in terms of the average number of photons detected p_{av} over a time interval τ and the quantum efficiency η of converting photons into electrons is

$$I_p = \frac{ep_{av}\eta}{\tau} \tag{16.19}$$

Using this in Eq. (16.18) and after a bit algebra, one obtains

$$p_{av} = \frac{2B\tau F(M)}{\eta}\frac{S}{N} \tag{16.20}$$

Example 16.9 *The average number of photons needed to ensure a BER = 10^{-9} may now be estimated by calculating $F(M) = kM + \left(2 - \frac{1}{M}\right)(1 - k)$, with $k = 0.02$ and assuming $\eta = 0.8$, and $B\tau = 0.6$ for a raised cosine filter following the APD and SNR = 144. The value $p_{av} = 866$ is obtained.*

16.6.4   Power limits of nanolasers

Ma and Oulton in their work used a typical value of 1,000 photons per bit as sufficient number to overcome thermal noise and to provide required sensitivity. It is straightforward to calculate the energy per bit and the optical power needed for different modulation speed. Such relationships, for a wavelength of 1,550 nm, assuming 100% conversion efficiency are shown in Fig. 16.4, in which the optical energy per bit and photon numbers as a function of bit rate are plotted. Different lines in the figure represent constant power.

Example 16.10 *Assuming that 10^4 photons of wavelength 1550 nm (energy = 0.8 eV) are incident on the photodetector, the energy per bit is $1.6 \times 10^{-19} \times 0.8 \times 10^4 = 1.12\,fJ$, and the average emissive power of the laser for a bit rate of 10 Gb/s is 11.2 μW.*

At a given bit rate, the upper limit of power emitted by a nanolaser should be less than 10 fJ/bit while the lower bound is determined by the thermal noise.

It should be noted that the curves given by Fig. 16.4 are somewhat idealistic. One must consider the external quantum efficiency and the energy needed to switch the parasitic capacitance $E = \frac{1}{2}CV^2$, where C is the parasitic capacitance and V is the operating voltage. For low power consumption, C must be as small as possible, preferably at fF level. For a 100-nm junction length and picosecond carrier recombination lifetime, the footprint of the laser should be less than 1 μm^2. The maximum modulation speed has an upper bound $(2RC)^{-1}$, where R is the resistance of the device.

The lower bound of nanolaser energy is limited by thermal noise and as noted already typically 1,000 photons/bit are needed to maintain proper SNR. External quantum efficiency values of 10% or more have recently been reported. The modulation bandwidth of nanolasers due to their small size may exceed a few hundred GHz.

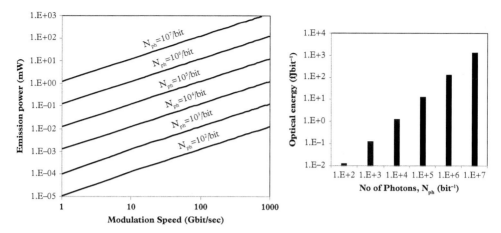

Figure 16.4 *Output power of a nanolaser as a function of modulation speed for different values of photon number/bit. $N_{ph} = 1000/bit$ sets the thermal noise limit. Optical energy in fJ/bit is shown in terms of photon number/bit in the right panel.*

16.6.5   Electrical injection

Most of the operating systems including optical interconnects require small lasers which are electrically driven and can operate in continuous mode at room temperature, Metallic contacts are needed to provide electrical connections as well as to serve as heat sinks. Room temperature CW operation has been reported with a small cavity volume of $0.67\lambda^3$ only by Ding et al (2013). In this connection, nanoLEDs are considered as a promising light source for optical interconnect. One such indium phosphide (InP)-based nano-pillar LED grown on Si has been reported by Dolores-Calzadilla (2017).

To date, however, electrically injected nanolasers meeting all the requirements for perfect optical interconnect suitable for practical applications have not been achieved. The requirements are high thermal stability, integration with high coupling efficiency with waveguides and bandwidths over tens of GHz for direct modulation, and low power consumption < 0.1 fJ/bit are needed for future data centres.

Schematic diagrams of current optical interconnects using VCSELs and optical fibres in present day supercomputers and future on-chip interconnections using nanolasers and waveguides are given by Ning (2019). We provide two such arrangements in Fig. 16.5, which are conceptually similar to the diagrams of Ning (2019).

16.7   Metal-based nanolasers

The nanolasers discussed so far rely on localized or propagating surface Plasmon po-laritons for their operation and are typically made of metal-semiconductor interfaces.

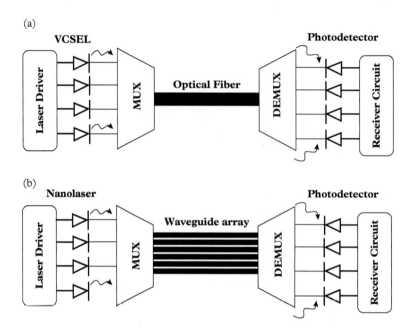

Figure 16.5 *(a) Schematic of present-day optical interconnect in supercomputers using VCSELs and optical fibre; (b) schematic diagram of future on-chip interconnect based on nanolasers and plasmonic waveguides.*

While such nanolasers have subwavelength dimensions, and fairly low threshold current, the pumping mechanism is mostly optical and therefore the devices have a large footprint. Electrically pumped nanolasers have not yet attained the level of maturity needed for system operation. Besides, the modulation speed of the nanolasers is still below the desired level. In some application areas, there is a need to have an in-built locally generating tiny optical emitter, an example of which is a nanoantenna to be discussed next.

In such applications, some alternate light sources, based purely on metals, are being investigated. In this context, light emission by inelastic electron tunnelling has been identified as a promising area of study. In the following subsections, we shall discuss the basic principle and a few recent demonstrations and one application area as an on-chip interconnect using nanoantenna.

16.7.1   Light emission by inelastic electron tunnelling

Generation of light by the process known as inelastic electron tunnelling (IET), was first discovered in a metal-insulator-metal (MIM) tunnel structure by Lambe, and McCarthy (1976).

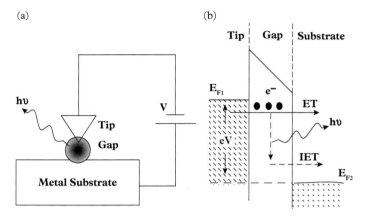

Figure 16.6 *(a) Schematic of a tunnel structure consisting of a metal tip, and a metal substrate separated by a small gap acting as a barrier. Electric fields are more or less confined within the shaded region; (b) band diagram of the structure under a bias voltage V showing elastic (ET) and inelastic (IET) electronic tunnelling. In IET, energy is given up as a photon or a SPP.*

The process of light generation by IET is illustrated by Fig. 16.6. On the left side of Fig. 16.6(a) a schematic structure employed is shown. It consists of a metal tip separated from another metal substrate by a very narrow gap, which may be a vacuum or an insulator. Under application of a bias, the Fermi levels of the tip metal and the substrate will be misaligned and a potential barrier will develop across the thin gap as shown in the band diagram in Fig. 16.6(b). Under this situation, electrons will tunnel elastically (ET) from the metal tip through the barrier to the metal substrate, without losing energy. However, when the applied bias is large enough to make $eV = h\upsilon$, a quantum of light energy, electrons may tunnel through the barrier inelastically by the IET process to reach the substrate. The energy is thus given up as a photon of energy $h\upsilon$. Alternatively, the energy may be given up by exciting localized surface plasmon polariton.

Unfortunately, however, the IET process has a very poor efficiency as one photon is generated per million electrons. On the other hand, the process is intrinsically fast as tunnelling is involved. The tunnelling time is about h/eV, and for $V \approx 1V$, the value is 4.16×10^{-15}s, allowing ideal modulation bandwidth to cross 200 THz. In practice however, the bandwidth is limited by the RC time constant. Methods to improve this and further details may be found in Parzefall and Novotny (2018) and Braun et al (2018). Qian et al (2018) using nanocrystals assembled into MIM junctions, reported ~2% efficiency of far-field free-space light generation, indicating at least two orders of magnitude improvement over previously reported values. Zhang et al (2019) reported three orders of magnitude enhancement of efficiency over the values for planar structures.

Uskov et al (2016) proposed a novel idea of enhancing manifold the efficiency of light emitters utilizing IET by introducing the concept of resonant inelastic electron

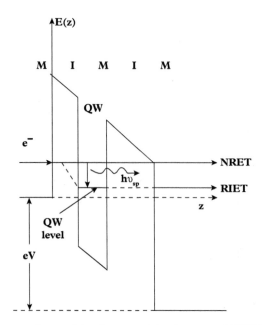

Figure 16.7 *Band diagram of a metal-insulator-metal-insulator-metal (MIMIM) structure to illustrate non-resonant elastic tunnelling and resonant inelastic electron tunnelling. The middle metal layer acts as a quantum well, the quantized level of which is resonant with the surface plasmon quantum.*

tunnelling (RIET). Basically the structure is a MIMIM structure. The process may be understood by following the band diagram shown in Fig. 16.7. As usual a bias eV is applied between the first (injector) and the last (collector) metal layers. The middle metal layer acts as a quantum well (QW) and has bound energy states. Electron from the injector may cross the structure by non-resonant elastic tunnelling (NRET) as well as by RIET as shown in the diagram. The emission of surface Plasmon with energy $h\nu_{sp}$ occurs when the propagating electron makes a downward transition to the QW level, and then tunnels through the second insulating barrier to be collected by the collector at the other side. The two insulating layers in the MIMIM multilayer structure drastically reduce the NRET process, but the RIET process is facilitated by the presence of resonant electron state at the middle QW layer. The idea is similar to the resonance effect in photon assisted resonant tunnelling in semiconductor multilayer structures.

The idea has been used by Qian et al (2021) to increase the emission efficiency for IET emitters. They used conducting indium tim oxide (ITO) as the injector and metallic titanium nitride (TiN) as the collector. Silver nanorods (AgNRs) were grown on top of metal QWs. An ultrathin metal film (~ 1.4-nm TiN) was grown with atomically flat interfaces in-between dielectrics (~ 10 nm for each Al$_2$O$_3$) over a large area. Quantum size effect was visible in ultrathin TiN metal layer, in which the conduction band is split into several electron subbands, due to growth of high-quality dielectric/ultrathin-metal/dielectric sandwich heterostructures

Qian et al observed multiple peaks in the I–V characteristics signifying onset of resonant tunnelling to different QW states. They found external quantum efficiency(EQE) as large as 32% in their experiment. There are scopes for optimization.

16.7.2 Wireless interconnects using nanoantenna

Nanometer sized optical antenna, termed as nanoantenna, are currently being developed for different applications, such as, on-chip optical interconnect, nanoscale sensing, metrology, and as nanoscale light emitting devices, including single-photon emitters for quantum information processing. In this section, the use of nanoantenna for on-chip wireless interconnect and the basic principle of light generation by inelastic electron tunnelling employed in nanoantenna will be briefly discussed.

Transmitting antennas are used to radiate information carrying radio frequency (RF) signals for broadcasting. Similarly, receiving antennas are tuned to different transmitting signals and the weak received signals are amplified and processed to decipher the information. Antenna had been of use since the beginning of wireless radio transmission and reception. Instead of isotropic transmission and reception, antennas can be directional as used in TV network. The directional properties had been achieved by using suitable reflectors, active feed and director as in Yagi–Uda antennas. Such antennas proved to be crucial for enabling television broadcasting and had been installed on many rooftops.

The Yagi–Uda concept has been in the focus in recent years with a view to adapting it at optical frequencies. Two key benefits are: (1) dramatic increase in the bandwidth at much higher frequencies and (2) remarkable scale-down of the footprint to the nanometer scale. The applications envisaged are (1) development of an efficient link between electron-based IC chips used in computers and photonics-based fibre networks, and (2) realization of an on-chip data communication link. It is obvious that antennas outperform subwavelength plasmonic waveguides used for longer distances. In addition, the nanoantennas have an adaptable footprint, allow multiple beam crossings, and are free from Joule heating, as in metallic waveguides.

The challenges involved, the methods adopted to realize such Yagi–Uda optical nanoantenna, the results obtained with realized dimensions and other details, in particular, the values of directivity, have been described by Kullock et al (2020). They created a 1-nm tunnel gap by first fabricating the antenna structure having a 25 nm gap. Then gold (Au) particles were dropcasted on the sample. Using an atomic force microscope cantilever suitable particles were then pushed into the antenna gap.

Description of the growth and fabrication, precisely positioning Au nanoparticles, generating light locally by inelastic tunnelling mechanism, and many other details may be found in the paper.

Problems

16.1 *Consider Eq. (16.9) expressing variation of $\mathcal{J}_{th}$ with thickness d of the active layer. An approximate formula $\Gamma = \frac{\beta^2}{2+\beta^2}$, where $\beta = \frac{2\pi d}{\lambda_0}\sqrt{\eta_a^2 - \eta_{cl}^2}$ is used. Use the parameter values given for GaAs in Example 16.1. Calculate and plot the values of $\mathcal{J}_{th}$ vs. d $(0.05 \leq d \leq 0.4\mu m)$.*

16.2 *Modify the expression for SNR given by Eq. (16.18) by including the contribution of thermal noise. Prove that the SNR becomes maximum for an optimum value of multiplication factor M.*

16.3 *Express photon lifetime in terms of length and mirror reflectivities of a Fabry–Perot (FP) resonator.*

16.4 *Obtain the relationship between photon lifetime and Q of a cavity as given in Example 16.5.*

16.5 *Calculate the voltage needed to generate light of wavelength 870 nm by EIT. Calculate also the electric field in the air gap having 3 nm width.*

16.6 *Discuss the nature of variation of current with voltage in a MIM structure used for light emission by IET.*

16.7 *Experimental I–V characteristics for IET have been obtained by Zhang et al (2019). By using a theoretical model, they also found close agreement between theory and experiment as shown in Fig. 2a of their paper. Reproduce the theoretical curve by using the parameters given and compare with the experimental data.*

References

Altug H, Englund D, and Vuckovic J (2006) Ultrafast photonic crystal nanocavity laser. *Nature Physics* 2: 484–488.

Amano H, Asahi T, and Akasaki I (1990) Stimulated emission near ultraviolet at room temperature from a GaN film grown on sapphire by MOVPE using an AlN buffer layer. *Japanese Journal of Applied Physics* 29: L205.

Arakawa Y and Sakaki H (1982) Multidimensional quantum well laser and temperature dependence of its threshold current. *Applied Physics Letters* 40(11): 77–78.

Basu P K, Mukhopadhyay B, and Basu R (2015) *Semiconductor Laser Theory*. Boca Rator, FL, USA: CRC Press.

Bhattacharya P, Frost T, Deshpande S, Baten M Z, Hazari A, and Das A (2014) Room temperature electrically injected polariton laser. *Physical Review Letters* 112: 236802.

Braun K, Laible F, Hauler O, Wang X, Pan, A, Fleischer M, and Meixner A J (2018) Active optical antennas driven by inelastic electron tunneling. *Nanophotonics* 7(9): 1503–1516.

Chang S-W, Lin T-R, and Chuang S L (2010) Theory of plasmonic fabry-perot nanolasers. *Optics Express* 18: 15039–15053.

Chou Y-H, Hong K-B, Chang C-T, Chang T-C et al (2018) Ultracompact pseudowedge plasmonic lasers and laser arrays. *Nano Letters* 18:747–753.

Chou Y H, Wu Y-M, Hong K-B, Chou B-T et al (2016) High-operation-temperature plasmonic nanolasers on single-crystalline aluminium. *Nano Letters* 16: 3179–3186.

Cisco. (2017). The Zettabyte Era: Trends and Analysis. https://www.cisco.com/c/en/us/solutions/collateral/service-provider/visual-networkingindex-vni/vni-hyperconnectivity-wp.html

Coldren L A, Corzine S W, and Masanovic M (2012) *Diode Lasers and Photonic Integrated Circuits*, 2nd edn. Hoboken, NJ, USA: John Wiley.

Deen M J and Basu P K (2012) *Silicon Photonics: Fundamentals and Devices*. Chichester, UK: Wiley.

Ding K et al (2012) Room-temperature continuous wave lasing in deepsubwavelength metallic cavities under electrical injection. *Physical Review B* 85:041301(R).

Ding K et al (2013) An electrical injection metallic cavity nanolaser with azimuthal polarization. *Applied Physics Letters* 102: 041110.

Ding K et al (2013) Record performance of electrical injection sub-wavelength metallic-cavity semiconductor lasers at room temperature. *Optics Express* 21: 4728–4733.

Dolores-Calzadilla V et al (2017) Waveguide-coupled nanopillar metal-cavity light-emitting diodes on silicon. *Nature Commununications* 8: 14323.

Faist J, Capasso F, Sivco D L, Sirtori C, Hutchinson A L, and Cho A Y (1994) Quantum cascade laser. *Science* 264(5158): 553–556.

Galanzha E I (2017) Spaser as a biological probe. *Nature Commununications* 8: 15528.

Gu Q et al (2014) Amorphous Al2O3 shield for thermal management in electrically pumped metallo-dielectric nanolasers. *IEEE Journal of Quantum Electronics* 50: 499–509.

Hill Martin, Marell Milan, Leong Eunice, Smalbrugge Barry, Zhu Youcai, Sun Minghua, Van Veldhoven Peter J, Geluk Erik J, Karouta Fouad, Oei Yok, Noetzel Richard, Ning CunZheng, Smit Meint K (2009) Lasing in metal-insulator-metal sub-wavelength plasmonic waveguides. *Optics Express* 17:11107–11112.

Hill M T et al (2007) Lasing in metallic-coated nanocavities. *Nature Photonics* 1:589–594.

Hill MT and Gather MC (2014) Advances in small lasers. *Nature Photonics* 8:908–918.

Ho JF et al (2015) Low-threshold near-infrared GaAs–AlGaAs core–shell nanowire plasmon laser. *ACS Photonics* 2:165–171.

Huang B J et al (2019) Electrically injected GeSn vertical-cavity surface emitters on silicon-on-insulator platforms. *ACS Photonics* 6:1931–1938.

Iga K (1977) Laboratory notebook of P&I Laboratory 1977 Issue. *Tokyo Institite of Technology* Tokyo, Japan, Mar. 22.

Jewell J L, Harbison J P, Scherer A, Lee, Y H, and L T Florez. (1991) Vertical-Cavity surface-emitting lasers: Design, growth, fabrication, characterization. *IEEE Journal of Quantum Electronics* 27(6):1332–1346.

Johnson J C et al (2001) Single nanowire lasers. *Journal of Physics and Chemisry B* 105:11387–11390.

Khajavikhan M et al (2012) Thresholdless nanoscale coaxial lasers. *Nature* 482:204–207.

Kirstaedter N, Ledentsov N N, Grundmann M, Bimberg D, Ustinov V M, Ruvimov SS, Maximov M V, Kop'ev PS Z I, Alferov Zh I Richter U, Werner P, Gosele U, and Heydenreich J (1994) Low threshold, large T_0 injection laser emission from (InGa)As quantum dots. *Electronics Letters* 30(17):1416–1417.

Kullock R, Ochs M, Grimm P, Emmerling M, and Hecht B (2020) Electrically drive Yagi-Uda antennas for light. *Nature Communications* 11(115):1–7.

Kwon S H et al. (2010) Subwavelength plasmonic lasing from a semiconductor nanodisk with silver nanopan cavity. *Nano Letters* 10: 3679–3683.

Lakhani A M, Kim M K, Lau E K, and Wu, M C (2011) Plasmonic crystal defect nanolaser. *Optics Express* 19: 18237–18245.

Lambe J and McCarthy S L (1976) Light Emission from inelastic electron tunneling. *Physical Review Letters* 37: 923–925.

Lau E K, Lakhani A, Tucker R S, and Wu M C (2009) Enhanced modulation bandwidth of nanocavity light emitting devices. *Optics Express* 17(10): 7790–7799.

Lu Y J et al (2012) Plasmonic nanolaser using epitaxially grown silver film. *Science* 337: 450–453.

Lu Y J et al (2014) All-color plasmonic nanolasers with ultralow thresholds: Autotuning mechanism for single-mode lasing. *Nano Letters* 14: 4381–4388.

Ma R M and Oulton R F (2019) Applications of nanolasers. *Nature Nanotechnology* 14: 12–22.

Ma R M, Oulton R F, Sorger V J, Bartal G, and Zhang X (2011) Roomtemperature sub-diffraction-limited plasmon laser by total internal reflection. *Nature Materials* 10: 110–113.

McCall S L, Levi A F J, Slusher R E, Pearton S J, and Logan R A (1992) Whispering-gallery mode microdisk lasers. *Applied Physics Letters* 60:2 89–291.

Michalzik R and Ebeling K J (2003) Operating principles of VCSEL. In: *Vertical Cavity Surface Emitting Devices, Springer Series in Photonics*, vol.6. H Li and K Iga (eds.) pp. 53–98. Berlin: Springer.

Miller D A B (2000) Rationale and challenges for optical interconnects to electronic chips. *Proceedings of the IEEE* 88(3): 728–749.

Miller D A B (2009) Device requirements for optical interconnects to silicon chips. *Proceedings of the IEEE* 97: 1166–1185.

Nakamura S, Senoh M, Nagahama S, Iwasa N, Yamada T, Matsushita T, Kiyoku H, and Sugimoto Y (1996) InGaN-based multi-quantumwell-structure laser diodes. *Japanese Journal of Applied Physics B* 35(1): L74–76.

Nezha M P et al (2010) Room-temperature subwavelength metallo-dielectric lasers. *Nature Photonics* 4: 395–399.

Ni C -Y A and Chuang S L (2012) Theory of high-speed nanolasers and nanoLEDs. *Optics Express* 20: 16450.

Ning C -Z (2010) Semiconductor nanolasers. *Physica Status Solidi B* 247(4): 774–788. DOI 10.1002/pssb.200945436.

Ning C -Z (2019) Semiconductor nanolasers and the size-energy-efficiency challenge: A review. *Advanced Photonics* 1:014002.

Noginov M A et al (2009) Demonstration of a spaser-based nanolaser. *Nature* 460: 1110–1112.

Oulton R F et al (2009) Plasmon lasers at deep subwavelength scale. *Nature* 461: 629–632.

Painter O et al (1999) Two-dimensional photonic band-gap defect mode laser. *Science* 284: 1819–1821.

Parkhomenko R G, Kuchyanov A S, Knez M, and Stockman M I (2021) Lasing spaser in photonic crystals. *ACS Omega* 6(6): 4417–4422.

Parzefall M and Novotny L (2018) Light at the end of the tunnel. *ACS Photonics* 5: 4195–4202.

Purcell E M (1946) Spontaneous emission probabilities at radio frequencies. *Physical Review* 69: 681–681.

Qian H, Hsu S-W, Guruntha K et al (2018) Efficient light generation from enhanced inelastic electron tunneling. *Nature Photonics* 12: 485–488.

Qian H, Li S, Hsu S-W, Chen C-F, Tian F, Tao A R and Liu Z (2021) Highly-efficient electrically-driven localized surface plasmon source enabled by resonant inelastic electron tunneling. *Nature Communications* 12(3111): 1–7.

Schneider C, Rahimi-Iman A, Kim N Y, Fischer J, Savenko I G, Amthor M, Lermer M, Wolf A, Worschech L, Kulakovskii V D, Shelykh I A, Kamp M, Reitzenstein S, Forchel A, Yamamoto Y, and Höfling S (2013) An electrically pumped polariton laser. *Nature* 497(7449): 348–352.

Senior J M (1993) *Optical Fibre Communications: Principles and Practice*, 2nd edn. Hertfordshire, UK: Prentice_Hall International (UK) Ltd.

Shen K C et al (2018) Deep-ultraviolet hyperbolic metacavity laser. *Advanced Materials* 30: 1706918.

Tiwari P, Wen P, Caimi D, Mauthe S, Trivino N V, Sousa M, and Moselund K E (2021) Scaling of metal-clad InP nanodisk lasers: Optical performance and thermal effects. *Optics Express* 29(3): 3915–3927.

Uskov A V, Khurgin J B, Protsenko I E, Smetanin I V, and Bouhelier A (2016) Excitation of plasmonic nanoantennas by nonresonant and resonant electron tunneling. *Nanoscale* 8(30): 14573–14579.

van Exter M P, Nienhuis G, and Woerdman J P (1996) Two simple expressions for the spontaneous emission factor β. *Physical Review A* 54: 3553.

Wang R, Xu C, You D, Wang X, Chen J, Shi, Z, Cui Q, and Qiu T (2021) Plasmon–exciton coupling dynamics and plasmonic lasing in a core–shell nanocavity†. *Nanoscale* 13: 6780–6785.

Wang S et al (2017) Unusual scaling laws for plasmonic lasers beyond diffraction limit. *Nature Communications* 8: 1889.

Wang S, Chen H-Z, and Ma R-M (2018) High performance plasmonic nanolasers with external quantum efficiency exceed 10%. *Nano Letters.* 1: 7942– 7948.

Xu L, Li F, Liu Y, Yao F, and Liu S (2019) Surface plasmon nanolaser: Principle, structure, characteristics and applications. *Applied Science* 9:861. DOI:10.3390/app905086.

Yablonovitch E (1987) Inhibited spontaneous emission in solid-state physics and electronics. *Physical Review Letters* 58(20): 2059–2062.

Yablonovitch E, Gmitter R J, and Bhat R (1988) Inhibited and enhanced spontaneous emission from optically thin AlGaAs/GaAs double heterostructures. *Physical Review Letters* 61(22): 2546–2549.

Yamamoto Y, Machida S, and Björk G (1991) Microcavity semiconductor laser with enhanced spontaneous emission. *Physical Review A* 44(11): 657–668.

Yariv A and Yeh P (2007) *Photonics: Optical Electronics in Modern Communication*, 6th edn. Oxford, New York: Oxford University Press.

Yu K, Lakhani A, and Wu M C (2010) Subwavelength metal-optic semiconductor nanopatch lasers. *Optics Express* 18: 8790–8799.

Zhang C, Hugonin J-P, Coutrot A-L, Sauvan C, Marquier F, and Greffet J-J (2019) Antenna surface plasmon emission by inelastic tunneling. *Nature Communications* 10:(4949): 1–7.

Zhang Q et al (2014) A room temperature low-threshold ultraviolet plasmonic nanolaser. *Nature Communications* 5: 49–53.

Index